REVIEW

OF

RESEARCH

CONTRIBUTORS

TOM BIRD
C. A. BOWERS
DAVID K. COHEN
MICHAEL COLE
MARTIN ENGEL
BARBARA FINKELSTEIN
NATHALIE J. GEHRKE
PETER P. GRIMMETT
MICHAEL S. KNAPP
JAMES S. LEMING
ALLAN M. MacKINNON
HUGH G. PETRIE
WILLIAM F. PINAR
EMILY ROBERTSON
TONY SCOTT
KENNETH A. SIROTNIK
JAMES P. SPILLANE
GARY SYKES

IN EDUCATION

18

1992

GERALD GRANT
EDITOR
SYRACUSE UNIVERSITY

PUBLISHED BY THE
AMERICAN EDUCATIONAL RESEARCH ASSOCIATION
1230 Seventeenth Street, NW
Washington, DC 20036-3078

Contents

Introduction

The essays brought together in this volume, which were commissioned by the Board of Editors of the *RRE*, often originated in the form of a simple question. At the start, some of them were also a bit saucy, as "Which John Dewey do we need now?" which led to the magnificent review of Dewey's thought by Emily Robertson. Or, "What was the upshot of the revisionist debate?" which became the stimulus for Barbara Finkelstein's expansive review of work by historians of education. Most of these queries came from my colleagues on the editorial board, but occasionally we stumbled across a suggestion from afield that was too good to turn down.

One such encounter at Stanford University led to what is now chapter 4. I was having lunch with a distinguished senior scholar at the School of Education there who had recently returned from a conference in São Paulo. It was a three-day affair that opened with short speeches by various academic stars. Participants then selected which of these scholars they wished to have seminars with for the remainder of the conference. My luncheon partner told me the conference organizers were embarrassed because only a handful had signed up for his seminar while a great many were interested in something called critical pedagogy. He had never paid much attention to it. What was it all about, and had these critical theorists produced anything anyone in education should pay any attention to? William F. Pinar and C. A. Bowers accepted his challenge and produced the essay that appears here on the politics of the curriculum and the significance of critical perspectives.

Pinar and Bowers admirably fulfill the expectations we have for a review. We do not seek mere summaries of research, no matter how encyclopedic. A good review is one that synthesizes and evaluates scholarship in a given field in a style that can be understood by a wide audience while also providing fresh perspectives to experts within the field. It appraises what has been done, tells us what controversies exist, and suggests fruitful avenues for further exploration. At their best, *RRE* reviews have redirected research and set new policy agendas, as I think many of those in this volume will do. Volume 18 brings together an extraordinary set of reviews of research and scholarship in the fields of curriculum, history and philosophy, and teacher education.

David Cohen and James Spillane open the volume with a profoundly pessimistic view of the possibilities for major reform in the curriculum. They were asked to assess whether how we organize our systems of education makes any difference in what children learn. For example, do highly centralized educational systems with an extensive apparatus of instructional guidance produce better outcomes than more fragmented or decentralized systems of education? They organize their essay around two paradoxes. The first is that, especially in the last decade, the United States has launched intensive national movements for reform, with proponents arguing for the creation of state and national curricula as well as national tests to assess them. But the political system in the United States, unlike many other nations, was specifically designed to frustrate the exercise of such centralized authority in education. Educational policy is made in 110,000 schools, 15,000 school districts, and thousands of state and federal agencies. It is further fragmented by organizations set up to bypass existing agencies in order to coordinate policies mandated in categorical programs. Policy networks that grew up around such programs as P.L. 94-142, education for disabled children, were ingenious, "for they support state and national efforts to solve local problems in a system that was designed to frustrate such efforts" (p. 10).

Compared to other nations such as Japan or Australia, systems for instructional guidance are weak and inconsistent in the United States. Whether one looks at instructional frameworks or curriculum guides, assessment practices, or systems for monitoring instruction, guidance is diffuse. Teachers hear a babble of voices. Although many persons and agencies issue advice about instruction, few take account of each other's advice. There is enormous variance on instructional guidance from state to state and among nations, but little research on its effects. Cohen and Spillane also recognize that demands for more consistency run against the grain of bottom-up reform efforts such as those advocated by Theodore Sizer. They agree with those who argue that major reforms must combine bottom-up with top-down strategies.

The second paradox is that education reform has never been so intellectually ambitious while education capacity is so weak. In contrast to earlier reforms for increased access or a return to the "basics," contemporary reformers are much more educationally ambitious. They insist that all students must become more active and engaged, and must learn to become independent thinkers and problem solvers. But these ideals collide with the reality of instructional practice that is dominated by lectures, recitations, and worksheets. Intellectual demands are light and the work is dull. Students have few incentives to work hard; most students will be granted admission to higher education, with the exception of a few elite colleges, no matter how little effort they make, and employers in the United States seldom pay attention to actual school achievement apart from easily obtained credentials. Only a modest fraction of public school teachers have deep knowledge of any academic subject, and past efforts to transform teaching into a more ambitious enterprise have had small effects at best:

Neither new exams nor new curricula would work unless teachers understood them, and as things now stand, most teachers would not. This problem might be solved in part if teachers were extensively involved in building new frameworks, curricula, and assessments, and in grading students' work . . . [which] could be extraordinarily educative. . . . [But] these activities would reach only a fraction of the teaching force and would touch only part of the reeducation need. (p. 38)

Cohen and Spillane believe that a genuine transformation that would achieve reformers' intellectually ambitious aims requires a decades-long effort to reeducate not only teachers, but university faculties and parents and policymakers. They fear, however, that reformers are better at addition than subtraction. The most likely outcome is that new reformers will have their day, adding more mandates on top of what exists, creating more clutter in an undemanding and inconsistent system.

Although they might not agree with Cohen and Spillane's conclusions, Gehrke, Knapp, and Sirotnik corroborate their account in many respects. In chapter 2, Gehrke and her colleagues ask what is known about what schools really teach. They look at the elementary and secondary school curriculum in social studies, language arts, mathematics, and science. They surveyed textbooks and formal statements of curriculum in state and district guidelines, but focused primarily on the curriculum as enacted by teachers in the classroom. Teachers in all areas see themselves as having considerable control over the curriculum that they teach and believe they are teaching what should be taught. Scholars disagree. There is virtually no content area where scholars agree that the curriculum being taught is the curriculum that should be taught. While teachers tend to believe that discrete skills should be mastered before being combined for use in real experiences, many scholars see that view of sequencing as a barrier to needed curriculum change.

In social studies, the curricula are characterized as teaching students what they already know and focusing on the transmission of factual knowledge to the exclusion of conceptual understanding and problem solving that are the object of current reforms. The curriculum is disconnected from real experiences, authentic writing efforts, real problems of citizenship, or serious questions about the earth's environment.

The authors distinguish three models of instruction in the language arts: The mastery model closely associated with instructional objectives and basic skills, the heritage model that is oriented to understanding the traditional canon, and the process model that is concerned with the way each individual constructs knowledge from experience. While the process model has made some headway, the mastery model dominates most elementary schools and the heritage model in high schools. The skills of speaking and listening are taught to few students. Most high school students study the major literary works through a single anthology, and they do less reading than in the elementary grades. Note taking is the most common form of writing; extended essays are rare. Courses are marked by a strong demand to memorize information, and most students are not encouraged to be creative or to develop a personal voice.

Although mathematics teachers have developed consensus about reforms that should be implemented, the curriculum as currently enacted is lamentable. It is described as an underachieving curriculum that expects less of students than it should or could. Mathematics instruction is seen as exhibiting a narrow range of instructional formats with teacher presentations followed by seatwork.

Both mathematics and science are heavily tracked, but science teaching is more diverse. Consensus on what ought to be taught is growing, however, partly as a result of the work of the American Association for the Advancement of Science. Its Project 2061 recommends a thematic approach to science content, interdisciplinary thinking, an emphasis on connections between science and technology, and the inculcation of scientific habits of mind.

Gehrke and her coauthors conclude that we do not need more large-scale studies of the curriculum such as those conducted by the National Science Foundation. Researchers should ask why the curriculum is so slow to change and understand more about why teachers make the choices they do in selection of learning activities, sequences, and assessment. They should also focus on studies of the middle schools, where we know too little about either the curriculum that is actually taught or the way that students make choices that will influence whether they go to college or vocational schools. The curriculum should be studied from both the students' and the teachers' points of view.

Looking beyond the core areas of the curriculum, in chapter 3 James Leming examines the expanding edge of what is taught in schools in the United States. Americans tend to see their schools as a kind of universal social solvent, and most proposals for social reform usually find their way into the curriculum in some form. Leming defines his review of "contemporary issues curricula" as those "areas of formal instruction . . . that have as a primary goal the amelioration of pressing problems facing youth and/or society" in the last decade. Only secondarily are "these curricula intended to achieve more traditional academic objectives" (pp. 111–112). Leming reviews research on 11 of these contemporary issues curricula. He includes programs to lower rates of teenage pregnancy, reduce drug and alcohol abuse, prevent suicide, counter racial prejudice and gender stereotypes, help students cope with death, and expand global awareness.

The general finding is that success rates are rather low, although two thirds of these programs achieve some knowledge gains. However, knowing more translated into desired changes in attitude in only about a third of the cases and changes in behavior in 28% of the cases. Some of the programs are ill defined, poorly conceptualized, and apparently poorly taught. So-called "death education," which attempts to help students examine their feelings about death and cope with losses, is taught in about 11% of schools in the United States. It has no agreed-upon curriculum, and its outcome is often the opposite of what is intended, namely, to increase anxiety about death. Similarly, many alcohol abuse programs inculcate a more positive attitude toward drinking. Leming finds that peace education, also referred to as nuclear war education or peace studies, "has no agreed-upon conceptualization regarding its goals, is surrounded by contentious controversy, and has almost no research regarding its impact on students," although it is estimated to be a part of the curriculum in 12% to 15% of public schools. (p. 142). One study of a suicide prevention program that analyzed the responses of those who had previously attempted suicide with those who had not found that the attempters had more negative reactions to the program. None of the three major evaluations of suicide prevention programs reviewed found effects on either knowledge or attitudes.

In some areas, such as programs to reduce pregnancy rates, the failure to find any significant impact has led to the conclusion that goals of changing strong sexual drives are simply unrealistic. However, Leming points out that too many reviews of the research may have lumped good and bad programs together. Schools seem to do likewise, uncritically accepting all varieties of sex education programs and paying no attention to research evidence. Hence good and bad programs may be implemented at the same rate.

If one examines some of the most successful programs, such as efforts to reduce racial prejudice through collaborative learning groups, or to increase environmental awareness, one finds that these programs incorporate active, engaged learning strategies. For example, the best results in the environment education programs were achieved when teachers involved students in action strategies such as investigating the causes of stream pollution in their community and then developing a plan for combating it.

The authors of the first three chapters have taken the curriculum mostly as given, as formalized and enacted, whether at the center or the periphery. In chapter 4, William Pinar and C. A. Bowers survey the history of scholarship about the hidden curriculum, what some have called the unintended consequences of the way the school is organized and socially structured. That is, schools not only teach mathematics and social studies but the way they teach it and track students through course sequences stamps some students as losers and others as winners, or "teaches" them to be docile and conformist members in a preordained social order. Although they credit Philip Jackson as one of the first to explicate the idea of a hidden curriculum, Pinar and Bowers are primarily interested in that field of curriculum study, originally rooted in a politically left or Marxist analysis, that has been identified as "critical pedagogy" or "critical perspectives."

Their essay traces the evolution of the field from Marxist theories of schools as reproducing the social order to more complex models that incorporate the notion of resistance to what dominant groups would like the schools to reproduce. The authors' broad assessment of the domain of critical scholarship is that it succeeded in raising the consciousness of scholars about the way that the curriculum can be viewed as a political text or as part of a system that protects the privileged social position of some members of society. The controversies revolve around criticisms that scholars in this field have tended to view the schools *only* as a political text. They have walled themselves off and are conceptually isolated. They have failed to engage meaningfully with other traditions or with the schools. Some see their perspective as utopian because critical pedagogy "fails to challenge any identifiable social or political position, institution or group" (p. 178).

In chapter 5, Tony Scott, Michael Cole, and Martin Engel provide a stunning survey of the uses of computers in education since John Nevison invented the term "computer literacy" in a paper in 1976. In an essay that will appeal to both the novice and the expert, the authors are candid about the shortcomings of the computer revolution. The computer cart (of drill and practice routines) too often comes before the subject-matter horse. Alas, these are the programs that seem to sell, perhaps because teachers feel safer with these more routinized uses of computers than they do with more holistic or creative approaches. Many of the so-called Integrated Learning Systems have been developed and targeted for inner-city schools where superintendents are often desperate to get the scores up quickly. And at a superficial level, the drill-and-skill programs do that. But a tracking system has developed in which poor pupils in big cities are more likely to use computers as substitute workbooks while students in more affluent suburbs are using the new technology for tutoring and simulations and other creative activities.

If you are interested in assessing your own creative computer skills, chapter 5 provides a scale of 22 competencies developed by J. D. Ellis for computer-using science educators. Scott and his colleagues discuss a variety of exciting developments, such as students linked in computer networks using hypermedia to produce self-portraits of their schools. In the Apple Classroom of Tomorrow experiments in classrooms saturated with computers, students have become spontaneous peer tutors, and the adults have become coaches and orchestrators. Teachers employing advanced applications of microcomputers in K–12 classrooms have changed their teaching styles so radically that they find that some administrators don't know how to evaluate them!

Section II of the volume turns to the domains of history and philosophy, and opens with Barbara Finkelstein's chapter on "Education Historians as Mythmakers." By "mythmakers," Finkelstein wishes to underscore the role of historians as "cultural authorities carrying messages from the past" whose narratives also "formulate myths that reveal education purposes and practices." She draws out the visions that are projected in these narratives and seeks to analyze the role of historians as "social messengers." Chapter 6 explores "education history as an evolving series of narrative visions that . . . have revealed ever more nuanced and complete aspects of the U.S. educational past" (p. 257). (Do I hear an echo here of a myth of continual progress among historians of education?)

As noted earlier, Finkelstein was originally asked to do a review that would assess the outcome of the revisionist debate among historians. She chose to place that question in the larger context just indicated, although most of her essay does review the field as a series of revisions or assaults on the traditional canon laid down by the first generation of historians of education such as Ellwood P. Cubberly, R. Freeman Butts, and a young Lawrence Cremin. She portrays them as establishing the "myth of benevolence," or as focusing on the "empowering qualities of public schools for individual mobility and civic nurture" (p. 261). Those who came later to demythologize the canon also created new myths of their own as they attempted to uncover the "hidden injuries of class, race, and gender and . . new modes of social control" (p. 258). Finkelstein argues that recent scholarship has sought to reconcile contradictions between "imperial traditions of history writing in education" and later revisions of it. But she concludes that education historians must develop new forms of discourse that capture the history of human relationships, the history of communication, and cross-cultural understandings.

More than 15 years ago, Hugh Petrie wrote a seminal essay on the topic of interdisciplinarity in education and hence we asked him to revisit the topic in light of a multitude of educational reforms stressing interdisciplinary approaches. In chapter 7, Petrie sorts out the conceptual differences among disciplinary, interdisciplinary, multidisciplinary, and transdisciplinary forms of inquiry. Multidisciplinary work he sees as group problem solving in which members do not cross the bounds of their own disciplines, whereas in true interdisciplinary work key elements of the disciplinarians' use of their concepts and tools change. A typical school curriculum is an example of the former, whereas a subject like immunopharmacology would be an example of the latter. His review is guided by three questions: (a) What is the relation of theoretical and practical wisdom in interdisciplinary efforts? (b) Does integration of knowledge take place in individuals, or is it a group or social phenomenon? and (c) How do we overcome the paradox that interdisciplinary approaches can be justified only from the perspective of established disciplines?

John Dewey certainly fostered interdisciplinary forms of inquiry, and some of his critics would say he was overly optimistic, if not naive, about the possibility of solving our problems through inquiry, dialogue, and persuasion. In chapter 8, Emily Robertson asks whether Dewey's educational vision is still viable. In assessing the work of Dewey and his critics—recently adumbrated in a 50-page bibliography of work published only in the last decade—Robertson focuses on his theory of radical democracy and his commitment to democratic education.

Robertson believes that Dewey is now neglected in professional philosophy. Within education, some argue that Dewey's version of progressive education had little lasting impact on public schools, and one scholar notes that to understand the history of education in the United States one must realize that "Edward L. Thorndike won and John Dewey lost." Dewey's voice is muffled, if not silenced, in many discussions of contemporary education reforms because he believed that substantial change in schools required major transformations in the culture of the broader society. At heart, Dewey had a radical faith in the possibilities of a truly participatory democracy that the schools must both model and help to produce.

Robertson's lucid review explores recent interpretations and criticisms of the social ends that Dewey sought. She grounds the reader in two central strands of Dewey's philosophy—experimentalism and radical democracy. She examines critiques from the right that he was too much the social engineer and from the left that he was naive about the uses of power in distorting dialogue and frustrating change. Her aim is to be neither judge nor advocate but to show us the living edges of Dewey's thought. She argues that his theory of ethics, his proposals for resolutions of social conflict, and his conception of social science continue to offer promising avenues for further scholarship. And Robertson suggests how his voice could fruitfully enter current debates on teacher empowerment, feminist theories, and multiculturalism. Of the latter, she argues that Dewey is basically right in posing the central issue as the problem of how Americans achieve "genuine assimilation *to one another*" (p. 361).

The volume concludes with two related essays on teacher education, the first on the development of craft knowledge and the second on the use of case studies. Peter Grimmett and Allan MacKinnon begin chapter 9 with a poem to make their point that craft knowledge—learning how rather than learning that—is not something that can be adequately conveyed in usual research report formats. They argue that it is primarily knowledge acquired "at the elbows" rather than in books, noting that their essay promoting craft knowledge is therefore something of an oxymoron in a volume that reviews research. Their review focuses on teaching as a craft rather than as a profession that is built on propositional knowledge. They believe that teacher education should make more room for craft without denigrating codified knowledge. Following Herbert Kohl, they define craft knowledge as "balancing teacher-initiated ideas with student-initiated ones, . . . doing research as a teacher, and observing and responding to what actually succeeded with the students" (p. 392). Teacher education programs would give more weight to growth of craft knowledge if they incorporated more texts (including films, poems, narratives of teaching and research on teaching) authored by teachers. Student teachers should be encouraged to develop their own metaphors and write their own credos of teaching. Methods courses should be school-based, with university faculty teaching alongside their student teachers so that they could both learn at each other's elbows.

In the last chapter, Gary Sykes and Tom Bird explore the way craft and other kinds of knowledge can be conveyed through imaginative use of cases. Although they do not neglect criticisms of case methods, they are sympathetic to Lee Shulman's view that the field of teaching "is itself a body of cases linked loosely by working principles, and case methods are the most valid way of representing that structure in teaching" (pp. 513–514). They review a growing literature on the use of cases with a focus on their application to the reform of teacher education. Although case methods have long been a feature of professional education in law, medicine, business, and other fields, they have not found a central place in teacher education. A small chorus has been promoting their use, but thus far little research has been done to compare the effectiveness of case methods with what they describe as the traditional foundations-methods-clinical sequence of teacher preparation.

Sykes and Bird provide an excellent guide to the literature on cases produced by proponents of the method. They furnish many examples of "encounters with cases" through texts both literary and scientific, videotapes of actual teaching situations, computer-based interactive hypermedia, and simulations. They doubt, however, that the field is even close to having produced a substantial "curriculum of cases." Much development needs to be done before alternatives can be evaluated. Their review helps sort out the kinds of questions we need to ask about the use and effectiveness of case methods, beginning with, What counts as a case and what does it represent? Why is a case significant? And what understandings about cases allow a community to use them fruitfully?

Let me conclude this introduction by offering my thanks to the editorial board, who provided me with much delightful instruction, and to the contributing editors, who played a most significant role in bringing these essays to an optimal state. They are acknowledged at the beginning of the volume. I would like to add here my gratitude to Judith Torney-Purta and Harold Stevenson, who offered valuable consultation. Finally, let me acknowledge my special appreciation of John Covaleskie and Rajeswari Swaminathan for their editorial assistance.

January 1, 1992
Syracuse University

Gerald Grant
Editor

I.
CURRICULUM

Chapter 1

Policy and Practice: The Relations Between Governance and Instruction

DAVID K. COHEN and JAMES P. SPILLANE
Michigan State University

Ours is a time of remarkable ferment in U.S. education. The recent school reform movement initially focused on the "basics," but then took off in a dramatically new direction in the late 1980s. Reformers started to demand more thoughtful and intellectually ambitious instruction. Leaders in politics and business argued that students must become independent thinkers and enterprising problem solvers. Educators began to say that schools must offer intellectually challenging instruction that is deeply rooted in the academic disciplines.

These ideas are a dramatic change. For most of this century, politicians and businessmen ignored public education or supported only minimum programs for most students. And most leaders in education long have been inclined to the view that most students need basic and practical education rather than more high-flown and demanding stuff. These tendencies were entirely representative. Though the American people have been enthusiasts for schooling, few have been keen on intellectually ambitious education.

More unusual still, recent reformers have proposed fundamental changes in politics and policy to achieve the new goals. They argue for the creation of state or national curricula, to push instruction to new heights. Or they advocate state or national tests or examination systems, to pull instruction in the same direction. Or they propose to link examinations and curricula so as to gain even more leverage on teaching and learning. Prominent politicians, businesspeople, and professors have en-

We are grateful to Carol Barnes, Linda Darling-Hammond, Robert Dreeben, Robert Floden, Susan Fuhrman, Harry Judge, James Kelly, Magdalene Lampert, George Madaus, Barbara Neufeld, Andrew Porter, Daniel Resnick, Brian Rowan, Lauren Sosniak, Marshall Smith, Gary Sykes, Teresa Tatto, Suzanne Wilson, and Rona Wilensky for comments on earlier drafts of this essay. Gerald Grant and Linda Darling-Hammond offered especially helpful suggestions on the next to last draft.

dorsed one or another of these proposals. Several state and national agencies have begun to implement them. Major efforts are under way to mobilize much more consistent and powerful direction for instruction from state or national agencies.

These developments seem hopeful to some and unwise to others. But everyone agrees that they mark an astonishing reversal, and many therefore wonder whether the new proposals are attainable. One set of problems concerns politics. Power and authority have been extraordinarily dispersed in U.S. education, especially in matters of instruction. Could state or national agencies actually mobilize the influence required to steer teaching and learning in thousands, or hundreds of thousands, of far-away classrooms? That would require extensive new state or national infrastructure in education, as well as a radically new politics of education. Are such things possible?

A second set of problems concerns instructional practice. The new proposals envision much more thoughtful, adventurous, and demanding instruction. But most instructional practice in the United States is quite traditional: Teachers and students spend most of their time with lectures, formal recitations, and worksheets. Intellectual demands generally are modest, and a great deal of the work is dull. Only a modest fraction of public school teachers have deep knowledge of any academic subject. Hence, even if state or national agencies accumulated the infrastructure and influence required to steer teaching and learning, could they be steered so sharply away from long-established practice?

To answer these questions about how things might change, one must ask others about how they now work. How do instructional policies made in state and national agencies play out in local classrooms? What are the relations between policy and practice? What might it take to change them? Have central agencies ever tried to promote innovative and adventurous teaching? If so, with what results? These seem crucial issues for America today and tomorrow, but our knowledge about them is limited by what we did yesterday. The dispersed organization of American education rendered the connections between policy and instruction inconsequential for most of our history. The topic barely entered educational inquiry because it seemed so distant from educational reality. There is little American evidence about the structure or consequences of much greater state or national control. Similarly, American disdain for intellectually challenging education has left us with only modest evidence on how such education might turn out in this nation's schools. In order to learn much about such matters we must look beyond the U.S. education mainstream, and to studies of other national school systems.

We tackle the issues in four chunks. First, we probe the relations between state and national government on the one hand and instruction on the other. We explore how the structure and activities of central government affect classroom practice. But in some systems, key decisions about instruction, like what texts to read or what tests to use, are made by no central agency. Hence, in the second chunk of the essay we identify the specific sources of guidance for instruction, including tests, texts, and other things. We explore how such things interact with governance structures, and we probe their effects on classroom practice.

In the third chunk of the essay we scrutinize change in classroom practice. The recent U.S. reforms propose very ambitious shifts in instructional purposes, processes, and content: We inquire about the prospects for such change in teaching.

Finally, we consider nongovernmental influences on instruction. Recent reformers have proposed radical changes in policy, politics, and instructional guidance, seeing these as potent influences on classroom work. Yet studies of schooling here and abroad often suggest that social and cultural influences may be no less significant. For instance, some researchers report that Japanese families tend to support children's hard work and academic achievement, while Americans tend not to. Such differences may account for many of the effects often ascribed to policy and institutions.

GOVERNMENT STRUCTURE AND POLICY MAKING

The formal institutions of government are widely supposed to shape the relations between education policy and instructional practice. In France and many other nations, central agencies have enormous authority and power (Holmes, 1979; Lewis, 1985). Ministries of education make most policy for local education, and they often do so in great detail. But the U.S. political system was specifically designed to frustrate central power. Authority in education was divided among state, local, and federal governments by an elaborate federal system, and it was divided within governments by the separation of powers. These divisions were carefully calculated to inhibit the coordinated action of government, and they gained force from the country's great size and diversity (Kaufman, 1969).

The U.S. federal government thus has had relatively weak influence on education, as a matter of both law and tradition. But since World War II the central government has accumulated increasing influence on state and local decisions about funding, education for disadvantaged groups, civil rights and civil liberties in schools, research, and curriculum improvement. Despite these changes, direct federal governance of education is marginal. Federal agencies directly operate few schools and contribute only a little more than 6% of school operating budgets, on average (U.S. Bureau of the Census, 1989).

State governments are the constitutional center of U.S. education. But most states delegated most authority to localities, for most of their history. States supported the establishment of public schools with enabling statutes and, sometimes, a bit of money in the 19th century, but most of the pressures to establish public schools lay outside of state government. There has been some variability in states' influence in education. Hawaii has no local districts, and southern states have tended to be stronger than those elsewhere (Wirt & Kirst, 1982). But until 15 years ago the general pattern was extensive delegated state power. Most state agencies were small and weakly staffed (McDonnell & McLaughlin, 1982; Murphy, 1974). State governments have begun to exercise more power during the last decade (Cantor, 1980), but most are still far from what, in world perspective, could be called central control.

Such weakness in higher level agencies is quite unusual. In many nations the national ministry is the senior and often sole partner, managing all educational programs and paying most or all operating costs. In modern France, the schools have until recently been a creature of the national government in Paris, not of local or departmental governments (Cameron, Cowan, Holmes, Hurst, & McLean, 1984b; Holmes, 1979; Lewis, 1985). Even state or provincial governments in other federal systems have much greater power and authority. Australian state governments hold most constitutional authority in education, as they do in the United States. But the six Australian states also are the basic operating units in education (Boyd & Smart, 1987; Cameron, Cowan, Holmes, Hurst, McLean, 1984a). Each state operates all the public schools within its boundaries, performing all the functions that Americans associate with both state and local school government.

The United States thus has a remarkably fragmented governance system. Many important educational decisions are made in the nation's roughly 110,000 individual schools (U.S. Bureau of the Census, 1989). These include decisions about educational programs, student assignment, teacher assignment, and resource allocation among students (Wirt & Kirst, 1982). One result is remarkable variation across schools (Cusick, 1983; Powell, Farrar, & Cohen, 1985). Recent efforts at local "restructuring" and "school-based management" will almost certainly enhance the influence of many schools.

Local districts are the fundamental governance agencies, by tradition and practice. There are some 15,000 local districts (U.S. Bureau of the Census, 1989), and their influence is extraordinary in world perspective. Despite the recent growth of state and national power, these districts make a great range of decisions, including those that bear on levels of funding, the nature of educational program, and the teachers to be hired (Travers & Westbury, 1989). Financial support for most U.S. schools is still tied to local tax bases and taxation decisions, which produces enormous variation in educational resources and, thus, instructional programs. The key role of local districts builds many differences into U.S. education (Firestone, 1989).

Individual schools and districts have had much less influence in many other nations (Travers & Westbury, 1989). The French and Singaporean ministries of education have until recently monopolized decisions about educational programs, teacher assignment, and resource allocation (Cameron, Cowan, Holmes, Hurst, & McLean, 1984b, 1984d). Local schools have had little leeway within central guidelines, a condition that some nations have begun trying to change (Cohen, 1990a; Resnick & Resnick, 1985, 1989). And many nations simply have no local districts. Australian state education departments deal directly with each school (Cameron et al., 1984a), although some use regional offices for some administrative purposes. Funding decisions typically are made by national or state agencies, greatly reducing or eliminating fiscal and programmatic variation among schools. Some nations with strong central governments do have local jurisdictions that are supposed to play a large role in education. The postwar Japanese constitution guarantees local authority in such educational decisions as teacher hiring and curriculum (Cameron, Cowan, Holmes, Hurst, & McLean, 1984c). But the influence of local prefectures is constrained both by the broad authority of national agencies and by centuries-old habits of deference to the center. The result limits educational variation of many sorts (Cameron et al., 1984c).

In most nations, the relations between policy and practice are framed by systems of central power or by a small number of powerful state or provincial governments. The authority of the state is immense and, in many cases, theoretically unlimited. Schools are creatures of the nation-state or the province, and usually were created in the process of consolidating those entities (Meyer, 1983; Ramirez & Rubison, 1979; Ramirez & Boli, 1987).

The connection between central power and public education is a world pattern to which the local mobilization of schooling in the United States is one of the few great exceptions. Despite growing state and federal power, local government still is the key element in U.S. schooling. And the relations between policy and practice are framed by sprawling government structures in which fragmented power and authority express a considered mistrust of government.

If government structure frames the formal relations between central policy and classroom practice, policy-making fills that frame with specific content. The two are often at odds. While the design of American government incarnates a deep mistrust of state power, the design of most education policy expressed an abiding hope for the power of government and a wish to harness it to social problem solving. Collisions between the two were precipitated by the proliferation of state and federal education policies and programs in the last three decades. These included federal efforts to improve curriculum and instruction in the 1950s and early 1960s and to eliminate the racially dual school system throughout the South in the 1960s and 1970s. They also included federal and state efforts to improve education for disadvantaged students, to reform the education of handicapped students, to provide bilingual education for non-English-speaking students, and to ensure sex equity in schools across the nation. Nearly all of these policies and programs sought to solve problems that crossed jealously guarded jurisdictional boundaries among and within governments.

To speak of the relations between policy and practice in the United States is thus to speak both of collisions between policy and governance and of the consequences in educational institutions. Those collisions have affected the relations between policy and practice in several ways. New educational policies expanded central authority and drew the agencies of policy and practice closer together. But these policies did not commensurately reduce the autonomy of "lower level" agencies. The flood of state and federal policies and programs coursed through a large and loosely jointed governance system, and agencies throughout the system retained much of their operating independence. For instance, the states depend on localities for political support and policy execution, as any higher level agent depends on subordinates. State governments, therefore, should be constrained by what localities will accept. Yet the states often act with remarkable independence. The state education reforms of the last 10 years have in some respects been quite offensive to local educators, but many still have been enacted with little difficulty (Fuhrman, Clune, & Elmore, 1988). Similarly, the national government has only a modest constitutional role in education, and it has long deferred to state and local authorities. Nonetheless, federal agencies have taken various dramatic initiatives designed to greatly change state and local education. Many were taken over local and state opposition, some over fierce and even violent opposition (Orfield, 1969). Despite the constraints that lower level agencies can impose on their superiors, agencies above have regularly pushed far beyond the presumed limits.

The same phenomenon obtained in reverse: State and local autonomy has been only modestly constrained by higher level policy. Researchers have documented the states' great flexibility in responding to the dramatic federal policies and programs of the 1950s and 1960s (Murphy, 1974). Researchers also have shown that local schools and districts retain considerable latitude in coping with state and federal policies (Berman & McLaughlin, 1977; McLaughlin, 1987). Despite the increasing flow of higher level requirements, advice, and inducements, lower level agencies have much room to interpret and respond. Relations among state, federal, and local agencies therefore remain quite attenuated despite decades of effort to bring them closer together. Centers of organization and governance are widely dispersed and weakly linked, despite the growth of policy. Central agencies can make serious demands on others with relative ease; they need only mobilize the political resources to enunciate a policy or begin a new program. But the costs of enforcing demands are much greater. A great distance remains between state or federal policy-making and local practice (Firestone, 1989).

Yet policy-making has complicated educational organization. In order to make contact with local educational organizations, state and federal agencies have had to bridge vast political chasms artfully designed to frustrate central power. To increase general governance authority in education was politically unthinkable for the federal government. What is more, federal agencies were weak. They had no general capacity in curriculum, instruction, school personnel, or assessment, since both the Constitution and political practice were thought to forbid it. State agencies had much more authority, but with a few exceptions they had little more capacity. A majority of states had delegated most operations to local governments and private test and text publishers. Traditions of decentralization, suspicions about central power, and deference to local authority meant that higher level authority could only grow by way of individual, freestanding programs, each of which promised to solve a specific educational problem (Bankston, 1982; Meyer, 1983). But these individual programs were located in agencies that had little general operating capacity in the "technical core" of education.

Hence, when weak federal and state agencies tried to implement such ambitious programs as Head Start and Title I of the Elementary and Secondary Education Act (1965), each program had to be outfitted with its own minimum core of administrative operations (budget, personnel, evaluation, and the like). Furthermore, each program had to coordinate operations across many levels of government, owing to the lack of general administrative capacity above the local level. And each policy or program had to do so in ways that did not require much capacity in such key areas of education as curriculum or instruction, since such things were regarded as off limits to central government. Lacking general central authority and capacity, leaders of each program had to establish their own systems. How else could they hope to mobilize tens or hundreds of thousands of educators, in hundreds or thousands of jurisdictions, across several levels of government?

Work in such policies and programs therefore was confined within specialized administrative subunits organized around oversight tasks within each program (Wise, 1979). Administrative capacity grew, but within programs rather than across entire governments. Administrative burdens therefore multiplied as the same or similar administrative work was repeated across programs (Bankston, 1982; Cohen, 1990a; Meyer, 1983; Rowan, 1982, 1983). Central agencies grew, but in a fragmented fashion (Clark, 1965; Scott & Meyer, 1983; Stackhouse, 1982). And the administrative expansion added little to central capacity in the core areas of education such as curriculum and instruction. The collisions between optimistic policies and cautiously designed government have produced fractured and duplicative administration.

These fractures were reflected in the organization of agencies outside of government. As policies and programs took shape, networks of interested agencies—advocacy organizations, professional groups, and special purpose research and development agencies, among others—grew up around them. Examples include the loose network that helped to build support for the legislative proposals that became PL 94-142 and Title I of the 1965 ESEA (now Chapter I). Each network has helped to coordinate and stabilize program operations and mobilize support for programs across governments and among many sorts of agencies (Cohen, 1982; Peterson, 1981; Peterson, Rabe, & Wong, 1986). Like the programs and policies that they grew up around, these policy networks are ingenious, for they support state and national efforts to solve local problems in a political system that was designed to frustrate such efforts (Kaufman, 1969). But these clever inventions also encourage political fragmentation and multiply administrative work (Bankston, 1982; Cohen, 1982; Meyer, 1983; Rogers & Whetten, 1982). For they support fractured authority within education agencies, as managers in each program attempt to build their own bridges across great political chasms. The ingenious devices that cope with fragmentation among governments tend to exacerbate fragmentation within them.

Collisions between cautious designs of government and hopeful designs for policy also complicated local educational practice, because administrative work grew as localities coped with increasing state and federal policies and programs. Since higher level authorities are so distant from local practice, they are rarely held accountable for their actions there. Hence state and federal initiatives were generated with little regard for the relations among them, or for their cumulative local effects (Kimbrough & Hill, 1981; Kirst, 1988; Wise, 1979). Indeed, some of the most potent local effects of state and federal programs or policies had no intended programmatic content. The best case in point is underfunded mandates. Federal legislation for handicapped students placed unaccustomed procedural and substantive burdens on local education agencies, but the legislation carried less than half the estimated costs of compliance. Initially, it was thought that full funding would soon follow, but it never did. Federal requirements were never relaxed, though. Local and state school agencies had to allocate their own funds to this area of program support, often with grave results for other educational activities.

Yet requirements have limits. State and federal officials rarely can effectively oversee local program implementation. No state or federal education agencies have the inspectorates found in Britain, France, and their former colonies. At best, U.S. state and federal agencies use oversight-at-a-distance, such as written program evaluations, grant recipients' reports on operations, and the like. Such things multiply work without producing fruitful contacts among public servants at different levels of government (Bardach & Kagan, 1982). And local schools retain considerable autonomy. Administrators and teachers usually can tailor higher level programs to local purposes and conditions if they have the will and take the time (Berman & McLaughlin, 1977). Often they can cope with higher level directives simply by ignoring them. Inattention is a ubiquitous management tool (Kiesler & Sproull, 1982), and it can be especially efficient in a fragmented governance system.

These patterns contrast sharply with many foreign education systems. The ministries of education in France and Singapore deal with schools on a broad range of educational matters, as do the state departments of education in Australia and provincial governments in Germany. There are administrative subunits in these agencies, but they are broadly defined by the key areas of schools' operation (i.e., curriculum, instruction, personnel, and the like). The subunits have extensive general authority, and new initiatives typically subsist within them rather than being set aside in independent units, because the operating units make the key decisions about education and have the resources. That is what might be expected in nations founded on *etatist* traditions. Policy initiatives are not organized as though they were at war with government, or on the assumption that they can have little to do with the core operations of education.

The collisions between rapidly expanded policy-making and fragmented governance are a hallmark of U.S. education. Few nations have such dispersed authority and power in education, yet few have such intense higher level policy-making. Americans complain more than any other people about state interference with education and centralizing forces in schools, but authority and power are more dispersed here than in nearly any other nation. Perhaps that is why we complain more.

INSTRUCTIONAL GUIDANCE

State and federal governments have made many efforts to improve instruction. They offer financial aid to local districts, sponsor child health and nutrition programs, and support efforts to improve education for the disadvantaged. Yet such policies rarely make broad or close contact with instruction. Teaching and learning are more directly affected by the texts that students and teachers use, the examinations that assess students' academic accomplishments, the standards teachers must satisfy in order to secure a post, and the like. These instruments comprise the means so far invented to guide classroom work. We lump them under the rubric of instructional guidance and sort them into five categories: (a) instructional frameworks, (b) instructional materials, (c) assessment of student performance, (d) oversight of instruction, and (e) requirements for teacher education and licensure.

Nations use these instruments very differently. In some cases guidance is designed and deployed by governments, while in others private agencies play a large role. Additionally, the arrangement of government-sponsored guidance varies greatly across nations (Broadfoot, 1983). And while all school systems adopt some stance toward guiding instruction, often that stance includes offering little advice.

Instructional guidance also mediates the effects of other policies that seek to affect practice, because the effects of all government policies that try to influence instruction—including those that do so by offering extra aid to the disadvantaged or holding schools "accountable"—are mediated by such things as instructional materials, teachers' professional capacities, and methods of student assessment. Intentionally or not, the aggregate of instructional guidance is a medium in and through which many other educational policies and programs operate.

In what follows we compare instructional guidance in the United States with its counterparts in other national school systems. We focus on the instruments of guidance; while these are governed in many different ways in national school systems, we do not try to describe that variety here. Instead we use a few key categories that describe variations in instructional guidance. These variations can be produced by many different governmental and administrative arrangements (see Porter, Floden, Freeman, Schmidt, & Schwille, 1988):

Consistency: Given different domains of guidance, one important issue concerns the extent of agreement within and among domains. In some systems instructional frameworks are consistent internally, and consistent with texts or teacher education. But in other systems they are not.

Specificity or prescriptiveness: Teaching and learning are complex enterprises, and there are many different ways to enact them. Teachers and students are offered clear and detailed guidance about content coverage or pedagogy in some systems, while in others guidance is very general or vague.

Authority and power: To offer guidance is not to decide what weight it carries. Advice for instruction is presented in ways that have great authority with students and teachers in some systems, but in others such advice is presented in ways that carry little weight.

Instructional Frameworks

Instructional frameworks are general designs for instruction (i.e., broad conceptions of the purposes, structure, and content of academic work). Frameworks can set the terms of reference for the entire enterprise. In some school systems they guide course structure and content, the nature of textbooks, the purposes and content of examinations, and the like. They can be quite prescriptive: In some former French and British colonies such frameworks offer extensive and focused guidance about instructional content, and in some cases approaches to teaching as well. In France many curriculum decisions are made in the national ministry of education (Horner, 1986), which often details the topics to be studied, the teaching materials and methods to be used, and even time allocations (Beauchamp & Beauchamp, 1972; Lewis, 1989). The Japanese central ministry issues frameworks for each subject (Kobayashi, 1984), prescribing content and detailing the sequence of topics (Kida, 1986; Organization for Economic Cooperation and Development, 1971).

Such guidance often seems to carry great authority and power. In France, many curriculum decisions are made by the national assembly, while others are ministry decrees. But authoritative guidance need not be governmental. In Holland it is offered by autonomous agencies that are supported by government but are not part of it.

Frameworks have been unusual in the United States. The most common instructional designs have been bare listings of course requirements by states or localities. Apart from the New York State Regents, it was long uncommon for state agencies to offer advice about the material to be covered within particular subject areas, or about the structure of courses. This passivity was not unique to state governments. Until quite recently, few local systems seemed to prescribe topics within courses or curricula. And guidelines about pedagogy have been even more rare. Relatively weak state and local guidance concerning course content and pedagogy has meant that students and teachers have had great latitude in shaping the content and purposes of their courses (Cusick, 1983; Porter et al., 1988; Powell, Farrar, & Cohen, 1985; Schwille et al., 1983, 1986; Sedlak, Wheeler, Pullin, & Cusick, 1986).

A few states recently have moved more aggressively into instructional design. Florida, South Carolina, and a few other southern states instituted statewide basic skills curricula in efforts to improve students' performance during the past decade. These included guidance for content coverage and pacing, and at least implicitly for teacher education. Several states have published evaluations that claim gains in student achievement, although no independent evaluations seem to have been done. At the same time, several other states have pressed guidance for a radically different sort of content. In 1985, California issued the first of a series of curriculum frameworks that were intended to make teaching and learning intellectually much more ambitious and demanding. Arizona and Michigan have taken some similar steps, as has Connecticut.

Some local school systems also began to move toward instructional frameworks in the 1970s and 1980s, with the news that test scores were declining and mounting demands that schools get "back to the basics." Local districts came under unfamiliar pressure to improve performance, and some began to devise minimum instructional programs in response— Washington, DC, Chicago, and Philadelphia among them. There is little systematic research on these matters, so we cannot gauge the depth or extent of these changes. Additionally, several cities that adopted such schemes recently announced their demise. But officials in a few districts that we recently visited reported a move to greater central control. Schools can no longer determine their own instructional programs, and central offices have written rudimentary curriculum frameworks, usually blueprints for "essential skills."

Instructional Materials

Texts and other materials are found in all systems, but the extent of guidance for their content and use varies enormously. In many systems, the national ministry sets the terms of reference for text content, and/or authorizes the textbooks to be used based on curriculum frameworks (Kida, 1986; OECD, 1973). In such cases, there is a good deal of consistency between the guidance teachers receive from textbooks and from national curriculum frameworks. In some nations ministries actually publish texts, while in others texts are privately published. But in either event, materials are closely tied to curriculum frameworks.

Decisions about instructional materials have been much more fragmented in the United States. Since there have been few instructional frameworks until recently, publishers had little or no consistent, content-oriented guidance. Instead, they were guided by what had been done before, by official and unofficial expressions of state or local preferences, and by their own sense of the market. Texts have improved in many ways over those that were available in the 1920s, but most commentators regard most texts as intellectually shallow. Many states and localities officially adopt textbooks, and Americans often have thought this to be highly prescriptive for instruction. But lacking much official guidance for topic coverage within texts, save for such matters as evolution, these texts seem not to have been highly prescriptive for topic coverage (Floden et al., 1988). Researchers report that many texts mention many more topics than could be dealt with, which leaves open extensive topic choice by teachers (Tyson-Bernstein, 1988). Additionally, there seem to be appreciable inconsistencies in content coverage among the different texts for most subjects at most grade levels (Freeman et al., 1983). Hence texts have offered many opportunities for teachers and students to vary the content they cover (Freeman & Porter, 1989; Porter et al., 1988; Schwille et al., 1983).

As several states recently moved toward more explicit instructional designs, they tried to make them count for textbooks. California used its new curriculum frameworks in mathematics, and in literature and language arts, to press publishers to revise texts. Publishers were told that if they did not make satisfactory revisions their texts would not be approved for adoption. But the state's guidance still was general. The mathematics framework, for instance, offered little specific guidance about topic coverage. Casual comparisons of the new and old literature and language arts texts with the revised framework suggest that the state has won some significant changes, although systematic analysis remains to be done. Studies of the mathematics texts and framework suggest only modest change thus far (Putnam, Heaton, Prawat, & Remillard, in press).

Some local districts also have begun trying to promote greater consistency between instructional frameworks and materials. Several that devised such frameworks also specified the knowledge and skills that students and teachers should cover in texts and other materials, often doing so in compilations of "essential skills." In at least one case, local officials tied their guidance to recently published texts that seemed to fit with the local instructional frameworks. The district specified the material to be covered in the common text, and when it should be covered. This constitutes an extraordinary change for U.S. schools, but we have found no studies that gauge its breadth or depth.

Assessment of Results

Assessment of instructional results is an essential element of instructional guidance in most school systems. Though assessment practices are changing in European systems (Kellaghan & Madaus, 1991; Madaus, 1991), many nations tie assessment closely to curriculum. In France and many former French and British colonies, examinations are referenced to national curricula, instructional frameworks, or both. The examinations thus provide both a visible target for instruction and a means of checking its results (Madaus, 1991; Resnick & Resnick, 1985, 1989). The nature of assessment in these cases varies greatly among nations, but it all differs from American approaches. The examinations probe students' performance in specific curricula. Many systems mix multiple-choice questions with extended essay or problem-solving performances, though some— Japan, for instance—rely entirely on multiple-choice questions (Cheney,

1991). In contrast, U.S. schools employ standardized tests that are referenced to national norms and are designed to be independent of curricula. Performance has been limited to answering multiple-choice questions (Noah & Eckstein, 1989).

In France, Great Britain, and Japan, examinations count in very specific ways. Students' promotion and further education depend partly or entirely on their exam performance (Eckstein & Noah, 1989). Indeed, many school systems that employ examinations are highly selective, and the exams are the key agent of selection (Kellaghan & Madaus, 1991). In Singapore, exams are used to make nearly irrevocable decisions about streaming in both the primary and secondary grades and thus to influence decisions about students' careers and further education. This use of examinations sharply limits students' opportunities to recoup earlier poor performances. The United States lacks such a selective examination system, which is one reason why students here have more "second chances" than they do in any other nation. But the use of examinations for student selection does enhance the examinations' authority (Madaus, 1988, 1991; Madaus & Kellaghan, 1991; Resnick & Resnick, 1985, 1989). In New South Wales, Australia, for example, students' performance on the school leaving exam determines their opportunities for further education. Differences of a tenth of a point in exam scores can be crucial. In Japan, scores on both national secondary school leaving exams and university extrance exams determine which high school students will go on to university and determine the quality and prestige of the universities that students will attend (Ohta, 1986; OECD, 1973; White, 1987). Secondary schools' prestige also is tied to students' success in examinations for prestigious universities (OECD, 1971). The social and economic significance of exam performance offers many incentives for students and teachers to take exams seriously.

Matters are very different in the United States. There is a great deal of assessment, but it has an uncertain bearing on instruction. One reason is that most tests have been designed to minimize their sensitivity to specific curricula (Madaus, 1989; Resnick & Resnick, 1985; Smith & O'Day, 1990). What is more, many different tests are designed, published, and marketed by many different private testing agencies. And most decisions about which test to use have been made by thousands of local and state school agencies, each of which adopts tests of its own liking independent of the others' decisions. All of this has made for inconsistent guidance from assessment.

Variation in content coverage has been another source of inconsistency in the guidance that U.S. tests offer for instruction. Standardized tests often have been seen as interchangeable, but one of the few careful studies of topical agreement among tests raised doubts about that view. Focusing on several leading fourth-grade mathematics tests, the authors observed that "our findings challenge . . . th[e] assumption . . . that standardized achievement tests may be used interchangeably" (Freeman et al., 1983). They maintain that these tests are topically inconsistent and thus differentially sensitive to content coverage.

Inconsistency has been further enhanced by the widespread local practice of using one publisher's test in one grade and others in other grades. This problem has been magnified by the increase in testing during the past several decades, as local and state-sponsored minimum competency and essential skills tests have spread. American students are now tested much more often than they were 20 years ago, but they are tested with more different tests.

Thus, established U.S. approaches to assessment would have impeded consistency among the elements of instructional guidance, had consistency been sought. Until recently, however, it was not. The guidance for instruction that tests offered was general, and probably more a matter of the form of knowledge (i.e., it exists in multiple-choice formats and is either right or wrong) than its content. This guidance also was vague, since the test results were rarely known. Test results were kept from teachers, partly on the designers' view that they were not designed to guide instruction.

Indeed, decisions about test design, marketing, and adoption typically have been made apart from knowledge of specific school curricula, teacher education, and the like. Test theory and practice have held that such independence is crucial to test validity, but this has further weakened consistency between tests and instructional materials. Research seems to bear out the weak relations between the subject matter content of standardized tests and of texts. Several investigators concluded: "If a fourth-grade teacher limits instruction to one of the four books analyzed, students will have an adequate opportunity to learn or to review less than half of all topics that will be tested" (Freeman et al., 1983).

To the extent that tests have guided instruction, then, they have done so inconsistently. This has weakened the instructional authority of the tests. It is thus not surprising that many teachers report they rarely take test results into account in instruction (Floden, Porter, Schmidt, & Freeman, 1978; MacRury, Nagy, & Traub, 1987; Ruddell, 1985; Salmon-Cox, 1981; Sproull & Zubrow, 1981).

But two qualifications are in order. First, there have been a few exceptions to these patterns, notably the New York State Regents exams and the Advanced Placement Program (AP). The AP program is a special subsystem within public education in which high-achieving students take advanced courses. The AP exams seem to strongly influence instruction, in part because they are tied to a suggested curriculum and readings. The exams also seem to be taken seriously by most students and teachers, partly because the scores count for college entrance as well as college course taking. But these have been anomalies in American education (Powell, 1991).

Second, the patterns described above have begun to change. Rising public interest in testing and other political pressures led many states and localities to begin publishing scores in the early 1970s, after decades of secrecy. By now, many do so as a matter of course and often conviction. State and local school agencies also increasingly turned to tests in efforts to improve instruction. The favored method was to institute "accountability" schemes, often based on minimum competency tests. Many of these schemes included only a high school graduation requirement, but some also included tests for promotion. Some were hastily contrived under political pressure. The tests often were adapted from standardized norm-referenced tests designed for other purposes.

State and local use of tests to guide instruction marked a dramatic turn in assessment practices. But the fragmentation characteristic of U.S. education was evident here as well. Many minimum competency tests were unrelated to other elements of instructional guidance, such as curriculum. The tests effectively became the curriculum in some cases (Darling-Hammond & Wise, 1985). Recently, however, that has begun to change as well, as some publishers have brought out text series that are accompanied by criterion-referenced test systems. These link curriculum and instruction to testing. In several cities that we have studied, these test and text series are the heart of the instructional program. Students' performance is monitored by regular testing keyed to text pages, and sometimes students are retested until they achieve "mastery." We have discovered no studies that probe the frequency of such practices, though they seem to be found chiefly in cities with many disadvantaged students. In such cases, tests offer much more specific and prescriptive guidance than ordinarily has been the case in the United States.

How does such testing affect instruction? There has been surprisingly little research on the issue. Several researchers assert that the tests have had a powerful effect on teaching (Darling-Hammond, 1987; Darling-Hammond & Wise, 1985; Resnick & Resnick, 1989; Romberg, Zarinnia, & Williams, 1989). Competency tests are said to drive instruction in a mechanical and simplistic direction. Teachers orient instruction to the test items, and if students do poorly on the test, remediation consists of drill on the items they do not know (Kreitzer, Madaus, & Haney, 1989; Madaus, 1988). A recent U.S. Department of Education report claims that "accountability systems . . . are very powerful policy tools that have changed school-level planning and teaching activities" (OERI, 1988, p. 31).

But it also is often said that these claims only hold for situations in which the tests carry "high stakes" (i.e., that they count for students' academic progress, or for schools or teachers). This condition does not hold for many minimum competency testing programs (Ellwein, Glass, & Smith, 1988), or for many students in high stakes testing programs. It also seems to be accepted that such tests are much more likely to affect poor and minority group children, since more advantaged students pass the tests with little effort. These considerations suggest that the effects of minimum competency testing have been quite uneven, and are salient for a particular segment of the school population. Additionally, we do not know how salient the tests have been, because there have been no observational studies of teachers' responses to tests. The little research on competency testing thus far is based on interviews with teachers who describe the effects of testing in rather global terms, and the evidence they present is very mixed (OERI, 1988; Romberg, Zarinnia, & Williams, 1989).

The effects of testing have been complicated by recent reforms. Minimum competency testing has come under sharp attack, and standardized testing itself is the object of unprecedented criticism. Several states recently have begun to use novel testing programs in efforts to strengthen and radically change guidance for instruction. The California state education department has begun revising its statewide testing program in an effort to "align" the state's tests with its ambitious new curriculum frameworks. State officials hope that if the tests are changed to assess thinking and understanding rather than facts and memorization, they will "drive" instruction in the new directions. Connecticut has been making similar changes, although it seems to rely on tests much more than on instructional designs. Florida has dropped its minimum competency testing program in favor of a radically different approach to reform. Proposals for authentic assessments and performance assessments have become common, and many educational agencies claim to be implementing them. This ferment is quite unprecedented, but the developments are so recent that little is known about the operation of innovative assessments, let alone their effects.

Monitoring Instruction

The inspection of students' work, the observation of teaching, and other sorts of monitoring constitute a fourth type of instructional guidance. Monitoring also varies dramatically among nations. French and British central school agencies long included inspectorates, whose duties extended to checking on the topics that teachers covered, their pedagogy, and the materials they used. British inspectors visited schools to maintain standards of work and offer advice on content and pedagogy, though this role has fallen into disuse in Britain (Lawton & Gordon, 1987). They still publish reports and conduct continuing professional education for teachers (Lawton & Gordon). Such arrangements are common in one form or another in many former French and British colonies.

Monitoring has been extremely modest and inconsistent in the United States. Few states and localities systematically monitored either teachers' coverage of curriculum or the quality of classroom work. There were no education inspectorates, nor was it common for principals to keep tabs on students' and teachers' academic work (Schwille et al., 1983). Indeed, it was uncommon for students to keep the detailed records that would permit such monitoring. Even if such records were kept, few principals involved themselves in instruction. Hence there have been few checks on what materials are used, how they are used, or what instruction is provided. In this respect, U.S. teachers have had quite extraordinary autonomy.

Many observers believe that U.S. teachers nonetheless teach more or less the same thing anyway. They often point to the use of textbooks, believing that the text determines instruction in most classrooms. If teachers use the same texts, it is expected that they will teach the same thing. Though there has been little research on this matter, the assumed homogeneity of content coverage is unsupported by the available evidence. Even when teachers use the same texts, their content coverage seems to vary greatly (Putnam et al., in press; Schwille et al., 1983). The authors of one study concluded that "this investigation challenge[s] the popular notion that the content of math instruction in a given elementary school is essentially equal to the textbook being used" (Freeman & Porter, 1989, p. 418).

There are some recent signs of change. Many state and local systems attempt to monitor instruction with minimum competency tests, though the evidence suggests that these efforts are quite inconsistent and often ineffective. But at least one local school system that we visited went further: As it adopted more centralized instructional guidance, the district also devised a way to monitor teachers' coverage. Teachers fill out forms that report chapter and page coverage in required texts, and the forms are read by principals and central office officials. Some states also have begun monitoring of a sort. South Carolina has used test scores to identify both low-performing schools and districts that need special attention, and high-performing schools and districts that can be released from various state requirements. But there are few studies of these schemes, and we could find no investigations of their effects on instruction.

Teacher Education and Licensing

Guidance for instruction in this realm also varies greatly among nations. In many countries, the guidance offered by teacher education is quite consistent with other sorts of guidance. One key connection is with the schools' curriculum; for instance, in Singapore, teachers' professional education is closely tied to the curriculum of the schools. Additionally, in many nations the requirements for licensure are national rather than local, and teacher education is consistent across institutions. That seems to be true at the national level in France, partly because the ministry's inspectors play a central role in the preparation of elementary school teachers in the Ecole Normale (Lewis, 1985). This tends to create consistency in the professional education of teachers and in the messages they receive from different elements of the system.

Such guidance is more of a hodgepodge in the United States. States are the agency for licensure of virtually all occupations; however, unlike medicine, teacher certification requirements are inconsistent across states and often within them. Chicago and New York City, for instance, have different certification requirements than do the states in which they are located. The interstate differences are so considerable that one recent study concluded that "a teacher certified in one state is unlikely to meet the certification requirements in another" (Haggstrom, Darling-Hammond, & Grissmer, 1988, p. 12).

Most requirements for certification focus on teachers' education, and virtually all concern higher education. But the state agencies that set certification standards are remote from the colleges and universities that conduct most teacher education. Moreover, certification agencies usually have little connection with the state agencies that govern colleges and universities. Certification agencies, in addition, have tended to act purely in terms of course requirements rather than course content or students' performance. Hence there is room for considerable variation in how colleges and universities interpret the same requirements.

Another source of inconsistency is the loose relation between college and university requirements for teacher education and the schools' curriculum. Schools' curricula vary within states, as well as within local districts. The variety of local instructional programs cannot be accounted for by teacher education departments. And in many cases members of those departments regard the schools' curriculum as a collection of errors that intending teachers must learn to avoid.

Against this background, the mere idea of consistent guidance for teacher education and licensing seems revolutionary. Yet, recently there have been moves in that direction. Most notable is the National Board for Professional Teaching Standards (NBPTS), which has begun efforts to develop a voluntary national examination system for teachers. If successful, this could lead to a partial national system for teacher certification, which could profoundly affect teacher education.

Instructional Guidance: An Overview

Instructional guidance in the United States has been inconsistent and diffuse. Many private and public agencies issue advice for instruction, but few take account of each other's advice. Hence much guidance for instruction has been unrelated, divergent, or contradictory. It also has been largely decoupled from government. Public agencies have extensive authority to guide instruction, but they delegate most of it to private firms or local schools. The influence of U.S. school governments pales when compared with central or provincial agencies elsewhere.

Instructional guidance also filters the effects of other initiatives that aim to influence classrooms. Prolific and inconsistent guidance in the United States has muffled and diffused such initiatives. Since government officials could not turn to an established system of guidance in efforts to shape instruction, individual programs or policies could not exert a powerful and consistent influence on instruction. Each program or policy was on its own, each competing with a buzz of other advice. Federal and state policymakers dealt with this problem by trying to mobilize special arrangements (e.g., program guidelines, evaluation, and technical assistance). But these are ancillary to the core instruments of guidance, and have been no more than modestly influential.

The result is paradoxical. Public and private agencies prolifically produce guidance, more than in societies with much more potent advice for instruction. But this does not press instruction in any consistent direction, because when guidance is inconsistent and diffuse, no single test, curriculum, or policy or program is likely to have a broad or marked effect. Many teachers and students are aware of different sorts of advice, but few are keenly aware of most of it. Many know that most guidance is either weakly supported or contradicted by other advice and that much can safely be ignored. The din of diverse, often inconsistent, and generally weak guidance opens considerable latitude to those who work within it.

Teachers' habits and decisions are important in any system of instruction. But absent plain and strong guidance, they become unusually important. The result in U.S. classrooms is curiously mixed. The forms of instruction are generally traditional, and the intellectual level usually is low, but the specific content is remarkably variable. There are many reasons for the variation, including differences in students' inclinations and teachers' judgment. But one additional reason is that students' and teachers' preferences are not informed by a plain system of common purposes and content. Classrooms around the world are of course traditional in form as well, often much more so than in the United States. But classrooms here exibit a distinctive sort of diffuse, academically relaxed traditionalism. Yet the content is highly variable. Teachers' work is guided more by inherited practices and individual decisions than by any clear and common view of what is to be covered, how it is to be covered, and why. In this sense, American schools have the worst of both worlds.

Our point is not that instructional guidance has been irrelevant in U.S. schools. Rather, it has been relevant only when someone chose to notice it and to do something about it. In a sense, this is true anywhere: Teachers in Singaporean or French schools must notice guidance and choose to do something about it before it can shape instruction. But the consistency, prescriptiveness, and authority of instructional guidance in such places increases the chances that teachers will notice the same advice. In contrast, teachers' and students' autonomy have been enhanced in the United States because they work in such a diffuse system of instructional guidance. The classroom doors behind which teachers labor are no thicker here than elsewhere, but teachers in the United States receive fewer strong and consistent messages about content and pedagogy. Hence, they and their students have found it relatively easy to pursue their own preferences once the doors have closed behind them.

The situation has begun to change as recent school reformers seek to cure the ills of U.S. education by mobilizing more consistent guidance for instruction. We know little about the effects of these efforts, but the cures bear an uncanny resemblance to the disease. Several states are trying to promote some form of consistent guidance, but quite naturally do so independently of each other. Many localities are trying to promote some version of consistent guidance, but do so with no reference to each other or to state policy. Federal education officials recently have begun trying to create more consistent guidance for instruction, but their efforts so far have been independent of many state and local endeavors. Several national groups—the National Governors Association, NBPTS, the National Council of Teachers of Mathematics, and others—also are trying to promote more consistent guidance. Each, of course, is carrying on independently of the rest, and none are much related to state and local efforts. Some professional associations also have taken up the idea, as have several academic disciplines; however, there is modest contact among these endeavors as well, and little relation to state and local initiatives. We live in a blizzard of different, divergent, and often inconsistent efforts to create more consistent guidance.

There also are deep divisions over the content of the recent reforms. Proposals for more lively and demanding instruction are circulating in various political, disciplinary, and educational circles, but there are many versions of the new ideas. These novel schemes also compete with established ideas and practices, for "back to basics," "effective schools," and "direct instruction" all are alive, well, and firmly rooted in school and classroom practice.

All of this is par for the American course. Government structure has not been changed by recent reforms, nor has political practice. The power of our ingeniously fragmented political system is evident even in efforts to cure fragmentation. Some attack fragmentation as a barrier to more effective instruction, but others celebrate it as a source of vitality in American institutions. Similarly, today's disagreements about the aims and methods of education are only the most recent expression of old tensions between our practical and anti-intellectual bent and our occasionally more elevated aspirations. The dispute has deep roots in both popular culture and the institutions of education, and it would be astonishing if it were settled easily or soon.

Effects of Instructional Guidance

If instructional guidance is worth noticing, it must be because it makes a difference to teaching and learning. Does it?

Many educators around the world would think the answer obvious and affirmative. That guidance affects instruction is the working assumption of many European and Asian school systems. But many U.S. social scientists argue that it is difficult or impossible to steer education toward consistent practices or results, owing to weak knowledge of educational processes and other uncertainties (Berlak & Berlak, 1981; Floden & Clark, 1988; Jackson, 1968; Lampert, 1985; Lortie, 1975). John Meyer and his associates contend that school systems therefore create elaborate rituals, building a "logic of confidence" to replace evidence of rational relations between educational resources and processes on the one hand, and results on the other (Scott & Meyer, 1983). School systems "buffer" themselves by offering evidence on attendance and degrees instead of evidence on performance. Oddly, there is little evidence on these contending assumptions. For all the variation in instructional guidance, there is little research on its effects.

Teaching

Many scholars assert that guidance affects teaching. In writing of the effects of the French Baccalauréat examinations, for instance, Patricia Broadfoot notes that "examination questions virtually become the [schools'] syllabus" (Broadfoot, 1984). But guidance from one source can be offset by guidance from another. Hence we put the issue more specifically: Is teaching more consistent in school systems with more consistent instructional guidance? The only direct way to answer this question would be to connect evidence on the structure and content of guidance in education systems to evidence on teaching within them. The only study that permits such comparisons is the Second International Mathematics Study (SIMS). And while SIMS contained evidence on math teaching and curriculum for 15 nations, it offered little data on system structure. David Stevenson and David Baker compiled such data, focusing on the degree of central curriculum control. They tied this to SIMS data on the consistency of topic coverage among teachers within nations. They found that cross-national differences in the degree of central curriculum control were positively related to consistency in the topics that teachers reported they taught. Teachers in nations with more centralized curriculum control reported greater agreement on topics taught than did teachers in systems with less central control. More centrally controlled systems also had fewer teachers who reported teaching little of the prescribed curriculum. There was less within-system variation in the amount of mathematics instruction in systems with more national curriculum control than in those with local or provincial control. Finally, teachers in more locally controlled systems were more likely to report that they adjusted instruction to local conditions, including their perceptions of students' ability and mastery of mathematics (Stevenson & Baker, 1991). While modest, these differences all suggest an effect of consistent guidance. But Stevenson and Baker point out that they had no direct measures of consistency in guidance.

SIMS seems to be the only data set in which system-level effects can be explored, but instructional guidance operates at many levels of education. Many recent studies of school effectiveness have focused attention on school-level consistency in guidance. While the studies are of varying quality, they show that schools differ widely with respect to consistency in instruction. Some adopt a laissez-faire style, permit diverse offerings and approaches, and thus create many choices for teachers about what to teach and how, and for students about what to study and how much. Others offer more consistent instructional guidance, thus limiting both instructional offerings and faculty and student choices (Bryk, Lee, & Smith, 1990; Cusick, 1983; Powell et al., 1985).

What explains the effects of instructional guidance on teaching? Researchers who study individual schools offer varied answers to the question. Some point to school heads' leadership in forging consensus about goals and methods. Others focus on school "climate" or shared norms for instruction among faculty and students (Bryk et al., 1990). Levels of faculty collegiality and cooperation sometimes are offered as another sort of guidance for teaching (Purkey & Smith, 1983). But other analysts point as much to structural as cultural factors; that is, some schools are committed to less differentiation in the curriculum and thus to fewer choices for students and teachers, creating more consistency by organizing the curriculum around a common core of courses (Powell et al., 1985). Not surprisingly, such schools tend to be smaller (Bryk et al., 1990), which suggests another influence on consistency. Researchers who study school systems offer a different sort of answer: More central control of curriculum produces more consistent topic coverage (Stevenson & Baker, 1991). But it is possible that such consistency only expresses what teachers learned as students. If elementary and secondary schools are the prime agencies of teacher education, as many scholars argue, then the curriculum that teachers present may reflect their earlier school learning, rather than current official directives. The difference could be consequential for reform. If official directives are a potent influence on teachers' actions, then recent state and national reforms might quickly affect classroom work. But if consistency is more the result of inattentive curricular hand-me-downs, then changes in policy could take much longer to find their way into classroom practice.

Learning

Our interest in the effect of instructional guidance on teaching is partly instrumental: We want to know whether it affects learning. There is, unfortunately, no cross-national evidence on this issue, nor do we expect anything persuasive very soon. For researchers would have to connect evidence on the large structure of educational systems with evidence on the fine structure of teaching, and connect both of those to learning. Furthermore, they would have to do so across many different nations with different school systems. It would be an immensely complex task to make those connections while also taking other salient influences into account. If the prior history of research on school effects is any precedent, knowledge will grow slowly.

But many U.S. schools have tried to improve learning by increasing guidance for instruction, and many researchers have investigated the effects. One body of evidence arises from studies associated with the movement for "effective schools." Researchers reported that students' achievement improved, or was higher than expected, in schools in which leaders focused on common goals and faculty had high expectations for students (Purkey & Smith, 1983; Rowan, 1990). But these studies usually involved only a few schools, and most offered very limited data on school organization and culture (Purkey & Smith).

More systematic evidence on the effects of school-level guidance arises from reanalyses of the High School and Beyond data set. Bryk and Driscoll (1988) probed the relations between various measures of schools as "communities" and students' performance. Community included shared values, common curriculum and other activities, and an ethos of "caring" for students. Schools that were high on these dimensions had significantly lower dropout rates and absenteeism and slightly higher gains in mathematics achievement. Lee and Bryk (1989) used the same data set to probe differences in schools' constraint of curricular choices. Schools that channeled most work into a common curriculum created consistency by increasing the amount of work that students did in common. Lee and Bryk argued that such schools tended to reduce performance differences among students over time, particularly for minority-group students. In a later article they wrote that schools can "minimize the normal differentiation effects that accompany wide latitude in course choices. . . . Initial differences among students' [performance] can be either amplified or constrained" (Bryk et al., 1990, p. 178).

John Chubb and Terry Moe also reanalyzed High School and Beyond and stressed consistent instructional guidance even more. They argued that high-performing high schools are marked by "coherence," in which principals "provide a clear vision of where the schools are going . . . [and] encourage . . . cooperation and collegiality." They opine that these attributes add up to "organizational coherence" (Chubb & Moe, 1990, p. 91). They also found that students performed better when school staffs had a coherent vision of academic goals and were collegial and cooperative, although the magnitude of the effect was quite modest.

What can we conclude about the effects of instructional guidance? For one thing, consistency means somewhat different things, or is a construct with quite different dimensions. One line of thought focuses on culture and values, another on the organization of curriculum choice, and a third on leadership. For another, most research on the effects of instructional guidance is recent, and the evidence is modest. One cross-national study seems to show that more central curriculum control is modestly associated with greater topical consistency in teaching, and various U.S. school studies claim that more consistent instructional guidance is associated with more consistent instruction. But there are no field studies that make a convincing case for the causal power of guidance, and no cross-system studies connect consistency at the system level with student performance. Both are crucial gaps. A diverse body of research seems to show that more consistent instruction and instructional guidance in schools are associated with higher student achievement. But the causal ambiguities remain, and there are significant problems in inferences from schools to systems. Additionally, most studies reveal only modest effects, yet scholars argue fiercely about them (Witte, 1990).

Even if the studies were more extensive and convincing, there is another problem: The measure of student achievement in all this research has been traditional standardized tests. These tests entail a version of academic accomplishment that is said to depend heavily on recall of isolated facts and mastery of routine mental operations. This is just the sort of work that recent reformers wish schools to put aside in favor of more sophisticated endeavors. Can we assume that a positive effect of consistent guidance on such tests would hold for more challenging versions of achievement? It seems doubtful. Some would argue that the ambitious academic work recent reformers seek would be inimical to consistent guidance. With Theodore Sizer, they would say that if schools are to cultivate sophisticated and independent instruction, they must be sophisticated and independent.

Do we conclude that instructional guidance affects teaching and learning? Plainly it does, somehow. But how? Are the effects of guidance fragmentary or systemic? Specifically, are teaching and learning more consistent in systems that have more consistent guidance for instruction? Evidence on this question is thin. There is some support for the idea in one cross-national study, as well as in many smaller studies of schools. But these studies are limited in many ways, and the authors of the cross-national study caution their readers against making too much of the results (Stevenson & Baker, 1991). There is, for example, no evidence that would permit us to distinguish the effects of formal guidance from teachers' earlier learning. Research on this matter does not offer much support for recent U.S. efforts to use instructional guidance to press teaching and learning toward greater consistency.

CHANGE IN TEACHING

Uncertainty about the effects of instructional guidance looms even larger when we consider the content of guidance that reformers wish to offer teachers and students. For they propose to transform teaching and learning from relatively dull and routine practices into exciting and intellectually demanding ones. To this end, many argue for novel assessments that are tied both to new curriculum frameworks and to radically revised instructional materials. The combination is seen as a way to dramatically change learning and teaching. Would that happen? The studies discussed thus far have little to say on this point, for they all concern the present and past operations of schools and school systems. What do we know about how teachers might change in response to more consistent and ambitious guidance for their work?

Precious little, if we want a direct answer. We have found no studies of school systems that attempted to shift from local autonomy and traditional teaching to more centrally controlled and intellectually ambitious instruction. None of the national school systems that currently exhibit great consistency suddenly changed from fragmented to consistent operations. Some evolved over the course of several centuries, while others were hastily created in the wake of decolonization. But in neither case were teachers required to change from well-established traditional practice to novel and much more adventurous practice.

There have been some studies of efforts to turn teaching in a much more adventurous direction. Larry Cuban found that American classrooms remained traditional despite progressive reforms (Cuban, 1984). He argued that teaching changes at a glacial pace and in fragmentary fashion. In most cases, teachers borrowed bits and pieces of progressive ideas and practices and integrated them into standard classroom formats. That conclusion fits with the work of other investigators in the United States and the United Kingdom who studied various efforts to push teaching in more ambitious and adventurous directions. All concluded that efforts to make teaching more ambitious produced change at the margins but little else (Goodlad, Klein, & Associates, 1974; Popkewitz, Tabachnick, & Wehlage, 1982; Stevens, 1912).

It might be objected that progressivism was only a program. There were many ideas, books, articles, and pamphlets, some professors teaching courses, and even a few teacher education agencies devoted to the "new education." But there were no curricula, no assessments, and no instructional frameworks that might have helped teachers to learn a different pedagogy. From this perspective, the 1950s curriculum reforms were an improvement, for teachers had new texts as well as opportunities to learn about the new curricula. Some of the new texts were widely adopted, and many teachers took advantage of opportunities to learn. But reports of great change in teaching were few and far between. Some teachers seem to have dramatically changed their approach to instruction in the early years of reform, but many more struggled to understand and change (Sarason, 1977). Most teachers seemed to make only marginal changes, grafting bits of reform ideas and practices onto established, traditional teaching. There is indirect evidence that these were major changes for the teachers involved (Cohen, 1990b; Cohen & Ball, 1991). But the difficulty of such change was not appreciated by most of those involved (Sarason, 1981). Measures that might have supported more change thus were not contemplated, much less taken. And changed educational priorities soon swept away opportunities for teachers to learn more. A subsequent NSF-sponsored study found few classroom traces of the curriculum reforms (Stake & Easley, 1978).

Would not the recent reforms be much more potent? Instead of new texts and opportunities for further education, there would be an entire guidance system—new instructional frameworks that were reflected in novel sorts of assessment, in new curriculum materials, and in new approaches to teacher licensing and education. Would not "systemic reform" (Smith & O'Day, 1990) offer much more structure for teaching, much richer opportunities for teacher learning, and a chance for professional community in teaching? Would not more direction offer more support and pressure for change?

The idea has some appeal. But if greater structure and consistency would offer a more substantial basis for change in teaching, it does not follow that change would be easy or swift, for the greater structure would frame new and ambitious purposes, content, and methods. The agenda for teacher change would be vast, even with greater guidance. Consider, for example, studies of the "new math" in Europe. Some European school systems that adopted the new math had more consistent guidance for instruction than did others or the United States. But those differences did not seem to affect change in teaching. The research is spotty, but the most detailed study argued that the processes of reform were strikingly similar across systems with very different structures (Moon, 1986). Reports about change in teaching also were quite homogeneous across systems. Participants and researchers reported that classroom practice changed only a little, and for the most part in fragmentary ways (Damerow, 1980; Howson, 1980; Moon; Oldham, 1980a, 1980b; Van der Blij, Hilding, & Weinzweig, 1980). The new math seemed to fare little differently in the French system of consistent guidance structures than in the less consistent British or U.S. systems (Welch, 1979).

We are inclined to think that some versions of systemic reform could offer more support for radical change in teaching than purely decentralized arrangements. But there is no evidence on the relative rates or depth of change under various organizational conditions. More important, there is growing evidence of several fundamental obstacles to the changes that reformers currently urge, none of which are structural in nature. One concerns teachers' knowledge. The recent reforms demand a depth and sophistication in teachers' grasp of academic subjects that is far beyond most public school teachers. For instance, while math is a leading area in the current reforms, most elementary school teachers have a very modest understanding of the mathematics they teach (Post, Taylor, Harel, Behr, & Lesh, 1988; Thompson, 1984). They would need to learn a great deal more if the reforms were to have any chance of success. More important, teachers would have to shed established modes of understanding and adopt more modern, constructivist versions of knowledge. Such change is not just a matter of learning more—it could fairly be termed a revolution. Scholarship in several fields has shown that intellectual revolutions are very difficult to foment (Cohen, 1990b; Cohen & Ball, 1991; Fiske & Taylor, 1984; Kuhn, 1970; Markus & Zajonc, 1985; Nisbett & Ross, 1980).

Another obstacle lies in teaching. Even if teachers knew all that they needed, the reforms propose that students become active, engaged, and collaborative. If so, classroom roles would have to change radically. Teachers would have to rely on students to produce much more instruction, and students would have to think and act in ways they rarely do.

Teachers would have to become coaches or conductors and abandon more familiar and didactic roles in which they "tell knowledge" to students (Lampert, 1988; Newmann, 1988; Roehler & Duffy, 1988; Scardamalia, Bereiter, & Steinbach, 1984; Sizer, 1984). Researchers have studied only a few efforts at such change, but they report unusual difficulty, for teachers must manage very complex interactions about very complex ideas in rapid-fire fashion. The uncertainties of teaching multiply phenomenally, as does teachers' vulnerability (Cohen, 1988; Cuban, 1984; Lampert, 1988; Newmann, 1988; Roehler & Duffy, 1988).

Since the recent reforms would require much teacher learning, they would require many changes in teachers' opportunities to learn. That is a third obstacle to change. Those who presently teach would need many educational opportunities on the job, as well as off it, in colleges, universities, and other agencies. Yet few schools now offer teachers many chances to learn on the job, and what they do offer is generally deemed weak at best. Most continuing education in universities has a dismal reputation among teachers and researchers. In addition, intending teachers would require fundamentally revamped undergraduate disciplinary and professional education. Few intending elementary teachers can major in an academic subject, and few intending teachers of any sort can learn new approaches to subject matter or pedagogy, since college and university educators rarely teach as reformers now intend (Boyer, 1987; Cuban, 1984; Cohen, 1988; McKeatchie, Pintrich, Lin, & Smith, 1986).

More consistent guidance for instruction could not solve these problems. Under some conditions, too complex to spell out here, such guidance might help to solve them. But fundamental change in teaching also would require fundamental reform of the education of intending and practicing teachers, and equally fundamental changes in schools and universities to support such learning. Even with those reforms, deep change in teaching probably would be slow and difficult.

BEYOND FORMAL STRUCTURE

Guidance for instruction never stands alone. School systems consist not only in rules and formal structures, but also in beliefs about authority, habits of deference and resistance, and knowledge about how things work. Culture and social organization intertwine with formal structure in these systems. Many school systems that offer consistent guidance for instruction are situated in societies in which culture and other social circumstances seem to support academic effort. In contrast, U.S. society and culture seem to undermine academic effort. The success of school systems in Europe and Asia thus may owe more to the influences of culture and society than government or system structure. If so, the nearly exclusive attention to system structure in the current U.S. reform movement may be misplaced.

Social Circumstances of Schooling

Higher education and business firms are the two largest consumers of schooling in most societies. Hence their consumption patterns send signals concerning the qualities and accomplishments that they find desirable in students. The consumption patterns of American colleges and universities send mixed but generally weak signals about the importance of strong academic performance. Only a small group of highly selective colleges and universities has demanding admissions standards. A much larger fraction has very modest requirements: Students need only a thin record of academic accomplishment in high school, often only a C or low B average, to be acceptable for admission. Only high school graduation is required for admission in still another large group of institutions. And not even high school graduation is required in another large group. There is something to celebrate in this, for many students can have a second or third chance to make good despite previous failures. But these arrangements also signal that high school students need not work hard in order to get into college or university (Bishop, 1989; Powell et al., 1985; Trow, 1961, 1988). It is thus irrational for most students who aspire to higher education to work very hard in high school, for only a few have a chance to enter a highly selective college or university. Their opportunities lie instead at less selective institutions, where much less high school work is required for admission (Powell et al., 1985; Trow, 1988). It therefore is irrational for high school teachers to press those students to try hard and do their best work, for the students need not push themselves in order to push ahead.

A similar situation holds for the employment practices of most U.S. businesses. Few firms seem to ask for students' high school transcripts or references from teachers when considering them for employment. And even when firms do request transcripts, only a tiny fraction of schools supply them (Bishop, 1989). The lack of employer interest deters students from thinking that grades, effort, or behavior count for jobs and deters teachers from thinking that their judgments about students can make a difference (Rosenbaum & Kariya, 1989). Hence it would be irrational for students who intend leaving high school for work to do their best in school. Thinking deeply is difficult, and only a small fraction of students seem intrinsically motivated to do it. If students can get jobs without even presenting evidence about their grades, school behavior, and teachers' evaluation of their work, why should they work hard?

These patterns are unusual. Colleges and universities in Japan, France, and many other nations lay great weight on students' performance in high school or on high school leaving and university entrance exams. If students wish to enter university, it is essential to work hard in school and get good grades, prepare for the exams, or both (Rosenbaum & Kariya, 1987). There are many troublesome features of such systems, including the exclusion of able students who do not do well on exams. But these systems leave no doubt about the importance of hard work and good school performance.

Employers in many nations also pay close attention to students' secondary school records in hiring decisions. This is true in Japan; New South Wales, Australia; Singapore; and West Germany (Bishop, 1987, 1989; Clark, 1985; Kariya & Rosenbaum, 1987; Rosenbaum & Kariya, 1989). Employers routinely review transcripts and teacher references when high school graduates or early school leavers apply for jobs. In some cases, schools and employers work closely in placing students in apprenticeship or regular work situations. Teachers know these things, as do students. It is understood that students who do not apply themselves and behave decently in school will have difficulty finding good jobs. There are important rewards for academic effort and good behavior, even for students who have no ambitions for further education.

Culture

Incentives are not the whole story. The values attached to learning and teaching differ among societies, as do attitudes toward authority and habits of child rearing. Such beliefs, values, and habits may support the guidance that issues from formal agencies in some cases and subvert it in others.

Americans have long been ambivalent about academic work. Anti-intellectualism is a prominent feature in American culture (Hofstadter, 1963), and we are inclined to value experience over formal education. Americans also value practical rather than intellectual content within formal education (e.g., learning to "get along") and job-related knowledge and skills (Cusick, 1983; Lynd & Lynd, 1929; Powell et al., 1985). Eighty-one percent of the respondents in a recent Gallup poll said that the "chief reasons" people want their children to get a formal education are job opportunities, preparation for a better life, better-paying jobs, and financial security. Only 15% said that the chief reason was to become more knowledgeable or to learn to think and understand (Elam & Gallup, 1989). Relatively few American mothers report working closely with their children on academic tasks or offering support for hard work and success in school (Stevenson et al., 1985; Stevenson, Lee, & Stigler, 1986).

Intellectual work and academic accomplishment appear to be more highly regarded in other societies. In Japan and China, for instance, parents take education very seriously and hold teachers in high esteem. Investigators report that Japanese mothers play a central role in their children's academic success (Holloway, Kashiwagi, Hess, & Hiroshi, 1986; Lebra, 1976; Shimahara, 1986; White, 1987). They encourage children and work closely with them on assignments, creating an environment conducive to learning (Holloway et al.; Stevenson et al., 1985; Stevenson & Lee, 1990; White, 1987). Similar practices are found among Chinese parents (Stevenson & Lee). Japanese and Chinese mothers also seem to hold higher standards for their children and to have more realistic evaluations of their achievement than American mothers (Stevenson & Lee).

Family life and values thus seem to support successful schooling in Japan while impeding it in the United States. One researcher noted that "it would be quite impossible to take account of Japanese formal education without recognising that—in many ways—it lives in close symbiosis with that culture" (King, 1986, p. 75; see also White, 1987). American commentators often have offered complementary explanations of weak work in school here (Coleman, 1961; Cusick, 1983; Lynd & Lynd, 1929; Powell et al., 1985).

Habits of association and attitudes toward authority also may help to explain why formal guidance for instruction seems to be treated more seriously in some societies than in others. Since Alexis de Tocqueville, observers have noted Americans' distinctive individualism, their preoccupation with personal autonomy, and their focus on individual expression and development. These qualities often have been contrasted with more cooperative and deferential behavior in other societies, in which people seem more preoccupied with how they can fit in, work with others, and advance collective values.

For instance, Japanese teachers carefully foster cooperative work on common tasks, build habits of collaboration and conflict resolution, and teach accommodation to group preferences. They exercise great patience in encouraging students to work with groups, and use groups to regulate behavior, manage conflict, and support desired attitudes. In the process, Japanese teachers accommodate "discipline" problems that would be intolerable to most Americans. But they build many centers of support for the values and behavior they wish to inculcate rather than assuming the entire burden themselves. Many "discipline" problems are therefore managed by other students rather than the teacher alone (Boocock, 1989; Peak, 1989).

American teachers instead foster individual work on individual tasks. They cultivate little or no group activity and rarely build group strength. They do not support accommodation to group preferences but tend to impose their own preferences. They manage all discipline problems themselves and have little patience with misbehavior. American students learn little about alternative ways to manage conflict or about collaborative work.

Thomas Rohlen (1989) framed these comparisons in a broad analysis of differences in organizational life. He viewed Japanese classrooms as marked by more respect and deference than those in the United States but as less hierarchical and teacher centered. The Japanese emphasis on accommodation to group values and cooperative work helps to explain the coexistence of two things that strike Americans as inconsistent: deference to authority and enormous capacity for productive work in decentralized organizations. Rohlen notes that these qualities are found in all sorts of organizations, from primary classrooms to business and government.

The result is an overall social structure that is in many respects centrifugal in terms of affiliation and the capacity to order events. Social contexts and organizations are built up from the bottom (or the outside), so to speak, in a way that invests the peripheral entities with great stability. The locus of social order is in the lower-level, subordinate groupings. . . . These entities gain a degree of autonomy from the fact that internally they are strengthened by the pattern of attachment we are considering. (Rohlen, 1989, pp. 31–32)

Americans have few alternatives between individualism and imposed authority. We often fluctuate between centralized hierarchies and decentralization. The result makes it difficult for central authority to succeed, while also precluding the development of alternatives.

How does this bear on our analysis? The remarkable consistency in Japanese education may owe as much to deference to authority, habits of accommodation, and extraordinary pressures for cooperation at all levels as to formal guidance. Rohlen's account also suggests caution about the prospects for much more potent guidance for instruction in the United States, for it might be crippled by our habit of alternatively embracing central authority and fiercely resisting it.

CONCLUSION

Most schemes for fundamental change present a paradox. They offer appealing visions of a new order but therefore also contain a devastating critique of existing realities. If pursued, these critiques reveal the lack of many capacities that would be required to realize and sustain the new vision. Reformers can imagine a better world in which those capacities would be created, but their problem is more practical—how to create the new world when those capacities are lacking?

Recent reform proposals offer a version of this puzzle, for they entail two dramatic departures from American political and instructional practice. One is that schools should promote a new instructional order marked by deep comprehension of academic subjects, in which students are active learners rather than passive recipients and in which teachers practice a much more thoughtful and demanding pedagogy. The other departure is radical reform in school governance and instructional guidance to produce the desired changes in classrooms. These reforms include a national examination or testing system, national curricula, a national system of teacher certification, and many equally dramatic reforms at the state level. Although different in important ways, all of these plans and proposals move sharply toward greater state, national, or federal control of education. All seek to realize new and ambitious sorts of teaching and learning in ordinary classrooms. Hence all represent an effort to much more powerfully guide instructional practice with policy.

These are astonishing, even revolutionary, proposals and are appealing in many respects. But we have pointed to weak capacities for change in several crucial departments.

One is politics. Reformers seek much greater state, national, or federal control of education and a consequent tightening of the links between central policy and local practice. But the entire fragmented apparatus of American government weighs against such ventures. Past efforts at tightening the links between policy and instruction by increasing central control have met with extremely limited success and produced organizational side-effects that have greatly complicated governance and administration. The reforms sketched above seem unlikely to succeed unless the governance and organization of U.S. education is either greatly streamlined or simply bypassed. Streamlining has much appeal, including relief from the burdens consequent upon past efforts to reform local practice with state or federal policy. But streamlining would entail an unprecedented reduction of existing policies and programs, and thus a reduction in existing governmental authorities at all levels. It would spell the end of many state and local government functions in education, even though it could ease many administrative and organizational problems. What would induce local and state officials to accept the diminishment or demise of their domains? Visions of a better school system? Barring fiscal catastrophe or a sustained mass movement for fundamental change in education, we see no sign of the requisite inducements.

Bypassing government appeals to many partly because the prospects for streamlining seem so bleak. The creation of nongovernmental or quasi-nongovernmental authorities that may design a national examination system already is under way. A similar course of action has been taken by NBPTS in its efforts to create a national system of teacher examination and certification. Such bypass operations have great short-run appeal, for avoiding government sponsorship and operations could greatly ease the work of designing and developing national education systems of one sort or another. But the systems thus created would only work in the medium and long run if government were streamlined, for national curricula, examinations, or teacher certification systems could operate efficiently only if many extant policies and programs regarding testing, curriculum, instruction, and teacher licensing fell into disuse. Of course, that would require many existing state and local government authorities to fade away (i.e., streamlining), the difficulties of which we just touched upon.

Instructional practice is a second realm in which the capacity for change is weak. Reformers seek much more thoughtful, adventurous, and demanding teaching and learning, and they envision new instructional guidance to produce it. But nearly the entire corpus of instructional practice weighs against it. Teachers and students spend most of their time with lectures, recitations, and worksheets. Intellectual demands generally are modest, and a great deal of the work is dull. Only a modest fraction of public school teachers have deep knowledge of any academic subject. Research and experience both show that past efforts to fundamentally change teaching have had modest effects at best. Most often, they have resulted in fragmentary adoption of new practices, translation of new practices into old ones, or both.

Solutions to these capacity problems would require fundamental redevelopment in education. An intellectually ambitious system of instructional guidance would be one key element, but few Americans have had the education or experience that would prepare them to understand such guidance or put it to appropriate use. To build new capacities for education would be to reeducate many Americans. That is obviously true for teachers, but teachers' efforts would not prosper if parents and political leaders did not understand and support their work. Additionally, few teachers work in schools that could support radically different approaches to instruction, let alone teachers' efforts to learn such things. Building new capacity would require that schools become places in which teachers could learn and teach very differently.

Such redevelopment would be an immensely ambitious endeavor. The creation of new instructional guidance arrangements would be an extraordinary research and development task, surely the largest ever in U.S. education. For example, new examinations would have to be invented, to assess a broader range of academic knowledge and skills than conventional tests. The exams also would assess students' skill and knowledge in more diverse ways (e.g., writing essays in English, explaining and justifying answers in chemistry, and offering nonnumerical representations of mathematical problem solving). Because examinations of this sort would invite students to use and display a broad range of knowledge and skill (Nickerson, 1989), the results would be difficult and time-consuming to evaluate, especially for a large and diverse population. Such things are possible, and approximations can be found here and there in the United States and some other nations (Resnick & Resnick, 1989). But the approach has been little tried in the United States, and Americans have little experience. Specialists are just beginning to invent examples, and a few states are beginning to experiment with them (California State Department of Education, 1985). A few problems with such exams have been suggested (Porter, 1990), but many others are likely to appear if they actually are developed and widely used.

New curricula also would be needed for any guidance system keyed to deep understanding of academic subjects. Instructional frameworks would have to be devised, along with curriculum guides that focused attention on key elements of each academic subject. Texts and other materials that effectively opened access to those topics would have to be composed. These materials would have to be accessible to a large and diverse population of teachers and learners, and they would be most useful if designed in a way that teachers could learn from them while teaching, preparing to teach, and reconsidering their teaching. Such materials would be most helpful to teachers and students if subject coverage was integrated across the grades. Though such things seem possible, they are entirely unfamiliar in the United States. A few states are just beginning to develop more demanding and thoughtful curricula. It seems reasonable to expect that such a novel endeavor would take a long time to develop and longer to mature.

Neither new exams nor new curricula would work unless teachers understood them, and as things now stand, most teachers would not. This problem might be solved in part if teachers were extensively involved in building new frameworks, curricula, and assessments, and in grading students' work. Such activities could be extraordinarily educative if they were designed with that end in view, but to do so would greatly complicate the development tasks. Additionally, these activities would reach only a fraction of the teaching force, and would touch only part of the reeducation need. Teacher education itself would have to be greatly improved, which would require fundamental changes in college and university education. Both professional and subject matter education would have to be deepened and focused much more closely on the content of schooling, and university teaching would have to radically improve. Such changes are daunting to contemplate, but they might be encouraged by the sort of examination system that the National Board for Professional Teaching Standards has proposed. If intending teachers' grasp of subject matter and pedagogy could fairly be assessed, then the exams might offer college and university programs sensible targets for their educational efforts. If the targets were accepted, if teacher education programs were revised accordingly, and if school systems used the exam results as hiring criteria, the quality of teaching might be greatly improved. But note that such changes would be immense: The examinations do not now exist, and it would take colleges and universities at least a generation to make the required instructional reforms. The National Board is just beginning to develop some exams, and current estimates are that it will take at least 3 years to produce an initial prototype in a single subject for a grade or two. The NBPTS staff hopes that a full set of examinations might be developed by the year 1997, though no one really knows how long it will take. Here too Americans are relatively inexperienced. Even if the exams were developed roughly on schedule, it would be prudent to assume that many adjustments would be required as the exams came into use. And only a handful of professors have given any thought to the reforms of higher education that might dramatically improve the education of intending teachers.

It would be no mean feat to solve any one of these research and development problems by itself. But the recent reforms are "systemic"; that is, they seek to link assessment of students' performance to the content and form of curriculum guides and course materials, and to tie both of them to teacher education. Hence the research and development tasks sketched above should be undertaken jointly. That would be an extraordinarily demanding and time-consuming effort. It is another reason we believe that devising a guidance system to support deep understanding of academic subjects would be a huge endeavor.

If changes in instructional guidance are crucial, they would not work all by themselves. Americans are well used to local control of education, and they have been less and less inclined to defer to teachers. Radical reform of instruction would be unlikely to get very far unless parents and political leaders supported it. Yet to do so these Americans would have to embrace very different conceptions of knowledge, teaching, learning, and schooling than they currently do. That is possible. For instance, administrators, political leaders, and parents could learn about new examinations by participating in their development, in field trials, and in revisions. Though such work probably would increase conflict in the short run, it might increase the long-run chances that the finished exams would be understood, accepted, and used appropriately. But the learning would be a great change for parents and political leaders, no less large or difficult than for teachers and students. And to give administrators, parents, and politicians opportunities to learn would complicate, slow down, and alter the development of new exams.

Finally, the reforms that we have been discussing would require changes in individual school operations and organization. One reason is that teachers and administrators would have a great deal to learn. It is unlikely they could offer the intellectually ambitious instruction that reformers seek unless they had ample time to learn on the job. Another reason is that the new instruction would be much more complex and demanding than the common fare in schools today. It is unlikely that teachers could do such work unless they had the autonomy to make complicated decisions, to work with colleagues, and to revise as they went. Still another reason is that teachers could hardly contribute to the development of a common instructional system unless they had much more time and opportunities to work with others in education beyond their school. These considerations suggest radical changes in schools and school operations, so that they offered more opportunities to learn on the job and greater autonomy for school professionals. But how could that be done in the context of reforms that entail much greater central authority and power? Not easily, unless the reforms were carefully designed to enhance such autonomy, and unless the capacities to exercise it were nurtured at all levels of education. Marshall Smith and Jennifer O'Day (1990) have sensibly argued that systemic reform would require a combination of "bottom-up" and "top-down" change. But given the present organization of U.S. schools and school governments, and the work habits of policymakers, teachers, and administrators, that would be a great change indeed.

These observations suggest that the recent reforms might have more chances of success if the entire venture were conceived and executed as a great educational enterprise, one in which state and national leaders had as much to learn as teachers and students. That seems appropriate to a set of proposals that would require such radical change in individuals and institutions, and the cultivation of so many new capacities. But the recent reforms began to catch American's imagination just as economic and social problems began to further constrain the capacities for change. States and localities are struggling with a massive fiscal crisis, and many confront staggering social problems.

What happens when grand visions of change collide with limited capacities? The most common consequences are incremental alteration at the margin of institutions and practices, or self-defeating results, or both. For example, education reformers could relatively easily add streamlining mandates and a layer of streamlining agencies to the existing accumulation of mandates and organizations. But they would find it much more difficult and costly to replace the present cluttered and fragmented structure with one that was much simpler and more powerful. Similarly, bypass operations could easily add complexity rather than reducing it. For governments would not sit still. Experience suggests that they would respond by regulating the bypassing agencies, or by finding roles for themselves in interpreting and managing the functions generated by the bypassing agencies, or by taking over the bypass agencies without closing down the authorities that were to have been bypassed, or some combination of the above. In this event efforts to increase simplicity and clarity might well further complicate and confuse matters.

Similarly, it would be relatively easy and cheap for reformers to add mandates for more thoughtful teaching and learning on top of extant mandates for teaching basic skills, informing students about drugs and AIDS, not to mention remedial education, bilingual education, and programs for cooperative learning and improving students' self-concepts. It would be much more difficult and costly to replace the present cluttered and fragmented accumulation of instructional guidance with a system that was simpler, more focused, and more powerful. It would be even more difficult to redevelop education in ways that would enable most educators to take good advantage of such changes.

But uncommon results are always possible. For instance, the recent reforms might succeed by a sort of osmosis. If reformers kept up the pressure for several decades, much more consistent and demanding instruction might result. Indeed, the extraordinary fragmentation of American institutions may create a porosity that permits such change. Something like that sort of osmosis seems to have occurred in the spread of basic skills instruction during the 1970s and 1980s. More time and pressure would be required for the more difficult reforms that we have discussed here, and most reform movements in education are notoriously brief, but the fragmentation of American government could open many opportunities to persistent reformers.

We also may underestimate the ingenuity of policymakers and educators. Perhaps they will seize on the growing social and financial crisis to turn schools in the direction that reformers wish. Streamlining, simplification, and consistency could be appealing slogans in an era of falling budgets and rising problems. Perhaps the crippling legacy of the Reagan years in public finance and the economy will become an opportunity to press ahead with its nationalizing legacy in education.

No one knows how the story will turn out. But in most cases, today is the best guide to tomorrow. If American politics and education run true to form, reformers will do better at addition than subtraction. They will introduce many different schemes to make education more consistent, but they will be less able to produce consistency among those schemes, to greatly reduce the clutter of previous programs and policies, or to fundamentally change teaching. If so, current efforts to reduce fragmentation would only add several new and unrelated layers of educational requirements and instructional refinements on top of many old and inconsistent layers. The new ideas would have their day, but only at the expense of further clutter and inconsistency.

REFERENCES

Bankston, M. (1982). *Organizational reporting in a school district: State and federal programs*. Stanford, CA: Institute for Finance and Governance, School of Education, Stanford University.

Bardach, E., & Kagan, R. A. (1982). *Going by the book: The problem of regulatory unreasonableness*. Philadelphia: Temple University Press.

Beauchamp, G., & Beauchamp, K. (1972). *Comparative analysis of curriculum systems*. Wilmette, IL: The Kagg Press.

Berlak, A., & Berlak, H. (1981). *The dilemmas of schooling: Teaching and social change*. London: Methuen.

Berman, P., & McLaughlin, M. (1977). *Federal programs supporting educational change*. Santa Monica, CA: RAND.

Bishop, J. (1987). *Information externalities and the social payoff to academic achievement*. Ithaca, NY: Cornell University, Center for Advanced Human Resource Studies.

Bishop, J. (1989). *Incentives for learning: Why American high school students compare so poorly to their counterparts overseas*. Ithaca, NY: Cornell University, Center for Advanced Human Resource Studies.

Bishop, J. (1989). Why the apathy in American high schools? *Educational Researcher, 18*(1), 6–10.

Boocock, S. (1989). Controlled diversity: An overview of the Japanese preschool system. *Journal of Japanese Studies, 15*, 41–66.

Boyd, W., & Smart, D. (1987). *Educational policy in Australia and America: Comparative perspectives*. Philadelphia: The Falmer Press.

Boyer, E. (1987). *The undergraduate experience in America*. New York: Harper and Row.

Broadfoot, P. (1983). Assessment constraints on curriculum practice: A comparative study. In M. Hammersley & A. Hargreaves (Eds.), *Curriculum practice; Some sociological case studies* (pp. 251–269). New York: Falmer.

Broadfoot, P. (1984). From public examinations to profile assessment: The French experience. In P. Broadfoot (Ed.), *Selection, certification, and control* (pp. 199–219). London and New York: Falmer Press.

Bryk, A., & Driscoll, M. (1988). *The high school as a community: Contextual influences and consequences for students and teachers*. Madison, WI: National Center on Effective Secondary Schools.

Bryk, A., Lee, V., & Smith, J. (1990). High school organization and its effects on teachers and students: An interpretive summary of the research. In W. Clune & J. Witte (Eds.), *Choice and control in American education* (Vol. 1, pp. 135–226). New York: Falmer.

California State Department of Education. (1985). *Mathematics framework*. Sacramento: Author.

Cameron, J., Cowan, R., Holmes, B., Hurst, P., & McLean, M. (Eds.). (1984a). *Austrialia* (International Handbook of Educational Systems). Chichester, England: Wiley.

Cameron, J., Cowan, R., Holmes, B., Hurst, P., & McLean, M. (Eds.). (1984b). *France* (International Handbook of Educational Systems). Chichester, England: Wiley.

Cameron, J., Cowan, R., Holmes, B., Hurst, P., & McLean, M. (Eds.). (1984c). *Japan* (International Handbook of Educational Systems). Chichester, England: Wiley.

Cameron, J., Cowan, R., Holmes, B., Hurst, P., & McLean, M. (Eds.). (1984d).

Singapore (International Handbook of Educational Systems). Chichester, England: Wiley.

Cantor, L. (1980). The growing role of states in American education. *Comparative Education, 16*(1), 25–31.

Cheney, L. (1991). National tests: What other countries expect their students to know. Washington, DC: National Endowment for the Humanities.

Chubb, J., & Moe, T. (1990). *Politics, markets, and American schools*. Washington, DC: Brookings Institution.

Clark, B. (1965). Interorganizational patterns in education. *Administrative Science Quarterly, 10*, 224–237.

Clark, B. (1985). The high school and the university: What went wrong in America, Part II. *Phi Delta Kappan, 66*, 472–475.

Cohen, D. K. (1982). Policy and organization: The impact of state and federal educational policy on school governance. *Harvard Educational Review, 52*, 474–499.

Cohen, D. K. (1988). *Teaching practice: Plus ça change*. East Lansing: National Center for Research on Teacher Education, Michigan State University, 1988.

Cohen, D. K. (1990a). Governance and instruction: The promise of decentralization and choice. In W. Clune & J. Witte (Eds.), *Choice and control in American education* (Vol. 1, pp. 337–386). New York: Falmer.

Cohen, D. K. (1990b). A revolution in one classroom: The case of Mrs. Oublier. *Educational Evaluation and Policy Analysis, 12*, 311–330.

Cohen, D. K., & Ball, D. (1991). Relations between policy and practice: A commentary. *Educational Evaluation and Policy Analysis, 12*, 331–338.

Coleman, J. (1961). *The adolescent society*. New York: Free Press.

Cuban, L. (1984). *How teachers taught*. New York: Longman.

Cusick, P. (1983). *The egalitarian ideal and the American high school*. New York: Longman.

Damerow, P. (1980). Patterns of geometry in German Textbooks. In H. G. Steiner (Ed.), *Comparative studies of mathematics curricula—Change and stability, 1960–1980* (pp. 281–303). Haus Ohrbeck: Institut fur Didaktik der Mathematik der Universitat Bielefeld.

Darling-Hammond, L. (1987, April). The over-regulated curriculum and the press for teacher professionalism. *NASSP Bulletin*, pp. 22–29.

Darling-Hammond, L., & Wise, A. E. (1985). Beyond standardization: State standards and school improvement. *Elementary School Journal, 85*, 315–336.

Eckstein, M., & Noah, H. (1989). Forms and functions of secondary school leaving examinations. *Comparative Education Review, 33*, 295–316.

Elam, S. M., & Gallup, A. M. (1989). The 21st Annual Gallup Poll of the public attitudes toward the public schools. *Phi Delta Kappan, 71*, 41–54.

Ellwein, M. C., Glass, G. V., & Smith, M. L. (1988). Standards of competence: Propositions on the nature of testing reforms. *Educational Researcher, 17*(8), 4–9.

Firestone, W. (1989). Educational policy as an ecology of games. *Educational Researcher, 18*(7), 18–24.

Fiske, S., & Taylor, S. (1984). *Social cognition*. Reading, MA: Addison-Wesley.

Floden, R. E., & Clark, C. (1988). Preparing teachers for uncertainty. *Teachers College Record, 89*, 505–524.

Floden, R., Porter, A., Alford, L., Freeman, D., Irwin, S., Schmidt, W., & Schwille, J. (1988). Instructional leadership at the district level: A closer look at autonomy and control. *Educational Administration Quarterly, 24*, 96–124.

Floden, R. E., Porter, A. C., Schmidt, W. H., & Freeman, D. J. (1978). *Don't they all measure the same thing? Consequences of selecting standardized tests* (Research Series No. 25). East Lansing: Michigan State University, Institute for Research on Teaching.

Freeman, D., Kuhs, T., Porter, A., Floden, R., Schmidt, W., & Schwille, J. (1983). Do textbooks and tests define a national curriculum in elementary school mathematics? *Elementary School Journal, 83,* 501–514.

Freeman, D. J., & Porter, A. C. (1989). Do textbooks dictate the content of mathematics instruction in elementary schools? *American Educational Research Journal, 26,* 403–421.

Fuhrman, S., Clune, W., & Elmore, R. (1988). Research on educational reform: Lessons on the implementation of policy. *Teachers College Record, 90,* 237–257.

Goodlad, J., Klein, M., & Associates. (1974). *Looking behind the classroom door.* Worthington, OH: Charles Jones.

Haggstrom, G., Darling-Hammond, L., & Grissmer D. (1988). *Assessing teacher supply and demand.* Santa Monica, CA: Rand Corporation.

Hofstadter, R. (1963). *Anti-intellectualism in American life.* New York: Vintage Books.

Holloway, S., Kashiwagi, K., Hess, R., & Hiroshi, A. (1986). Causal attributions by Japanese and American mothers and children about performance in mathematics. *International Journal of Psychology, 21,* 269–286.

Holmes, B. (1979). *International guide to education systems.* Paris: UNESCO.

Horner, W. (1986). Curriculum research in France and Luxembourg. In U. Hameyer, K. Frey, H. Haft, & F. Kuebart (Eds.), *Curriculum research in Europe* (pp. 91–107). Strasbourg: The Council of Europe.

Howson, A. (1980). Some remarks on the case studies. In H. G. Steiner (Ed.), *Comparative studies of mathematics curricula—Change and stability, 1960–1980* (pp. 502–508). Haus Ohrbeck: Institut fur Didaktik der Mathematik der Universitat Bielefeld.

Jackson, P. (1968). *Life in classrooms.* New York: Holt, Rinehart, and Winston.

Kariya, T., & Rosenbaum, J. (1987). Self-selection in Japanese junior high schools: A longitudinal study of students' educational plans. *American Journal of Sociology, 60,* 168–180.

Kaufman, H. (1969). Administrative decentralization and political power. *Public Administration Review, 29*(1), 3–15.

Kellaghan, T., & Madaus, G. (1991). *Proposals for a national American test: Lessons from Europe.* Unpublished manuscript, Boston College.

Kida, H. (1986). Educational administration in Japan. *Comparative Education, 22,* 7–12.

Kiesler, S., & Sproull, L. (1982). Managerial responses to changing environments: Perspectives on problem sensing from social cognition. *Administrative Science Quarterly, 27,* 548–570.

Kimbrough, J., & Hill, P. (1981). *The aggregate effects of federal education programs.* Santa Monica, CA: RAND.

King, E. (1986). Japan's education in comparative perspective. *Comparative Education, 22,* 73–82.

Kirst, M. W. (1988). *Who should control our schools: Reassessing current policies.* Stanford, CA: Center for Education Research.

Kobayashi, V. (1984). Japanese and U.S. curricula compared. In W. Cummings, E. Beauchamp, S. Ichikawa, V. Kobayashi, & M. Ushiogi (Eds.), *Educational policies in crisis* (pp. 61–95). New York: Praeger.

Kreitzer, A. E., Madaus, G. F., & Haney, W. (1989). *Competency testing and dropouts*. Unpublished manuscript, Boston College.

Kuhn, T. (1970). *The structure of scientific revolutions*. Chicago: University of Chicago Press.

Lampert, M. (1985). How do teachers manage to teach? Perspectives on problems in practice. *Harvard Educational Review, 55*(2), 178–194.

Lampert, M. (1988). *Teachers' thinking about students' thinking about geometry: The effects of new teaching tools*. Cambridge, MA: Educational Technology Center.

Lawton, D., & Gordon, P. (1987). *The HMI*. London: Routledge.

Lebra, T. (1976). *Japanese patterns of behavior*. Honolulu: University Press of Hawaii.

Lee, V., & Bryk, T. (1989). A multilevel model of the social distribution of high school achievements. *Sociology of Education, 62*, 172–192.

Lewis, H. (1985). *The French education system*. London: Croom Helm.

Lewis, H. (1989). Some aspects of education in France relevant to current concerns in the U.K. *Comparative Education, 25*, 369–378.

Lortie, D. (1975). *Schoolteacher: A sociological study*. Chicago: University of Chicago Press.

Lynd, R., & Lynd, H. (1929). *Middletown: A study in modern American culture*. New York: Harcourt, Brace and World.

MacRury, K., Nagy, P., & Traub, R. E. (1987). *Reflections on large-scale assessments of study achievement*. Toronto: Ontario Institute for Studies in Education.

Madaus, G. (1988). *The influence of testing on the curriculum*. In L. Tanner (Ed.), *Critical issues in curriculum* (pp. 83–121). Chicago: University of Chicago Press.

Madaus, G. (1989). *The distortion of teaching and testing: High-stakes testing and instruction*. Unpublished manuscript, Boston College.

Madaus, G. (1991, June). *The effects of important tests on students: Implications for a national examination or system of examinations*. Paper presented at the American Educational Research Association Invitational Conference on Accountability as a State Reform Instrument: Impact on Teaching, Learning, Minority Issues and Incentives for Improvement, Washington, DC.

Madaus, G., & Kellaghan, T. (1991). *America 2000's proposal for American achievment tests: Unexamined issues*. Unpublished manuscript.

Markus, H., & Zajonc, R. (1985). The cognitive perspective in social psychology. In G. Lindzey & E. Aronson (Eds.), *Handbook of social psychology* (3rd ed., Vol 1). Hillsdale, NJ: Erlbaum.

McDonnell, L. M., & McLaughlin, M. W. (1982). *Education policy and the role of the states*. Santa Monica, CA: RAND.

McKeatchie, W., Pintrich, P., Lin, Y., & Smith, D. (1986). *Teaching and learning in the college classroom: A review of the research literature*. Ann Arbor, MI: National Center to Improve Postsecondary Teaching and Learning.

McLaughlin, M. (1987). Learning from experience: Lessons from policy implementation. *Educational Evaluation and Policy Analysis, 9*, 171–178.

Meyer, J. (1983). Centralization of funding and control in educational governance. In J. W. Meyer & W. R. Scott (Eds.), *Organizational environments: Ritual and rationality* (pp. 179–198). Beverly Hills, CA: Sage.

Moon, B. (1986). *The new math curriculum controversy: An international story*. Philadelphia: The Falmer Press.

Morone, J. (1991). *The democratic wish.* New York: Basic Books.

Murphy, J. (1974). *State education agencies and discretionary funds: Grease the squeaky wheel.* Massachusetts: Heath.

Newmann, F. M. (1988). Higher order thinking in the high school curriculum. *NASSP Bulletin, 72,* 58–64.

Nickerson, R. (1989). New directions in educational assessment. *Educational Researcher, 18*(9), 3–7.

Nisbett, R., & Ross, L. (1980). *Human inference: Strategies and shortcomings of social judgement.* Englewood Cliffs, NJ: Prentice-Hall.

Noah, H., & Eckstein, M. (1989). Tradeoffs in examination policies: An international comparative perspective. *Oxford Review of Education, 15*(1), 17–27.

Organization for Economic Cooperation and Development. (1971). *Reviews of national policies for education: Japan.* Paris: Author.

Organization for Economic Cooperation and Development. (1973). *Educational policy and planning: Japan.* Paris: Author.

Office of Educational Research and Improvement State Accountability Study Group. (1988). *Creating responsible and responsive accountability systems.* Washington, DC: U.S. Department of Education.

Ohta, T. (1986). Problems and perspectives in Japanese education. *Comparative Education, 22*(1), 27–30.

Oldham, E. (1980a). Case studies in geometry education: Ireland. In H. G. Steiner (Ed.), *Comparative studies of mathematics curricula—Change and stability, 1960–1980* (pp. 326–346). Haus Ohrbeck: Institut fur Didaktik der Mathematik der Universitat Bielefeld.

Oldham, E. (1980b). Case studies in algebra education: Ireland. In H. G. Steiner (Ed.), *Comparative studies of mathematics curricula—Change and stability, 1960–1980* (pp. 395–425). Haus Ohrbeck: Institut fur Didaktik der Mathematik der Universitat Bielefeld.

Orfield, G. (1969). *The reconstruction of southern education.* New York: Wiley.

Peak, L. (1989). Learning to become part of the group: The Japanese child's transition to preschool life. *Journal of Japanese Studies, 15,* 93–124.

Peterson, P. (1981). *City limits.* Chicago: University of Chicago Press.

Peterson, P., Rabe, B., & Wong, K. (1986). *When federalism works.* Washington, DC: Brookings Institution.

Popkewitz, T., Tabachnick, B., & Wehlage, G. (1982). *The myth of educational reform: A study of school responses.* Madison: University of Wisconsin Press.

Porter, A. (1990, October). *Assessing national goals: Some measurement dilemmas.* Paper presented at the 1990 Educational Testing Service Invitational Conference, The Assessment of National Education Goals, New York.

Porter, A., Floden, R., Freeman, D., Schmidt, W., & Schwille, J. (1988). Content determinants in elementary school mathematics. In D. A. Grouws, T. J. Cooney, & D. Jones (Eds.), *Effective mathematics teaching* (pp. 96–113). Reston, VA: National Council of Teachers of Mathematics.

Post, T., Taylor, B. R., Harel, G., Behr, M., & Lesh, R. (1988). *Intermediate teachers' knowledge of rational number concepts.* Unpublished manuscript, University of Wisconsin.

Postlethwaite, T. N. (Ed.). (1988). *The encyclopedia of comparative education and national systems of education* (Vol. C). Oxford, England: Pergamon Press.

Powell, A. (1991). *Private schools.* Unpublished draft essay, Cambridge, MA.

Powell, A., Farrar, E., & Cohen, D. K. (1985). *The shopping mall high school.* Boston: Houghton Mifflin.

Purkey, S., & Smith, M. (1983). Effective schools: A review. *Elementary School Journal, 83,* 428–452.

Putnam, R. T., Heaton, R., Prawat, R., & Remillard, J. (in press). Teaching mathematics for understanding: Discussing case studies of four fifth-grade teachers. *Elementary School Journal.*

Ramirez, F., & Boli, J. (1987). The political construction of mass schooling: European origins and worldwide institutionalization. *Sociology of Education, 60,* 2–11.

Ramirez, F., & Rubison, R. (1979). Creating members: The political incorporation and expansion of public education. In J. W. Meyer & M. Hannan (Eds.), *National development and the world system* (pp. 72–82). Chicago: University of Chicago Press.

Resnick, D. P., & Resnick, L. B. (1985). Standards, curriculum, and performance: A historical and comparative perspective. *Educational Researcher, 14,* 5–20.

Resnick, L. B., & Resnick, D. P. (1989). *Assessing the thinking curriculum: New tools for educational reform.* Pittsburgh, PA: Learning Research and Development Center.

Roehler, L., & Duffy, G. (1988). Teachers' instructional actions. In R. Barr, M. Kamil, P. Mosenthal, & P. Pearson (Eds.), *Handbook of reading research* (Vol. 2, pp. 861–884). New York: Longman.

Rogers, D., & Whetten, D. (1982). *Interorganizational coordination.* Ames: Iowa State University.

Rohlen, T. (1989). Order in Japanese society: Attachment, authority and routine. *Journal of Japanese Studies, 15,* 5–40.

Romberg, T. A., Zarinnia, E. A., & Williams, S. R. (1989). *The influence of mandated testing on mathematics instruction: Grade 8 teachers' perceptions.* Unpublished manuscript, University of Wisconsin—Madison.

Rosenbaum, J., & Kariya, T. (1987). Self-selection in Japanese junior high schools: A longitudinal study of students' educational plans. *Sociology of Education, 60,* 168–180.

Rosenbaum, J., & Kariya, T. (1989). From high school to work: Market and institutional mechanisms in Japan. *American Journal of Sociology, 94,* 1334–1365.

Rowan, B. (1982). Organizational structure and the institutional environment: The case of public schools. *Administrative Science Quarterly, 27,* 259–279.

Rowan, B. (1983). Instructional management in historical perspective: Evidence on differentiation in school districts. *Educational Administration Quarterly, 18,* 43–59.

Rowan, B. (1990). Commitment and control: Alternative strategies for the organizational design of schools. In C. Cazden (Ed.), *Review of research in education* (Vol. 16, pp. 353–389). Washington, DC: American Educational Research Association.

Ruddell, R. B. (1985). Knowledge and attitude toward testing: Educators and legislators. *The Reading Teacher, 38,* 538–543.

Salmon-Cox, L. (1981). Teachers and standardized achievement tests: What's really happening? *Phi Delta Kappan, 62,* 631–633.

Sarason, S. (1977). *The culture of the school and the problem of change.* Boston: Allyn & Bacon.

Scardamalia, M., Bereiter, C., & Steinbach, R. (1984). Teachability of reflective processes in written composition. *Cognitive Science, 8,* 173–190.

Schwille, J., Porter, A., Alford, L., Floden, R., Freeman, D., Irwin, S., &

Schmidt, W. (1986). *State policy and the control of curriculum decisions*. East Lansing: Institute for Research in Teaching, Michigan State University.

Schwille, J., Porter, A., Floden, R., Freeman, D., Knappen, L., Kuhs, T., & Schmidt, W. (1983). Teachers as policybrokers in the content of elementary school mathematics. In L. Shulman & G. Sykes (Eds.), *Handbook of teaching and policy* (pp. 370–391). New York: Longman.

Scott, R., & Meyer, J. (1983). The organization of societal sectors. In J. W. Meyer & W. R. Scott (Eds.), *Organizational environments: Ritual and rationality* (pp. 129–154). Beverly Hills, CA: Sage.

Sedlak, M., Wheeler, C., Pullin, D., & Cusick, P. (1986). *Selling students short: Classroom bargains and academic reform in American high schools*. New York: Teachers College Press.

Shanker, A. (1989). *Asking the right questions*. Washington, DC: American Federation of Teachers.

Shimahara, N. (1986). The cultural basis of students achievement in Japan. *Comparative Education, 22*, 285–303.

Sizer, T. (1984). *Horace's compromise*. Boston: Houghton Mifflin.

Smith, M., & O'Day, J. (1990). Systemic school reform. In S. Fuhrman & B. Malen (Eds.), *The politics of curriculum and testing* (pp. 233–267). Philadelphia: The Falmer Press.

Sproull, L., & Zubrow, D. (1981). Standardized testing from the administrative perspective. *Phi Delta Kappan, 62*, 628–630.

Stackhouse, E. (1982). *The effects of state centralization on administrative structure*. Unpublished doctoral dissertation, Stanford University.

Stake, R., & Easley, J. (1978). *Case studies in science education Vol. 1, The case reports* (Stock No. 038-000-00377-1). Washington, DC: U.S. Government Printing Office.

Stevens, R. (1912). *The question as a measure of efficiency in instruction*. New York: Teachers College.

Stevenson, D., & Baker, D. (1991). State control of the curriculum and classroom instruction. *Sociology of Education, 64*(1), 1–10.

Stevenson, H., & Lee, S. (1990). Contexts of achievement: A study of American, Chinese and Japanese children. *Monographs of the Society for Research in Child Development, 55*(1–2) (Serial No. 221).

Stevenson, H., Lee, S-Y., & Stigler, J. (1986). Mathematics achievement of Chinese, Japanese, and American children. *Science, 231*, 693–699.

Stevenson, H., Stigler, J., Lee, S., Lucker, G., Kitamura, S., & Hsu, C. (1985). Cognitive performance and academic achievment of Japanese, Chinese and American children. *Child Development, 56*, 718–734.

Thompson, A. (1984). The relationships of teachers' conceptions of mathematics and mathematics teaching to instructional practice. *Educational Studies in Mathematics, 15*, 105–127.

Travers, K., & Westbury, I. (1989). *The IEA Study of Mathematics I: Analysis of mathematics curricula*. New York: Pergamon.

Trow, M. (1961). The second transformation of American secondary education. *International Journal of Comparative Sociology, 2*, 144–166.

Trow, M. (1988). American higher education: Past, present, and future. *Educational Researcher, 17*(3), 13–23.

Tyson-Bernstein, H. (1988). The academy's contribution to the impoverishment of America's textbooks. *Phi Delta Kappan, 70*, 194–198.

U.S. Bureau of the Census. (1989). *Statistical abstract of the United States, 1988*. Washington, DC: U.S. Department of Commerce.

Van der Blij, F., Hilding, S., & Weinzweig, A. (1980). A synthesis of national reports on changes in curricula. In H. G. Steiner (Ed.), *Comparative studies of mathematics curricula—Change and stability, 1960–1980.* Haus Ohrbeck: Institut fur Didaktik der Mathematik der Universitat Bielefeld.

Welch, W. W. (1979). Twenty years of science curriculum development: A look back. In *Review of research in education* (Vol. 7). Washington, DC: American Educational Research Association.

Westbury, I. (1980). Conclusion to conference proceedings "Reflection on case studies." In H. G. Steiner (Ed.), *Comparative studies of mathematics curricula—Change and stability, 1960–1980* (pp. 509–522). Haus Ohrbeck: Institut fur Didaktik der Mathematik der Universitat Bielefeld.

White, M. (1987). *The Japanese educational challenge.* New York: Free Press.

Wirt, F., & Kirst, M. (1982). *The politics of education: Schools in conflict.* Berkeley, CA: McCutchan.

Wise, A. (1979). *Legislated learning.* Berkeley: University of California.

Witte, J. F. (1989). *Choice and control in American education: An analytic overview.* Madison: University of Wisconsin—Madison.

Witte, J. (1990, August). *Understanding high school achievement.* Paper presented at the meeting of the Political Science Association, San Francisco.

Chapter 2

In Search of the School Curriculum

NATHALIE J. GEHRKE, MICHAEL S. KNAPP, and KENNETH A. SIROTNIK
University of Washington

Euclid may have been among the first to note that "the whole is the sum of its parts." But surely he was not describing "the curriculum." Given our experience over the several months spent on this project, and judging by what has been written by others attempting to explain the "curriculum field," we are reminded again why the field is at once so fascinating and frustrating: One seems to get a general sense of what "the curriculum" is without knowing quite how to define it in all its detailed parts; yet, once having made inferences at this level of generality, there remain nagging concerns that much remains to be discovered.

Put simply, the challenge we have taken on is to review what is known about what schools really teach. "What schools really teach" is a matter squarely in the domain of study called "curriculum," a field that has generated enormous activity and scholarship. It is a field, however, that remains contentious in terms of definition and delineation. After perusing all the curriculum texts on our collective shelves, we rediscovered what we and others have known for some time: "A quick survey of a dozen curriculum books would be likely to reveal a dozen different images or characterizations of *curriculum*. It might even reveal more, because the same author may use the term in different ways" (Schubert, 1986, p. 26).

This observation, tired and old as it is, weighs heavily on those who undertake a review of a domain that resists definition. This resistance, of course, is what makes curriculum so fascinating and the field so prolific (Eisner & Vallance, 1974). Yet it puts reviewers like us between the proverbial rock and hard place. No one feasible slice through the extant literature could reveal what schools really teach, when for some it is found in the content of tests, for others, in the personal phenomenologies of teachers and learners, and so on.

We had to make hard decisions, decisions that probably affected what we found and what is left to find. In this review, therefore, we need to accomplish three things. First, we share with readers our view of the complexity of the curriculum field and in that context indicate how we chose to find our way through this territory for purposes of the present review. Second, we present our findings—our attempt to provide answers, if partial, to the question of what schools really teach. Finally, we summarize and discuss the search for what schools really teach in four subject areas: social studies, language arts, mathematics, and science.

APPROACHING THE TASK

The only word in the phrase "what schools really teach" that would be minimally contentious is "schools." Here we restrict our review to those places typically referred to as elementary and secondary public schools. We did not deliberately seek out special or alternative public schools (e.g., magnets, continuation schools).

"What," "really," and "teach" are much more problematic. "Teach" conjures up the controversy over whether "instruction" is fundamentally part of, or sufficiently separate from, "curriculum." It forces the question: When we consider "what," must we also consider "how"? Judging from textbooks on curriculum, it appears to be impossible not to devote substantial space to process issues and, therefore, to issues pertaining directly to teaching. A "curriculum" on "critical thinking" could end up being delivered as a deductive, problem-solving unit in mathematics or as a unit on dialectical argument and competent communication in social studies. Each of these units, and their attendant conceptions of "critical thinking," could be handled in a variety of instructionally interesting ways. Yet it is not likely that considerations of instruction could be independent of considerations of the differing conceptions of "critical thinking." The act of teaching shapes what is taught, and what is to be taught shapes how it is taught.

Perhaps separations between "curriculum" and "instruction" are useful in curriculum inquiry and instructional design theory, and certainly there are large bodies of work in both domains. But from the standpoint of making inferences about "what schools really teach," ignoring instruction is about as fruitful as ignoring *knowing* when considering the nature of *knowledge* (Dewey, 1916). We cannot consider one in absence of the other.

The problematic nature of "what," therefore, becomes increasingly apparent. It includes not only "how" (instructional strategies, nature of teacher-student interactions, and the like), but also many other features that characterize what really goes on in schools and classrooms, not the least of which are content and subject matter and the way these are organized (e.g., scope and sequence issues). Although the array of definitions for *curriculum* is diverse (e.g., see Beane, Toepfer, & Alessi, 1986, pp. 30–31), one clear commonality is that whatever it is, it is multifaceted. Implicit in Tyler's (1949) seminal work was the notion that content was not the whole of curriculum. To begin understanding what goes on in schools and classrooms requires appraising such features as intended goals and objectives; breadth, depth, and organization of content and subject matter; instructional strategies; learning activities; use of human and material resources; use of time and space; grouping patterns; and assessment of learning (Goodlad & Associates, 1979).

But through what "eyes" are these appraisals conducted? When we ask what schools *really* teach, about whose "reality" are we asking? The phenomenology of curriculum suggests an amalgam of perceptions and interpretations manifested in what teachers say they do, students say they get, and observers say they see. Given the putative influence of official state, district, or local curriculum guides, course syllabi, adopted textbooks, standard examinations, and the like, the interpretations and inferences made by reviewers of these curriculum documents and of similar documents-in-use in classrooms must be added to the phenomenological mix. Moreover, given the putative influence of "those in the know"— the curriculum theorists, the subject matter experts, the educational reformers, and so forth—the analyses and interpretations of what the curriculum *should be* must also be added to the "Rashomon" of what the curriculum presumably is.

To complicate our understanding of "really" even further, what does *not* get taught is also part of the curriculum. This null curriculum, as it is often called, is powerful by virtue of its absence. Consider, for example, a curriculum devoid of higher level thinking activities. The null curriculum often suggests a hidden (or implicit) curriculum, for example, a basic skills emphasis versus a critical thinking emphasis. The hidden curriculum, however, is implied not only by what is not present, but by much of what is. What is implicit can become rather explicit once one looks for it. Consider, for example, Bowers's analysis of technology (1988), McNeil's analysis of structure and control (1986), and Oakes's analysis of tracking (1985). Finally, the null, the hidden, and the not-so-hidden all attest to the normative or ideological content of schooling and therefore the curriculum. What schools are for, what they should be for, what values and beliefs are operant, whose interests are being served—these are issues that are always implicit (if not explicit) in structures and functions of schools and schooling.

Such is the nature of "the curriculum." The rather straightforward question "What do schools really teach?" is not so straightforward after all. For the purposes of our review, we did not attempt to develop a grand theory of curriculum that would direct us in defining, sorting, and selecting our way through the field. We felt that this would be inappropriate anyway, considering the current state of the art. Instead, we opted for a pragmatic course, following two primary paths through the territory that would reveal a sizable literature base for inferring much about what schools really teach.

First, of all the possible curricular commonplaces, subject matter appears to be among the most comprehensive organizers. Locating relevant literature by focusing on such features as learning objectives, use of time, or assessment strategies would probably be less productive than first focusing on a selection of major content areas and then looking for evidence pertaining to all relevant commonplaces. In an attempt to fairly represent content and to build upon our own working knowledge, we selected the following subject matters: social studies, language arts, mathematics, and science. Since much that is written about subject matter is tempered by level of schooling, we deliberately sorted by elementary and secondary levels.

It might be argued that focusing on instructional strategies and classroom interaction would be an equally viable path through the territory. However, given the broad level of generality inherent in our task, the modal instructional patterns appearing and reappearing for nearly a century represent a sufficient summary of what we would probably find were we to take this tack again (e.g., see the reviews and studies by Amidon & Hough, 1967; Cuban, 1984; Dunkin & Biddle, 1974; Goodlad, Klein, & Associates, 1970; Hoetker & Ahlbrand, 1969; Jackson, 1968; Peterson & Walberg, 1979; Sirotnik, 1983; and Stevens, 1912). Much of what really goes on in schools is suggested by persistent patterns of teachers lecturing to students in primarily direct instruction, total class formats. The vast majority of teachers' questions are at the factual recall and literal comprehension levels; affect-neutral classroom climates are the rule of thumb; there is a very small likelihood of students working together on common tasks in small groups; and so forth. Much of what does not go on is also suggested by these patterns (Sirotnik, 1988). Rather than repeat these kinds of analyses, however, we decided to take the subject matters as alternative high-yield paths through the curriculum territory.

Second, given the multiplicity of perceived realities discussed above, it seems that the clearest distinction was between what the curriculum *ought* to be and what it *appears* to be. The former is found in writings by content experts, reform commissions, and the like; the latter is found in reports of research relying on one or more data sources and concomitant perceptions and inferences (teachers, students, observers, course materials, textbooks in use, etc.). This dichotomy, although inadequate for dealing with the important issues of the ideological, null, and hidden curricula, served our purposes here because we were not focusing on these more implicit features of curriculum.

What the curriculum "appears to be," however, covers a lot of territory, and as a consequence a further set of distinctions proved useful, although not as a systematic organizer for analysis or discussion. Some insight into what is apparently taught can be gleaned from studying the *planned* curriculum—that which resides in district or state frameworks, curriculum guides, textbooks, and ultimately teachers' minds as they formulate what they will do. Further insight comes from evidence about the curriculum as *enacted* by teachers in the classroom, although one is left with the quandary over whose account of enactment—students', teachers', or observers'—to believe. Finally, neither the planning nor the enactment of curriculum fully captures the curriculum as received or *experienced* by students, although there are a few systematic and convincing accounts of the students' experiences. Where appropriate, in our discussion, we indicate whether the evidence we summarize refers primarily to the planned, enacted, or experienced curriculum. We do so to explain more fully the aspects of curriculum that are—and are not—made visible by existing research.

Several reentry points into the territory were selected for each of the four major content areas: major reform reports and the empirical bases (if any) ostensibly supporting the need for reform; reviews in major handbooks of research on the subject matter domain; an ERIC search using key words such as curriculum and curriculum change delimited by such phrases as K–12 education, educational reform, subject matter titles, and teaching and instructional practices; an on-line search through the University of Washington library acquisitions using similar key words for sorting; and other eclectic sources with which we were already familiar (large-scale survey studies, comprehensive multiple case studies, etc.). The list was further augmented with suggestions made by consulting reviewers.

Our findings, then, are presented separately for social studies, language arts, mathematics, and science, although salient points of comparison and contrast are noted. Each is organized in five parts: introduction and overview, views on what should be taught, patterns in what apparently is taught, differences between elementary and secondary levels, and summary comments. We found that any more systematic and detailed organization and interpretation of the literature was not possible because of the lack of consistency in type of information, commonplaces, perspectives, and levels of inference. In what follows, therefore, we note consistently what data sources are involved and whose reality is being reported and relied upon.

Finally, readers will note that although we report exceptions to general trends when we find them, we did not set out to search for the unusual. Indeed, our task was quite the opposite. And even though we have found many modal patterns to be similar across content areas, we have reported them separately in order to document adequately our conclusions.

WHAT IS KNOWN ABOUT THE NATURE OF THE SOCIAL
STUDIES CURRICULUM

The social studies curriculum is generally agreed to be aimed toward education for citizenship, but both content and approaches toward achieving that end have been much affected by the social and political agendas of each age. The area of the curriculum that came to be known as social studies began to take shape around 1900. History had been a part of the curriculum long before that, but, as the century turned, other disciplines were introduced. By 1916, the Social Studies Committee of the Commission on the Reorganization of Secondary Education had made recommendations that secondary social studies include history, economics, political science, sociology, and civics (Hertzberg, 1981). At both the secondary and elementary level, however, history has maintained the central position—even today, as historians complain about the decline in substance of the curriculum.

As in most curricular areas, spokespersons in social studies education bemoan the lack of rigorous research into many facets of the curriculum. They also complain about how little is known about the way it is enacted and about the effects it has on students in the short and long term. In the November/December 1990 issue of *Social Studies*, several authors expressed dissatisfaction of this kind while writing in response to the just-published contents of the enormous (661 pages) *Handbook of Research on Social Studies Teaching and Learning* (Shaver, 1991). For all the complaints, the handbook includes 53 separate chapters summarizing and critiquing research in the social studies field. Curiously, no section or chapter therein claims to address the nature of the curriculum per se, although 12 chapters are included in a section called "Teaching for and Learning Social Studies Outcomes." These chapters are organized around several commonly agreed-on outcomes for learners (e.g., thinking and decision making, empathy and moral development, concept development, social and political participation) and around the content of the several contributing disciplines (e.g., history, political science, economics, anthropology, sociology, psychology, geography). One chapter, strangely grouped in a section on contexts, does focus on "Scope and Sequence, Goals and Objectives" (Joyce, Little, & Wronski, 1991). However, the authors quickly admit that these matters "have not been thoroughly researched" (p. 321), meaning, apparently, not researched, although the authors point to "a variety of philosophical, historical, and conceptual writings" that add to understanding. A second relevant chapter on teachers as curricular-instructional gatekeepers in social studies (Thornton, 1991) reminds the reader of the powerful role that teachers play in the curriculum experienced by students. Thornton reviews those studies relating teacher planning and definitions of social studies to the curriculum and to instructional strategies. It is this role of the teacher in determining the contents of the social studies curriculum to which we will return later in this section, for the degree to which one can describe a modal curriculum seems to be directly related to the degree to which teachers in that field are in agreement on the aims, content, and methods of instruction.

When social studies educators describe the curriculum, they do so primarily on the basis of large-scale studies, mostly done in the 1970s, that surveyed social studies educators (Morrissett, 1977; National Assessment of Educational Progress [NAEP], 1980; Weiss, 1978) or textbook publishers (Schneider & Van Sickle, 1979) or analyzed textbooks, curriculum guides, and documents from states and local districts (Wiley, 1977). Two of these studies (Weiss, 1978; Wiley, 1977) along with a third—an extensive multiple case study (Stake & Easley, 1978)—were part of a single project directed by Helgeson and supported by the National Science Foundation (NSF). The trio presented the most comprehensive picture of the field that had been painted to that time.

All three concurrent studies, reported in nine major documents, have had a profound effect on the social studies literature and research. A vast portion of the inquiry in social studies done since the appearance of the reports references one or more of the three studies, the NSF report, the Project Span reports (1982) that depended heavily upon them, or a summary and interpretation of them by Shaver, Davis, and Helburn (1980), also supported by NSF. The recent effort by Jenness (1990), supported by the National Commission on Social Studies in the Schools, draws on all of the above as well as many others to present not only the history of social studies, but also a current picture of the curriculum. That picture focuses on the secondary curriculum. The elementary curriculum is treated peripherally, perhaps because social studies in the elementary school is accorded only about 20 minutes a day, according to Jenness (1990, p. 19), but it becomes a significant portion of the required curriculum at the secondary level.

Views on What Should Be Taught in Social Studies

Although nearly all in the field agree that the main purpose of social studies is education for citizenship, this field is most tolerant of alternative definitions of how one prepares for citizenship, and what the content of the field should be. It appears equally tolerant of alternative approaches to teaching and, possibly, to research. Demonstrating this tolerance, the National Council for the Social Studies has offered not one, but three, different suggested scope and sequence patterns based on three different points of view on the field's main goals (*Social Education*, October 1989). The organization had entertained as many as five patterns at one time

(*Social Education*, November/December 1986). Perhaps the reason for this tolerance is that social studies draws from and is taught by individuals from a wide array of disciplines. Perhaps it is because the focus on education for democratic participation that pervades the field requires the modeling of consideration of alternative points of view; or perhaps it is because this field is simply as confused about citizenship and education for it as is the greater culture in which it finds itself. In any case, there is little consensus among professionals on what should be taught to make one a good citizen.

Because of a diversity of points of view about the main purposes of social studies beyond education for citizenship, social studies educators have spent a considerable amount of energy identifying categories (at least three, but no more than seven) and refining descriptions of the major traditions in approaches to social studies (Barr, Barth, & Shermis, 1977; Brubaker, Simon, & Williams, 1977; Engle & Ochoa, 1988; Hass, 1979; Morrissett, 1977). Perhaps labeling gives a sense of control over the field. Nearly all the categorizers agree that although these several approaches are represented in the various curricula that teachers plan and enact, the most common approach is that of conservative cultural continuity. Conservative cultural continuity essentially means teaching social studies as transmission of citizenship-related information with emphasis on inculcation of traditional values (Morrissett & Hass, 1982). The three NSF studies, working from different data sources and using different inquiry methods, all concluded that most social studies curricula, in their planned and enacted forms, were aimed primarily at transmitting information and at socialization aimed to prepare students to enter adult life "without a ripple of discontent or change" (Stake & Easley, 1978, preface).

Periodically, the conservative cultural continuity approach to social studies is challenged by two other approaches, according to Hass (1979). One, "the intellectual aspects of history and the social sciences," often comes from academics in one of the social studies contributing disciplines (e.g., Ravitch, 1987). The recent restructuring of the California social studies curriculum framework has been heavily influenced by those representing the intellectual aspects of the disciplines, particularly history. The second challenger approach, which focuses on the "process of thinking reflectively," emerged with Dewey's work in the 1920s and resurfaced among social studies education academics. The "New Social Studies" projects of the late 1960s and 1970s represented a continuation (Sanders & Tanck, 1970; Oliver & Shaver, 1974; Wiley & Morrissett, 1972). Most of the curricula developed at that time revolved around discovery, inquiry activities, and social problems. More recently, the proposal of Parker (1990) and the ongoing work of Newmann (1990) represent this point of view. Social studies educators claim that this approach is the one that has the largest body of research evidence supporting its merits for reaching the goal of good citizenship (Evans, 1989; Parker, 1991).

Although there is little consensus on what should be taught in the social studies curriculum, there appears to be agreement on the shortcomings in what is being taught, how it is being taught, and the preparation of those who teach it (Morrissett, 1981). Complaints are common about the vacuous, redundant nature of the curriculum, especially as presented in textbooks (Ravitch, 1987); about the flattened content (Goodlad, 1984) and the inattention to the relationship of the content to the developmental levels of students (Armento, 1986; Egan, 1982; Levstik, 1986); and about the limited and decreasing inclusion of formal social studies learning activities, especially in the primary grades (Gross, 1977; Hahn, 1985; Jarolimek, 1981). Critics have not agreed, however, on how the shortcomings should be remediated. That, of course, depends on the point of view.

Patterns in the Social Studies Taught Across the Grades

Social studies educators conclude that the conservative cultural continuity approach has been maintained over some 80 years, primarily because of teacher and parent preference (Stanley, 1985) and because of the contents of textbooks (Wiley, 1977). Scholars further conclude that the formal social studies curriculum in the United States is characterized by an emphasis on the transmission of information selected from the various disciplines, but mostly history; by an avoidance of controversial or sensitive topics; and by a topical organization that leaves wide latitude to the teacher for selection of content for emphasis, and subsequently allows enormous diversity in the topics actually covered. There is evidence of differential curricula for students in schools in different kinds of communities (Hemmings & Metz, 1990) and for students within separate tracks in the same schools (Sanders & Stone, 1987). Whether this diversity is appropriate or objectionable is hotly debated. But it exists, even though formal curriculum documents suggest great uniformity in course titles, requirements, and sequencing.

Teachers use a read-recite format for social studies instruction but supplement their transmission of information activities with a few activities for reflective thinking and experiential learning; they claim they aim for higher level thinking, but most do not appear to pose questions or create activities that would seem to exercise such thinking. The typical class is teacher centered, dependent on a single hardback textbook from one of a handful of major publishing companies, and unlikely to include much if any student choice of content or activity. These generalizations hold true across the elementary and secondary grades, according to those writing at the end of the 1970s (Lengel & Superka, 1982; Mehlinger & Davis,

1981), those writing in the mid-1980s (Tye, 1985), and those writing at the end of the 1980s (Brophy, 1990; Shaver, 1991).

The Social Studies Curriculum in the Elementary Grades

We can turn to the three concurrently completed NSF studies (Stake & Easley, 1978; Weiss, 1978; Wiley, 1977) and the related writings, especially the Project Span report (1982), to draw a sketch of the social studies curriculum within the elementary grades.

Wiley's analysis of curriculum guides, textbooks, and other materials used between 1955 and 1975 has been the primary basis for statements about the formal, planned curriculum within the United States. Lengel and Superka (1982, pp. 85–88) summarized the main topics generally taught in each elementary grade level as gleaned from Wiley's study, elementary social studies textbooks, and Project Span staff experience. Their topical summary is as follows:

Kindergarten—Self, Home, School, and Community: Discovering myself (Who am I? How am I alike and different from others?), school (my classroom, benefits of school), working together, living at home, community helpers, children in other lands, rules celebrating holidays, working and playing safely.

Grade 1—Families: Family membership, recreation, work, cooperation, traditions, families in other cultures, how my family is alike and different from others, family responsibilities, my senses and feelings, the family at work, our school and other schools, and national holidays.

Grade 2—Neighborhoods: Workers and services in the neighborhood, food, shelter and clothing, transportation, communication, living in different neighborhoods, my role within the neighborhood and community, changes in my neighborhood, neighborhoods and communities in other cultures, farm and city life, and protecting our environment.

Grade 3—Communities: Different kinds of communities, changes in communities, community government, community services, communities in other countries, cities, careers in cities, urban problems, business and industry, pioneers and American Indians, and communities past and present.

Grade 4—Geographic Regions & State History: Different world regions, people of the world, climatic regions, physical regions, population, food, problems of our state, our state government, state history, people of our state, state laws, roles of state workers.

Grade 5—U.S. History: The first Americans, exploration and discovery, colonial life, westward movement, War Between the States, immigrants, the Roaring Twenties, lifestyles in United States, values of the American people, our neighbors to the North and South, United States as world power, great American leaders.

Grade 6—World Cultures/Hemispheres: Political and economic systems, land and resources, people and their beliefs, comparative cultures; Western Hemisphere—Early cultures of South America, the ABC Countries, Central American Countries, Canada, Mexico, historical beginnings of Western world; Eastern Hemisphere—Ancient Greece and Rome, Middle Ages, Renaissance, Middle East, Europe, Africa, India, and China. (Lengel & Superka, 1982, pp. 85–87)

This framework has come to be termed the "expanding environments" pattern, that is, one that begins from the immediate environment of the child and moves outward to ever-enlarging contexts. The framework is traced back to a report in 1916 done in conjunction with the secondary-level Report of the Committee on Social Studies (Dunn, 1916, as reported by Morrissett, 1981). Hanna (1963) is credited with elaborating and most clearly explaining the expanding communities pattern, although most curricula are faulted with following the letter but not the spirit of his recommendations.

A quick review shows that in addition to the expanding environments topics, one may also find interspersed traditional topics that do not fit the pattern, such as holidays and famous persons. In addition, special topics of interest are added by the local district or the teacher. Lengel and Superka (1982) mention the addition of law-related studies, ethnic heritage, women, and careers; Jarolimek (1981) added economic education and global education. Wiley had concluded that the first-grade scope had become progressively cosmopolitan and multicultural and less rurally oriented, that the focus on neighborhoods in second grade had become more urban, and that there was an increasingly comparative orientation to community studies in third grade (1977, p. 30). She further explained that one can find considerable diversity in topics in curriculum documents. The appendix to her report contains a nine-page-long table of topics found in state and local curriculum guides.

While the K–3 pattern had been extraordinarily stable across time, the 4–6 level had shown more changes in both curriculum documents and textbooks. State history became the main focus of Grade 4 courses by 1975, replacing geography. Grade 5 had become more "presentist" in its orientation to U.S. history by 1975, according to Wiley. Grade 6 content had changed from a predominantly Western hemisphere focus to an emphasis on world cultures, although some sixth–seventh-grade sequences were exploring Western and Eastern hemispheres, one a year, in either order. The latter pattern may be connected to the concurrent increase in middle schools with a sixth–eighth-grade configuration replacing the junior high schools (Grades 7–9). Patterns of curricula in particular states vary as well. For example, New York and Texas show considerable differences at certain middle grade levels, because Texas has "an unusual history, and distinct constitutional-legal legacy" (Jenness, 1990, p. 26), whereas New York's relationship to the union makes its state history feasible to teach as part of national history.

Although critics of the expanding environments pattern abound, Lengel and Superka, and later Brophy (1990), contended that the pattern has survived because the topics are generalized and nonspecific and allow teachers to incorporate new content and adapt the topics to local needs and teacher interests. In addition, all the major textbook series in the

1970s and the 1980s held to this pattern, as did methods textbooks used in preservice programs (e.g., Kaltounis, 1969). The problems inherent in the expanding environment framework are, according to some, the lack of attention to students' interests (Egan, 1982) and the delayed engagement of students with content likely to shape their attitudes on important issues. Mitsakos (1978), for example, noted that children's attitudes toward people of other cultures could be positively shaped by early (K–3) introduction of a global education program offered in place of the usual sequence. Others counter that the expanding environment pattern could accommodate such global education content.

Most troublesome to social studies educators in the early 1980s was not the stability or change in content of the elementary curriculum, but a decided decrease in the amount of social studies being taught, particularly at the primary (K–3) level. The survey of practitioners by Weiss (1978), the Stake and Easley multiple case study (1978), and the survey of educators done by Gross (1977) all pointed to a trend that found social studies losing time in the school day to mathematics and reading as part of a back-to-basics movement. There had been a decline in sales of early-grades social studies textbooks as well. This same decrease in time allocated and in textbook purchasing was not in evidence in Grades 4–6, where social studies seemed to have a stable place, though still of less importance in the curriculum than mathematics and reading. Jenness's update claimed "most experts agree" that today less time is being given to social studies and science and more to reading and mathematics (1990, p. 19).

Some contended that although social studies may have lost time allocated to its exclusive study in Grades K–3, many materials in reading and language arts include social studies content (Stake & Easley, 1978), so it is being taught indirectly. This indirect teaching tends to substantiate the claims that social studies content is conceived by teachers as what is known or factual (pp. 13–32). In one of the few "close-up" studies of elementary curriculum, J. White (1985) further reported that, no matter what the content of the materials read, the elementary teacher whose first-grade class she studied closely for a year tended to use the stories and descriptions to point out the implications of the material for students' own values and behaviors. White claimed that when the first-grade teacher told a story about modern-day travelers in the Saudi Arabian desert, she took the opportunity to get across the message that children should always go to the bathroom before they go on a long bus trip; when the black plague was discussed, the virtue of cleanliness was affirmed (1985, p. 234). This, then, represents the second part of the conservative cultural continuity approach to social studies, that is, the socialization of students to important community values.

The Social Studies Curriculum in the Secondary Grades

It should come as no surprise that the secondary social studies curriculum pattern can be traced back to 1916, just as the elementary pattern has been. Instead of the expanding environments organizer, however, the secondary pattern is said by social studies educators to exhibit two cycles of "contracting environments" (global to national) heavily weighted with history.

Grade 7:   World History/Cultures/Geography
Grade 8:   U.S. History
Grade 9:   World Cultures/History or Civics/Government
Grade 10: World Cultures/History
Grade 11: U.S. History
Grade 12: American Government or Sociology/Psychology
            (Lengel & Superka, 1982, pp. 88–89)

These two cycles are said to have been planned for the early-20th-century curriculum on the basis of the school-leaving pattern of the day (with large numbers leaving after sixth grade, and again after eighth or ninth). By including world history in seventh grade, U.S. history in eighth, and world culture/history (earlier, European history) or civics in ninth (having given U.S. history to fifth graders earlier), one could give a first exposure to both national and world history before most students had left school for good. Those who stayed to graduate—about 20% at the time—received another opportunity for learning through a world culture course (previously European history) in 10th grade, U.S. history in 11th, and American government in 12th. In more recent times, the 12th graders may have taken sociology/psychology or economics in place of American government. In some places, students are freer to take the courses in a different sequence and, in many states, where only some of the above courses are required, students may add other disciplinary and interdisciplinary electives.

While the elementary social studies expanding environments organizer can still be justified, however tenuously, on the basis of child development, the secondary social studies organizer is clearly based on outdated assumptions about school leaving (better than 75% of all students stay to graduate from high school). Still, Wiley's study (1977) and a survey of secondary social studies teachers (Fontana, 1980) confirmed the Grade 7–12 sequence outlined. Furthermore, it appears that little or no articulation is attempted among the various courses taken in this series. The new California framework claims greater articulation, but the subject matter still looks amazingly close to the traditional pattern (Jenness, 1990, p. 26). This lack of articulation is often exacerbated by the purchase of texts for each course completely independent of the purchase of other social studies texts (Patrick & Hawke, 1982, p. 109).

Aside from the descriptions of the course labels, social studies in the middle grades (6–9) seems to have received very little research attention, even though it is a designated course in the middle-grades curriculum. One source of insight into middle-grades social studies is the four separate grade-level shadow studies (Lounsbury & Clark, 1990; Lounsbury & Johnston, 1988; Lounsbury & Marani, 1964; Lounsbury, Marani, & Compton, 1980) of multiple students supported and published by the National Association of Secondary School Principals. Each followed about 150 students throughout a single school day. The reports highlight a few students and summarize and discuss the reports on the total group. We thus can see the experiences of several students at each grade level in their social studies and other classes through the eyes of their observers. Each student's experience is somewhat different, but one analyst's comment on the eighth-grade reports seems telling:

> With few exceptions, curriculum content in these schools is locked into preparation for life 25 years ago. Observers' comments on course content evidenced no attempt to deal with emerging demands upon this nation and our planet in terms of shifting realities and problems already facing humanity. (Lounsbury & Clark, 1990, p. 35)

Although the sequence shown above is typically offered in high schools across the country, the requirement to take such course work varies by region. While the mean number of courses required across the nation in 1988 was just under three (2.9), high schools in the North Atlantic region required more (3.2) and high schools in the Southeast region required less (2.7; U.S. Department of Education Center for Educational Statistics, cited in Jenness, 1990). Across the country, states have tended to require more social studies than mathematics or science (Jenness, 1990; Weiss, 1978). The most common patterns are those requiring U.S. history and civics/government, plus either world history, economics, or world geography (Jenness, 1990, p. 37). Educational reforms have seen increases in mathematics and science requirements in the last several years, but apparently not at the expense of social studies requirements (Grossman, Kirst, & Schmidt-Posner, 1986). Electives in social studies such as sociology, psychology, economics, and international relations have shown declines based on a California course offering analysis (Catterall & Brizendine, 1985) and on a national high school transcript analysis (U.S. Department of Education, 1987); history courses have increased.

The problem inherent in using requirements, transcripts, and course descriptions to understand the nature of the curriculum is that they often do not reflect the actual course content. Earlier as well as recent explorations show that course titles and district documents can be quite misleading and/or vague. Examination of textbooks and materials, syllabi, and class activities is necessary to get at the nature of the curriculum. The large-scale Study of Schooling conducted by Goodlad and colleagues (1984) not only examined formal curriculum documents, but also "snapshots" of life in classrooms. Tye (1985) separately reported the summary and analysis of these high school data by content area. Her comments highlight the "enormous differences between schools in the range of social studies courses offered" (p. 224), from one school that offered only the required courses to another that offered 55 quarter courses, any 16 of which would make up the students' 4-year social studies requirement. Most of the other 11 schools fell somewhere in between. Yet the actual course content of the basic and required courses was, Tye (1985) claimed, nearly identical across all 13 schools. For example:

U.S. History begins with the colonial era and moves through certain topics (more or less predictably skimming very lightly over others, or omitting them altogether), up to the 1960's. U.S. Government includes the Constitution and the three branches of government, and often covers state and local government as well as the federal system. Yet students never visit a local courtroom to observe the judicial process, nor does the local congressperson ever come to speak to the class. (pp. 224–225)

She concluded that, based on observations, the students in required courses were not only covering the same general topics, but also doing basically the same thing in their classes: listening to the teacher, having class discussions, writing answers to questions, taking tests, and occasionally seeing films or filmstrips. Curiously, students in social studies talked and worked in small groups less than in any other core content area (Tye, 1985, pp. 214, 221).

Here we come to a basic descriptive crisis. On one hand, we have several large-scale studies, such as the NSF studies and the Study of Schooling as interpreted by Tye, that point to a modal secondary social studies curriculum containing a basic set of course offerings, each with a relatively defined set of topics, "predictably" emphasizing some topics and skimming others, and each most typically using a predictable set of instructional practices regardless of the student audience. These studies find some confirmation in the close-up studies of McNeil (1986), who looked at several high school social studies programs, and of McKee (1988), who examined seven U.S. history courses emphasizing thinking skills. This should suggest that most students are getting a relatively common, although often uninspiring, curriculum in social studies across the country.

On the other hand, an equally large-scale study—of 26 high schools in California (Sanders & Stone, 1987)—seems to show that the modal curriculum is not truly the most common: Students on different tracks through high school experience very different curricula, with different goals, textbooks and materials, homework demands, and classroom activities. All the 26 representative California high schools tracked students in English, mathematics, and science, but not in social studies, so at first glance it would seem that students would have similar exposure to content in this curricular area. However, the study reported that the highest-achieving students have the fewest choices because academic sequences planned for them fill almost all of the available time. High-achieving students end up in required courses, and then in only a very few additional social studies courses that match their schedules. At the same time, most other social studies electives are likely to be populated with students from lower tracks, which is what Stake and Easley's researchers found in the case study of a high school that was heavily tracked (1978). The freedom to select electives, coupled with tracked versions of the required courses, sets up different opportunities and demands for different students.

How can we reconcile these two seemingly polar positions on the uniformity or diversity of the curriculum? It appears that the factors examined by the various studies may contribute to the differences in conclusions about differentiation of the curriculum, from a third-party point of view. When factors such as homework assignments, textbook used, and access to electives are included in descriptions, more variability is seen; when the factors are curriculum documents, or in-class teacher behavior or classroom activities, less variability is described. We may, in other words, have an example of the blind men examining different parts of the elephant—and doing so beginning from certain preconceived notions about what will be found.

Before we leave the secondary social studies curriculum, however, and to further illuminate the above issue, we must mention briefly those studies in the area that shed light on the particularistic presentations of the social studies curriculum. All were done recently, depended on qualitative methods, and sought less to generalize than to interpret small numbers of cases. They include studies done to examine the role of context, or teacher background and knowledge, in the shaping of the curriculum. In the process of reporting on these important elements, glimpses at the curriculum are also offered.

At least three studies are representative of this close look at enacted curricula, all examining history teaching. In the first (Wilson & Wineburg, 1988), we see history curricula being taught by four beginning history teachers who selected different content, used textbooks differently, and planned significantly different learning activities. The differences were related by the researchers to the distinct disciplinary backgrounds of the four, not all of whom were history majors. In the second (Thornton, 1988), three 10th-grade U.S. history teachers in a single school were observed and their practices described. The wide variety of conceptions about the purposes of the course and, therefore, what was offered were related, according to Thornton, not to conceptions of how students learn but to context and management concerns. Similarly, McNeil (1986) reported a case study of history teachers in different high schools in which she found teachers in certain contexts making curriculum choices more often to ensure control rather than inspire learning. Teachers' typical choices of approach and content selection are described, revealing something of the history curriculum as well as the nature of the schools.

With their focus on the teachers, these last three studies begin to illustrate why courses that may bear the same title often cover highly divergent topics, have highly variable teacher-stated goals, and yet often, underneath, converge on transmission of similar values.

Making Sense of the Evidence Regarding the Social Studies Curriculum

From the evidence presented in large-scale studies done over a decade ago, we find social studies to be a curriculum area where there is the appearance of uniformity in basic topics, sequences, course titles, and textbook contents. There is also considerable uniformity in the kinds of instructional practices in which teachers engage in the classroom. But there appear to be incredible differences as well, and those differences have some disturbing implications for those concerned about educational equity. The differences appear to lie in the decisions teachers make in adjusting the topics, sequence, textbook contents used, and homework assignments. Teachers make these adjustments as a result of their own background knowledge and beliefs, to meet the needs of particular groups of students, or to conform to school and community values. The social studies curriculum is very much in the hands of the individual teacher.

Because there is no single national professional organization that speaks to and for social studies teachers, but rather an array of competing organizations and disciplinary groups, there are many nudges and tugs but no consensus and no clarion calls to bring teacher-level action toward a kind of curriculum that shows greater consensus. In addition, social studies is viewed by educators outside the field and by lay people as a second-tier content area. Although it is given increasing time in the school curriculum as students move from kindergarten through high school, it seems more tolerated than touted and more tangential than core. With national leadership fractured and local leadership disinterested, it is no wonder that teachers look primarily to themselves, in the most isolated sense, for choices about what really gets taught in the social studies curriculum.

We do not yet know well, and may never be able to articulate, what the subject matter of the modal social studies curriculum is, although Jenness's work is the most recent valiant attempt to characterize the history and current status of its study at the secondary level. This is not due so much to the shortcomings of research methods, lack of studies, or lack of interest, but rather to the tremendous variability of subject matter any one student may experience because of the teacher's knowledge and the influences of the context. Recent work by Stodolsky (1988; Stodolsky, Salk, & Glaessner, 1991) is beginning to give a picture of the social studies curriculum as experienced by students and how their experiences differ in other subject areas. More of this work is needed, for there appear to be considerable differences in what students claim to experience and what teachers and even third-party observers claim is offered.

We do not yet have many interpretive portraits of social studies in the elementary and middle grades, nor in the many different social studies courses students may take in the secondary grades. Some pictures are emerging of history courses, but all the other electives and required courses have been given less examination. What, for example, gets taught in high school world geography courses, or in sociology, psychology, economics, anthropology, Black studies, third-world studies, or any number of other electives?

The three new proposed scope and sequence patterns offered by the National Council for the Social Studies in 1989 (*Social Education*, October 1989) may provide some of the structures missing in the formal documents of the planned curriculum and perhaps in the enacted curricula of the teachers. Cautious optimism is necessary, however, because in social studies, as in other content areas, teachers are not viewed as likely to turn to the literature and research as a guide to practice. Unlike mathematics or science teachers, social studies teachers typically believe they know all they need to know to teach it (Goodman & Adler, 1985; Stake & Easley, 1978). Given that they also tend to view social studies primarily as a vehicle for transmitting cultural values and facts that the community believes and the textbook supports, the curriculum is likely to be slow to change.

WHAT IS KNOWN ABOUT THE NATURE OF THE LANGUAGE ARTS CURRICULUM

Until the present century, elements of language arts (e.g., reading, writing, spelling, grammar, and rhetoric) were typically taught separately. These aspects have gradually merged, at least as they are viewed by scholars, but their presentation in the K–12 grades still exhibits much of the separateness of an earlier era (Squire, 1991). For example, only during the last decades has writing been viewed by some as the process in which spelling, grammar, usage, and other skills learning should be embedded. Although scholars in the field may recognize the relationships among the elements, their own special interests have been questionable contributors to further merger. Each of the four language arts (reading, writing, speaking, and listening) has attracted more or less separate groups of researchers and spokespersons who have formed separate organizations, hold separate conferences, and seem to write for different audiences through different journals. The recent publication of another mega-handbook, the *Handbook of Research on Teaching the English Language Arts* (Flood, Jensen, Lapp, & Squire, 1991), attempts to compile research from all the specialized areas. But at the same time, the equally large second volume of the *Handbook of Reading Research* (Barr, Kamil, Mosenthal, & Pearson, 1991) has appeared, to perpetuate the separation. Oral language is the stepchild in all of this.

Language arts seems to be one area of the curriculum where scholars in the field do not bemoan the sorry state of research. Several research and development centers and major projects have focused resources on language arts questions over the past 30 years (Early, 1991). These projects include the well-known First Grade Studies (Bond & Dykstra, 1967), Looking in English Classrooms (Squire & Applebee, 1968), and the National Writing Project directed by Graves (Graves & Myers, 1978).

In spite of all of the research that has apparently been carried out on the English language arts in recent decades, only a limited amount has been focused on describing the state of the curriculum. Authors write as though there is agreement about the nature of current language arts curriculum and that it is not good, then proceed to study other things. Scardamalia and Bereiter (1986) listed nine major focuses in writing research, only one of which was "classroom practices" (p. 780). There are hints that quite a number of dissertation studies have examined the curriculum as presented in different basal reading series or grade-level texts, but these have not been reviewed here.

Research in language arts often centers instead on the learners, particularly on their development in reading, writing, speaking, and listening, in that order of importance. Several single or multiple case studies have received considerable attention as English teachers and scholars have observed their own children or a small number of other children learning to read or write, or both (e.g., Bissex, 1980; Calkins, 1983). Another popular kind of study examines the alternative outcomes of different kinds of experimental curricula, but the curricula described are the exceptional, not the commonly found (e.g., Atwell, 1987). A more recent research focus, comparable to that found in social studies, is on the nature of teacher knowledge and beliefs and their relationship to teachers' curricular decisions (e.g., Grossman, 1990). Through this last group we can see, as we also did in social studies, some of the particularistic ways that the language arts curriculum is enacted in variance from any modal description that might be given.

Interestingly, when scholars in the field wish to present a picture of the curriculum in language arts, they frequently reference studies across the general curriculum such as Goodlad's Study of Schooling (1984) and Boyer's Carnegie Study of High School (1983). The most often cited status studies done within the field appear to be those of Squire and Applebee (1968) and later of Applebee (1981, 1989), but these are focused mostly on high school language arts, not the entire K–12 curriculum. Squire and Applebee's (1968) report of the National Study of High School English Programs gathered data from 158 high schools with reputations for excellence in English. Applebee's (1981) study focused on writing in the high school. No recent studies of large scale are found, suggesting that, as in social studies, pictures drawn in the past are believed to hold true today, except in the most innovative classrooms.

In elementary programs, the notable reference points include the ethnographic studies of Heath (1983, 1986) on the development of language skill in school and in the family, Cazden (1988) and Dillon and Searle (1981) on classroom discourse, Durkin (1978–1979, 1987) on instruction and testing emphases of teachers, and Florio and Clark (1982) on uses of writing. For our purposes here, we will also use the 1988 NAEP survey, interpreted by Langer, Applebee, Mullis, and Foertsch (1990) in reading and Applebee, Langer, Jenkins, Mullis, and Foertsch (1990) for writing. These surveys gathered data from 4th-, 8th-, and 12th-grade teachers and students (a summary of the 1984 data was provided earlier by LaPointe, 1986). We make assumptions that grades not covered in the survey are more like than unlike those included (e.g., second and third grade are little different from fourth, and so on).

Views on What Should Be Taught in Language Arts

As with social studies, multiple traditional approaches can be described in the teaching of language arts, and they have each had periods of ascendency over the past several decades (Applebee, 1974). The approaches are near counterparts to those in social studies, springing from similar philosophical orientations. There are three approaches, as named by Farrell (1991): the mastery model, the heritage model, and the process model (p. 65). The *mastery model* is closely associated with competency assessment, instructional objectives, and basic skills learning. In the form endorsed by scholars (e.g., Bloom, 1981), students experience carefully controlled, sequenced learning activities until mastery of the skill/knowledge is exhibited; until mastery the student does not proceed to other skills. With some looseness, the mastery model is similar to the conservative cultural continuity approach in social studies. Language arts educators do not, as a rule, speak from this point of view. Instead, this is a model expressed more in practice and through curriculum materials and textbooks.

The second language arts model championed by scholars is the *heritage model*. Like its academic discipline-oriented counterpart in social studies (the intellectual aspects of history and the social sciences), the heritage model looks to the traditionally accepted canon for literature of enduring worth that embodies the values and great ideas of the larger culture. Farrell cites Hirsch (1987), Cheney (1987), and Ravitch and Finn (1987) as advocates of the heritage model, scholars who are active voices in the academic disciplines social studies approach as well; Suhor (1988) names Fancher (1984) and Adler (1982) as additional spokespersons.

The last, *process model* is considered the most student centered. It is especially concerned with the way each individual constructs knowledge from experience. One can relate it to the reflective thinking process model in social studies. Scholars working from this model see language use as developmental, encourage creativity and writing from an early age, and focus on students' understanding of the relationship between the writer's aim and the reader's response. Rosenblatt (1938), Britton (1972), and Moffett and Wagner (1983) have been among the primary spokespersons for the process model, which seems to have captured the center stage in recent practitioner-oriented journals and curriculum reform literature.

In characterizing the strength of each approach, one could say that the mastery model is pragmatically focused on identifying the necessary skills for human communication and then figuring out efficient ways to transmit them to the novice; whether successful or not, the aim is efficiency. The heritage model is most concerned with how, and particularly on what, those skills are put to use in intellectual activity, once they are learned. The process model is focused on language arts activity (usually, though not exclusively, writing) at a developmentally appropriate level as a means to teach the necessary skills, as well as to encourage creative and more frequent use. The process model has had its greatest effect on elementary language arts programs, some influence in the middle grades, and very little effect in the high schools, according to Farrell (1991, pp. 73–74).

There is some evidence that a few scholars are developing arguments for a model that incorporates the concerns of the heritage advocates for experiences with great ideas (usually from literature) with the concerns of the process advocates for developmentally appropriate experiences (e.g., Suhor, 1988). The focus on children's oral language development finds a more welcome home in this emergent model. Key to this way of approaching language arts are ideas such as the dialectical interaction of learner and knowledge, engagement in authentic purposeful language activities, the close relationship of language form to function, and the integration of new with existing knowledge in language and thought (Hansen-Krening, personal communication, July 11, 1991).

The most apparent problem in newer integrated models in language arts is the high demand they place on teachers. The recent research on the status of the curriculum seems to show that some teachers are attempting to act on the process aspects, while some newer textbooks are providing idea-rich, though not necessarily classical, literature bases, combined with teacher guides that assist the teacher in enacting a process model (e.g., the Silver-Burdett & Ginn *World of Language* series).

Patterns in the Language Arts Taught Across the Grades

Despite the influence of the process model and the emergent integrated model on some recently published curriculum materials and in some classrooms, observers of language arts are quick to admit that what one sees in schools is closest to the mastery model in most elementary and middle schools, with overtones of the heritage model in high schools. Language arts, generally taught from this skills and information perspective, assumes a central position in the general curriculum, taking as much as 100 minutes a day in the K–3 curriculum and somewhat less in upper elementary grades (National Science Foundation, 1978). In the middle grades and high school, more language arts is required than any other subject area (U.S. Department of Education, as reported in Jenness, 1990). Learning information *about* language and literature predominates. Writing receives scant, though apparently increasing, emphasis. Within the allotted time, K–12, a negligible amount is spent in developing skills in oral language. Speaking and listening, the language arts most frequently used in daily life (Pinnell & Jaggar, 1991), are nearly missing in the curriculum for most students.

Language Arts Curriculum in the Elementary School Grades

When understanding what elementary schools teach in the language arts, we must look first to the textbooks used, for in this subject area, in particular, the textbook is highly influential in determining the curriculum. Furthermore, the accompanying workbooks constitute almost the whole of instructional programs in Grades 3–6, and nearly the same in K–2, according to the findings of observation-based research in multiple classrooms (Durkin, 1978–1979, 1987). Teachers' selection of content and its sequencing is directly tied to a particular text and its order of presentation (Durkin, 1978–1979, p. 518).

As of the early 1980s, decoding appeared to be the central skill taught in most basal reading series, and it was taught through a skill-based drill-

and-practice program (Calfee & Drum, 1986). Little vocabulary instruction was in evidence in several popular basal reading series (Beck, McKeown, McCaslin, & Burkes, 1979); that which could be found was dependent mostly on rote practice and was disconnected from the text being read. Writing requirements in workbooks accompanying basal readers follow a similar pattern. Writing for personal interest and satisfaction, or "expressive writing" (Britton, 1982), was not generally called for in published materials. New series published recently (e.g., the Silver-Burdett & Ginn series) appear to be different in several respects, including more writing and vocabulary instruction and a higher proportion of "literature of quality" (Chall & Squire, 1991, p. 125). Some time will pass, however, before these new materials reach the majority of classrooms.

With the aforesaid characteristics of curricular materials in mind, we can then offer a sketch of the reading and writing curriculum as enacted in classrooms. Reading instruction for fourth graders is given, on average, about an hour each day, but lower ability students are likely to spend somewhat more time in reading than higher ability students (Langer et al., 1990, p. 40). Fourth-grade teachers claimed that their colleagues in Grades 1, 2, and 3 placed emphasis in beginning reading instruction on either phonics-based or what was rather generously termed "eclectic" approaches (Langer et al., 1990, p. 38). Only a tiny percentage of the teachers (3.6%) reported that their peers used language experience approaches (a process model).

Reading instruction for fourth graders is typically based on a single basal reader, with other books and magazines available for supplementary reading. Seventy-nine percent of the students are in ability-based reading groups, and while in those groups they report never or rarely being asked to discuss their readings or write about them. Writing receives little or no direct instruction, but is mainly worked into the curriculum by making it a part of other activities—it is practiced but not taught (Florio & Clark, 1982). According to the fourth-grade teachers, half of the students receive daily instruction on comprehension skills, somewhat fewer receive daily instruction on vocabulary, and even fewer receive daily instruction on reading aloud and word attack (p. 44). Better than half receive workbook or skill-sheet assignments each day. Lower ability students receive not only more workbook and skill-sheet assignments, but also more homework, than their higher ability counterparts (p. 46).

What the teachers report as instruction is open to considerable question, however. The "study" of reading or any other component is frequently characterized not by instruction (Duffy & McIntyre, 1982), but by what Durkin calls "mentioning" (1978–1979, p. 505). Teachers mention concepts in brief introductory bursts, then assign workbook pages or worksheets in which the students put the mentioned concepts to use. The teachers then question the students to assess whether they have acquired the concept and its use. Thus, mentioning, assignment giving, and interrogating for assessment purposes appear to constitute the primary ways teachers spend time in presenting the language arts curriculum. These activities reinforce the orientation of the curriculum toward skills and information memorization.

When it comes to looking at the curriculum specifically for development of oral language, the whole of classroom discourse can be viewed as the object of study. This being the case, studies such as those done by Cazden (1988), Heath (1983, 1986), Duffy and McIntyre (1982), and Mehan (1979) tell us something about the interaction patterns between student and teacher and the opportunities they provide for students to become orally competent. The unhappy situation, according to Strickland and Feeley (1991), who have summarized research on classroom discourse, is that in typical primary classrooms students have so little time to speak and such restricted formats in which to speak that schools do not offer a productive "oral language curriculum." If students do not acquire certain language-use patterns at home, they are unlikely to acquire them at all, especially if they are in typical classrooms where teachers see "mentioning" as instruction.

Language Arts Curriculum in the Secondary Grades

High school English has been dominated by the study of major literary works. Junior high/middle schools have spent considerable time on the study of grammar, but literary study has received attention there as well. Composition has been a lesser part of the secondary curriculum until recent decades, while speaking and listening have never successfully made it into the fold, because, according to Squire (1991), those interested in speech lost power to influence English teachers when they withdrew from the National Council of Teachers of English to form their own association (now the Speech Communication Association) in 1922. Speech electives are offered in larger high schools but rarely in smaller schools.

In the secondary schools, most students work in a single literary reader composed of short stories, poetry, essays, plays, and longer-length works of fiction. They may also use a grammar and usage text. Some supplemental full-length books are read by most (Squire & Applebee, 1968). These vary with the track in which the student is placed (Applebee, 1989). High school courses are generally a mixture, then, of literature study, grammar, and mechanics. The amount of reading done is, on average, less than in the elementary grades, according to the 1988 NAEP survey (Langer et al., 1990). The amount of writing of anything beyond short answers tends to be limited; note taking is the most common form of writing (Applebee et al., 1990). The courses are marked by a strong demand to memorize information and exhibit discrete skill in reading, grammar, and usage, and they seek little or no creativity or personal voice—all in line with the mastery model. The most critical learnings for the majority of secondary language arts teachers are those related to subject matter mastery (Tye, 1985, p. 210).

According to teachers of eighth graders, three quarters of them offer an hour or less of writing instruction each week (Applebee et al., 1990). This instruction focuses primarily on grammar or writing skills, uses a writing process approach, and integrates reading and writing. High-ability classes are apparently experiencing more writing process instruction than low-ability classes. According to teacher self-reports, two opposite models (skill-based and process) seem to coexist in most teachers' classrooms. Most of the writing requested is of essays held to one or two paragraphs, although students report writing more and longer assignments than do their teachers. Teachers differentiate writing assignments for eighth graders, giving more analytic, imaginative, and literary papers to higher ability classes than to others. Most of the students are asked to follow a structured writing process that includes getting information before writing, making notes or an outline, defining purpose and audience, and revising the paper at least once (Applebee et al., 1990, pp. 33–58).

More course work is required in English/language arts than in any other subject area in the high school, in most states 4 full years (Pipho, 1991). Although a wide array of electives was fashionable in the early 1970s in some large high schools, such electives declined in the late 1970s as a result of the back-to-basics movement. Budget constraints in the late 1970s and early 1980s caused further reductions in language arts electives in states such as California (Catterall & Brizendine, 1985). Smaller high schools have been likely to offer few if any choices all along; instead, they require a series that usually includes American literature in the 11th grade and either British or world literature in the 12th (Applebee, 1989). Students in college preparation tracks have few choices among language arts courses, while students in general and remedial tracks show transcripts with widely varying course work to fulfill requirements. If courses must be eliminated because of budget constraints, however, advanced courses are likely to remain and electives are not (California State Department of Education, 1984).

Most required English courses make heavy use of a single textbook of collected works (Purves, 1975; Squire & Applebee, 1968). Higher ability classes are likely to use more book-length works in addition to the basic text. Applebee's (1989) survey of most frequently used book-length works showed that a short list of books dominates public school, Catholic school, and independent school lists. Drawn from 322 schools, the public school top 10 includes the following: *Romeo and Juliet, Macbeth, Huckleberry*

Finn, Julius Caesar, To Kill a Mockingbird, The Scarlet Letter, Of Mice and Men, Hamlet, The Great Gatsby, and *The Lord of the Flies*. With slight changes in order, the list is the same for Catholic schools; independent schools are more likely to drop *Of Mice and Men* and add the *Odyssey*. All of the books can be found at any grade level, 9–12, but some most typically show up at 11th (e.g., *The Great Gatsby, The Scarlet Letter, Huckleberry Finn*) and others at 12th (e.g., *Macbeth, Hamlet*, and, with some frequency, *Oedipus Rex*).

Applebee noted, with some disappointment, that the list includes few 20th-century writings, no works by authors who are members of a minority group, and only one, *To Kill a Mockingbird*, written by a woman (1989, p. 5). The list is drawn from a narrower group of authors than a list constructed from a similar survey 25 years earlier (Anderson, 1964). Applebee further notes that students in higher tracks read somewhat different books from those in low tracks, and there is much more variability in the books read in lower tracks. Young-adult literature, such as *The Outsiders* and *The Pigman*, is likely to appear on the lower track list; it does not appear on any upper track list (this news will be cheered or denounced, depending on one's championed position, i.e., the heritage model or the process model).

Although the book-length works provided above appear in a large percentage of schools surveyed, the list may be slightly misleading concerning the uniformity of the literature base for English courses in the United States. In the appendix to his report, Applebee lists seven double-column, single-spaced pages of books required in less than 5% of the schools.

Lest we begin to think that literature and its analysis constitutes the secondary language arts curriculum, Tye's (1985) comment, based on a review of packets of materials from English/language arts teachers in the Study of Schooling, is important to remember:

> As far as course content is concerned, the most common English course was a combination of grammar, language mechanics, and literature study. The second most common course seemed to be literature study only, and the third was grammar study only. These formed the core of required English courses. The remainder were special-interest offerings such as speech, journalism, and creative writing. We found a strong emphasis on the basics of grammar and composition reflected in the teachers' lists of topics and skills taught in our sampled English classes. On the other hand, we found little or no fiction writing, etymology study, or media literacy being taught in these classes. (p. 210)

The most recent NAEP data (1990) add to Tye's portrait of high school English. These data suggest that literature study has a somewhat more prominent role than the Study of Schooling data showed earlier. Teachers were not surveyed, but 12th-grade students in the 1988 NAEP study reported that more than half of their time in English class was spent on literature (p. 41); students in academic tracks reported more time spent on literature and writing than those in general or vocational tracks. We know less well the way in which that time "on literature and writing" was spent. Evidence collected in 1968 showed that honors students were more likely to be asked to analyze what they read, while "less-gifted peers" were asked to report or summarize (Squire & Applebee, 1968, p. 174). That differential seems to be apparent in more recent looks into the classroom as well (e.g., Oakes, 1985). Newkirk speaks of this as a "painfully consistent" practice in high school English (1991, p. 333).

Although a considerable amount of time is apparently spent on literature, about a third of 12th-grade students reported that half of their time in English was spent in learning to write. Most students do spend regular time talking with their teachers about the work in progress, but students in high-ability classes were more likely to be asked to discuss work in progress with peers and teachers than were students in lower ability classes (Applebee et al., 1990, p. 51).

Making Sense of the Evidence on the Language Arts Curriculum

As in social studies, there does not appear to be a loud cry for updates on the research done earlier that revealed a skills- and information-based curriculum both at the elementary and secondary level. One must suppose that the experts are not convinced that much has changed since the Study of Schooling or Applebee's examination of secondary English classrooms, both done over 10 years ago. There are indications that writing instruction is changing, at least to the extent that more writing is being asked of students and that the language experience model of early reading instruction is gaining ground. But the overwhelming sense from the literature is that there has not been great change in most classrooms, or in most curriculum materials in use in those classrooms. Some new texts are appearing that may begin to shift the model from mastery to process, but these texts will take time to reach the classroom, especially in financially constrained districts.

The instruction as "mentioning" that Durkin (1978–1979) encountered when looking for comprehension instruction is an important key to examining the curriculum as enacted in the classroom and experienced by students. This is so because the conception of instruction as mentioning confines the kind of knowledge taught in schools to what Cuban called "artifactual knowledge" (1984); it relies on previous knowledge taught somewhere outside the schools for actual completion of school assignments (Heath, 1991), and it puts students from certain cultural backgrounds at a distinct disadvantage in the process. Explaining why this form of instruction is prevalent and powerful in shaping students' encounters with knowledge is a matter of import. It may be that teachers, whether elementary or secondary, do not have sufficient grounding in the subject matter to do more than mention concepts (Gehrke & Sheffield, 1985), or it may be a lack of pedagogical knowledge (Grossman, 1990). It may also be, as Durkin hypothesized (1978–1979), because of pressures for coverage and completion or because mentioning and then assigning seatwork offers greater control of student behavior. Hillocks and Smith (1991) proposed some reasons as they commented on the persistence of traditional grammar instruction:

> Many explanations have been adduced, some not so flattering: it is easy to teach by simply assigning page and exercise numbers; it is easy to grade; it provides security in having "right" answers, a luxury not so readily available in teaching writing or literature; it helps to demarcate social distinctions, at the same time raising the teacher to a higher class by virtue of at least seeming to possess a prestige dialect; and it is an instrument of power by which one may hold those unable to learn in thrall. The more generous explanations for the persistence of grammar have to do with the lingering legacy of the Middle Ages—the belief that grammar is the key to something else. At the very least, teachers say, grammar provides the basis for "correctness" in writing, for putting commas and periods in the conventional places. (p. 600)

Our sketches of the enacted curriculum are based mostly on survey data from students and teachers at the 4th-, 8th-, and 12th-grade levels. We can only infer what happens in the other nine grades. We have few status reports from third parties observing the curriculum in action in middle and high schools, except those from the Study of Schooling (Goodlad, 1984; Tye, 1985), Squire and Applebee (1968), and Applebee (1981). Mostly, we must infer what the curriculum is by examining data about instructional practices. Even studies of the curriculum as presented in texts and curriculum documents are rare, though they may be showing up as theses and dissertations that do not get published and circulated. We have almost nothing about the part of the language arts curriculum focused on speaking and listening in the upper grades. The reason could be that we have not looked in the right places, or that there is little research done in these two critical language areas, or that there is little on which to do research. Finally, we do not know much about the various language arts electives at the secondary level, and here, as in the social studies area, we know the least about the middle grades.

WHAT IS KNOWN ABOUT THE NATURE OF THE MATHEMATICS CURRICULUM

Perhaps more than any other area of the school curriculum, mathematics exhibits remarkable consistency across grades and classrooms.

What is more, there is widespread agreement about the nature of this consistent curricular pattern.

The empirical evidence to verify the consensus is varied and a little thin; questions have not been systematically addressed with all levels of the curriculum in mind. However, the following kinds of research provide data from students, teachers, and observers that help to construct a picture of what is taught: large-sample national surveys such as NAEP (Dossey, Mullis, Lundquist, & Chambers, 1988); international comparative studies, such as the Second International Mathematics Study (SIMS; Crosswhite, Dossey, Swafford, McKnight, & Cooney, 1985); examinations of reform effects in mathematics, such as transcript studies determining the effects of "first-generation" reforms on high school course-taking patterns (Clune, 1989) and research on implementation of new state frameworks (e.g., Peterson, 1990a); small-sample, intensive studies of curriculum in action drawing on yearlong patterns of teacher log data (Porter, 1989) or on detailed observations of activity structures in the mathematics classroom (e.g., Stodolsky, 1988); and research on the content of mathematics textbooks (e.g., Flanders, 1987; Porter et al., in press).

Other studies can also contribute to an understanding of what is taught, although indirectly. For example, research on the determinants of instructional content in mathematics (e.g., Barr, 1988; Porter, Floden, Freeman, Schmidt, & Schwille, 1988) helps to pinpoint what is being taught and why. The burgeoning research literature derived from cognitive psychology on the teaching and learning of mathematics (Romberg & Carpenter, 1986) illuminates further the nature of current curricula, even though such studies typically focus on the learner and the curriculum as it might be rather than as it is.

Despite the deficiencies of the research base, the available evidence permits reasonable inferences about the modal features of the mathematics taught in the nation's schools. On the basis of the research, we can describe the most common foci of instruction, the range of topics taught, the degree of attention to mathematical thinking and reasoning, and the articulation of content through the grades. In addition, we can characterize the dominant approaches to instruction and the attention paid to attitudes and beliefs.

As before, we will discuss the nature of the mathematics curriculum at different grade levels, but first we summarize themes that cut across grades. Finally, we close the section with some observations about what is and is not known about the mathematics curriculum. To provide context for the discussion, we start with a brief review of current conceptions of what should be taught, as these are premised on a view of typical practice in most mathematics classrooms.

Views on What Should Be Taught in Mathematics

There is considerable consensus among a broad spectrum of mathematicians, mathematics educators, and curriculum scholars about what should be taking place in mathematics classrooms nationwide. That consensus has been effectively communicated by documents such as *Everybody Counts* (National Research Council, 1989) and *Curriculum and Evaluation Standards for School Mathematics* (National Council of Teachers of Mathematics [NCTM], 1989).

In brief, these documents project a view of curriculum that emphasizes a broad range of mathematical topics, conceptual understanding of the mathematical ideas underlying operations or algorithms, ability to solve nonroutine mathematical problems, and application of mathematics to students' lives or other real-world situations. Reform proposals encourage teaching modes that support these curricular goals—for example, through the use of manipulatives (to represent mathematical ideas in various ways), extended discussions of mathematical ideas (to stimulate alternative viewpoints on problems and encourage students to develop their own solutions to problems), and group collaborative work (to encourage more complex problem-solving processes).

This consensus has been carefully orchestrated over a long period of time. Leaders in the mathematics education community have engaged in a process across much of the past decade to develop the ideas presented in these reform documents and to promote them across the nation. For example, mathematics teachers are becoming familiar with the NCTM *Standards*, perhaps to the point that the rhetoric of this document is becoming the "correct" view of what mathematics should be in the schools. However, visions of the ideal are not constant over time: The current vision of ideal mathematics is noticeably different from the "new math" on which the mathematics education community focused three decades ago.

The current consensus about good mathematics curriculum corresponds to an equally consistent view of what is being taught in the typical American classroom. As best we can tell from available evidence, that perception is accurate, at least at the level of modal patterns described below.

Patterns in the Mathematics Taught Across Grades

One of the most widely cited descriptions—and indictments—of contemporary mathematics curriculum in the United States describes it as an "underachieving curriculum," one that expects less of students than it should or could (McKnight et al., 1987). Derived from international comparative research, that characterization captures a number of features of the curriculum about which there is general agreement, at least among the experts interested in promoting the new vision of mathematics curriculum. In this view, the curriculum does the following:

- Emphasizes algorithmic and procedural skills while deemphasizing conceptual understanding of the mathematical ideas underlying these skills
- Places relatively little emphasis on mathematical reasoning, thinking, or problem solving (especially as applied to novel or unfamiliar problem situations)
- Downplays or ignores applications of mathematics to students' lives or other "real-world" situations, and instead concentrates on mathematical abstractions taught without context
- Focuses on a narrow range of topics (e.g., arithmetic in the lower grades, algebra and geometry at the secondary level) while exposing students to other topics (e.g., probability and statistics, estimation) in a generally superficial way, if at all
- Repeats material in an increasingly redundant manner as students progress through the grades

Two other points of consensus—one about the way mathematics is taught and the other concerning the attitudes and beliefs embedded in mathematics curriculum and instruction—round out the indictment. Both properly belong in a discussion of the curriculum because they are so inseparable from the enactment of curriculum. By comparison with other subjects, mathematics instruction exhibits a narrow range of instructional formats—teacher presentation or lecture followed by individual seatwork predominate (Dossey et al., 1988; Stodolsky, 1988). This fact affects the nature of the students' learning experiences and, in effect, shapes the content of that experience. Across their years of exposure to mathematics, students form an increasingly negative image of the discipline, which they believe to be about applying the right algorithm and getting correct answers (Dossey et al., 1988). Although initially enthusiastic about mathematics, most come to believe that the subject has relatively little to do with their lives.

These patterns are believed to be true of the planned as well as the enacted curriculum. There is some evidence that intentions for the curriculum—as expressed in mathematics textbooks, state frameworks, district scope-and-sequence charts, and the like—are beginning to change. Still, there are substantial gaps between new intentions and classroom practice.

The planned curriculum in mathematics is elusive, as in many subject areas, because in any particular setting at least three different intentions are at work: the district's (and state's, where states proactively seek to define what should be taught), the textbook publisher's, and the teachers'. Ultimately, of course, the individual teacher merges these three in the way that she or he enacts the curriculum within the classroom. Because in mathematics teachers appear to rely heavily on the textbook (Dossey et al., 1988), an examination of the textbook's contents provides a first approximation of what is likely to be taught.

Systematic reviews of textbook content are few and far between, but those that have been done underscore the redundancy of the mathematics curriculum, at least at the elementary school level. One comparative review documented this pattern, along with the surprising and sudden disjuncture between elementary and secondary level content:

A relatively steady decrease occurs in the amount of new content over the years up through eighth grade, where less than one-third of the material is new to the students. This decrease is followed by an astounding increase in the amount of new content in the texts of the most common ninth-grade course, algebra. (Flanders, 1987, p. 18)

The reasons for the disjuncture are not obvious and are probably rooted in long-established and seldom-questioned traditions of the introduction of algebra and other forms of "higher" mathematics.

More extensive research on mathematics textbook content at the elementary school level (Porter et al., in press) confirms other general beliefs about the curriculum, such as the primacy of skill development and the cursory treatment of large numbers of mathematical topics not having to do with arithmetic computation.

Research on the way mathematics textbooks are used suggests that, important as they may be as a source of lesson material, the content of instruction is in no sense dictated by the textbook, at least at the elementary school level (Freeman & Porter, 1989). Teachers are selective about what they use from the textbook, the order in which they teach it, and the relative time devoted to particular topics in the text. Their decisions about what to emphasize, what to leave out, and how to organize their students' use of the textbook are driven by a variety of forces ranging from their own convictions about what should be taught to a variety of external forces (Porter et al., 1988).

Beyond the big picture described above, the existing research base affords a more detailed picture of the mathematics curriculum for elementary through middle school grades (1–8) and for the high school years (Grades 9–12). Unlike social studies, the most relevant dividing line appears at the end of the middle or junior high school years rather than at the beginning, for reasons noted above regarding the introduction of algebra. In recent years, the elementary grades have drawn the most attention from researchers interested in describing the curriculum; their investigations parallel, and are often linked to, the efforts of reformers to change the mathematics to which children are exposed in their early years of schooling.

The Mathematics Curriculum in the Elementary and Middle School Grades

"Mathematics" in these years (Grades 1–8) appears to be synonymous with "arithmetic" and, more specifically, with arithmetic computation. As a result of the textbook pattern described above, widespread beliefs among teachers, testing pressures, and limitations on teachers' background knowledge, the curriculum typically proceeds through basic number concepts, followed by the four primary operations (addition, subtraction, multiplication, and division), first with whole numbers, then with fractions, mixed numbers, decimals, and percentages. The precise mixture of topics may vary somewhat, but the overall contours of the curriculum are familiar and relatively unvarying: addition and subtraction, with and without carrying, in the first and second grades; an introduction to multiplication and division in the third and fourth grades while solidifying addition and subtraction skills; more complex multiplication and long division in the fifth and sixth grades; and so on.

In principle, the mathematics curriculum in Grades 1–8 follows a "spiral" pattern, but as one observer points out, the fundamental premise of spiral curricula—that students continually revisit simple yet powerful ideas and explore them in greater depth each time—is not realized (McKnight et al., 1987). Instead, students repeat learning experiences that convey a fragmented picture of arithmetic skills, with little new insight conveyed each time; the simple yet powerful ideas—for example, that mathematics is a science of patterns, that events have varying probabilities of occurring, and that unknown quantities can be represented symbolically and manipulated—are all but lost in the process. One fourth-grade student, cited in McKnight et al. (1987), aptly summarized it all: "I didn't learn much this year that I didn't already know from last year. Math is my favorite class but we just did a lot of the same stuff we did last year" (p. 98).

With the focus on arithmetic skills predominating, other topics in elementary mathematics are given less play in the curriculum or ignored altogether. In particular, geometry, estimation, graphing, probability and statistics, and measurement play relatively small supporting roles in the elementary and middle school grades. Many mathematical topics are given very little time, based on the teachers' view that these things should be taught for "exposure": One study documented the fact that 70% of all topics taught by a sample of third- through fifth-grade teachers received

30 minutes or less total instructional time across the entire school year (Porter, 1989). This pattern of topical coverage in instruction parallels what is found in the most commonly used elementary mathematics textbooks (Porter et al., in press).

The relative lack of time and attention devoted to mathematical problem solving, reasoning, conceptual understanding, and applications is striking. Small-sample, intensive research on small numbers of teachers suggests this pattern. In one study, instruction aimed at either problem solving or conceptual understanding constituted less than 30% of the total instructional time across the year; the great majority of time went to the teaching of computational skills (Porter, 1989). As the researchers describe it:

What little attention is given to problem solving is largely limited to artificial story problems and . . . story problems presented in a repetitive format that tends to diminish their problem-solving character. Students are rarely, if ever, asked to formulate a problem for themselves, yet problem formulation may be the most difficult and important aspect of the kind of higher order thinking that students need. (Porter, 1989, p. 11)

Mathematics instruction at the elementary and middle school level appears to have a routine, rigid quality that parallels and supports the repetitive focus on arithmetic computational skills (Knapp, St. John, Zucker, Needels, & Stearns, 1987). As some small-sample research on fifth-grade urban classrooms suggests, the pattern distinguishes mathematics from other subject areas, such as social studies, in which a variety of activity structures are found: "The traditional math classes contain only one route to learning—teacher presentation of concepts followed by independent student practice" (Stodolsky, 1988, p. 123). Nationwide samples of students report the same basic pattern (Dossey et al., 1988).

The Mathematics Curriculum at the Secondary Level

The pattern at the secondary level (Grades 9–12) is similar to that described for the earlier grades, although with substantial differences in topics. The curriculum is well described by multiple case study research from the late 1970s:

In most of our sites, we found secondary school mathematics to be just as traditional and work-oriented as learning to compute in the elementary school. The judgement from mathematicians and other visiting site observers was that many of the courses were tedious. Several noted, however, that the students in high school mathematics classes did not appear to find them as dull, uninspiring, and irrelevant as the observers did. . . . It appeared obvious and acceptable to most people that the justification of the traditional high school content (algebra, geometry, trigonometry, and their continuation into analytic geometry and calculus) is that these courses prepare students for engineering, physics, economics, statistics. (Stake & Easley, 1978, p. 13-23)

As noted earlier, the curriculum suddenly shifts focus in the ninth grade from arithmetic to algebra, followed by blocks of time (typically, a half year to a year) devoted to geometry, trigonometry, and calculus, usually in that order. A transitional "prealgebra" course is also commonly offered, as are "general math" and "consumer math" courses emphasizing advanced arithmetic applications, which typically are reserved for lower track students who are unlikely to proceed through the full high school sequence. Computer programming (which we will treat here as a mathematics course) is an increasingly common feature of the high school curriculum.

Unlike the first eight grades of schooling, the mathematics curriculum at the high school level is a function of requirements for graduation. Students' choices of courses to meet these requirements, and the options available to them, vary by virtue of track placement. With each successive year in high school, a decreasing percentage of students persist in mathematics. For example, transcript studies reveal the following percentages of graduating seniors in 1987 having taken each type of college-track math course: algebra I (77%), geometry (61%), algebra II (46%), trigonometry (20%), precalculus (12%), and calculus (6%; Educational Testing Service, 1989). Track placement has a great deal to do with what math courses one takes. Few vocational track students are likely to take the courses just discussed (with the exception of algebra I, which the majority of such students take; Westat, 1988). Differentiation of the mathematics curriculum does not begin in the high school years, however. Informal tracking exists in middle and junior high school mathematics programs. One international comparative study distinguished four distinct mathematics curricula at the eighth-grade level resulting from varying mixtures of topics: arithmetic (e.g., fractions, ratio/proportion/percentage), geometry, and prealgebra or, for advanced students, algebra (McKnight et al., 1987).

Both course offerings and requirements have changed considerably in the last decade, and along with them, student enrollments. As a consequence, the sum total of the secondary mathematics curriculum as experienced by students is changing, even though individual courses may be taught as they have for years. Figures for both vocational and academic track students suggest that enrollments in mathematics courses increased across the 1980s (Westat, 1988). As far as we can tell from multiple-site case study research, the most noticeable changes in course offerings relate to lower track students:

The courses added [in response to increased graduation requirements during the 1980s] were overwhelmingly at the basic, general, or remedial level . . . apart from scattered evidence of watering down of course content, little is known about the quality of the courses. (Clune, 1989, p. v)

There is reason to believe that high school mathematics focuses on abstract, algorithmic skills in much the same way that the enacted curriculum in the elementary grades is preoccupied with arithmetic computation skills. For example, at least 30% of first-year algebra is spent on mechanical techniques such as factoring polynomials (Usiskin, 1985). It is not surprising, then, that students show themselves least able in measures of mathematical thinking ability, conceptual understanding, and ability to apply mathematics to practical problems.

The typical pattern of secondary-level mathematics instruction reinforces the limitations in curriculum just described. As far as we can tell from student self-report data, teacher presentation and individual seatwork are emphasized as much or more than is the case at the elementary level (Dossey et al., 1988). As in the earlier grades, textbooks are the predominant instructional resource, while alternatives such as the calculator and the computer have been used by less than half of all students. However, the presence of these technologies in high school mathematics is becoming more noticeable than in the earlier grades. As of 1986, a third of all high school graduates—between two and three times the percentage 8 years earlier—had taken a course in computer programming and had written at least one computer program to solve a mathematics problem (Dossey et al., 1988).

Making Sense of the Evidence Regarding the Mathematics Curriculum

The picture of the curriculum we have painted so far emphasizes the sameness of the mathematics curriculum and the relatively unvarying quality of mathematics instruction across grades. In brief, the curriculum is highly consistent across schools and grades and limited in scope—many important curricular goals receive short shrift in both the planned and enacted curriculum at present. While this portrayal of the mathematics curriculum helps to distinguish it from other subject areas at a global level, it masks some important variations and complexities.

First, there is reason to believe that the mathematics offered students in schools serving affluent populations differs from that available in schools serving poor populations (Zucker, 1991b). We know this both from research on the schooling offered students from different social class backgrounds (e.g., Anyon, 1981) and from recent work examining mathematics curricula in schools serving high concentrations of children from low-income backgrounds (Knapp et al., 1991). In such settings, the mathematics typically offered to students resembles the profile we have been describing (emphasis on procedural skill, lack of emphasis on conceptual understanding, redundancy across grades, etc.), only in somewhat greater degree than what is available under more affluent circumstances.

Second, the profile of mathematics curriculum we have offered is an average that obscures the variation across cases. Even though intensive research verifies the high degree of similarity that exists across mathematics classrooms (e.g., Stodolsky, 1988), there are still important differences that may have great importance for the students' learning experience. For example, recent research in a sample of relatively "effective" high-poverty elementary schools has uncovered substantial differences among classrooms in the teachers' emphasis on conceptual understanding and on topics other than arithmetic (Zucker, 1991a).

We do not yet understand the full extent and depth of variation within the mathematics curriculum. The question becomes increasingly pertinent as the current consensus among leaders within the mathematics education community becomes accepted as the new standard for practice. Textbooks are beginning to reflect this consensus, under market pressure from large states that have substantially revised their curriculum frameworks. To be sure, mathematics textbooks have a long way to go before they significantly alter the curricular patterns described above (Porter, 1989), but they have begun to change. The mandates and framework revisions that stimulate textbook changes are simultaneously affecting what teachers do, even though these effects are complex. Case study research indicates that teachers are considering and attempting to change their approach to mathematics, yet not necessarily reaching the goals envisioned by reformers (Cohen, 1990; Peterson, 1990b), while others are redefining reform approaches in terms that are variations on traditional practice (e.g., Wilson, 1990). The least that can be said is that, in many quarters, educators are no longer as complacent and unquestioning about the mathematics curriculum as they once were. Although traditions die hard (if they die at all), the present limitations in the U.S. mathematics curriculum are not a foregone conclusion.

WHAT IS KNOWN ABOUT THE NATURE OF THE SCIENCE CURRICULUM

If the mathematics curriculum is a story of consistency, then science is one of diversity. Both within and across schools, and across grades, there is a great deal of variation in what is taught. The pattern was well captured by multisite case study research over a decade ago:

As one reads through the case studies, one is struck by the diversity of the many opportunities to learn science. The diversity existed within buildings and departments, mostly as a difference among teachers. . . . Some differences were found at the district level. . . . Less obvious to statistical studies but greatly obvious to the observers were the extent to which course content, format and teaching method differed among individual science classes. Even though teachers might use the same texts, they improvised to such an extent that two otherwise seemingly identical courses looked greatly unalike. . . . One could not avoid the inference that teachers develop and follow their own guidelines. (Stake & Easley, 1978, p. 13-3)

By itself, diversity is neither good nor bad. On the one hand, it may represent healthy experimentation, responsiveness to local conditions and interests, and a sign of teachers seizing the initiative. On the other hand, it may simply reflect inattention to this subject area, lack of coherence in the curriculum, or the influence of insufficient preparation on the part of teachers.

Despite this diversity, there are still modal patterns regarding the content taught, types of activities in which students engage, and so on that can be derived from sources similar to those reviewed above regarding what is taught in mathematics: large-scale national surveys of various kinds (e.g., Mullis & Jenkins, 1988; Weiss, 1978, 1987; and others summarized in Gilford, 1987), international comparative studies (e.g., Jacobsen & Doran, 1987), case studies and other forms of intensive research (e.g., Glasgow, 1983; Stake & Easley, 1978), analyses of science textbooks and what has influenced them (e.g., Cohen, 1987; Quick, 1977; Yager, 1983), and analyses of reform attempts such as the NSF curriculum reforms of the 1960s (Jackson, 1984; Welch, 1979) or research on the effect of graduation requirements on course offerings (e.g., Clune, 1989). Mainstream curriculum research in science education (summarized in White & Tisher, 1986) and research on science teaching and learning (summarized in Linn, 1986) have some bearing on the nature of the current curriculum; however, these bodies of work focus primarily on issues pertaining to the nature of the learner, the learning process, and the development of more effective curricular approaches.

As with the other subject areas, we review what is asserted about the nature of the ideal curriculum, the primary patterns across grades, and the unique features of curriculum in the elementary versus secondary grades.

Views on What Should Be Taught in Science

Given the range, proliferation, and rapid change of the disciplines on which school science draws, it is little wonder that the science education community does not speak about what should be taught with a voice as unified as that of mathematics educators. Nonetheless, there is growing consensus on a number of points, especially concerning the science curriculum that is—and is not—available to the nation's young people. The proliferation of reform manifestos, over 100 by one analyst's count (Hurd, 1986), share many ideas about where science curriculum should be headed. Statements of "what should be," such as *Project 2061: Science for All Americans* (American Association for the Advancement of Science, 1989), project a vision for science education that reflects an increasingly wide spectrum of opinion among scientists, science educators, curriculum scholars, and teachers.

This vision of the science curriculum emphasizes the following:

- A thematic approach to the content of science, such that broad unifying ideas (e.g., systems, evolution, or constancy) are treated in considerable depth and repeatedly throughout the student's school experience
- Interdisciplinary thinking that goes beyond the traditional disciplinary boundaries that separate physics, chemistry, and other bodies of scientific thought
- The connections between science and technology and the implications of both for human society
- The inculcation of scientific "habits of mind"
- The development of skills in scientific inquiry, at least in part through direct experience with the process of scientific knowing

Whether this kind of vision can be translated into more concrete and widely accepted form, as in the case of the NCTM *Standards* in mathematics, remains to be seen. Numerous attempts to do so are under way.

The vision of reform in science education rests on a conception of what is currently taking place that resembles the pattern described above for mathematics, but with important differences, discussed below.

Patterns in the Science That Is Taught Across Grades

From national surveys of teachers and students, case studies of science education in action, and other sources, the following overall picture of the science curriculum in Grades 1–12 emerges.

The curriculum focuses on the coverage of factual knowledge rather than conceptual understanding. Numerous commentators have remarked on the tendency of the science curriculum to favor breadth of coverage at the expense of depth of understanding (Bybee et al., 1989; Knapp et al., 1987). The pattern is most obvious in the structure and content of science textbooks, which have a tendency toward encyclopedic coverage of topical material (Knapp et al., 1987; Weiss, 1986) and the introduction of technical vocabulary (Yager, 1983). Various forces drive science curriculum toward an emphasis on factual material—such an emphasis requires less understanding on the part of teachers, reflects the burgeoning of new scientific knowledge (Stake & Easley, 1978), and enables publishers to prepare textbooks that will appeal to states with different lists of key topics (Capper, 1987).

As currently taught, the science curriculum deemphasizes learning the processes and skills of scientific inquiry. Numerous curricula were built in the 1960s with scientific inquiry as a central focus—for example, the Elementary Science Study (ESS) and the Science Curriculum Improvement Study (SCIS) at the elementary level, the Earth Science Curriculum Project (ESCP) and the Intermediate Science Curriculum Study (ISCS) at the middle school level, and the Physical Science Study Committee (PSSC) and the Chemical Education Materials (CHEM) Study at the high school level. These emphases have not, however, become a central feature of science programs most commonly used today, even though a majority of science teachers indicate that the development of inquiry skills is an important objective (Weiss, 1987).

Given the lack of emphasis on the process of scientific inquiry, it is not surprising that opportunities to learn science experientially constitute a relatively small portion of the curriculum and may, in fact, be declining over time. In most instances, textbooks and teacher presentation are the primary resources for science learning. Despite decades of attention by reformers and curriculum developers to "hands-on" science instruction in the lower grades and laboratory-based or field-based science learning in the upper grades, the evidence suggests that these are not prominent features of science curriculum today (St. John, 1987; Weiss, 1987) Instead, the textbook is the dominant instructional resource at all levels (Bybee et al., 1989; Weiss, 1987). Teachers primarily teach through presentation and demonstration (Mullis & Jenkins, 1988).

The science curriculum is constructed across the grades in such a way that it best serves the needs of the "pipeline" for individuals pursuing scientific careers. Although there is considerable rhetoric to the contrary (e.g., National Science Board, 1983), it appears that science curricula are articulated in such a way that individuals most likely to pursue scientific or engineering careers are most effectively served.

In light of the kinds of students for whom science courses are typically designed, one might expect technology to figure prominently in the curriculum, yet the content of science courses places relatively little emphasis on "technology" either as a topic of study or as a tool. As a topic of study, technology is rarely treated separately, if at all; this is apparent from reviews of programs at the elementary school level (Bybee et al., 1989) and elsewhere, even though a movement to promote "science-technology-society" (STS) as a course theme has been active for a decade or more (Solomon, 1989). Commonly used textbooks at the elementary level, for example, make little attempt to expose students to the social implications of science and technology or to what these fields of endeavor imply for individuals' lives (Pratt, 1981). The computer is beginning to make inroads on science classrooms but has a long way to go before it is a common feature of science instruction (Weiss, 1987).

These overall patterns describe the typical case, but there are important variations on the theme, as illustrated by data regarding the incidence of different approaches to high school physics. While most high school physics courses do not emphasize the development of inquiry skills or conceptual understanding, a noticeable minority of the nation's physics classrooms make use of curricula developed with this goal in mind, such as PSSC and the Harvard Project Physics courses developed under NSF sponsorship in the 1960s and early 1970s. A decade later, 9% of high schools offered PSSC physics and 8% offered Harvard Project physics, while 54% offered *Modern Physics*, a textbook with a more conventional curricular emphasis (French, 1986).

The variation in what is taught in science has to do with more than the choice of text or program for a particular slot in the curriculum. Distinct patterns of curriculum appear at different levels of schooling. The amount of time and priority accorded science, for example, varies greatly across the grades, as does the range of choices about appropriate science content for each level of schooling. Grade-specific differences are described in more detail below.

Science in the Elementary and Middle School Grades

The nature of the science that is taught differs greatly by level of schooling within the first nine grades. The overriding story at the elementary school level, especially in the primary grades (1–3), is that very little science gets taught. The amount of time allocated to the subject tells part of the story—national estimates from a survey of teachers place the figure at 19 minutes per day for the lower elementary grades and 38 minutes a day for Grades 4 through 6 (Weiss, 1987). These figures are based on teachers' self-reports and are likely to overestimate the actual classroom time spent on science (in this sense, the data may tell us more about the planned than the enacted curriculum); other estimates, supported by a good deal of anecdotal evidence, suggest that the real figure may be closer to an hour a week (National Science Board, 1983). The more telling finding from national surveys is the relative position of science within the elementary school curriculum; according to one survey, math receives approximately twice and language arts nearly four times the number of instructional minutes allocated to science in fourth grade (Cawelti & Adkisson, 1986).

The science taught to most elementary school children rests on a static and somewhat limited view of the scientific enterprise, as one review aptly puts it:

Science [at the elementary school level] is primarily presented as a body of knowledge and only secondarily as a process for establishing new knowledge. The organization of science in current programs conveys the idea that science is disciplinary, cumulative, and largely independent of the processes used to develop new scientific knowledge. Learning scientific knowledge is the primary goal of science education in the elementary school. (Bybee et al., 1989, p. 13)

Not surprisingly, "topics, not concepts, are the dominant orientation for science in elementary schools. Scientific information, facts, and processes are presented within topics" (Bybee et al., 1989, p. 32).

Instructional approaches reflect and reinforce this conception of science content. Although a great deal of energy has been invested over the past two decades in promoting "hands-on" approaches to science instruction at the elementary school level, it seems this dimension of instruction is not apparent in the typical elementary school science lesson. Multisite case study research paints the basic picture:

When and where science was formally taught, the instructional material was usually taken directly from a textbook series. The method of presentation was: assign-recite-test-discuss. The extent to which the emphasis on reading and textbooks pervaded the elementary science program is illustrated by an episode observed in an elementary science class where the teachers opened a recitation period with the question: How do we learn? A chorus of students replied: *We learn by reading.* (Stake & Easley, 1978, p. 13-6)

A more recent review of evidence regarding elementary science curriculum and instruction reiterates the point:

Teachers frequently take a dominant role in which they deliver science information to children through lectures, demonstrations, and text readings. Typically, they choose topics with which they are comfortable, usually in the life sciences rather than the physical sciences. (Bybee et al., 1989, p. 72)

On the basis of a national survey of teachers, lecture and discussion are the mode of instruction 74% of the time in the lower elementary grades and 87% of the time in the upper elementary grades (Weiss, 1987).

Various forces encourage teachers to adopt this conception of the science content and the corresponding pattern of teaching. Elementary teachers feel chronically ill-prepared to teach science (Stake & Easley, 1978); they have little time and meager resources, on average, for setting up science lessons based on experiential activities; furthermore, teachers generally see the task of managing materials-based science lessons as too difficult and troublesome, no matter how much they believe this is the right thing to do (St. John, 1987; Weiss, 1987). Besides, elementary school teachers have many other things to teach, most of which have greater priority and a more established place in the academic program.

There are other questions to ask about the nature of what is taught in science to the nation's elementary school children. In an attempt to understand how female and ethnic minority students systematically lose interest in science, and over time perform less well in it, some research has begun to explore the assumptions about cognitive style and the nature of knowledge embedded in science curricular materials (Cohen, 1987). This work suggests that the planned curriculum (as represented by commonly used textooks designed for the upper elementary grades) heavily emphasizes the cognitive skills that are "used in standardized measures of intelligence and achievement, and to related concepts of time, space, and causality" (p. 56), and at the same time excludes material that invites the use of alternative conceptual styles. Although there is still much to learn about the way these emphases show up in classroom instruction, there is reason to believe that the textbook's assumptions are transferred into classroom instruction in a straightforward way, given teachers' tendency to rely on textbooks to provide the intellectual structure for their science teaching (Raizen & Jones, 1985).

Much of what has been said about the science curriculum in Grades 1–6—emphasis on factual or topical knowledge, lack of emphasis on inquiry skills, reliance on textbook-based instruction—applies to the next few grades as well, but there are several key differences. First, science occupies an established niche in the curriculum in most middle and junior high school programs. Accordingly, more time is spent on science instruction (e.g., four 45-minute class periods per week in many junior high schools), and the instruction is carried out by a teacher specializing in science (or science and mathematics). These teachers are more likely than their elementary school counterparts to have a room designated for science instruction, many of which have at least some equipment and facilities appropriate to science teaching.

Second, the content of instruction in the middle grades typically fits within a more established structure featuring some combination of four blocks of study: "general science" (an eclectic mixture of scientific topics drawn from many disciplines), "life science" (incorporating topics drawn from biology, botany, and, often, ecological studies), "physical science" (combining elements of physics and chemistry), and "earth science" (including material from geology, oceanography, atmospheric and space science, and studies related to the physical environment). These four content areas are represented in the curricula for Grades 7–9 in almost equal proportions (Weiss, 1986; Welch, Harris, & Anderson, 1984), although the first two account for a somewhat larger percentage of students and classes.

There is no agreement on what is included within each of the four content areas. As indicated by multisite case study research:

As with the elementary science programs, what actually happened was left to the discretion of the [junior high school] teacher. There was little agreement (as perhaps there should not be) on what should be included in a junior high school science course. . . . The philosophical orientation of the teacher played a great role in determining what content was taught and how it was taught. The junior high teaching ranks were composed of many former elementary teachers with a "whole child perspective" who had "moved up" and many "subject oriented" high school teachers who had "moved down" from or were waiting to "move up" to high school teaching. . . . How they approached [the special problems of this transition period between elementary and high school] was largely a matter of individual choice on the part of the teachers. There was very little attempt to articulate courses with the many different learnings different students have had in elementary school, nor with what they would be expected to take in high school. (Stake & Easley, 1978, p. 13-7)

Third, although a majority of science teachers at this level indicate that the development of inquiry skills is an important objective of science instruction, the percentage of classes devoted to active "hands-on" activities has decreased significantly (from 53% to 42% between 1977 and 1986) over the past decade, while the percentage devoted to lecture and discussion has increased (from 76% to 82%; Weiss, 1987). It is not clear what accounts for the trend, but declining resources, poorly prepared teachers, lack of administrative support, and teachers' own disillusionment with curricula that emphasize "hands-on" approaches are likely factors.

Science in the Senior High School Grades

As at the junior high school level, science in Grades 10–12 occupies an established niche in the curriculum, but the traditions are more firmly entrenched regarding what is taught, in what sequence, and for whom. The traditional sequence of three yearlong courses—biology, chemistry, and physics, offered in that order to college-track students—is firmly rooted in the curricular offerings of most of the nation's high schools (e.g., Stake & Easley, 1978).

A decade ago, the typical pattern was for large numbers of high school students to take the first in the sequence (biology) to satisfy the requirement for 1 year of high school science necessary for high school graduation. Thereafter, many fewer students enrolled in chemistry and fewer still in physics; both were typically offered as electives and were especially tailored to the most devoted science students. This pattern is still largely true, but under the influence of increasing graduation requirements the proportions of students who have taken the three courses have grown in the last decade: Among high school graduates in 1987, for example, 90% had taken a biology course, 45% a chemistry course, and 20% physics, figures that represent a 6% to 14% increase over the graduating cohort from 5 years earlier (Educational Testing Service, 1989).

As in the case of mathematics, the science curriculum available to students in high school depends heavily on the track to which one is assigned. Vocational track students, for example, are extremely unlikely to study chemistry or physics, as the figures for 1987 graduates indicate: Figures for enrollment in biology, chemistry, and physics courses among vocational track graduates in that year were 58%, 6%, and 1%, respectively (Educational Testing Service, 1989).

Unlike mathematics, however, the science available to students in high school is not limited to the "big three" sequence. Observers have documented the large number and popularity of "recreational science" courses available to students as electives.

[In the study schools] we did not find a single credit course on "recreational mathematics", but we found a number of recreational science courses. We found ornamental horticulture, mushrooms of the local forests, nature walks, nature camps, photography, and an electronics course toward graduation. (Stake & Easley, 1978, p. 13-8)

Such courses were often a vehicle for the introduction of "new scientific topics" that did not fit so neatly into the conventional course frameworks, although such topics (e.g., oceanography, ecology) were also squeezed into existing courses as additional units or digressions.

The manner in which high school science is taught follows the pattern of earlier grades: Lecture and demonstration predominates, and students have relatively little opportunity to "do" science. The laboratory component of high school science courses is apparently in serious decline, as suggested by national survey data regarding the percentage of science classes in which students have some chance to participate in "hands-on" activities. Thirty-nine percent of the most recent science lessons taught by 10th–12th-grade science teachers in 1986 featured this kind of activity as compared with 59% 8 years earlier; the corresponding figures for lecturing show a substantial increase across the 8-year period, from 68% to 84% (Weiss, 1987). This teaching approach is likely to reinforce student conceptions of science as a fixed body of knowledge rather than as an active process of inquiry into the nature of the physical world.

Making Sense of the Evidence Regarding the Science Curriculum

As in the case of mathematics, the overall patterns that describe the science curriculum obscure important differences for particular groups of students. It has long been known that certain groups—among them women and individuals with racial and ethnic minority backgrounds—are underrepresented in scientific careers and in science education, commencing with the time that they are able to make choices about courses (typically at the high school level, but also at earlier stages; Berryman,

1983). Recent data confirm that course-taking patterns differ by demographic groups, with women and some ethnic minorities (African Americans, Hispanics) less frequently enrolling in certain science courses such as chemistry and physics (Educational Testing Service, 1989). The figures are no doubt a reflection of tracking, which in some degree assigns minority students disproportionately to lower track programs, where "advanced" courses such as physics and chemistry are unlikely to be offered.

But apart from tracking effects, we do not understand sufficiently whether the curriculum, as enacted in the classroom, is different for these groups in any way. The provocative research on upper elementary science materials described above (Cohen, 1987) suggests that hidden assumptions in the way science content is presented and experienced by children may systematically limit the learning of certain kinds of children. But this is only one of the features of instruction as enacted in the classrooms that is poorly understood. The nature of curriculum in action is subtle and requires intensive forms of research, of which there has been relatively little over the years designed to illuminate the current state of science curriculum.

Finally, the broad-brush picture that can be constructed from available data sources does not permit us to see easily the gradual evolution in curriculum that is likely to be taking place under the influence of changes in society, the policy environment for schools, and the professional communities of which teachers are a part. Although there is little evidence that major or sudden changes in the science curriculum are taking place, the data we have reviewed provide hints of trends in course-taking patterns, in the availability of technology and materials, and in instructional approach. The direction of change does not suggest that the ambitious curricular reforms of the 1960s and early 1970s have taken root; if anything, it is clear that these have not—perhaps cannot—become standard practice in the majority of the nation's classrooms, at least in their original form (Jackson, 1984; Welch, 1979). But there are hints of possible long-term spillover effects—for example, in the design of the most common science textbooks, which began to incorporate many features of the NSF-sponsored curriculum projects a decade after the original development took place (Quick, 1977). Whether incremental changes such as these will accumulate to form a qualitatively different science in our nation's schools remains to be seen.

CONCLUSIONS: THEMES AND REMAINING QUESTIONS

As we expected, wending our way through the curriculum field using the subject matters as a first-order organizer yielded material from which inferences about modal patterns can be made. Also as expected, the major patterns are not inconsistent with those that have been suggested in the major studies and reviews of central tendencies in instructional practices and classroom interactions. Finally, consistent with the resistant nature of curriculum with respect to delineation and detection, we are left with the unsettling feeling that we are providing reasonable answers to the question of what schools really teach, while at the same time wanting more and deeper answers to the question. Following a brief review of general themes, we will return to this and other issues.

A few common themes and generalizations can be discerned across the four subject areas. First, there is no content area where scholars are agreed that the curriculum apparently being taught is the curriculum that should be taught. The distance between the idealized and the enacted is considerable; between the idealized and the experienced, even more. The curricula under criticism, whether language arts, social studies, mathematics, or science, are characterized as teaching students what they already know while omitting many things they should know, as focusing on the transmission of factual knowledge to the near exclusion of conceptual understanding and problem solving, and as being disconnected from real experiences—authentic writing efforts, real problems of citizenship, serious questions about the earth's environment, practical uses of mathematical thinking. In only one field, mathematics, is there consensus about the direction the curriculum in that area ought to take. In all others, conflicting viewpoints are exerting pressures, on the one hand, toward a discipline-based orientation and, on the other hand, toward a process-of-thinking orientation; some seek an integration of the two. These are not new conflicts but are traceable to earlier eras, and they hinge on the inherent tension between the society and the individual student as the primary source for the curriculum.

Studies reported in each subject area suggest that there are some classrooms in which the curriculum is very different from the criticized norm, generally in the direction of process-of-thinking models, but these classrooms are in the minority. Instead, a tendency to emphasize discrete skills and factual information pervades instruction in all of the subject areas. However, while there is considerable uniformity in the focus on the teaching of discrete facts, there is considerable diversity in the particular facts selected by the teacher for commitment to memory by the students. The topical nature of the curriculum in each content area as presented in texts and curriculum guides seems to reinforce the teachers' orientation to transmission of information. The structure of the school environment and the demands for teacher control over student behavior further influence teachers' choices.

In spite of the controls on teachers' choices described by third parties, teachers in all areas tend to see themselves as having considerable control over the curriculum they teach, that is, apparently, over the particular topics, information, and skills they will select and empahsize. More often than not, they believe they are teaching what they should. Typically, they do not see that there is a choice to teach something other than topical coverage information transmission or skill building. While the scholars in all the fields have raised serious questions about the current sequencing of content from discrete parts to application of the whole, practice in the classroom as reflected in the research suggests that, in general, teachers in all areas have a distinct and different notion about "proper" curriculum sequencing—that basic discrete skills and information must be mastered before being combined in use in real experiences. This practical theory held by teachers stands as a formidable barrier to curriculum changes espoused by experts in all subject areas.

How Well Do We Really Know What Schools Teach?

This paper has presented a modal description of what schools teach in the United States. The picture has been constructed from a variety of data sources, and even though there is limited research that addresses the question before us in a comprehensive way, inferences can be drawn on the basis of national surveys (both teacher and student self-report), closely observed case studies, analyses of textbooks, and carefully measured quantitative studies of small numbers of classrooms. These sources yield a picture of curriculum at a high level of generality in which the details, variations, and daily operation of the curriculum do not come into sharp focus. The result is that we can describe the typical case—the central tendency of the nation's K–12 curriculum, to put the matter metaphorically in statistical terms—but not the important variations that represent departures from the typical case.

In one sense, the outcome of our efforts could not be otherwise. The curriculum as practiced in the nation's 85,000 schools is many things at once. The attempt to describe it in all its manifestations and latencies lies beyond our capacities and may even call into question the importance of high-level generalizations about its nature. Nonetheless, as we have attempted to demonstrate, there is a portrait emerging from broad-brush statistical profiles and finely observed accounts of the curriculum in action.

Such portraits tend to be reaffirmed by personal experience. Here we have an enormous database. Anyone who has had a typical K–12 schooling experience will have been a participant observer in some 50 classes and will have accumulated some 14,000 hours of observational time. Add to this the insight many of us gain through the experiences of our children in schools, plus our own professional experiences as educators. To be sure, this collective corpus of conventional wisdom does not meet traditional canons of "scientific knowledge," nor has it been collected and codified in any rigorous way; however, it is a potent source of working knowledge. It seldom is at odds with the modal descriptions that arise from the various studies we have consulted for this review.

The congruence between experiences in schools and portraits based on careful research may explain, in part, why the educational research community has not engaged in definitive, large-scale inventories of what goes on in our schools and classrooms. We think we already *know* what such inventories would disclose, or at least that we *know enough*, and that more systematic study is simply not worth the enormous expense required to do so properly. The congruence may also explain, in part, why so many articles and books are written on the ideal or expert-determined curriculum, after a breezy introductory remark claiming that what currently goes on in schools is obviously *less* than ideal. Many curriculum reform reports are written in this manner, as are reports in handbooks of research and organization position papers.

Yet, consider for a moment some of the important variations that we know less well, or not at all, but are central to deeper understanding of the curriculum. First, the curriculum offered and received by different groups of children varies in important ways. Class-based differences in educational opportunities and instructional approaches have long been a matter of record, and in the wake of more recent examinations of curricular inequalities and tracking, we have begun to understand in more detail the divergence in actual subject matter taught to more and less "advantaged" groups within the nation's student population. But there is much more to learn about these differences and what they mean for the cognitive development, formation of attitudes, and inculcation of skills that is likely to follow from engagement in the curriculum.

Second, there are numerous attempts to change the curriculum being offered, some that go under the banner of one or another curricular reform movement, others undertaken more quietly by teachers who simply wish to be doing a better job in the classroom. We know enough to be skeptical about the widespread adoption or staying power of many reforms; innovations do not budge the state of typical practice easily, as Cuban (1990), Sarason (1990), and others remind us. We do not understand very well how far and wide currently popular reforms in mathematics, language arts, social studies, and science have spread, nor do we know well the range of experiments currently being tried. Crudely put, are these reforms being tried—however clumsily—in 5%, 20%, or more of the nation's classrooms? What residue of reform intentions, if any, are manifest in the way these teachers are adapting the "new ways" to their own situations and preferences? The answers have a bearing on the general state of the curriculum at the present time. In all likelihood, there is a wider range of curriculum in action, notwithstanding the modal tendencies we have described.

Third, teachers shape curriculum by the day-to-day decisions they make in preparing and delivering lessons. Across teachers, within and among schools, there is likely to be enormous variation in the nature of what is taught that is impossible to see at the level of generality employed in our review. These differences, often subtle and not always conscious, are the product of individual beliefs, knowledge, preferences, and responses to the constraints and opportunities of particular classroom settings. Research on teacher beliefs, decision making, and subject matter knowledge have begun to uncover the role and importance of these factors in determining the content of instruction and the learning of students. But there is still much to know about how, in aggregate, these beliefs, decisions, and knowledge alter what is taught and learned in particular subject areas.

Fourth, curricula change over time. Even though it is unlikely that the current generation of reform activity has brought about major changes in what is taught or how, there is reason to believe that the curriculum is different than it has been in the past. We know that some aspects of curriculum—what teachers and students do in classrooms, for example— are likely to vary little over many years. But the conceptions of appropriate reading and writing, the topics taught in science, the emphasis given to problem-solving strategies in mathematics, and the information base on which social studies draws are among the facets of curriculum that have evolved and are likely to evolve further, even though it may be difficult to detect the cumulative changes that are taking place. The most obvious changes are in the topics and information presented, but enduring change in the curriculum, based on different conceptions of knowledge and concomitant instructional practices, is probably slower and harder to see.

We do not understand well the evolution of curriculum that is going on now, if indeed such evolution is taking place. Benchmark data such as that available from NAEP suggest that some changes are taking place. Previously unknown technologies are slowly becoming a regular feature of classrooms; new topics are appearing in the curriculum and the overall topical mix of what is taught is shifting; new and decidedly different curriculum materials and textbooks are making their way into classrooms. To grasp the significance of these developments for the curriculum writ large is difficult, and there is much to understand about the way new ideas seep into professional discourse, assumptions about teaching or learning, and, ultimately, the act of teaching itself.

Thoughts on Further Inquiry

Our exploration of what schools really teach has led us to believe that more large-scale research, such as the Study of Schooling and the NSF studies of mathematics, science, and social studies education, is not the kind in which our limited resources should be invested at this time. While we believe that the NAEP studies are useful, and, if wisely planned, can continue to provide us with helpful information on trends in teaching within the various content areas, we do not think that other descriptions of modal patterns will reveal additional earthshaking or particularly helpful information at this time. Instead, we think there are three areas that deserve attention and could be more fruitfully pursued. These are all areas where qualitative, holistic approaches seem more appropriate and where massive resources are unnecessary.

First, more studies of teachers and their curriculum development and presentation seem useful, studies such as those done by McNeil, Heath, Florio and Clark, and Wineburg and Wilson. As we try to understand why the enacted curriculum is so slow to change, we must gain a better understanding of the reasoning of teachers and the circumstances surrounding their choices of subject matter, selection of learning activities, sequencing, assessment, and so forth. Only with this understanding can we hope to influence and modify what gets taught. We ought, by now, to have learned lessons from the curriculum-change efforts of the 1960s; teachers mediate the curriculum, and some would say they *are* the curriculum. Recent trends toward collaborative inquiry and action research are hopeful signs that teachers' perspectives will be better represented and understood as part of what schools really teach.

Second, students' experiences of the curriculum are important, too, and research that delves deeply into these experiences should be encouraged. While the NAEP survey of students gets us one kind of information—and it is useful—we also need the kind of close-up studies based on interview and observation such as Stodolsky has done. Just as all communicative acts must be understood in context and from both the producer's and the receiver's point of view, curriculum must be so studied.

A third area of inquiry should focus on what gets taught in the middle grades. Consistently, throughout our subject area searches, we found a dearth of descriptive inquiry on the middle grades, whether in K–8 schools, middle schools, or junior high schools. At this critical time in a child's education, where research suggests that students are making decisions about whether they will eventually attend college or whether they will eventually drop out, we seem to know the least about what we are teaching them. Several signs from both federal agencies and private foundations suggest that others have finally become aware of the need for study of the middle grades as well. Early adolescents and the curriculum offered to them must not continue to be the ignored stepchildren of the schools.

REFERENCES

Adler, M. (1982). *The paideia proposal: An educational manifesto.* New York: Macmillan.

American Association for the Advancement of Science. (1989). *Project 2061: Science for all Americans.* Washington, DC: Author.

Amidon, E. J., & Hough, J. B. (Eds.). (1967). *Interaction analysis: Theory research and application.* Reading, MA: Addison-Wesley.

Anderson, S. B. (1964). *Between the Grimms and "The Group": Literature in American high schools.* Princeton, NJ: Educational Testing Service.

Anyon, J. (1981). Elementary schooling and the distinctions of social class. *Interchange, 12,* 118–132.

Applebee, A. J. (1974). *Tradition and reform in the teaching of English: A history.* Urbana, IL: National Council of Teachers of English.

Applebee, A. J. (1981). *Writing in the secondary schools: English and the content areas* (NCTE ED 197347). Urbana, IL: Nation Council of Teachers of English.

Applebee, A. J. (1989). *A study of book-length works taught in high school English courses* (Report Series 1.2). Albany, NY: Center for the Learning and Teaching of Literature.

Applebee, A. J., Langer, J. A., Jenkins, L. B., Mullis, I. V. S., & Foertsch, M. A. (1990). *Learning to write in our nations' schools.* Washington, DC: U.S. Department of Education.

Armento, B. (1986). Learning about the economic world. In V. Atwood (Ed.), *Elementary school social studies: Research as guide to practice* (NCSS Bulletin No. 79, pp. 85–101). Washington, DC: National Council for the Social Studies.

Atwell, N. (1987). *In the middle.* Portsmouth, NH: Heinemann.

Barr, R. (1988). Conditions influencing content taught in nine fourth-grade mathematics classrooms. *Elementary School Journal, 88,* 378–411.

Barr, R. D., Barth, J. L., & Shermis, S. S. (1977). *Defining the social studies.* Arlington, VA: National Council for the Social Studies.

Barr, R., Kamil, M. L., Mosenthal, P., & Pearson, P. D. (1991). *Handbook of reading research Vol. II.* New York: Macmillan.

Beane, J. A., Toepfer, C. F., & Alessi, S. J. (1986). *Curriculum planning and development.* Boston: Allyn and Bacon.

Beck, J. L., McKeown, M. G., McCaslin, E. S., & Burkes, A. M. (1979). *Instructional dimensions that may affect reading comprehension: Examples from two commercial reading programs.* Pittsburgh: University of Pittsburgh Learning Resource & Development Center.

Berryman, S. E. (1983). *Who will do science?* New York: Rockefeller Foundation.

Bissex, G. (1980). *GNYS AT WRK: A child learns to write and read.* Cambridge, MA: Harvard University Press.

Bloom, B. (1981). *All our children learning.* New York: McGraw-Hill.

Bond, G. L., & Dykstra, R. (1967). The cooperative research program in first-grade reading. *Reading Research Quarterly, 2*(4).

Bowers, C. A. (1988). *The cultural dimensions of educational computing: Understanding the non-neutrality of technology.* New York: Teachers College Press.

Boyer, E. (1983). *High school: A report on secondary education in America.* New York: Harper & Row.

Britton, J. (1972). *Language and learning.* Harmondsworth, Middlesex, England: Penguin Books.

Britton, J. (1982). Spectator role and the beginnings of writing. In M. Nystrand

(Ed.), *What writers know: The language, process, and structure of written discourse* (pp. 149–169). New York: Academic Press.

Brophy, J. (1990). Teaching social studies for understanding and higher order applications. *Elementary School Journal, 90*, 351–417.

Brubaker, D. L., Simon, L. H., & Williams, J. W. (1977). A conceptual framework for social studies curriculum and instruction. *Social Education, 41*, 201–205.

Bybee, R. W., Buchwald, C. E., Crissman, S., Heil, D. R., Kuerbis, P. J., Matsumoto, C., & McInerney, J. D. (1989). *Science and technology education for the elementary years: Frameworks for curriculum and instruction.* Andover, MA: The Network, Inc.

Calfee, R., & Drum, P. (1986). Research on teaching reading. In M. C. Wittrock (Ed.), *Handbook of research on teaching* (3rd ed., pp. 804–849). New York: Macmillan.

California State Department of Education. (1984). *California high school curriculum study: Paths through high school* (ED 254 574). Sacramento: Author.

Calkins, L. M. (1983). *Lessons from a child: On the teaching and learning of writing.* Exeter, NH: Heinemann.

Capper, J. (1987). Updating the science curriculum: Who, what, and how? In A. B. Champagne & L. E. Hornig (Eds.), *This year in school science 1986: The science curriculum* (pp. 235–240). Washington, DC: American Association for the Advancement of Science.

Catterall, J. S., & Brizendine, E. (1985). Proposition 13: Effects on high school curricula, 1978–1983. *American Journal of Education, 93*, 324–351.

Cawelti, G., & Adkisson, J. (1986). *ASCD study reveals elementary school time allocations for subject areas: Supplement to ASCD update.* Alexandria, VA: Association for Supervision and Curriculum Development.

Cazden, C. B. (1988). *Classroom discourse: The language of teaching and learning.* Portsmouth, NH: Heinemann.

Chall, J. S., & Squire, J. R. (1991). The publishing industry and textbooks. In R. Barr, M. L. Kamil, P. Mosenthall, & P. D. Pearson (Eds.), *The handbook of reading research* (Vol. 2, pp. 120–146). New York: Macmillan.

Champagne, A., & Hornig, L. (1987). Critical questions and tentative answers. In A. B. Champagne & L. E. Horning (Eds.), *This year in school science 1986: The science curriculum* (pp. 1–12). Washington, DC: American Association for the Advancement of Science.

Cheney, L. V. (1987). *American memory: A report on the humanities in the nation's public schools.* Washington, DC: U.S. Government Printing Office.

Clune, W. H. (1989). *The implementation and effects of high school graduation requirements: First steps toward curricular reform.* New Brunswick, NJ: Consortium for Policy Research in Education.

Cohen, D. (1990). A revolution in one classroom: The case of Mrs. Oublier. *Educational Evaluation and Policy Analysis, 12*, 327–346.

Cohen, R. (1987). A match or not a match: A study of intermediate science teaching materials. In A. B. Champagne & L. E. Hornig (Eds.), *This year in school science 1986: The science curriculum* (pp. 35–60). Washington, DC: American Association for the Advancement of Science.

Crosswhite, F. J., Dossey, J. A., Swafford, J. O., McKnight, C. C., & Cooney, T. J. (1985). *Second international mathematics study: Summary report for the United States.* Champaign, IL: Stipes Publishing Co.

Cuban, L. (1984). *How teachers taught: Constancy and change in American classrooms 1890–1980.* New York: Longman.

Cuban, L. (1990). Reforming again, again, and again. *Educational Researcher, 19*(1), 3–13.

Dewey, J. (1916). *Democracy in education.* New York: Macmillan.

Dillon, D., & Searle, D. (1981). The role of language in one first grade classroom. *Research in the Teaching of English, 15,* 311–328.

Dossey, J. O., Mullis, I. V. S., Lundquist, M. M., & Chambers, D. L. (1988). *The mathematics report card: Are we measuring up?* Princeton, NJ: Educational Testing Service.

Duffy, G., & McIntyre, L. (1982). A naturalistic study of instructional assistance in primary grade reading. *Elementary School Journal, 83,* 134–139.

Dunkin, M. J., & Biddle, B. J. (1974). *The study of teaching.* New York: Holt, Rinehart & Winston.

Durkin, D. (1978–1979). What classroom observation reveals about reading comprehension instruction. *Reading Research Quarterly, 14*(4), 481–533.

Durkin, D. (1987). Teaching and testing in kindergarten. *Reading Teacher, 40,* 766–770.

Early, M. (1991). Major research programs. In *Handbook of research on teaching the English language arts* (pp. 143–158). New York: Macmillan.

Educational Testing Service. (1989). *What Americans study.* Princeton, NJ: Policy Information Center, Educational Testing Service.

Egan, K. (1982). Teaching history to young children. *Phi Delta Kappan, 63,* 439–441.

Eisner, E. W., & Vallance, E. (1974). *Conflicting conceptions of curriculum.* Berkeley, CA: McCutchan.

Engle, S., & Ochoa, A. (1988). *Education for democratic citizenship: Decision-making in the social studies.* New York: Teachers College Press.

Evans, R. W. (1989). A dream unrealized: A brief look at the history of issue-centered approaches. *Social Studies, 80,* 178–184.

Fancher, R. (1984). English teaching and humane culture. In C. Finn et al. (Eds.), *Against mediocrity* (pp. 49–69). New York: Holmes and Meier.

Farrell, E. J. (1991). Instructional models for English language arts, K–12. In *Handbook of research on teaching the English language arts* (pp. 63–84). New York: Macmillan.

Flanders, J. R. (1987). How much of the content in mathematics textbooks is new? *The Arithmetic Teacher, 39,* 18–23.

Flood, J., Jensen, J. M., Lapp, D., & Squire, J. R. (1991). *Handbook of research on teaching the English language arts.* New York: Macmillan.

Florio, S., & Clark, C. M. (1982). The functions of writing in an elementary classroom. *Research in the Teaching of English, 16,* 115–130.

Fontana, L. (1980). *Status of social studies teaching practices in secondary schools.* Bloomington, IL: Agency for Instructional Television.

Freeman, D. J., & Porter, A. (1989). Do textbooks dictate the content of mathematics instruction in elementary schools? *American Educational Research Journal, 26,* 403–421.

French, A. P. (1986). Setting new directions in physics teaching: PSSC 30 years later. *Physics Today, 39*(9), 30–34.

Gehrke, N. J., & Sheffield, R. (1985). Are core subjects becoming a dumping ground for reassigned high school teachers? *Educational Leadership, 42*(8), 65–69.

Gilford, D. M. (1987). Data resources to describe U.S. precollege science and mathematics curricula. In A. B. Champagne & L. E. Hornig (Eds.), *This year in school science 1986: The science curriculum* (pp. 67–106). Washington, DC: American Association for the Advancement of Science.

Glasgow, D. (1983, May). What research says: Identifying the "real" elementary school science curriculum. *Science and Children*, pp. 56–59.

Goodlad, J. I. (1984). *A place called school*. New York: McGraw-Hill.

Goodlad, J. I., and Associates. (1979). *Curriculum inquiry: The study of curriculum practice*. New York: McGraw-Hill.

Goodlad, J. I., Klein, M. F., & Associates. (1970). *Behind the classroom door*. Worthington, OH: Charles A. Jones.

Goodman, J., & Adler, S. (1985). Becoming an elementary social studies teacher: A study of perspectives. *Theory and Research in Social Education, 13*, 1–20.

Graves, J., & Myers, M. (1978). The Bay Area Writing Project. *Phi Delta Kappan, 59*, 410–413.

Gross, R. E. (1977). The status of the social studies in the public schools of the United States: Facts and impressions from a national survey. *Social Education, 41*, 194–200.

Grossman, P. L. (1990). *The making of a teacher*. New York: Teachers College Press.

Grossman, P. L., Kirst, M. W., & Schmidt-Posner, J. (1986). On the trail of the omnibeast: Evaluating omnibus education reforms in the 1980's. *Educational Evaluation and Policy Analysis, 8*, 253–266.

Hahn, C. L. (1985). The status of the social studies in the public schools of the United States: Another look. *Social Education, 49*, 220–223.

Hanna, R. (1963). Revising the social studies: What is needed? *Social Education, 27* (4), 190–196.

Harper, L. (1989). *To kill a mockingbird*. Philadelphia: Lippincott.

Hass, J. D. (1979). Social studies: Where have we been? Where are we now? *Social Studies, 72*, 249–253.

Heath, S. B. (1983). *Ways with words: Language, life, and work in communities and classrooms*. Cambridge, England: Cambridge University Press.

Heath, S. B. (1986). Sociocultural contexts of language development. In *Beyond language: Social and cultural factors in schooling language minority students* (pp. 13–24). Sacramento: Bilingual Education Office, California State Department of Education.

Heath, S. B. (1991). The sense of being literate: Historical and cross-cultural features. In R. Barr, M. L. Kamil, P. Mosenthal, & P. D. Pearson (Eds.), *Handbook of reading research* (Vol. 2, pp. 3–25). New York: Macmillan.

Hemmings, A., & Metz, M. H. (1990). Real teaching: How high school teachers negotiate societal, local community, and student pressures when they define their work. In R. Page & L. Valli (Eds.), *Curriculum differentiation* (pp. 91–111). Albany, NY: SUNY Press.

Hertzberg, H. W. (1981). *Social studies reform: 1880–1980*. Boulder, CO: Social Science Education Consortium.

Hillocks, G., Jr., & Smith, M. W. (1991). Grammar and usage. In *Handbook of research on teaching the English language arts* (pp. 591–603). New York: Macmillan.

Hirsch, E. D. (1987). *Cultural literacy*. Boston: Houghton Mifflin.

Hoetker, J., & Ahlbrand, W. P. (1969). The persistence of the recitation. *American Educational Research Journal, 6*, 145–167.

Hurd, P. D. (1986). Perspectives for the reform of science education. *Phi Delta Kappan, 67*, 353–358.

Jackson, P. W. (1968). *Life in classrooms.* New York: Holt, Rinehart & Winston.
Jackson, P. (1984). The reform of science education: A cautionary tale. *Daedalus, 112,* 143–166.
Jacobsen, W. J., & Doran, R. L. (1987). The second international science study: U.S. results. *Phi Delta Kappan, 66,* 414–417.
Jarolimek, J. (1981). The social studies: An overview. In H. D. Mehlinger & O. L. Davis, Jr. (Eds.), *The social studies* (The eightieth yearbook of the National Society for the Study of Education, pp. 3–18). Chicago: University of Chicago Press.
Jenness, D. (1990). *Making sense of social studies.* New York: Macmillan.
Joyce, W. W., Little, T. H., & Wronski, S. P. (1991). Scope and sequence, goals, and objectives: Effects on social studies. In J. P. Shaver (Ed.), *Handbook of research on social studies teaching and learning* (pp. 321–331). New York: Macmillan.
Kaltsounis, T. (1969). *Teaching elementary social studies.* West Nyack, NY: Parker Publishing Company.
Knapp, M. S., Adelman, N., Needels, M. C., Zucker, A. A., McCollum, H., Turnbull, B. J., Marder, C., & Shields, P. M. (1991). *What is taught, and how, to the children of poverty: Interim report from a two-year investigation.* Washington, DC: U.S. Government Printing Office.
Knapp, M. S., St. John, M., Zucker, A. A., Needels, M., & Stearns, M. S. (1987). *Opportunities for strategic investment in K–12 science: Options for the National Science Foundation.* Menlo Park, CA: SRI International.
Langer, J. A., Applebee, A. N., Mullis, I. V. S., & Foertsch, M. A. (1990). *Learning to read in our nation's schools.* Washington, DC: U.S. Department of Education.
LaPointe, A. (1986). The state of instruction in reading and writing in U.S. elementary schools. *Phi Delta Kappan, 68,* 135–138.
Lengel, J. G., & Superka, D. P. (1982). Curriculum organization in social studies. In *The current state of social studies: A report of Project Span* (pp. 81–102). Boulder, CO: Social Science Education Consortium.
Levstik, L. S. (1986). The relationship between historical response and narrative in a sixth grade classroom. *Theory and Research in Social Education, 14,* 1–19.
Linn, M. (1986). *Establishing a research base for science education: Challenges, trends, and recommendations.* Berkeley, CA: Lawrence Hall of Science, University of California at Berkeley.
Lounsbury, J. H., & Clark, D. C. (1990). *Inside grade eight: From apathy to excitement.* Reston, VA: National Association of Secondary School Principals.
Lounsbury, J. H., & Johnston, H. J. (1988). *Life in the three sixth grades.* Reston, VA: National Association of Secondary School Principals.
Lounsbury, J. H., & Marani, J. V. (1964). *The junior high school we saw: One day in the eighth grade.* Washington, DC: Association for Supervision and Curriculum Development.
Lounsbury, J. H., Marani, J. V., & Compton, M. (1980). *The middle school in profile: A day in the seventh grade.* Columbus, OH: National Middle School Association.
McKee, S. J. (1988). Impediments to implementing critical thinking. *Social Education, 52,* 444–446.
McKnight, C. C., Crosswhite, F. J., Dossey, J. A., Kifer, E., Swafford, J. O., Travers, K. T., & Cooney, T. J. (1987). *The underachieving curriculum: As-*

sessing U.S. school mathematics from an international perspective. Champaign, IL: Stipes Publishing Co.

McNeil, L. (1986). *Contradictions of control: School structure and school knowledge.* New York: Routledge & Kegan Paul.

Mehan, H. (1979). *Learning lessons: Social organization in the classroom.* Cambridge, MA: Harvard University Press.

Mehlinger, H. D., & Davis, O. L. (Eds.). (1981). The social studies. *Eightieth yearbook of the National Council for the Study of Education, Part 2.* Chicago: University of Chicago Press.

Mitsakos, C. L. (1978). A global education program can make a difference. *Theory and Research in Social Education, 6,* 1–15.

Moffett, J., & Wagner, B. J. (1983). *Student-centered language arts and reading K–13: A handbook for teachers* (3rd ed). Boston: Houghton Mifflin.

Morrissett, I. (1977). Curriculum information network. Sixth report: Preferred approaches to the teaching of social studies. *Social Education, 41,* 206–209.

Morrissett, I. (1981). The needs of the future and the constraints of the past. In H. D. Mehlinger & O. L. Davis (Eds.), *The social studies* (The Eightieth Yearbook of the National Society for the Study of Education, Part II, pp. 36–59). Chicago: University of Chicago Press.

Morrissett, I., & Hass, J. D. (1982). Rationales, goals, and objectives in social studies. In *The current state of social studies* (pp. 1–79). Boulder, CO: Social Science Education Consortium.

Mullis, I., & Jenkins, L. B. (1988). *The science report card: Elements of risk and recovery—Trends and achievements based on the 1986 national assessment.* Princeton, NJ: Educational Testing Service.

National Assessment of Educational Progress. (1980). *Citizenship and social studies objectives.* Denver: Author.

National Council of Teachers of Mathematics. (1989). *Curriculum and evaluation standards for school mathematics.* Reston, VA: Author.

National Research Council. (1989). *Everybody counts: A report to the nation on the future of mathematics education.* Washington, DC: National Academy Press.

National Science Board. (1983). *Educating Americans for the 21st century.* Washington, DC: Author.

National Science Foundation. (1978). *The status of pre-college science, mathematics, and social science education: 1955–1975.* Washington, DC: Author.

Newkirk, T. (1991). The high school years. In *Handbook of research on teaching the English language arts* (pp. 331–342). New York: Macmillan.

Newmann, F. M. (1990). Priorities for the future. *Social Education, 50,* 240–250.

Oakes, J. (1985). *Keeping track: How schools structure inequality.* New Haven, CT: Yale University Press.

Oliver, D. W., & Shaver, J. P. (1974). *Teaching public issues in the high school.* Logan: Utah State University.

Parker, W. C. (1990). A final response: Searching for the middle. *Social Education, 55,* 27–28, 65.

Parker, W. C. (1991). *Renewing the social studies.* Alexandria, VA: Association for Supervision and Curriculum Development.

Patrick, J. J., & Hawke, S. (1982). Social studies curriculum materials. In *The current state of social studies: A report of Project Span* (pp. 105–158). Boulder, CO: Social Science Education Consortium.

Peterson, P. (1990a). The California study of elementary mathematics. *Educational Evaluation and Policy Analysis, 12,* 257–262.

Peterson, P. (1990b). Doing more in the same amount of time: Cathy Swift. *Educational Evaluation and Policy Analysis, 12*, 277–296.

Peterson, P. L., & Walberg, H. J. (Eds.). (1979). *Research on teaching*. Berkeley, CA: McCutchan.

Pinnell, G. S., & Jaggar, A. M. (1991). Oral language, speaking and listening in the classroom. In *Handbook of research on teaching the English language arts* (pp. 691–720). New York: Macmillan.

Pipho, C. (1991). Centralizing curriculum at the state level. In M. F. Klein (Ed.), *The politics of curriculum decision-making* (pp. 67–97). Albany: State University of New York Press.

Porter, A. (1989). A curriculum out of balance: The case of elementary school mathematics. *Educational Researcher, 18*(5), 9–15.

Porter, A., Floden, R., Freeman, D., Schmidt, W., & Schwille, J. (1988). Content determinants in elementary school mathematics. In D. A. Grouws & T. J. Cooney (Eds.), *Perspectives on research on effective mathematics teaching* (pp. 96–113). Hillsdale, NJ: Erlbaum.

Porter, A. C., Kuhs, T. M., Freeman, D. J., Floden, R. E., Knappen, L. B., Schmidt, W. H., & Schwille, J. R. (in press). Elementary school mathematics textbooks. In R. W. Selden et al. (Eds.), *The textbook in American education*. New York: Ablex.

Pratt, H. (1981). Science education in the elementary school. In N. C. Harms & R. E. Yager (Eds.), *What research says to the science teacher* (Vol. 3, pp. 73–93). Washington, DC: National Science Teachers Association.

Project Span. (1982). *The current state of social studies*. Boulder, CO: The Social Science Education Consortium.

Purves, A. C. (1975). Culture and deep structure in the literature curriculum. *Curriculum Theory Network, 2*, 139–150.

Quick, S. K. (1977). *Secondary impacts of the curriculum reform movement: A longitudinal study of the incorporation of innovations of the curriculum reform movement into commercially developed curriculum programs*. Unpublished doctoral dissertation, Stanford University, Stanford, CA.

Raizen, S. A., & Jones, L. V. (Eds.). (1985). *Indicators of precollege education in science and mathematics: A preliminary review*. Washington, DC: National Academy Press.

Ravitch, D. (1987). Tot sociology or what happened to history in the grade schools. *American Scholar, 56*, 343–353.

Ravitch, D., & Finn, C. E., Jr. (1987). *What do our 17 year olds know? A report on the first national assessment of history and the nation*. New York: Harper & Row.

Romberg, T. A., & Carpenter, T. P. (1986). Research on teaching and learning in mathematics: Two disciplines of inquiry. In M. C. Wittrock (Ed.), *Handbook of research on teaching* (3rd ed., pp. 850–873). New York: Macmillan.

Rosenblatt, L. M. (1938). *Literature as exploration*. New York: Appleton-Century.

St. John, M. (1987). *An assessment of the school in the exploratorium program*. Inverness, CA: Inverness Research Associates.

Sanders, N. M., & Stone, N. C. (1987). *The California high school curriculum study: Paths through high school*. Sacramento, CA: Department of Education.

Sanders, N. M., & Tanck, M. L. (1970). A critical appraisal of twenty-six national social studies projects. *Social Education, 34*, 383–449.

Sarason, S. B. (1990). *The predictable failure of educational reform: Can we change course before it's too late?* San Francisco: Jossey-Bass.

Scardamalia, M., & Bereiter, C. (1986). Research on written composition. In M. Wittrock (Ed.), *Handbook of research on teaching* (3rd ed., pp. 778–803). New York: Macmillan.

Schneider, D. O., & Van Sickle, R. L. (1979). The status of the social studies: The publishers' perspective. *Social Education, 43*, 461–465.

Schubert, W. H. (1986). *Curriculum: Perspective, paradigm, and possibility.* New York: Macmillan.

Shaver, J. P. (Ed.). (1991). *Handbook of research on social studies teaching and learning.* New York: Macmillan.

Shaver, J. P., Davis, O. L., Jr., & Helburn, S. W. (1980). *An interpretive report on the status of precollege social studies edcuation based on three NSF-funded studies.* Washington, DC: National Council for the Social Studies.

Sirotnik, K. A. (1983). What you see is what you get: Consistency, persistency, and mediocrity in classrooms. *Harvard Educational Review, 53*, 16–31.

Sirotnik, K. A. (1988). What goes on in classrooms? Is this the way we want it? In L. E. Beyer & M. W. Apple (Eds.), *The curriculum: Problems, politics and possibilities* (pp. 56–74). New York: State University of New York Press.

Solomon, J. (1989). Science, technology, and society as a curricular topic. In T. Husén & T. N. Postlethwaite (Eds.), *Encyclopedia of education: Research and studies, supplementary Volume 1* (pp. 668–671). Oxford, England: Pergamon Press.

Squire, J. R. (1991). The history of the profession. In *Handbook of research on teaching the English language arts* (pp. 3–17). New York: Macmillan.

Squire, J. R., & Applebee, R. K. (1968). *High school English instruction today.* New York: Appleton-Century-Crofts.

Stake, R. E., & Easley, J. (1978). *Case studies in science education: Volume II— Design, overview, and general findings.* Champaign: Center for Instructional Research and Curriculum Evaluation, University of Illinois.

Stanley, W. (1985). Recent research in the foundations of social education: 1976– 1983. In W. Stanley (Ed.), *Review of research in social studies education: 1976– 1983* (pp. 309–399). Washington, DC: National Council for the Social Studies.

Stevens, R. (1912). *The question as a measure of efficiency in instruction: A critical study of classroom practice* (Contributions to Education No. 48). New York: Teachers College, Columbia University.

Stodolsky, S. S. (1988). *The subject matters: Classroom activity in math and social studies.* Chicago: University of Chicago Press.

Stodolsky, S., Salk, S., & Glaessner, B. (1991). Students' views about learning math and social studies. *American Educational Research Journal, 28*, 89–116.

Strickland, D. S., & Feeley, J. T. (1991). Development in the elementary school years. In *Handbook of research on teaching English in language arts* (pp. 286– 302). New York: Macmillan.

Suhor, C. (1988). Content and process in the English curriculum. In R. S. Brandt (Ed.), *Content of the curriculum* (pp. 31–52). Alexandria, VA: Association for Supervision and Curriculum Development.

Thornton, S. J. (1988). Curriculum consonance in United States history classrooms. *Journal of Curriculum and Supervision, 3*, 308–320.

Thornton, S. J. (1991). Teacher as curricular-instructional gatekeeper in social studies. In J. Shaver (Ed.), *Handbook of research on social studies teaching and learning* (pp. 237–248). New York: Macmillan.

Tye, B. B. (1985). *Multiple realities: A study of thirteen American high schools.* Lanham, MD: University Press of America.

Tyler, R. (1949). *Basic principles of curriculum and instruction.* Chicago: University of Chicago Press.

U.S. Department of Education, National Center for Education Statistics. (1987). *1987 high school transcript study.* Unpublished tabulations.

Usiskin, Z. (1985). We need another revolution in secondary school mathematics. In C. R. Hirsch & M. J. Zweng (Eds.), *The secondary school mathematics curriculum: The 1985 yearbook of the NCTM.* Reston, VA: National Council of Teachers of Mathematics.

Weiss, I. R. (1978). *Report of the 1977 national survey of science, mathematics, and social studies education.* Research Triangle Park, NC: Center for Educational Research and Evaluation.

Weiss, I. R. (1986). *Developing options for managing NSF's middle school science education programs.* Research Triangle Park, NC: Research Triangle Institute.

Weiss, I. R. (1987). *Report of the 1985–86 national survey of science and mathematics education.* Washington, DC: National Science Foundation.

Welch, W. (1979). Twenty years of science curriculum development: A look backwards. In D. C. Berliner (Ed.), *Review of research in education* (Vol. 7, pp. 282–306). Washington, DC: American Educational Research Association.

Welch, W. W., Harris, L. J., & Anderson, R. E. (1984). How many are enrolled in science? *Science Teacher, 51,* 14–19.

Westat. (1988). *Nation at risk study as part of the 1987 high school transcript study* (Tabulations for the U.S. Department of Education). Washington, DC: Author.

White, J. J. (1985). What works for teachers: A review of ethnographic research studies as they inform issues on social studies curriculum and instruction. In W. B. Stanley (Ed.), *Review of research in social studies education 1976–1983* (pp. 215–308). Boulder, CO: ERIC Clearinghouse for Social Studies Education and the National Council for the Social Studies.

White, R. T., & Tisher, R. P. (1986). Research on natural science. In M. C. Wittrock (Ed.), *Handbook of research on teaching* (3rd ed., pp. 874–905). New York: Macmillan.

Wiley, K., & Morrissett, I. (1972). Sincerely yours. *Social Education, 36,* 712–713.

Wiley, K. B. (1977). *The status of pre-college science, mathematics, and social science education: 1955–1975.* Washington, DC: National Science Foundation.

Wilson, S. (1990). A conflict of interests: The case of Mark Black. *Educational Evaluation and Policy Analysis, 12,* 309–326.

Wilson, S. M., & Wineburg, S. S. (1988). Peering at history from different lenses: The role of disciplinary perspectives in the teaching of American history. *Teachers College Record, 89,* 525–539.

Yager, R. (1983). The importance of terminology in teaching K–12 science. *Journal of Research in Science Teaching, 20,* 577–588.

Zucker, A. (1991a, April). *Mathematical problem solving: Opportunities and barriers in classrooms serving high concentrations of economically disadvantaged students.* Paper presented at the annual meeting of the American Educational Research Association, Chicago.

Zucker, A. (1991b). Review of research on effective curriculum and instruction in mathematics. In M. S. Knapp & P. M. Shields (Eds.), *Better schooling for the children of poverty: Alternatives to conventional wisdom.* Berkeley, CA: McCutchan.

Chapter 3

The Influence of Contemporary Issues Curricula on School-Aged Youth

JAMES S. LEMING
Southern Illinois University at Carbondale

A persistent characteristic of American education has been the conviction among both educators and the public that the school should play a role in the amelioration of important social problems. Even a cursory reading of the history of American education reveals that schools have been involved with much more than the three Rs. At various times in this century, many schools have assumed responsibility to teach about such diverse areas as war and peace, values and character, sex and drugs, race relations, adjustment to life, automobile safety, mental health, and the environment. These curricular initiatives have typically followed historical periods where the issues dealt with by the curriculum have galvanized the American public and the educational establishment. For example, peace education has received the greatest attention from educators during times of international instability. Likewise, education to reduce racial prejudice and multicultural education has penetrated the school curriculum most dramatically during times in our history when race and ethnic relations were in the forefront of the American consciousness.

In the 1920s, when, as a result of dramatic changes occurring within society, many traditional American values were being challenged, an extensive movement called character education arose (Yulish, 1980). In the tumultuous decade of the 1960s, a similar movement, called moral or values education, although with a different, more liberal spin to it, penetrated the school curriculum. Eventually, a more conservative emphasis was imparted to this movement during the Reagan administration. The American people have frequently expected the schools to assume responsibility for teaching about more than the three Rs.

Contemporary issues curricula will be defined as those areas of formal instruction that take place largely within classrooms that have as a primary goal the amelioration of pressing problems facing youth and/or society. Only as a secondary emphasis are these curricula intended to achieve more traditional academic objectives. The problem areas addressed by contemporary issues curricula typically have both a social and personal dimension. Within some problem areas, the emphasis of the curriculum is primarily at the personal level, as with the objective that the individual becomes less at risk to engage in behaviors harmful to the self (e.g., drug and AIDS education). With other problem areas, the curriculum may focus more on the study of the issue as a societal problem leading to the development in students of informed positions and the disposition to work toward the solution of the problem (e.g., justice in the society and environmental education). In some areas, such as nuclear education, the goals may be both personal and social (e.g., the reduction of anxiety and the creation of a safer world).

Because schooling is fundamentally a moral undertaking in that it contributes to the present and future development and well-being of youth, assessing the influence of curricula on youth is of vital significance. Such knowledge is worth having, for if it is found that there is no significant impact on students of a program of study, then schools may wish to reconsider the instructional approach or drop the area altogether to devote more time to other significant educational goals.

The purpose of this chapter, therefore, is to provide a succinct review of those areas of sustained interest and research regarding contemporary issues curricula and their influence on youth. The focus will be on research regarding curricular effects; however, information regarding the sources of public and educational interest, the nature of the content of the curricula, and any variables associated with differential curricular effects across studies will be reported. When recent quality reviews in a curricular area exist, the findings will be briefly reported and supplemented with any subsequent studies.

Because what a society considers a social problem, as well as how that society defines that problem, can change very quickly, and because schools are so susceptible to political influence and fads, I decided to limit this review to those curricular areas that received attention by the general public and were prevalent in schools in the 1980s. Furthermore, the research involved must have been conducted in classroom settings with school-aged youth (K–12). Although I made no effort to limit the review to experimental studies, this type of research provides the strongest evidence with regard to the influence of curriculum on students and was the dominant research methodology found in the studies identified.

A review of journals and annual meeting programs of professional educational associations, general educational journals, and research journals, as well as reports appearing in the news media, revealed a welter of potential curricular areas for possible review. Although the curricular areas identified as a result of the search process defy any rigorous classification, they will be organized under two broad headings for the purposes of this review. The first section will focus on curricula primarily concerned with the declining well-being of youth, including sex education, AIDS education, suicide prevention programs, drug education, and death education. The second section will focus on curricula concerned with the amelioration of national and international problems, including problems of racial and gender inequity and injustice, global education, nuclear education and peace studies, and environmental education.

THE DECLINING WELL-BEING OF YOUTH AND SCHOOL CURRICULA

These are not easy times for young people. By all indications, the well-being of youth is at an alarmingly low level. Educators and society, in addition to being concerned about low levels of academic performance, also are concerned about the increasing propensity for risk-taking behavior among youth, as reflected in statistics that reveal disturbingly high levels of adolescent injuries and homicide, suicide, delinquency, substance abuse, and sexual activity. Some of the negative trends in youth behavior may be attributable to the shifting values in society; however, the changing status of the American family has undoubtedly been the primary contributing factor in the decline of youth well-being.

The following data regarding the status of the American family highlight the instability many youth have faced over the past two decades. Since the 1970s, almost one of every two marriages has ended in divorce; between 1970 and 1989, the number of households maintained by women increased 98% from 5.5 million to 10.9 million; almost one fourth of the nation's children (under 18) lived with only one parent in the 1980s, as compared with only 12% (8.2 million) in 1970; and almost 20% of all children live below the poverty level, usually in households headed by women (Wentzel, 1990). The salience of marital dissolution for the well-being of youth is found in a recent study using longitudinal data from a nationally representative sample; Allison and Furstenberg (1989) found pervasive and lasting effects of early marital dissolution on youth problem behavior, psychological distress, and academic performance.

The remainder of this section will review the available research in four areas where the school curriculum was designed to influence the propensity of youth to engage in risk-taking behaviors: sex education, AIDS education, drug education, and suicide prevention. In addition, research on curricula where the intent is to alleviate anxiety related to the experience of death will also be briefly considered.

Sex Education

It has been estimated that 57% of teenagers are sexually active by the time they graduate from high school (Harris & Associates, 1986). One result has been that the United States today has the highest fertility rate among teenagers of any developed nation. More than 1 million teenage pregnancies currently occur each year—about 3,000 per day (Strasburger, 1985). The vast majority of these pregnancies are unplanned, and half are terminated through forced abortions. One response to these statistics has been the development and implementation of sex education programs in schools. One survey indicated that 75% of schools in urban areas had some form of formal instruction in sex education (Sonenstein & Pittman, 1984). Dawson (1986) reports that 68% of American women between 15 and 19 have received formal instruction in school about pregnancy and contraceptive methods.

Because there exist a number of recent reviews of the field of sex education (Dawson, 1986; Hofferth & Miller, 1989; Kirby, 1980, 1984; Reppucci & Herman, 1991; Stout & Rivara, 1989), only broad findings will be discussed here. The most common form of sex education is the sexuality unit typically taught within a health or physical education class. In a review of model sex education programs, Kirby (1984) identified three categories of goals contained in contemporary sex education programs: (a) an increase in student knowledge about topics related to sexuality, including sexually transmitted diseases, pregnancy and birth, abortion, contraception, anatomy, and physiology; (b) a clarification of values regarding dating, marriage, and drugs and alcohol and their relationship to sexual activity; and (c) improved decision-making and communication skills. Reppucci and Herman (1991) refer to such formal programs as secular sex education in that they attempt to provide information without specific religious or other value judgments being made about personal sexual decisions and behaviors.

There is nearly unanimous agreement among researchers in the field of sex education regarding the effects of traditional or secular sex education programs. The conclusions of Kirby (1980), in his seminal review of research in the field, have not been challenged in subsequent reviews. His findings may be summarized as follows: Knowledge about the nature of sexuality increases as a result of sex education instruction; where increases in knowledge are found, reproduction, contraception, and the nature of sexually transmitted diseases are typical subject matter areas. Sex education curricula have also been found to alter some student attitudes. Students become more tolerant of the sexual practices of others but do not change attitudes that are related to their own sexual behavior.

Relatively few of the programs involving adolescents have attempted to measure changes in behavior. Reviews have generally found these studies to be methodologically weak; nearly all involve nonrandomized designs, contain weak instruments, and make no attempt to assess long-term effects. More significant, the presumed link between behavior and knowledge and attitudes is speculative and largely unproven (Kirby, 1985).

An additional source of information regarding the effectiveness of sex education programs, other than studies using experimental designs, involves data from cross-sectional national databases. Whereas experimental evaluations typically involve the assessment of only immediate effects, cross-sectional studies that use an event history methodology permit speculation about long-term effects. Four cross-sectional studies were identified in which the investigators measured whether the respondent had been exposed to sex education as well as whether at least one of the following behaviors occurred: coitus, its onset or frequency, contraceptive behavior, and pregnancy rates.

Marsiglio and Mott (1986), using data from the National Longitudinal Survey of Work Experience of Youth, examined the impact of sex education on sexual activity, contraceptive use, and premarital pregnancy among American teenagers. They found a minimal impact of sex education on behavior. Adolescent women at ages 15 and 16 who indicated that they previously had taken a sex education course were more likely than those who have not had a sex education course to have initiated sexual activity. However, at ages 17 and 18, this pattern did not hold. Among 15- and 16-year-old girls, the effect of prior sex education was found to be weaker than virtually every other variable identified as being significantly related to the onset of sexual activity. Older sexually active females who have had a course in sex education are significantly more likely to use an effective contraceptive method (73%) than those who have never taken such a course (64%). No association was found between taking a sex education course and subsequently becoming premaritally pregnant before age 20.

Dawson (1986), using the 1982 Survey of Family Growth, analyzed data from 1,888 women aged 15 to 19 at the time of the interview. The respondents were surveyed regarding first intercourse, contraceptive use, reproductive experience, and demographic and socioeconomic characteristics. In addition, they were asked whether, by age 18, they had received any formal instruction in sex education and, if so, the nature of the topics covered. Dawson detected no consistent association between prior sex education and the probability of first intercourse. Young women who had had sex education were just as likely to initiate sexual activity at a given age as were those who had not had sex education. Sexually active adolescents who have had formal instruction report greater knowledge of contraceptive methods and are more likely to have practiced contraception at some time. When asked about current contraceptive practice, however, no differences appear between the two groups. Finally, no relationship between exposure to sex education and the risk of premarital pregnancy was found among sexually active teenagers.

Furstenberg, Moore, and Peterson (1985), using the second wave of the National Survey of Children conducted in 1981, report data on a nationally representative household sample of children who were between the ages of 7 and 11 in 1976. They found that the overall reported prevalence of sexual intercourse was over 50% higher among youth who did not have a sex education course (26% vs. 17%). Zelnik and Kim (1982) used data from two nationally representative surveys, one conducted in 1976 involving young women aged 15 to 17 and the other fielded in 1979 involving young women 15–19 and young men 17–21. In 7 of the 12 categories analyzed (determined by age, sex, race, and survey year), young people who had taken a sex education course were more likely to be sexually active than those who had not. In the remaining 5 cases, the reverse was true. It was concluded that there is no association between sex education (or its absence) and sexual activity. In addition, it was found that use of contraceptive devices was independent of prior sex education. Zelnik and Kim also tested the assumption that sexually active young people who have taken a course that contains information on contemporary contraceptive methods will experience fewer premarital pregnancies than those who have not taken such a course. It was found that in only one of eight age groups did the women who had taken a sex education course register lower pregnancy rates than women who had not had such a course (Black women aged 15–17 and surveyed in 1979).

These findings from both experimental and cross-sectional studies suggest that traditional approaches to sex education have little impact on sexual activity among teens or on teen pregnancy. The results of this research provide little insight into characteristics of curricula that can be linked to effectiveness with regard to cognitive and attitudinal outcomes. In his review, Kirby (1984) found that longer and more comprehensive programs yielded results little different from shorter programs. He did find that younger children learned more than older children, probably because older youth possess more knowledge prior to instruction.

The failure to find any significant impact of sex education on pregnancy rates has led some researchers to suggest that the effort to influence such a basic human activity in a specified direction is unrealistic (Stout & Rivara, 1989). Other authors have argued that the weak research findings are due not to the focus on values and behavior, but to the combining of good and bad programs in reviews of research of the field. It is argued that school districts uncritically accept all varieties of sex education programs without any research evidence, the result being that good and bad programs are implemented at about the same rate ("Sex Education," 1986).

Two additional approaches to sex education can be mentioned. Family planning clinics have been the preferred public policy strategy over the past two decades. Programs associated with these clinics, originally established to assist married couples in controlling their fertility, have been extended to serve adolescents by dispensing contraceptives in an effort to control sexually transmitted diseases and unwanted pregnancies. Because these clinics are not a part of the school curriculum and they are most frequently operated off campus and through referrals, the research findings will not be discussed here.

Value-based programs constitute the most recent development in the field of sex education. In these programs sexuality is placed within the context of human relationships and values such as respect for others, personal dignity, commitment, and self-control are stressed. All such programs emphasize and encourage the notion that sexual abstinence is a positive decision that each individual is free to make. Two recent programs, one short term and one long term, illustrate the nature and effectiveness of such an approach. The Responsible Sexual Values Program (Adamek & Thoms, 1991) involves the presentation of a 3-day instructional unit at the middle school level that integrates information about human sexuality, marriage, and parenting skills with group activities to develop positive self-esteem and wise decision making. After the curriculum has been presented, volunteer student organizations are formed to provide peer support for the norm of abstinence. In addition, parents attend a two-session workshop on the goals of the program and are involved in homework assignments that require their input. The results of the first-year evaluation of the program indicate that student knowledge of sexuality increased, as did attitudes supportive of the abstinent lifestyle (Adamek & Thoms, 1991). Data regarding pregnancy rates, sexually transmitted diseases, and precocious sexual activity will not be collected until the third year of the program.

A similar, but broader and more intensive, value-based program is reported by Vincent, Clearie, and Schluchter (1987). The School/Community Program for Sexual Risk Reduction Among Teens uses value-based sex education information and activities integrated, K–12, within regular school subjects. The program was implemented on a county-wide basis and involved not only parents, but also clergy, church leaders, local newspapers and radio stations, and special events to raise community awareness. Two years following the implementation of the program, it was found that estimated pregnancy rates declined for young women aged 14–17.

Value-based sex education that involves schools, parents, and the community in a common effort to encourage responsible sexual behavior appears to be a promising approach to changing adolescent attitudes and sexual behavior. The newness of these types of sex education programs and the relatively few evaluations conducted require that caution be used in interpreting the data; however, the results are encouraging and deserve careful attention and replication.

Evaluations of AIDS education curricula. Human immunodeficiency virus (HIV), the cause of AIDS, now infects more than 1 million people in the United States. In the eyes of many health care professionals and educators, it is the number one potential health problem facing youth today. The surgeon general of the United States has emphasized the importance of AIDS education and called AIDS a preventable epidemic that can be controlled by the modification of personal behavior (Koop, 1986). With current patterns of adolescent sexual activity and drug use—57% reporting having had sexual intercourse by age 17 (Harris & Associates, 1986) and 67% of high school seniors reporting they have experimented with drugs by graduation (Johnston, O'Malley, & Bachman, 1987)—large numbers of adolescents are at increased risk of HIV infection. Although it appears reasonable to expect that many schools have responded to this health threat by incorporating AIDS education into their curriculum, the time required to conduct evaluations of such curricula has been such that reports have only recently begun to be published. Three evaluation studies were identified, all published within the last 3 years (Brown, Fritz, & Barone, 1989; DiClemente et al., 1989; Miller & Downer, 1988).

All three studies used pretest-posttest research designs and involved treatments of from one to three class periods in length. The subjects used in the studies ranged from 7th grade to 11th grade. The goals of all three programs were similar: to increase student knowledge about AIDS, to increase tolerance toward persons with AIDS, and to modify attitudes toward the propensity to engage in AIDS-related risk behaviors. All three instructional units relied heavily on the presentation of information. The DiClemente et al. (1989) and Brown et al. (1989) studies also included class discussion of issues associated with contraction of AIDS.

DiClemente et al. (1989) found an advantage for the intervention classes, compared with control classes, with regard to knowledge about AIDS, and more positive attitudes reflecting less fear and a greater acceptance about having classmates who have AIDS. Brown et al. (1989) found changes in knowledge about AIDS and tolerance for AIDS patients. Attitudes toward AIDS-related high-risk behaviors changed only for the 7th-grade students, whereas change in coping behaviors for dealing with a friend with AIDS was found only for 10th-grade students. No comparison group was used in this study. Miller and Downer (1988) found gains in knowledge and attitudes among treatment subjects—a 13% increase in knowledge and a 19% increase in tolerant attitudes toward people with AIDS. These gains were maintained on an 8-week delayed posttest.

On the basis of the three studies cited, the research on AIDS education must be judged as somewhat encouraging at this time. It appears that with respect to AIDS education, the same pattern of findings as was found in traditional sex education is emerging. That is, knowledge and attitudes are found to change as a result of instruction. However, in the evaluation of AIDS education programs, researchers have not yet assessed the curriculum's influence on at-risk behaviors. It is probable that the same gap as exists between knowledge/attitudes and behavior as detected in the sex education research will also be found to exist with regard to the objectives of AIDS education programs, but this remains to be empirically established.

Drug/Substance Abuse Education

According to a recent national survey, substance abuse rates among teenagers in the United States are higher than those found in any other developed nation. Sixty-seven percent of high school seniors reported having used alcohol in the previous 30 days; 57% reported having tried an illicit drug, and 36% indicated lifetime use of an illicit drug other than marijuana; 5% reported daily use of marijuana, and 5% indicated daily use of alcohol; 37% reported an instance of heavy drinking within the previous 2 weeks; and 20% indicated they are daily cigarette smokers (Johnston et al., 1987). Recent polls indicate that drug abuse is seen both by the public and by teens as one of the most important problems facing America today (Coles, 1989; Elam, 1990). Given the widespread public concern regarding the epidemic of drug abuse and the perceived need to reach individuals in their formative years, it is not surprising to find some sort of drug abuse education in virtually every school in America. No other contemporary issue reviewed in this chapter has received such widespread attention by curriculum developers and evaluators. The research program regarding the effects of drug abuse education on youth is the most well-developed and extensive of any involving contemporary issues curricula.

In the past 30 years, there have been three broad shifts in the approach to drug education. Throughout the 1960s, the approach to drug abuse education largely involved providing information regarding the deleterious effects of the use of alcohol and other drugs and using scare tactics in an attempt to deter students from substance abuse. The "affective" or "humanistic" strategies of the early to mid-1970s focused more on teaching students personal skills such as problem solving, decision making, and developing positive health-related attitudes and self-concepts. The most recent approach to drug education, in the 1980s, involved a shift to the "social influences" strategy that emphasized making students aware of the factors in the social environment that create pressures to use drugs and developing the social skills to resist those pressures. Authors of the reviews of the research on the first two waves of drug abuse curricula are uniform with regard to their conclusions regarding the efficacy of these programs: They tend to be successful in increasing knowledge but less successful in changing attitudes, and they have little or no effect on drug and alcohol abuse (Berberian, Gross, Lovejoy, & Papanella, 1976; Kinder, Pape, & Walfish, 1980; Schapps, Dibartalo, Moskowitz, Palley, & Churgin, 1981). In programs that presented only knowledge or used scare tactics, some negative findings were noted. All of the authors of the reviews noted the generally poor quality of the existing research and cited the presence of inappropriate control groups, small sample sizes, sizable attrition rates, and poor measurement techniques. More recent reviews of drug education research have used meta-analytic techniques to assess the overall impact of programs and the characteristics of these programs. Three recent meta-analytic reviews are discussed briefly below.

In the most extensive review to date, Tobler (1986) conducted a meta-analysis of 143 adolescent drug prevention programs, including tobacco use prevention studies. The review focused on programs with students in Grades 6–12. Although the review was not limited to school-based programs, only 32 of 475 comparisons involved settings outside of schools. The period covered by the review was 1972 to 1984. Effect sizes were computed for 475 separate program effects grouped into five general outcomes identified in the studies: knowledge of characteristics of drugs and drug abuse, attitudes toward drug use, self-reports of drug use, skills (self-esteem, decision making, assertiveness), and behavior as measured by reports of others who were in a position to assess student drug use (e.g., parents, police, school personnel). The mean effect sizes found in Tobler's analysis were as follows: knowledge, 0.52; attitudes, 0.18; use, 0.24; skills, 0.26; and behavior, 0.27. Peer programs, programs that involved positive peer influence through peer teaching and peer counseling and helping, were found to show a definite superiority for the magnitude of the effect size obtained on all outcome measures. On the ultimate criterion of drug use, results of peer programs were significantly different from the combined results of the remaining programs.

Bangert-Drowns (1988), critical of Tolbert's unselective procedures regarding the methodological quality of the studies reviewed, excluded any studies that contained critical threats to internal validity. Originally, 126 reported studies were identified between 1968 and 1986, of which only 33 were finally reviewed. The average effect sizes were 0.76 for knowledge criteria, 0.34 for attitude criteria, and 0.12 for effect on behavior; however, the behavior effect was found to be not statistically different from zero. On measures of knowledge, no reliable relationships were found between study features and achievement. Use of class discussion and use of peers as leaders in instruction were found to be related to higher average effect sizes on student antidrug attitudes.

In a review of school-based programs intended to deter students from smoking and drinking, Rundall and Bruvold (1988) assessed the effects of 47 smoking and 29 alcohol school-based intervention programs published between 1970 and 1986 in which the researchers used a comparison or control group design. Pooled effect sizes were computed for immediate effects and measures collected at least 3 months after the end of the intervention. With regard to immediate and long-term behavioral impact, the ratios of desirable to total outcomes for alcohol were 18:27 (immediate) and 9:18 (long term). For smoking, the ratios were 28:35 (immediate) and 29:34 (long term). The pooled effect sizes for smoking and alcohol behavior, immediate and long term, were 0.15, 0.34, and 0.11, 0.12, respectively. The pooled effect sizes for smoking and alcohol knowledge, immediate and long term, were 0.59, 0.65, and 0.52, 0.38, respectively. Finally, with regard to attitude change, the pooled effect sizes for smoking and alcohol attitudes, immediate and long term, were 0.05, 0.13, and 0.18, -0.23, respectively. In other words, the long-term effect of alcohol programs on student attitudes was found to be a more positive attitude toward drinking. Generally, programs that emphasized other than awareness objectives, and involved the discovery of social norms and the development of social awareness, were more effective in both immediate and long-term outcomes.

The programs reviewed in the meta-analyses described above generally failed to include the third wave of program development and research in the field. This third wave is a more fully developed peer-based approach of preventative education that emphasizes identifying pro-drug pressures and acquiring strategies for resisting these pressures. In these programs students typically are taught to identify pressures to use drugs, to develop reasons not to use drugs, to recognize the benefits of not using drugs, and to develop the skills needed to resist drug usage. This last behavior is usually practiced through role playing in class. This peer approach involves group activity and the discussion of personal experiences in an effort to develop group norms against drug abuse.

Four recent program evaluations conducted with upper elementary or junior high school students suggest that this approach has promise. In one of the early evaluations of the social influence approach, Flay et al. (1985) found that a six-lesson unit concerned with smoking resulted in reduced smoking behavior, especially among those most "at risk" (i.e., with models at home of smoking behavior). Dielman, Shope, Leech, and Butchart (1989) found that a four-lesson social influence program reduced the rate of increase of alcohol use among sixth-grade students who had already begun to use alcohol. Ellickson and Bell (1990) found that their eight-lesson program reduced alcohol and tobacco use among junior high school students but did not sustain the reduction over time with regard to alcohol use.

In a longitudinal trial of a primary prevention program of cigarette, alcohol, and marijuana use, Pentz et al. (1989) reported data from the first 2 years of the Midwestern Prevention Project. This project is a comprehensive, community-based program that uses school study, parent involvement with homework, mass media, community organization, and health policy programming to combat adolescent drug usage. The entire adolescent population from 15 midwestern communities constitutes the sample. In the first 2 years, 22,500 sixth- and seventh-grade students received the school-based component. It was found, in the second year of what is to be a 6-year evaluation, that the use of all three target drugs was lower among students in the program for 1 year than among students just entering the program (17% vs. 24% for tobacco use, 11% vs. 16% for alcohol use, and 7% vs. 10% for the use of marijuana within the previous month).

The following conclusions appear warranted regarding the characteristics of efficacious drug education programs. Gains in knowledge were common to all such programs. With regard to attitude changes, lecture appears to have the smallest effect; peer programs have a greater influence. With regard to behavioral outcomes, social influence programs appear to be the most effective in reducing the incidence of drug usage; however, in cases where delayed measures were used, attrition of effects was usually found. This attrition of impact was found to be less substantial in social influence programs.

The history of drug education research demonstrates many of the characteristics found to be associated with a progressive research program: increasing sophistication with regard to research design, improved instrumentation, revision of instructional strategies based on prior research findings, and revision and reconceptualization of theoretical assumptions. The newer social influence and social skills approaches to drug use prevention programs appear to be more successful than earlier approaches in achieving the desired outcomes. Researchers in the field of drug education program evaluation appear to be self-critical and responsive to research findings. Even with this new approach, the findings to date regarding the efficacy of drug education programs have not been dramatic, and the long-term consequences of drug prevention programs are still indeterminate.

Suicide-Prevention Programs

The risk of secondary school students attempting to commit suicide is greater today than at any time in the past (Davis, 1988). Although a portion of this alarming high rate may be an artifact of improved reporting and record-keeping procedures, there exists considerable concern among educators and the public at large regarding this phenomenon. This concern is apparently warranted. A recent national poll by the Gallup organization found that 6% of teenagers reported having attempted, and 15% reported having contemplated, suicide ("Six Percent of U.S. Teenagers," 1991). One result of this concern has been the rapid growth of suicide-prevention programs for high school students. Wass, Miller, and Thornton (1990) estimate that one quarter of American secondary schools have some sort of suicide-prevention program. Garland, Whittle, and Schaffer (1989), in a survey of associations and agencies involved in suicide prevention, found that between 1984 and 1986 the number of schools having suicide-prevention programs increased from 787 to 1,709. The mean program duration was found to be 3.8 hours, with a modal duration of 1 hour. Instruction was typically conducted by agency staff or by a combination of agency and school personnel. Most programs were found to have a similar format: presenting facts about youth suicide to heighten awareness, identifying warning signs that enable youth to identify "at-risk" colleagues (case finding), and providing information about mental health resources and how to access them. A minority of programs set out to increase students' coping abilities through training in stress management. Only three systematic and controlled evaluations of school-based suicide-prevention programs were located.

Shaffer, Garland, Underwood, and Whittle (1987) studied approximately 1,000 students ranging from 13 to 18 years of age who were exposed to three different programs in six different high schools. A similar number of students not exposed to a program in six different schools served as controls. On a pretest of knowledge about suicide, most students (program and nonprogram) held what the researchers considered sound knowledge regarding suicide. The program did not change that knowledge. Between 5% and 20% of program and nonprogram students held attitudes toward suicide that the researchers considered inappropriate (e.g., suicide is a reasonable solution to personal problems, or they would not refer to authorities a friend who was contemplating suicide). It was found that the program had no effect on these attitudes.

In an evaluation of a 6-week high school suicide-awareness program by Sprinto, Overholser, Ashworth, Morgan, and Benedict-Drew (1988), student knowledge, attitudes, helping behaviors, hopelessness, and coping skills were assessed 10 weeks after the end of the program. The curriculum consisted of the following topics: attitudes about suicide, awareness of suicide as a problem, facts about suicide, and identification, intervention, and referral techniques. It was found that the curriculum had little impact on any of the desired outcomes. Where small positive effects were noted, they were generally associated only with the females in the sample.

In the only study located in which authors identified adolescents that had attempted suicide, Shaffer et al. (1990) conducted a controlled evaluation of two suicide-prevention curricula delivered to 606 ninth- and tenth-grade students in two suburban and rural high schools. Program effect was assessed using a researcher-constructed 48-item questionnaire. The question "Have you ever tried to kill yourself?" was used to classify students as attempters or nonattempters. Sixty-three, or 6.5%, of the sample indicated on both the pretest and posttest an affirmative response to the question and were classified as attempters; 35 of the attempters were in the program group and 28 were in the control group. No program effects were detected between program and control attempters on eight key questionnaire items. Both groups endorsed statements supportive of the goals of the program to the same degree. The attempters' reactions to the program were generally more negative than those of the nonattempters exposed to the program. Program exposure did not influence previous attempters' deviant attitudes. There was some evidence of unwanted effects in that attempters exposed to programs, compared with attempters not in the program, were less likely to recommend that suicide-prevention programs be presented to other students and more likely to indicate that talking about suicide in the classroom makes some kids more likely to try to kill themselves.

In the case of suicide-prevention curricula, none of the three evaluations reviewed found effects on knowledge or attitudes. The saliency of the issue of suicide to students was clearly apparent in the finding that the pretest alone may influence student knowledge and attitudes (Sprinto et al., 1988). In the one study that identified attempters, potentially negative effects of a suicide-prevention program were noted (Shaffer et al., 1990).

Death Education

Death education, unlike the curricular areas so far reviewed, lacks a widely accepted rationale tied to the enhancement of the well-being of youth. It is an inescapable fact of life that everyone must face the loss of a loved one at some point; however, one does not find, among the general public or the education profession, any widespread acceptance of the need for death education. The origins of the death education movement are best explained as a movement whose impetus springs from, and has been sustained by, professional organizations. The 1970s and the early 1980s appear to have been the period of greatest activity for death education in the nation's schools. The first national conference on death education was held in 1970. The Association for Death Education and Counseling, which publishes the journal *Death Studies*, is the most active group in the promotion of death education today. The movement appeared to be somewhat in eclipse in the last half of the 1980s, as judged by citations in *Current Index to Journals in Education*, which have fallen by 50% since the early 1980s.

At the most general level, the purpose of death education is to promote positive and realistic attitudes toward death. The rationale for including death education in schools is derived from the observation that death, suffering the death of another, and facing one's own death are significant sources of stress in our society that we are not adequately prepared to handle. Leviton (1977) has described three general purposes of death education. First, it can serve a primary preventative function. It can prepare individuals and societies for subsequent events and consequences. Students can learn that grief and bereavement, and even feelings of anger and guilt, are appropriate and normal responses to death. Second, death education can be interventive; that is, it can help the student to more humanely and meaningfully interact with a dying person. Finally, death education can have a rehabilitative effect; it can help the person understand the crisis and to learn from the experience. The most common outcome variable in death education research is death anxiety (fear of death—one's own or that of others), which usually is measured by such instruments as the Collett-Lester Fear of Death Scale (Collett & Lester, 1969) and the Templer Death Anxiety Scale (Templer, 1970). In addition, a few studies report data on behavioral outcomes of death education defined as self-reports of initiating discussion and seeking out additional information about the topic.

The topics covered in a typical death education course vary considerably, because there exists no nationally agreed upon guidelines or curriculum. Most courses are locally developed and somewhat idiosyncratic; however, there are certain commonalities that cut across most units of study. All available programs are designed so that, through a variety of experiences, students will gain knowledge about the various aspects of death and dying, examine their own feelings and attitudes toward death, develop the ability to cope with feelings related to death and dying, and interact with others to share feelings and thoughts about the phenomenon of death.

There are few data regarding the extent to which death education is present in American schools. A survey by Wass et al. (1990) provides the most recent and comprehensive look at this question. They looked at a stratified random sample of 423 public schools, from prekindergarten through 12th grade, and found that the national estimate for schools offering a course or unit on general death education was 11%. They also found that 17% offered a grief/support program and 25% had suicide-prevention/intervention programs. The death education programs were most often taught as part of a health education course and were usually of 2 weeks duration or less. The grief and suicide programs were usually crisis-oriented and conducted by counselors or other school staff members on an as-needed basis.

The majority of the research investigating the influence of death education curricula used college-aged subjects, most frequently nursing and other health-care students. A search of the death education literature over the past 15 years found six studies that used school-aged youth as subjects.

Bailis and Kennedy (1977) assessed the influence of a death education module in two midwestern high schools. The modules were a part of the 10th-grade English course of study that involved field trips, guest speakers, and selected readings. Interaction between activities and discussion was emphasized. It was found that the students participating in these modules increased with respect to their fear of death and dying. The controls experienced no such increase.

Noland, Richardson, and Bray (1980) reported the results of two studies with ninth-grade girls. In the first experiment, it was found that both treatment and control classes gained in knowledge about death issues at the posttest and follow-up. Students exposed to the death education unit were found to have lower death anxiety on the immediate posttest, but not on the delayed posttest. With regard to a willingness to discuss death topics, the treatment group demonstrated more of the target behaviors at the follow-up. In the second experiment, subjects consisted of 146 female students (aged 13 to 17) enrolled in health and physical education courses at two high schools. In this experiment the authors found that experimental students improved their cognitive scores, demonstrated positive attitude shifts, and attempted desired behavioral activities more than control students.

Rosenthal (1980) examined the impact of an 18-week elective course on death and dying on the death anxiety scores of parochial high school students in California. The treatment group consisted of 18 juniors and seniors who had enrolled in an elective course that met four times a week. The comparison group consisted of similar students in an elective course, of the same duration, on marriage and the family. On two of the four factors measured, death avoidance and reluctance to interact with the dying, the treatment group decreased in death anxiety relative to the comparison group. On the other two factors, no difference in death anxiety between treatment and control students was found. Edwards (1983) taught a 15-day death education course in a Catholic high school in Mankato, Minnesota. The treatment group's mean level of death anxiety was unchanged on both the immediate and delayed posttests. On the delayed posttest, it was noted that the death anxiety of the female subjects was lower.

Glass and Knott (1984) attempted to assess the effect of a 10-day unit on death in eight North Carolina high schools. In the schools that agreed to participate, volunteer teachers were trained by the researchers to teach the curriculum. No difference was found between the treatment and control groups' death anxiety scores before or after the 10-lesson unit. In the most recent study reviewed, Glass (1990) developed a death curriculum in which death and other losses, grief, and recovery were the foci for a 10-lesson unit. Eighteen guidance counselors volunteered to participate in the study and teach the curriculum to one class in their school and use another class as a control. No difference between the treatment and control groups was detected at the conclusion of the study, or on a 2-month delayed posttest.

In the seven experiments reviewed where a reduction in death anxiety was the desired outcome, an effect was found in only one case (Noland, Richardson, & Bray, 1980). In one case it was found that death anxiety increased following instruction (Bailis & Kennedy, 1977). The research on the influence of death education on death anxiety suggests that the field should reassess both its theoretical assumptions and instructional methods.

THE AMELIORATION OF SOCIAL PROBLEMS THROUGH SCHOOL CURRICULA

Not only is the well-being of youth at risk in contemporary society, but society itself is beset by a wide range of significant problems. In the section below, the research on the major curricular areas that have been developed to address contemporary social problems will be reviewed.

Reducing Racial Prejudice

Although there is considerable debate in this country regarding the forms and extent of racism in contemporary life, there is little disagreement that it is an important issue on the American agenda. The recent publicity regarding racist incidents on college campuses and racially motivated police violence in our major cities once again brings home to society, and educators, the importance of this issue and the failure of American society to realize some of its most noble ideals. One of the enduring legacies of *Brown v. Board of Education* (1954) has been the awareness that schools should be free of racism and should respect racial and ethnic diversity and provide equal opportunity for all students.

Both an interest in racism as an educational issue and research into the effects of schools in dealing with this social problem have been related to social and political events occurring in society. Within the last 50 years, there have been two major periods of interest in education regarding the reduction of racism. Immediately following World War II, the intergroup education movement was a response to the major exodus of Blacks to northern cities and the resulting racial conflicts that occurred. The second major period of concern regarding race relations was an outgrowth of the civil rights movement of the 1960s. I was unable to find any data regarding the extent of school efforts to combat racism during these two periods; however, it seems reasonable to assume that racially integrated schools were the most active sites.

Research on the schools' potential to develop positive racial attitudes has followed both of the above periods: a period of research activity in the late 1940s and early 1950s and another period from the mid-1960s through the 1970s. The 1980s, the era of the Reagan presidency, were quiet with regard to research on prejudice reduction in schools. Today, the reduction of racial and ethnic prejudice as an educational goal has been largely subsumed under the rubric of multicultural education. Although the goals of multicultural education may go far beyond the mere reduction of prejudice (e.g., improved academic performance and enhanced self-esteem as a result of racial or ethnic pride; Banks, 1991; Sleeter & Grant, 1987), this review will focus only on research regarding prejudice reduction. Typical of the dearth of research on multicultural education, a search of *Current Index of Journals in Education* between 1985 and 1989 identified 209 citations; not a single one was a research study on the effects of multicultural education on children. Sleeter and Grant (1987), after reviewing the literature, note: "We perceive one particularly disturbing gap in the literature we reviewed: There are virtually no research studies on multicultural education" (p. 438).

The research on this question during the era covered by this review may be summarized under two general headings: (a) curriculum lessons or units designed to develop positive racial attitudes through the introduction of content that presents positive and realistic views of minorities, and (b) the development of positive racial and ethnic relations through cooperative learning strategies. Five studies were identified in which the authors evaluated the effect of the presentation of multicultural information on the racial attitudes of youth.

Shirley (1988) investigated the effects of integrating multicultural activities into English, social studies, and reading curricula on the racial attitudes of students in racially integrated fifth- and sixth-grade classes. Lessing and Clarke (1976) reported the results of a study of the effect of an 8-week intergroup relations curriculum on racial attitudes. The study took place with sixth-, seventh-, and eighth-grade students in a nearly all-White suburban school. Koeller (1977) studied the effects of reading excerpts from stories about Mexican Americans on the racial attitudes of sixth-grade students in racially integrated schools in the Denver area.

Control groups heard excerpts from stories with nonethnic themes. Gimmestad and De Chara (1982) studied the potential effectiveness of dramatic plays in the reduction of prejudice, with a unit designed to increase fourth, fifth, and sixth graders' knowledge about, and improve their attitudes toward, four ethnic groups: Blacks, Puerto Ricans, Jews, and Chinese. Finally, Keats (1990) evaluated the impact of an information-oriented, one-semester high school Afro-American history course on the racial attitudes of White students.

In two of the above studies (Lessing & Clarke, 1976, and Koeller, 1977) no change was detected in racial attitudes as a result of the treatment. Keats found that students electing to take the course developed more egalitarian attitudes on only 3 of 13 subscales of a racial attitudes inventory. He also found that the students not electing to take the course were more egalitarian on the pretest than the students electing to take the course and that, at all times, female students were more egalitarian than male students. In the other two studies (Gimmestad & De Chara, 1982; Shirley, 1988), more positive racial attitudes were found for treatment students. In the two studies where positive racial attitude change was detected, students were involved in group activities in racially integrated classrooms.

Textbook publishers, under considerable pressure from educationists, large urban school districts, and states with substantial minority populations, made great efforts in the 1970s and 1980s to remove racial, ethnic, and gender bias from textbooks. Little is known, however, about whether students' reading of materials that have been determined by adults to be biased contributes to the development of, or reduces the extent of, prejudice in youth. In Banks's (1991) analysis of the research on the influence of biased materials on youth, he found the research to be of poor quality and the results equivocal. Research regarding the potential of reading materials to affect student attitudes is somewhat more informative with regard to the related issue of the learning, or unlearning, of sex roles. This literature is reviewed in the section on gender equity below.

Cooperative Learning Methods and Racial and Ethnic Prejudice

The exception to the pattern of a lack of recent research on multicultural education and prejudice reduction is the cooperative learning research program. One of the success stories of contemporary educational research has been the ability of the authors involved in the research program on cooperative learning strategies to consistently achieve their desired cognitive and affective objectives. Relevant to this review are studies in which researchers have examined the use of cooperative learning to improve race relations as measured by cross-racial friendship choices and the development of more positive attitudes toward people of other races. For cooperative learning methods to improve racial relations, racially and ethnically integrated schools are required. Desegregated schools and the increased contact between races alone, however, apparently will not break down all the personal and social barriers that exist among youth (Gerard & Miller, 1975).

In cooperative learning, students work in groups of four to five students made up of the same mix of race, gender, and level of academic ability as that of the class as a whole. The groups then receive rewards, recognition, and/or evaluation based on the degree to which they can increase the performance of each member of the group. Although there are many variations of cooperative learning strategies, the basic one is composed of a cooperative task structure (team practice) and a cooperative reward structure among the teams (the team competition for recognition).

The research on cooperative learning strategies has been extensively reviewed elsewhere (Slavin, 1990). Of interest here are the 10 studies in which the researchers have examined the impact on the improvement of intergroup relations. In 8 of the 10 studies, students in the cooperative learning groups, compared with control (individualistic reward structure) groups, increased in the number of friendship choices outside of their own ethnic group (D. Johnson & Johnson, 1981, 1982; D. Johnson, Johnson, Tiffany, & Zaidman, 1984; Oishi, 1983; Sharan et al., 1984; Sharan & Shachar, 1988; Slavin, 1979; Zeigler, 1981). In three studies using delayed measures of friendship choices, the positive effect of cooperative learning was found to persist up to several months later (D. Johnson & Johnson, 1982; Slavin, 1979; Zeigler, 1981). Three of the above studies found that positive attitudes toward minorities extended to students outside of the cooperative learning group (Oishi, 1983; Sharan et al., 1984; Sharan & Shachar, 1988). Two studies found that cooperative learning strategies had mixed results on intergroup relations. Weigel, Wiser, and Cook (1975) found that White relations improved with respect to Mexican Americans, but not to Blacks. DeVires, Edwards, and Slavin (1978) found improved intergroup relations for only 7 of the 13 classrooms studied. It should be noted that the last two studies were the only ones conducted at the secondary level. All of the other studies involved elementary students, usually at the fourth-, fifth-, and sixth-grade levels.

Evaluation of Holocaust Curricula

More than any event in recent history, the Holocaust has drawn attention to the potential horrific consequences of racial, ethnic, and religious intolerance. Concerned educators throughout the United States have seen the teaching of the Holocaust as an opportunity to teach students greater tolerance of others. Although there are no exact data regarding the extent of teaching about the Holocaust in U.S. schools, Totten

(1991) reports that over the past decade, more than 30,000 educators from 46 states, Canada, and abroad have taken part in Facing History and Ourselves workshops and that, annually, over 450,000 students in public, private, and parochial schools have been taught using materials from these workshops. Dawidowicz (1990), in a recent review of the field of school-based Holocaust studies, found that 7 states and many large school districts had required Holocaust curricula. Only two Holocaust curricula have been formally evaluated with controlled studies. These studies are discussed below.

Facing History and Ourselves is a Holocaust curriculum originally developed for use in junior high schools (Strom & Parsons, 1982). An early evaluation of the curriculum focused on two outcomes: knowledge of the Holocaust, defined as the ability to correctly define terms and concepts, and development in interpersonal understanding (Lieberman, 1981). On Selman's (1980) measure of interpersonal understanding, it was found that Facing History classes, compared with control group classes, developed greater capacity for sociocentric thinking. With regard to performance on the content test, the proportion of students in the Facing History classes correctly answering items on the posttest was greater than the proportion answered correctly by a control class. In a more recent evaluation of the Facing History curriculum (Lieberman, 1991), pre-post gains were compared for eight geographically diverse classrooms exposed to the curriculum during the 1989–1990 school year. As with the earlier evaluation, it was found that knowledge-level outcomes were greater at the end of the curriculum. The data on the growth of interpersonal understanding have not, as of this date, been scored. On two additional items not included in the first evaluation report, the response of students was not consistent with the objectives of the program. In only three of the eight Holocaust curriculum classrooms did students change in their opinion that the Holocaust could have been prevented, and in only two of the classes did the students change positively in their perceptions related to their responsibilities to be active citizens (i.e., to protest injustice). The Facing History and Ourselves research does not include an assessment of the impact of the curriculum on student racial or ethnic prejudice.

A recently developed Holocaust curriculum, Life Unworthy of Life, was developed to be used in high school world history classes. The curriculum consists of short historical lectures, analysis of primary documents, videotapes of survivor testimony, and participation in authentic simulations. The unit requires 2 to 4 weeks of in-depth instruction. A careful and systematic evaluation of the program indicates that it is successful in achieving its three student outcome objectives: increased ability to express in writing the consequences of indifference toward the mistreatment of others, reductions in level of prejudice, and increased historical knowledge of the Holocaust (Nagourney, Bolkosky, Ellias, & Harris, 1990). Fifteen classrooms, one from each of 15 high schools in Oakland County, Michigan, volunteered to participate in a nonequivalent control group quasi-experimental study. Comparison classrooms were 14 comparable classrooms in the same schools. On an essay test scored for students' ability to express the consequences of indifference toward the mistreatment of others, posttest means favored the treatment groups. On two scales measuring attitudes toward minority groups (the Jews and Blacks scales), treatment subjects were found to be less prejudiced on the posttest. No such change was noted in the control groups. Finally, on a 36-item knowledge-based, goal-referenced unit test, treatment classes demonstrated gains, whereas the comparison classes experienced no such gains.

Modification of Sex Role Stereotypes

One of the many results of the women's movement of the past two decades has been the emergence of the question of whether or not it is beneficial for children to be socialized in accordance with traditional sex role expectations. The belief that traditional sex role images presented in curricular materials often have detrimental effects on female academic and occupational achievement has gained wide acceptance among the general public and the educational establishment. On the basis of a belief that sex role typing is now outdated and possibly even harmful in a culture where adult role options are becoming more diverse, considerable attention has been paid to the issue of elimination of gender bias in schools and to the cultivation of sex role identities in youth that are more androgynous and therefore, it is argued, less harmful to the future development of youth.

Although the issue of sexism in school has received considerable attention in the general media and among professional educators, there exists little actual research regarding the extent to which sex role identity is changeable within school environments. In this section, the research on this question will be reviewed. Two general types of studies were encountered in this sparse body of research: studies that assessed the impact of the experience of reading materials that presented men and women in nontraditional roles, and broader curricular experiences that attempt to expose children to nontraditional sex role models.

The most complete review to date of the research of the effect of instructional materials on sex role thinking is that of Schau and Scott (1984). Their findings will be briefly summarized below and supplemented by more recent research. Considerable attention has been directed toward the eradication of male generic language at all levels of written discourse in this country. Does gender-neutral or balanced writing result in changes in children's gender associations with occupational roles? Schau and Scott (1984) report that when "he" is used in reading materials, readers typically think of the role as filled by a male. The typical reaction to gender-unspecified language is for readers to also associate the role in a more gender-balanced manner.

Schau and Scott (1984) identified 21 studies that examined the effect of sex-equitable materials on sex role attitudes. Typically, in these studies students were exposed to narrative accounts of sex role portrayals such as female firefighters or male nurses. The authors found that exposure to sex-equitable story materials results in students' attitudes about themselves and others becoming less sex role steryotyped; however, these changes were not found to generalize to contents not in the materials. In the studies that compared sex differences in gender attitudes, it was found that there was no consistent pattern; that is, in some studies only girls changed, in some studies only boys changed, and in some studies both changed. The evidence does not permit other than speculation regarding the persistence of effects, change by age groupings, or effect of amount of exposure on gender attitudes. Although the research is very sketchy, it appears that students show a preference for same-sex characters regardless of the sex typing of the role content. Gender-equitable reading materials apparently have no positive or negative effect on comprehension, although the research is not consistent on this question. In a later study, Scott (1986) found that after reading nontraditional gender narratives, students in Grades 4, 7, and 11 were more likely to hold gender-neutral conceptions of occupational roles, with younger students holding the most flexible role attitudes.

Three studies were identified that assessed the effects of curricular interventions on students' sex role attitudes that went beyond the reading of narratives. Scott (1984) provides data on the effects of an intervention on middle school pupils' sex role flexibility. A yearlong curricular intervention designed to increase sex role flexibility, the Fair Play: Developing Self-Concept and Decision-Making Skills in the Middle School (1983) program, served as the intervention. In addition to increasing sex role flexibility, the program is intended to involve students more in decision making and to improve academic achievement. Activities for each lesson involved readings, discussion questions, small group work, role play, and evaluation exercises. An effect was found with regard to occupational role steryotypes, with treatment students more likely to endorse egalitarian roles as acceptable for either males or females. No main effect for treatment was found with regard to sex role self-concept; girls, however, did express more androgynous self-concepts.

An evaluation of the experiential economic education program Mini-Society (Kourilsky & Campbell, 1984) examined the program's influence on children's entrepreneurial and occupational stereotyping. The Mini-Society experience consists of the development of a microcosmic society where children experience and resolve various social and economic problems. Treatment lasted for 45 minutes a day, three times a week, for 10 weeks, and involved 938 children in Grades 3–6. A pretest-posttest single group design was used. A gain was detected, pretest to posttest, with regard to the samples' listing of female owners of economic enterprises in the class. Also, a gain pretest to posttest was found on a measure that assessed children's attitudes toward the appropriateness of female participation in "traditional" male occupations.

In the late 1970s, a television series, "Freestyle," was developed with funding from the National Institute of Education and shown nationally on the Public Broadcasting Service. Targeted at upper elementary students and designed to reduce the influence of sex and ethnic group sterotyping on children's career interests, the series consisted of 13 half-hour dramas that presented to children a view of an idealized world where gender and race were irrelevant considerations in career decisions. The series was evaluated at seven sites across the United States and involved 7,000 fourth, fifth, and sixth graders (J. Johnson & Ettema, 1982). In the treatment classrooms, each episode was viewed, and then the teacher led a class discussion with supplementary materials and activities to reinforce the theme of the episode. The series was found to be effective in changing girls' sterotypes and perceptions about the sex appropriateness of such activities as mechanics and athletics. Boys were found to develop more positive attitudes toward nurturing activities. Children's own occupational interests were less influenced, and long-term changes were more limited than short-term ones. Similar effects were noted, but to a lesser extent, for students who watched the series but did not engage in class discussion or activities. A similar effect was also found for students who watched the program at home, but this effect was less than that for either of the other two conditions.

The pattern of research findings for studies that have exposed children to gender-equitable language and role portrayals is consistent: The treatments, no matter what duration (usually they involve a single class period) and no matter what the nature of the treatment (whether it involves simply reading, viewing a video, or other class activities), result in changes in children's sex role attitudes and associations. The majority of studies have involved elementary aged students; the effect has been found to be weaker with older students. It has been found that the move to less sex-biased role characterizations has weakened over time but has not completely evaporated. The detected changes, most notably with regard to occupational choices, do not generalize to other situations and roles. Since the research reported above assessed primarily short-term effects, it is not known whether these changes lasted beyond the immediate period following the intervention. Finally, a consistent pattern of gender differences was found regarding changes in sex bias. Girls were consistently found to manifest greater change and are more likely to express a preference for egalitarianism. It is likely the case that in much of the research, shifts in female responses masked male intransigence when mean treatment effects were reported.

. Cooperative learning, which has positive effects on facilitating intergroup interaction and interracial friendship choices, appears to have much less of an impact on gender relations and stereotypes. As Sadker, Sadker, and Klein (1991) noted in their review in this volume last year, the limited research suggests that cooperative learning does not lead to a more sex-equitable learning environment or promote cross-sex interaction. Their view is that cultural patterns of male aggressiveness and dominance, coupled with traditional patterns of female inhibition and deference in social settings, may result in traditional patterns of gender interaction emerging in cooperative groups. In this respect, racial and ethnic stereotypes may be more changeable than gender stereotypes using cooperative learning experiences.

Future research into the reduction of gender stereotypes in school settings could benefit from assessing the relationship between changes in sex role attitudes and gender-related behaviors in spontaneous, real-world contexts. There is no evidence in any of the research that changes resulting from educational experiences have a meaningful effect on the decisions and actions of youth in their later development. Longitudinal or cross-sectional data with behavioral outcomes could provide insight as to the value of these changes. Additionally, initial levels of preferred sex role traditionality, or androgyny, need to be taken into account in interpreting the results. There are a number of mediating variables that influence individuals responses to gender-equitable language and portrayals. For example, Morgan (1982) found that IQ and social class affect the extent of acquisition of sex role stereotypes acquired from television viewing; those students with the greatest deficit with regard to gender-equitable attitudes were found to be the most resistant to change. Identification of the factors associated with gender attitude change and the development of more differentiated treatments are logical next steps in this evolving research field.

Global Education

The impetus for global education as a part of the school curriculum is based on the observation that profound changes have occurred in the world's structure over the past two decades (Anderson, 1991). Increased capacity for communication has led to the creation of an international economy and an increased awareness of the interconnectedness of the peoples of the world. Advocates of global education argue that the United States' effective participation in this new world environment requires a different form of education, one that does not emphasize U.S. dominance and control. Many educators have called for citizens to be educated so that they have the skills and attitudes that will enable them to participate as citizens as America moves into this new world environment.

In spite of the general acceptance of the need for global education, the field lacks a clear conception of purpose. In 1983, Kobus noted that "the issue of definition continues to baffle both the proponents of the field and the uninitiated alike surfacing over and over again in surveys of the related literature" (p. 21). Little improvement in this state of affairs was noted in a review of the literature 8 years later. According to Massialas (1991):

> The need for a defensible conceptual framework in education for international understanding has not been met by those who specialize in the field. . . . The student attitudinal or behavioral outcomes that are to issue from exposure to global education or similar programs are rarely specified. (pp. 449–450)

This lack of clarity regarding the purposes of global education may be more of a problem for academics and researchers than for public school personnel. Traditional social studies curricula have always included the study of world history. It is within the world history course that most schools situate the opportunity to achieve the goals of global education. According to Jenness (1990), 80% of high schools offer a course in world history and 53% of high school seniors report having taken such a course. Today, some 40 states have mandated an infusion of global education or global studies into the social studies curriculum. The goals of these programs, as judged by statements from schools and the outcomes assessed in project evaluations, are relatively simple and direct. A typical goal statement is that of Dade County, Florida (cited in O'Neil, 1989), where global education is defined as "the process that provides students with the knowledge, skills, and attitudes which are necessary for them to meet their responsibility as citizens of their community, their state, and their nation in an increasingly interdependent and complex global society" (p. 2). When educational researchers operationalize such statements, two types of outcomes are typically found: knowledge regarding the international environment and attitudes toward other peoples and nations.

Published research on the effectiveness of global education curricula is sparse. This is very puzzling given the broad range of scholarship and curriculum development covered under the rubric of global education. Torney-Purta (1982) noted that, in 1980, the U.S. Department of Education collected data on international projects funded from 1956 to 1977;

only 15 of some 500 in the category of "Teaching and Learning about Other Countries" had any evaluation of learning effectiveness. In 1991, the Association for Supervision and Curriculum Development published the monograph *Global Education from Thought to Action*, which contains nine chapters written by leading authorities in the field (Tye, 1991). Not a single chapter discusses curricular research or cites a study that might illuminate the question "What impact does global education have on youth?"

In one of the early studies of the effectiveness of a global education program, Mitsakos (1978) evaluated the impact of the Family of Man Social Studies Program on third-grade children's views of foreign people. The curriculum emphasized cultural universals, diversity, and interdependence among people with a focus on the value of human dignity. The treatment group consisted of 21 classes that had studied the Family of Man program for 3 years. It was found that the Family of Man students viewed foreign people and nations more favorably than comparison students who had taken no formal social studies course work but not more favorably than students who had taken other social studies course work.

In the mid-1970s, the Social Studies Development Center at Indiana University developed a global education curriculum as a part of the Global Studies Project. This curriculum consisted of six units designed for use with junior high school youth. Each unit was to take approximately 5 weeks of instructional time. The goals of the units were to increase student understanding of global issues, to reduce ethnocentricism, and to promote empathy and understanding for people of other nations. Activities involved the study of information, simulations, role playing, and the discussion of moral dilemmas focusing on global issues. Three doctoral dissertations conducted at Indiana University involved evaluations of the Global Studies Project (Armstrong, 1979; Smith, 1977; Soley, 1982). In all three studies, junior high school teachers were recruited from presentations at social studies conferences. These teachers volunteered to teach at least two units in their classrooms and to recruit a control class from the same school. No training was provided to the volunteer teachers. The teaching of two units usually took one semester. In the Armstrong (30 treatment classrooms) and Soley (38 treatment classrooms) studies, no effect was detected for the curriculum on student ethnocentrism. Smith (using 15 classrooms) found no effect on student attitudes toward other nations. It was found that certain cognitive outcomes were achieved as a result of the curriculum; that is, in the Armstrong and Smith studies, students in the Global Studies classes developed a greater understanding that people of other cultures have a preference for their own homeland.

Three doctoral dissertations were identified that assessed the impact of global education programs specially developed for the researcher's dissertation. McAlvin (1989) assessed the comparative influence of two approaches to global education on sixth- and ninth-grade students: the World Bank's The Developing World and an area studies class. The author found a knowledge gain for classes that studied issues-oriented content, but neither approach resulted in any change in interest in, concern for, or identification with peoples of the world. Schloss (1989) evaluated the relative effectiveness of three different approaches to teaching global studies (didactic, media focused, and inquiry) on cognitive growth (course content) and attitudes toward groups. No true control group was used. No change was detected in cognitive or affective growth as a result of instruction. Finally, Rose (1988) studied the influence of "twinning," bringing elementary school students from diverse cultures together through the exchange of letters, reports, and projects. It was found that treatment classes generally developed more empathetic feelings toward the twinned culture; however, this did not generalize to other cultures. This study must be interpreted cautiously in that only 38% of treatment students completed the evaluation instrument.

In all of the above studies, the treatment lasted at least 6 weeks and in many cases an entire semester. In a 2-day lesson based on the Universal Declaration of Human Rights, Kehoe (1980) attempted to get eighth-grade students to endorse and apply principles of basic human rights. One treatment condition consisted of a teacher-led discussion on cases involving the application of basic human rights. In the other treatment condition, the students reacted in writing to written case studies. In a control group, students watched filmstrips on the United Nations. Both treatment groups were more willing than the control group to reject cultural relativism; however, no difference between the groups was found regarding their willingness to say a contravention of international law was wrong.

Three reports of research identified and evaluated existing global education programs that were being taught as a regular part of school social studies curricula. Hamilton (1982) investigated the effects of three globally focused courses taught at an Atlanta area magnet high school on students' global attitudes. No differences between treatment and control classes were found with regard to understanding of and regard for other peoples or on a measure of willingness for international cooperation attitude measures. Yocum (1989) assessed whether participation in a global education course would increase global mindedness without adversely affecting appreciation of one's own country. One global education class within each of four different Michigan high schools was compared with a sample of students enrolled in a social studies course without an international focus. Global education was not found to affect either global mindedness or the level of patriotism of the students.

Torney-Purta (1985), in a study that drew students from eight states from different regions of the United States, compared students who had participated in global education programs with students who had not. The author, a well-known researcher in the field of global education, drew upon her nationwide contacts to assemble a sample of 16 schools that both taught global education and could provide a comparison group. The majority of the sample consisted of 11th- and 12th-grade students of above-average academic performance. Two instruments were used: a knowledge measure that assessed information about global issues (Global Awareness) and a measure of concern for world problems and interest in peoples of other cultures (Global Concern). Students enrolled in global education programs were in five different types of programs: an extra-curricular program that featured such activities as clubs and a model United Nations, programs developed as a result of federally funded curriculum efforts, traditional global studies approaches within world history courses, programs developed as a result of statewide curriculum efforts, and programs developed as a result of regional school district efforts. Approximately the same number of students in each school served as controls. Only in the extracurricular group and the area-wide program group were higher levels of global awareness detected. Only the students in the extracurricular global education group demonstrated higher global concern when viewed against the comparison groups. In the other four types of programs, no differences were noted on this variable.

Educational movements have never required a solid research base to achieve popularity or space within the school curriculum. The global education movement of the 1980s stands out as one of the more recent manifestations of the gap that often exists between practice and research. Three general types of outcomes were assessed in the 10 studies identified: knowledge of global issues, attitudes toward peoples of other nations, and ethnocentrism. In the only study to detect an impact of global education curricula on knowledge of global issues, the program featured extracurricular experiences. Eight studies assessed the impact of global education curricula on student attitudes toward people of other nations: Only in the extracurricular program was a significant influence detected. In the two studies where ethnocentrism served as an outcome measure, researchers found that global education curricula had no impact.

Environmental Education

It is almost impossible to pick up a newspaper or watch the evening news without reading about or seeing a news report that deals with the environment. Such issues as global warming, industrial and nuclear waste disposal, deforestation, oil spills, endangered species, acid rain, and holes in the ozone layer, to name a few, have been constantly in the forefront of the American consciousness throughout the 1980s and into the 1990s.

The educational community has also been deeply concerned about the environment. Within the field of science education, a subfield of environmental education has grown to the point that it has its own professional association, the National Association of Environmental Education, and its own journal, the *Journal of Environmental Education*.

The 1980s saw great activity in the field of environmental education. A search of the *Current Index of Journals in Education* yielded 1,275 citations on environmental education in the 11-year period between 1980 and 1990. This rate of publication, averaging over 100 articles per year, indicates a high degree of interest among professional educators. Only 5 of the 1,275 journal articles identified in the search involved controlled evaluations of treatment effects. I was unable to identify any data regarding the extent to which environmental education has become a part of the regular school curriculum in America.

Three studies were identified that report the effectiveness of the issue investigation approach of Hungerford, Litherland, Peyton, Ramsey, and Volk (1988). In this approach, students investigate an environmental issue (e.g., stream pollution in the local community), research the possible causes and solutions, develop a realistic action plan for improving the situation, and take some action intended to ameliorate the environmental problem.

Volk and Hungerford (1981), using the issue investigation approach to environmental education at the junior high school level, sought to measure the effect of instruction on the problem-solving skills of eighth-grade students. A posttest-only design was used. Treatment students received instruction in definition, classification, evaluation, and uses of sources of information as they related to finding solutions to environmental problems. The treatment was a separate unit lasting 6 weeks of class time; student written responses, scored by a panel of judges, provided data on students' issue investigation skills. It was found that the treatment students were better able to name issues, state opposing sides to issues, and provide rationales for a side to an issue than were control students.

Two studies conducted by John Ramsey evaluated the effects of the issue investigation approach to environmental education on the environmental behaviors of junior high school students. In the first study (Ramsey, Hungerford, & Tomera, 1981), eighth-grade students received one of three treatments: instruction on how to take environmental action, environmental awareness instruction in a case study format, and content-oriented science instruction (the control group). A pretest-posttest design was used. Treatment duration was not specified. It was found that the action group demonstrated more self-reported remediative environmental behaviors than either of the other two groups. The treatment subjects also demonstrated more knowledge of action strategies than the other two groups.

The second Ramsey study involved 8 seventh-grade classes that received a one-semester course consisting of issue investigation skill development and 4 seventh-grade classes that served as controls (Ramsey & Hungerford, 1988). It was found that the treatment classes achieved higher mean scores than the controls on the following variables: overt environmental behavior, individual locus of control, group locus of control, knowledge of environmental action skills, perceived knowledge of environmental action skills, and perceived skill in the use of environmental action skills. No differences between groups were found with regard to environmental sensitivity.

Two studies were identified that evaluated the effect of traditional didactic methods of science instruction on environmental orientations of youth. Stapp, Cox, Zeph, and Zimbelman (1978) evaluated the effects of a 6-week unit on the impact of local transportation on the global environment on the environmental knowledge, attitudes, and problem-solving skills of gifted sixth- and seventh-grade students. A knowledge increase was found for the treatment classes, but not for the controls. Treatment students also expressed more pro-environmental responses, were better able to state the nature of an environmental issue, knew where to obtain information, and suggested alternative courses of action. The effect of a 15-day environmental issues unit on the environmental attitudes of fifth graders was reported by Jaus (1982). The unit consisted of the traditional study of selected topics related to the environment (e.g., water pollution, air pollution, balance of nature). It was found that the group of fifth graders that received instruction possessed more positive attitudes toward the environment than the control students.

Given the interest in environmental education found in the literature, the low number of experimental studies was surprising. There exists considerable research that can be classified as environmental education research, but it is largely nonexperimental in nature. In a recent meta-analysis of research in the field, Hines, Hungerford, and Tomera (1987) reviewed 128 studies that examined only variables associated with environmental behavior. Only 3 of the studies reviewed involved classroom instruction in environmental education as an independent variable.

The studies cited above suggest that environmental education that features investigation into environmental issues may be effective in developing issue investigation skills, positive environmental attitudes, and participation in environmentally responsible behaviors. Some caution is warranted, however, with regard to these findings. Two studies reviewed above that focused on knowledge-level outcomes also reported achieving more positive environmental attitudes. Also, the research on the issue investigation approach is primarily based on single-item variables and self-report data, both potential sources of low reliability and validity. Finally, all five studies reviewed used classes only at the junior high school level.

Peace Education

Peace education, sometimes referred to as nuclear war education or peace studies, is a movement in contemporary education that has no agreed-upon conceptualization regarding its goals, is surrounded by contentious controversy, and has almost no research regarding its impact on students. Yet, in spite of these apparent liabilities, it is by all accounts an educational movement that has possessed considerable vibrancy and has appealed to many educational leaders, professional organizations, teachers, and school districts.

Since the advent of nuclear weaponry, educators have been concerned about the possibility of nuclear holocaust. For example, between 1945 and 1947 there were 120 articles on peace education published in education journals. Interest in peace education waned in the 1960s and 1970s as the nation was involved with more immediate and pressing social issues; however, as attention turned to the consequences of living in a nuclear age on mental health, interest revived. In addition, because of increased international instability and increasing public concern, the 1980s became a period of great interest in nuclear and peace issues. In the past decade, such highly regarded journals as the *Harvard Education Review, Phi Delta Kappan*, the *Teachers College Record, Educational Leadership*, the *International Journal of Mental Health, Social Studies, Social Education*, the *American Journal of Orthopsychiatry, Physics Today*, the *Bulletin of the Atomic Scientists*, and many others have published special issues about nuclear war education. Also in the 1980s, the National Congress of Parent-Teacher Associations and the National Education Association passed resolutions and encouraged actions in favor of nuclear education in American schools.

No national surveys were identified that indicate how widely peace education may be a part of the curricula in American schools today. On the basis of a survey of 300 of the largest school districts in the United States, London (1987) estimates that only 12% to 15% of schools have a formal curriculum, at either the elementary or secondary level, on nuclear war education; however, all districts responding to London's survey stated that at least one unit was devoted to the study of nuclear weapons, usually in a high school social studies course. Christie (1990) maintains that there are more than 35 commercially produced peace curricula available today. Ryerson (1986) contends that 50 teaching guides for peace education now exist. Reardon (1988a), in a survey conducted in 1984–1985 of K–12 curricula, identified 130 examples of peace curricula. La

Farge (1988) is convinced that all of these numbers are low and that, today, many more curricula and teaching guides exist for peace education.

Betty Reardon, one of the leading proponents for peace education, has recently noted that "there are as yet no clear and precise limits to, nor standards for, what is to be included in peace education" (Reardon, 1988b, p. xix). In her analysis of topics contained in over 100 curriculum guides sent to her by teachers, she and her research team identified nine topical areas that constitute the foci of contemporary peace education curricula (Reardon, 1988b):

- Peace (concepts, models, processes)
- Conflict, conflict management, conflict resolution, war, weapons
- Cooperation and interdependence
- Nonviolence (concepts, practices, cases)
- Global community, multicultural understanding, comparative systems
- World order, global institutions, peacekeeping (methods, models, cases), alternative security systems
- Human rights, social justice, economic justice, political freedom
- Social responsibility, citizenship, stewardship, social and political movements
- Ecological balance, global environment, world resources

There are two major types of student objectives emphasized in the peace education literature. The first is global citizenship. Reardon's (1988a) understanding of the goals of peace education is as follows:

Stated most succinctly, the general purpose of peace education, as I understand it, is to promote the development of authentic planetary consciousness that will enable us to function as global citizens and to transform the present human condition by changing social structures and the patterns of thought that have created it. (p. x)

This objective emphasizes the development of the attitudes and dispositions necessary for students if they are to work as citizens for peace and to prevent nuclear war.

In addition to the emphasis on citizenship, there exists a focus in the movement that is therapeutic in nature and concerned about the mental well-being of young people as it relates to their awareness regarding the possibility of nuclear war (Chivian, Robinson, Tudge, Popov, & Andreyenkov, 1988). Typical of the goals of the therapeutic approach to nuclear education are those found in Beardslee and Mack's (1982) description of the implications of their work with children and adolescents: "We need to educate our children to the realities of nuclear . . . weaponry so that they can be helped to overcome at least that aspect of fear which derives from ignorance and which leaves them feeling so powerless'' (p. 91).

In spite of what appears to be a curricular area of considerable educational interest, only three studies were identified that examined the impact of peace education on youth. None of these studies have been published as research reports in refereed journals. The obvious reason is that the research in this field is of such poor quality that it undoubtedly could not survive a critical review process. In spite of this poor quality, because of the interest in this area and the sometimes exaggerated claims made by proponents of the field, the three studies are reported below.

Christie and Hanley (1988) report the results of an evaluation of the curriculum Choices: A Unit on Conflict and Nuclear War, developed by the Union of Concerned Scientists and the National Education Association. The purpose of the Choices curriculum, as described in the introduction to the teachers guide, is ''to help students understand the power of nuclear weapons, the consequences of their use, and most importantly, the options available to resolve conflict among nations by means other than nuclear war'' (p. 7). The 2-week unit was taught in April and May of 1986 in 67 classrooms involving 1,518 students at Grades 6, 7, and 8 by volunteer teachers who had been trained by the authors. Eight classrooms where Choices was not presented were used as controls. A reduction on a measure of nuclear fear was noted in 42 of the 67 treatment classrooms. In 30 classrooms, a decrease was detected on student sense of futurelessness. Treatment students' sense of powerlessness decreased in 33 classrooms. Among the control classrooms, 6 of 8 decreased on nuclear fear, 4 of 8 decreased on futurelessness, and 3 of 8 decreased on powerlessness. Using change scores, an effect for condition was detected on the nuclear fear measure but not the other factors. On 7 of the 17 items contained in a political opinion questionnaire, changes were noted in the treatment sample. Students were more disposed to believe that nuclear war can be prevented, worried less about the possibility of nuclear war, talked more to others about nuclear issues, attributed more responsibility for nuclear weapons to the United States than to the Soviets, and developed more positive attitudes toward the Soviets.

Berman (1986) evaluated the impact of a 3-week unit on nuclear weapons and the arms race on 10th- and 11th-grade students in the Pittsburgh Public Schools. From the 4,000 students in these grades, 341 pretests and 644 posttests were returned to the author. The 130 students who completed both the pretest and the posttest constituted the sample for this study. There was no control group. The unit consisted of five major lessons covering the following topics: the history of nuclear weapons, the arms race, current status of arms proliferation, the effects of nuclear war, and what can be done to prevent nuclear war. The unit was developed so as to be informational only and not to influence student opinion in any particular direction. The survey (instrument) grouped student responses into five subscales: concern about the threat of nuclear war, sense of powerlessness or empowerment with regard to nuclear war, conservative or liberal attitudes toward the arms race, image of the Soviet Union, and cynicism or optimism toward the future. A pretest to posttest change was found on only one of the 31 items on the questionnaire: "I can do something to prevent nuclear war."

Zolik and Nair (1987) report the results of an evaluation of Decision Making in a Nuclear Age, a curriculum developed under the auspices of the Educators for Social Responsibility organization. The curriculum has a left-of-center bias and emphasizes social justice as prerequisite for peace and disarmament as preferable to deterrence as a national policy. The sample consisted of four intact ninth-grade social studies classes in the Brookline School District comprising 84 students. A nonequivalent pretest-posttest control group design was used. The length of the course was not specified. On knowledge of nuclear issues measures, the treatment subjects showed an increase. However, no changes were found in the treatment classes on measures of nuclear anxiety or with regard to attitudes toward disarmament. Zolik and Nair did find that the treatment subjects experienced increased personal efficacy in relation to the prevention of nuclear war.

The findings from the three studies must be interpreted cautiously. With regard to fear of the possibility of war, both Berman and Zolik and Nair found no change. Christie and Hanley found a decrease in nuclear anxiety among peace education classes, but also reported a decrease of fear in an equal percentage of the control classes. All three studies reported a decrease in powerlessness or an increase in sense of political efficacy. However, in Berman's questionnaire the decrease in powerlessness was the only change found in 31 item comparisons in the questionnaire. The inappropriate statistical procedures, coupled with the remarkably high mortality in the study and the failure to include a control group, make this finding not credible. Christie and Hanley did find that 41% of the treatment classes decreased in feelings of powerlessness; however, 38% of the control classes also decreased in powerlessness. With regard to optimism regarding the future, neither Christie and Hanley nor Berman found a curriculum effect. In all three studies, no curriculum impact on attitudes toward the arms race (on a liberal/conservative scale) or attitudes toward the Soviets was noted. Finally, the Zolik and Nair study found a significant impact on student knowledge of nuclear war issues.

Overall, the quality of peace education research is quite poor. Among the problems are a failure to randomly assign subjects to treatments, reliance on volunteer teachers, a lack of training for teachers, instruments of low reliability, a failure to report simple means and standard deviations for all variables, and inappropriate statistical tests. Generally, researchers tended to report the results very informally and without complete descriptive and inferential statistics. The overall judgment, based on fragmentary and weak research, must be that peace education fails to achieve its desired outcomes to any significant extent. The only comfort that can be derived by peace educators is that no iatrogenic effect was noted, as some critics have charged; that is, students apparently did not become more fearful and anxious about nuclear war. In spite of the rather bleak research outlook, the existing data have repeatedly been used to argue for the efficacy of peace education. For example, in a recent review of peace education curricula, Valett (1991) cites the Christie and Hanley (1988) and Zolik and Nair (1987) studies as evidence of the effectiveness of peace education in increasing students' sense of efficacy and decreasing their nuclear anxiety.

CONCLUSION

One means of assessing the effectiveness of contemporary issues curricula is the ratio of achieved outcomes to desired outcomes. This ratio was computed for knowledge, attitudinal, and behavioral outcomes across all the curricular areas reviewed. It was found that, with regard to knowledge goals, the outcomes were achieved in 66.6% of the cases (20 of 30). If the 3 of 11 success rate of global education is removed, 89% of the studies remaining reported achieving knowledge-level outcomes. In 32.6% of the cases (27 of 82), desired attitudinal outcomes were achieved; in 27.5% (11 of 40), desired behavioral outcomes were achieved. If the findings of cooperative learning strategies are removed from the data on behavioral outcomes, the desired outcomes were achieved in only 10% of the cases (3 of 30). The changing of student attitudes and behavior associated with the goals of contemporary issues curricula appears to be a much more formidable task for school curricula than the teaching of knowledge regarding those same issues. Given that no clear relationship between increased knowledge and changes in attitudes and behavior was detected, the overall educational and social significance of the knowledge gains achieved must be questioned.

Below, the characteristics of studies associated with the achievement of desired outcomes will be summarized, some possible reasons for the relative inability of the curricula to change attitudes and behavior will be discussed, and potential obstacles and opportunities for future research will be presented. The discussion regarding the characteristics of research studies associated with curricular effectiveness will be organized around three characteristics common to all educational settings: the learner, the nature of the curriculum, and the teacher.

Age of the subjects was found to be associated with the achievement of the behavioral outcomes of sex education; junior high school programs find a greater effect on sex-related attitudes and behaviors than high school programs. Age (elementary school) was also found to be a significant factor related to the achievement of desired attitudinal outcomes in the areas of gender and racial equity education; however, the relatively few studies in these areas at the secondary level make this association speculative. The gender of subjects was a salient characteristic with regard to attitude change in studies that evaluated sex and racial equity curricula. Female subjects were found to be significantly more egalitarian than male subjects and to generally become even more so as a result of the experience of the curriculum. The existing attitudes of subjects may need to be given more attention by researchers. It is possible that mean treatment effects may be the result of a shift only among some of the subjects, from a somewhat positive to a very positive position, whereas the attitudes of other students remain unchanged. It is unlikely that attitudinal uniformity exists among students around any of the topics contained within contemporary issues curricula. Whether a class mean changes needs to be supplemented with data on who changes, how much, and under what conditions. Only in the sex equity research, where gender is frequently included as a factor in data analysis, does the research provide insight into these questions.

Another potentially important characteristic of the learner is the interest, or motivation, for the subject area that the individual brings to the classroom. One likely reason for the low level of impact of contemporary issues curricula on student attitudes and behaviors may be that the topic is viewed by students as personally irrelevant. The potential for any curriculum to have an impact on students is increased by the degree to which the subject is one about which students have a genuine and immediate concern. The majority of the curricular areas reviewed in this chapter are topics about which youth apparently do not share the same level of concern as adults. Robert Coles (1989) recently noted, in analyzing the results of 5,012 field interviews conducted by Louis Harris & Associates (Girl Scouts Survey on the Beliefs and Moral Values of America's Children): "One finding emerges clear and strong: The issues most cited in newspaper headlines and most aggressively voiced by experts and activists as the source of youth crises are simply not an immediate concern to most children" (p. 34). According to the survey, the five most important concerns to youth today are pressure to do well in school or sports (24%), what to do with your life (17%), drug usage (13%), lack of close friends (8%), and teenage pregnancy (6%). Only 2 of the 11 contemporary issues identified in this review appear as areas of major concern to youth.

Two additional characteristics of students not identified as variables in the research reviewed may be significant with regard to understanding the impact of these curricula: intellectual ability and social class. One could reasonably expect that differences in intellectual ability might be related to knowledge gains and that social class, through the presence of class-related attitudes as a mediating variable, might be related to attitude changes (Morgan, 1982).

The characteristics of the curriculum associated with program effectiveness are relatively clear. The didactic presentation of information was generally effective with regard to the acquisition of knowledge among students. However, it enjoyed no special advantage over the other approaches used; they also were found to be effective means of transmitting knowledge. Generally, curricula that emphasized the presentation of information were not found to have any significant impact on attitudes or behavior. Exceptions to this generalization were two environmental education studies and the Life Unworthy of Life Holocaust curriculum. The latter curriculum, like many reviewed, used a mix of instructional strategies that make it difficult to attribute efficacy to any single instructional strategy. A clear advantage was found for cooperative learning strategies with regard to racial and ethnic intergroup attitudes and interactions; however, no such positive effect was noted when the goal was the promotion of sex equity attitudes and behaviors (Reppucci & Herman, 1991).

Curricula that involved peer interaction, most often through group discussion and activities where students were actively involved in the collective exploration of attitudes and values in an open and democratic atmosphere, were found to be consistently effective in producing attitudinal change. The use of instructional strategies that permit the exploration of meaning and facilitate the evolution of moral norms in groups has most recently been extended through community-based programs, especially with regard to drug and sex education programs. This extension to involve a broad range of significant others in the clarification and reinforcement of norms against careless sexual activity and substance abuse, based on early reports, appears to have great potential. Finally, the evidence regarding the duration and intensity of curricular implementation was not found to be consistently associated with positive results in any of the areas reviewed.

The research on the effectiveness of contemporary issues curricula offers little insight into teacher characteristics that are associated with attitudinal and behavioral change in students. Given the crucial role that teachers play in the delivery of instruction in American schools, this lacuna in the research is puzzling. There are clear attributes that teachers must have in order to effectively present information to students and to implement approaches that are not based on didactic presentation of information. For example, characteristics such as open-mindedness, flexibility, curiosity, and reflectiveness are required if a teacher is to effectively lead open discussions in the classroom. The instructional skills necessary to lead open and productive discussions are among the most difficult to learn and complex to implement in any teacher's repertoire. Classrooms where discussions of contemporary issues occur on a regular basis are rare. Recent reviews of the history of the teaching of all subject matters, and social studies in particular, have found that the dominant instructional pattern of lecture-recitation has remained relatively unchanged over the past 50 years (Cuban, 1984, 1991; Jenness, 1990). Previous attempts to disseminate curricula that involve teachers and students in the exploration of social issues have been unsuccessful (Shaver, 1989). If most teachers' views of the subject matter, their educational philosophy, and the context in which they practice their craft are inimical to open classroom discussion, this promising finding of the efficacy of this curricular approach will at best have only a limited impact on the teaching of contemporary issues because few teachers have the motivation and skills to use the methods in the classroom.

It is also unlikely that teachers exercise sufficient moral authority to directly influence student values through persuasion or moral authority. Not only do most students not share the same concerns as teachers with regard to the importance of contemporary issues, but most adolescents do not view teachers as significant others in their lives. When high school juniors and seniors were recently asked to list whom they turn to when making important decisions (Schultz, 1989), they listed, in order, friends (55%), parents (47%), relatives (10%), and teachers/advisors at school (5%). This is not to suggest that teachers cannot be significant moral authorities who, through their salience and moral persuasion, have a significant impact on students. One of the earlier and classic evaluations of character education curriculum (Hartshorne & May, 1928), while finding that schools were generally ineffective in changing students' character, did detect among students in a very limited number of classrooms a significant and lasting impact on character. The authors attributed this finding to the effectiveness of the individual teacher as moral authority.

The majority of the curricula reviewed in this chapter contain goals with a clear normative focus. That is, they have attempted to transmit to youth values that embody preferred states of being for the individual and the society in which he or she lives. At the most general level, these norms can be understood as moral in nature in that they incorporate the values of the existing society and, through their transmission, attempt to ensure the well-being of youth and the stability of the existing culture. Two seminal contributions to the understanding of how culture transmits values to youth are those of Durkheim (1925/1973) and Piaget (1932/1965). The dynamics identified by these scholars and an attempted synthesis of their work by Kohlberg may be useful in interpreting the above pattern of research findings associated with curricular impact on attitudes and behavior.

The debate over the nature of moral development in this century has its historical roots in the seminal work of the French sociologist Emile Durkheim. Durkheim (1925/1973) understood moral development as a process of cultural "impression" upon the child accomplished by an affective transmission of society's values to the child. The salient features of this transmission were its emotional nature through the use of authority and discipline in the early years, and later in childhood through the dynamic of the child's natural attachment to groups. This perspective was sociological in nature, but the dynamics and outcomes are highly consistent with the social learning analysis that was to follow later in the century. Piaget (1932/1965) rejected the view of Durkheim and argued instead that moral development was best understood as a cognitive understanding constructed by the individual as a result of experiencing conflicts of social interaction natural to growing up in any society. Piaget's initial perspective was based on his observations of children's games that took place removed from supervision by authorities and where they had to solve real social conflicts. Piaget placed the child and his or her interactions with the social environment as the dynamic that explained moral development. The child was not a passive recipient of the socialization process, but instead was the creator of moral meaning.

Kohlberg, like Piaget, in his earlier versions of his theory of moral education, rejected the perspective that moralization was a matter of individual accommodation to society. Instead, he viewed moral development as the development of reasoning around a sense of justice that enabled the individual to isolate the legitimate moral claims of individuals in a situation and to balance these perspectives in a way that takes into account the perspectives of all the individuals in the situation. He later came to the realization that there was only a weak association between the development of cognition and behavior in moral situations. This realization, coupled with the very real and immediate demands placed on schools with regard to responsibility for student values and behavior, led to the revision of his theory to incorporate the process of the evolution of group norms within the just community framework (Power, Higgins, & Kohlberg, 1989). In the just community environment, students and teachers, as representatives of society's values, discuss, develop, and enforce compliance with the group norms necessary for the functioning of community. It is through the environment of the just community, with integration of the moral authority of the adult, the attachment to the peer group, and the cognitive development resulting from the open discussion of societal norms, that a synthesis of the positions of Durkheim and Piaget was accomplished. The contemporary issues curricula that were found to have the greatest impact on attitudes and behavior incorporated the dynamics identified above through open discussion and exploration of questions of morals and values within the peer group. Additionally, when the moral authority and concern of adults, parents, and community members were included as a part of the curricula, positive effects were noted.

If, as suggested at the beginning of this paper, the decline of the family is a significant obstacle to the transmission of values to youth, and added to this the moral pluralism and relativism that is perceived by many youth to pervade contemporary society, then curricular reform that involves the reestablishment of moral community with clear standards within an environment of shared concern and mutual respect is on the right track. In cases where schools have been able to enlist the cooperation of family and community in the teaching of norms, most notably drug and sex education, those approaches have shown promise with regard to attitude and behavior change.

If schools are to become more effective agents in the future with regard to the normative socialization of youth, certain obstacles to productive research regarding contemporary issues curricula must be overcome. There exists wide variation among the contemporary issues curricular areas with respect to extent and quality of research. Two of the curricular areas, sex and drug education, have well-developed and evolving research programs with a 20-year history. In each of the other areas reviewed, 10 or fewer studies were identified. One important reason for the differences in quality and breadth of research programs may be related to the importance attached by society to the problem area. Sexual promiscuity and the use of drugs by youth command public attention and the attention of youth much more than the other topics discussed in this review. These areas clearly have the most developed research programs of any of the areas reviewed.

How, then, are we to explain the presence of some apparently energetic curriculum movements in the absence of widespread support? A possible explanation is that the attention these areas have received may be more the result of committed and aggressive individuals who, through their own writing and professional activity, have kept attention focused on the issue and been able to convince some teachers and schools that the school curriculum should include the topic. Such a dynamic results in a high level of interest within the educational profession independent of a genuine and deep public concern regarding the problem. A number of times in the history of American education, a single individual and a small cadre of associates or a professional organization have created an educational movement that attracted national attention. A case in point is the mental hygiene movement, where the National Committee for Mental Hygiene was able to effectively introduce personality development as an educational goal into schools without significant concern for their agenda among the general public or the educational profession (Cohen, 1983). Typically, leaders of such popular educational movements lack the interest, time, or expertise to engage in careful and systematic curricular research.

A second and related reason for the lack of developed research programs in some curricular areas is that program development, implementation, and evaluation is expensive. To the extent that the public and the political establishment support a particular educational program, two essential conditions for program implementation and evaluation are met. The expenses associated with the careful implementation, supervision, and evaluation of a significant curricular program and a large school district may be too expensive for local budgets to assume. In the case of drug and sex education, sustained political interest at all levels, concern among the local community, and the availability of local, state, national, and foundation monies have combined to produce a climate where research is seen as desirable, needed, and, because of the availability of funds, possible.

All of the contemporary issues curricular areas reviewed in this chapter have, at one time or another, been embroiled in public controversy. A third potential explanation for the low level of research in some curricular areas may again be due to the level of political support for curriculum development and evaluation in these areas. It is impossible to teach about any of the areas reviewed without running the risk of in some way bringing into public debate differing ideological positions. Curricular and research access to the schools has always been a sensitive political issue for superintendents and school boards. The difficulties many of the researchers reported in the above studies have in achieving access to samples in schools indicate that careful research into the effects of some contemporary issues curricula will continue to be a difficult task unless school district administrations perceive that it is politically in their interest to permit such access. Administrative timidity that forced researchers to modify their research design contributed to many of the weaknesses in the research reviewed in this chapter.

A fourth obstacle to the development of effective research programs in these areas is the elusive and inveterate nature of their objectives. To attempt to change fundamental characteristics of human beings that have been shaped in the crucible of the early family and community environment is a daunting task. Shedler and Block (1990) report the results of a longitudinal study in which they found that adolescents who used drugs most frequently were maladjusted and characterized by interpersonal alienation, poor impulse control, and manifest emotional distress. They concluded that the problem of drug usage is a symptom, not a cause, of personal and social adjustment, and they argue that the meaning of drug use can only be understood in the context of individual personality and developmental history. They suggest that current efforts at drug prevention are misguided in that the roots of drug usage do not lie in the peer group, school, or community, but can be traced to the earliest years of childhood and are related to the quality of parenting. This perspective is given some support in the finding that in spite of extensive and well-intentioned efforts, levels of drug use by youth have showed no substantial reductions in the past two decades (Johnston et al., 1987).

A fifth possible explanation as to why some curricular areas fail to develop research programs may be a result of the *Weltanschauung* of the individuals involved in the movement. One notices, in some of the writings of individuals in peace, environmental, and global education, a sense of alienation from a technological and scientific world view. Science and technology are viewed as sources of war, international friction, and environmental degradation. The values endorsed within these movements are presented as humanistic and affective rather than scientific. These individuals are more prone to eschew experimental evaluations and value instead impressionistic and anecdotal reports.

Finally, the methodological requirements for research in many of the contemporary issues curricular areas are among the most difficult faced in any area of behavioral research. Quality research in these curricular areas will require longitudinal designs and accurate assessment of behaviors that involve privacy issues. Research that includes these characteristics is only occasionally found in the educational literature.

The underlying question this review has examined is the assumption that the schools, through explicit curricular interventions, can contribute to the amelioration of individual and social problems facing this society. Schooling from this perspective is, in a very fundamental sense, a moral enterprise. In order for this goal to be achieved, the positive values of the culture must be passed on to youth, assuming that those values are consistent with the principle of justice and respect the fundamental dignity of all citizens. Just as educators have a responsibility to formulate and implement curricula consistent with these goals, they also have a moral responsibility to monitor the effects of those curricula on youth. Times are tough. There are far more demands on the schools than can possibly be met, and the resources for carrying on the functions of schools are limited.

In the past decade, Americans have become increasingly results oriented regarding the achievements of their schools. Americans have traditionally viewed themselves as preeminent in the world—economically, militarily, technologically, and culturally. For long periods of our history, acceptance of this sanguine and perhaps arrogant perspective has been warranted. In the last few decades, challenges have arisen that have questioned this optimism, most notably the loss of the technological edge that permitted the United States to maintain economic supremacy in world markets. The blame, justly or not, has been directed at the schools for this weaker America. Studies of comparative academic achievement between American schools and those of competing nations have become a major area of public and political interest. In 1969, the Congress of the United States established the National Assessment of Educational Progress to assess and report back to the American people on how our schools were doing. More and more states are turning to statewide assessment to ensure that schools are accountable. Increased state and federally mandated programs, more frequent and extensive assessment of student performance, and increasingly tough high school graduation and college entrance requirements have exerted tremendous pressures upon the school curriculum. The limited resources of time, energy, and money make it impossible for the schools to respond to all demands for space within the curriculum. For all practical purposes, the school curriculum has become a zero-sum game for administrators and teachers in most schools; there is simply no room for additional subjects without removing some other part of the existing curriculum. Part of the responsibility of educators is to do all they can to ensure that the decisions they make regarding the expenditure of educational time, energy, and money, all finite and scarce commodities, are ones that maximize schools' effectiveness. Hard choices have to be made. One such hard choice is the inclusion of contemporary issues curricula in the schools.

It is unlikely that there will ever be a time when individuals and groups, concerned for the well-being of youth and society, will not attempt to project their specific concerns into the schools. As Kliebard (1987) has recently documented, the curriculum in this country has taken shape largely as a result of the struggle between interest groups. If past history is any indication, some groups will be more successful than others in gaining access to the school curriculum, but evidence regarding the impact of these curricula will, in all likelihood, continue to be a minor concern when compared with the political struggles involved. One important function of educational research should be to assist the educational establishment and policymakers in determining the extent to which the best interests of children are being served by schools and in providing insight into how best to carry out the moral imperative placed upon educators by society. Some small insight into these important questions is provided by the research reviewed in this chapter. Let's hope that in the future the obstacles to careful research cited above will be overcome and that in the future greater clarity will emerge on these important issues.

REFERENCES

Adamek, R. J., & Thoms, A. I. (1991). Responsible sexual values program: The first year. *Family Perspective, 25*, 67–81.

Allison, P. D., & Furstenberg, F. F. (1989). How marital dissolution affects children: Variations by age and sex. *Developmental Psychology, 25*, 540–549.

Anderson, L. F. (1991). A rationale for global education. In K. A. Tye (Ed.), *Global education: From thought to action* (pp. 13–34). Alexandria, VA: Association for Supervision and Curriculum Development.

Armstrong, P. M. (1979). The effect of global studies instructional materials on dimensions of the global attitudes of middle school students (Doctoral dissertation, Indiana University, 1979). *Dissertation Abstracts International, 40*, 3731A.

Bailis, L. A., & Kennedy, W. R. (1977). Effects of a death education program upon secondary school students. *Journal of Educational Research, 71*, 63–66.

Bangert-Drowns, R. L. (1988). The effects of school-based substance abuse education—A meta-analysis. *Journal of Drug Education, 18*, 243–264.

Banks, J. A. (1991). Multicultural education: Its effects on students' racial and gender role attitudes. In J. P. Shaver (Ed.), *Handbook of research on social studies teaching and learning* (pp. 459–469). New York: Macmillan.

Beardslee, W. R., & Mack, J. E. (1982). *The impact on children and adolescents of nuclear developments* (Task Force Report No. 20). Washington, DC: American Psychiatric Association.

Berberian, R., Gross, C., Lovejoy, J., & Papanella, S. (1976). The effectiveness of drug education programs: A critical review. *Health Education Monographs, 4*, 377–398.

Berman, S. (1986). *Measuring change in student attitudes on the nuclear war issue.* Unpublished manuscript.

Brown, L. K., Fritz, G. K., & Barone, V. J. (1989). The impact of AIDS education on junior and senior high school students. *Journal of Adolescent Health Care, 10*, 386–392.

Brown v. Board of Education of Topeka, 347 U.S. 483 (1954).

Chivian, E., Robinson, J. P., Tudge, J. R. H., Popov, N. P., & Andreyenkov, V. G. (1988). American and Soviet teenagers' concerns about nuclear war and the future. *New England Journal of Medicine, 319*, 407–413.

Christie, D. J. (1990, August). *The measurement of psychological constructs in peace education.* Paper presented at the annual meeting of the American Psychological Association, Boston.

Christie, D. J., & Hanley, C. P. (1988). *The psychological impact of an educational unit about conflict and nuclear war on adolescents.* Marion: Ohio State University, Marion. (ERIC Document Reproduction Service No. ED 305 274)

Cohen, S. (1983). The mental hygiene movement, the development of personality and the school: The medicalization of American education. *History of Education Quarterly, 23*, 123–149.

Coles, R. (1989). *Girl Scouts survey on the beliefs and moral values of America's children.* New York: Girl Scouts of the United States of America.

Collett, L. J., & Lester, D. (1969). The fear of death and the fear of dying. *The Journal of Psychology, 72*, 179–181.

Cuban, L. (1984). *How teachers taught: Constancy and change in American classrooms 1890–1980.* New York: Longman.

Cuban, L. (1991). History of teaching in social studies. In J. Shaver (Ed.), *Hand-

book of research on social studies teaching and learning (pp. 197–209). New York: Macmillan.

Davis, J. M. (1988). Suicide and the schools, intervention and prevention. In J. Sandoval (Ed.), *Crisis counseling, intervention and prevention in the schools* (pp. 187–203). Hillsdale, NJ: Erlbaum.

Dawidowicz, L. S. (1990). How they teach the Holocaust. *Commentary, 90*(6), 25–32.

Dawson, D. A. (1986). The effects of sex education on adolescent behavior. *Family Planning Perspectives, 18*, 162–170.

DeVires, D. L., Edwards, K. J., & Slavin, R. E. (1978). Biracial learning teams and race relations in the classroom: Four field experiments using Teams-Games-Tournament. *Journal of Educational Psychology, 70*, 356–362.

DiClemente, R. J., Pies, C. A., Stoller, E. J., Straits, C., Oliva, G. E., Haskin, J., & Rutherford, G. W. (1989). Evaluation of school-based AIDS education curricula in San Francisco. *Journal of Sex Research, 26*, 188–198.

Dielman, T. E., Shope, J. T., Leech, S. L., & Butchart, A. T. (1989). Differential effectivenes of an elementary school-based alcohol misuse prevention program. *Journal of School Health, 59*, 255–263.

Durkheim, E. (1973). *Moral education: A study in the theory and application of the sociology of moral education.* New York: Free Press. (Original work published 1925)

Edwards, M. I. (1983). *The effect of a three-week death education course on the death anxieties of high school sophomores.* Unpublished master's thesis, Mankato State University, Mankato, MN.

Elam, S. M. (1990). The 22nd annual Gallup poll of the public's attitudes toward the public schools. *Phi Delta Kappan, 72*, 41–55.

Ellickson, P. L., & Bell, R. M. (1990). Drug prevention in junior high: A multisite longitudinal test. *Science, 24*, 1299–1305.

Flay, B. R., Ryan, K. B., Best, J. A., Brown, K. S., Kersell, M. W., d'Avernas, J. R., & Zanna, M. P. (1985). Are social-psychological smoking programs effective? The Waterloo study. *Journal of Behavioral Medicine, 8*, 37–59.

Furstenberg, F. F., Moore, K. A., & Peterson, J. L. (1985). Sex education and sexual experience among adolescents. *American Journal of Public Health, 75*, 1331–1332.

Garland, A., Whittle, B., & Schaffer, D. (1989). A survey of youth suicide prevention programs. *Journal of the American Academy of Child and Adolescent Psychiatry, 28*, 931–934.

Gerard, H. B., & Miller, N. (1975). *School desegregation: A long-range study.* New York: Plenum Press.

Gimmestad, B. J., & De Chara, E. (1982). Dramatic plays: A vehicle for prejudice reduction in the elementary school. *Journal of Educational Research, 76*, 45–49.

Glass, J. C. (1990). Changing death anxiety through death education in the public schools. *Death Studies, 14*, 31–52.

Glass, J. C., & Knott, E. S. (1984). Effectiveness of a lesson series on death and dying in changing adolescents' death anxiety and attitudes toward older adults. *Death Education, 8*, 299–313.

Hamilton, B. S. (1982). Assessing the effects of planned curricular interventions on high school students' attitudes toward and knowledge of global education (Doctoral dissertation, Georgia State University, 1982). *Dissertation Abstracts International, 43*, 3159A.

Harris & Associates, Inc. (1986). *American teens speak: Sex, myths, TV and birth control.* New York: Planned Parenthood Federation of America.

Hartshorne, H., & May, M. (1928). *Studies in the nature of character: Vol. 1. Studies in deceit.* New York: Macmillan.

Hines, J. M., Hungerford, H. R., & Tomera, A. N. (1987). Analysis and synthesis of research on responsible environmental behavior: A meta-analysis. *Journal of Environmental Education, 18,* 1–8.

Hofferth, S., & Miller, B. (1989). An overview of adolescent pregnancy prevention programs and their evaluations. In J. J. Card (Ed.), *Evaluation programs aimed at preventing teenage pregnancies* (pp. 25–40). Palo Alto, CA: Sociometrics Corporation.

Hungerford, H. R., Lithreland, R. A., Peyton, R. B., Ramsey, J. M., & Volk, T. L. (1988). *Investigating and evaluating environmental issues and action: Skill development modules.* Champaign, IL: Stipes Publishing Company.

Jaus, H. H. (1982). The effect of environmental instruction on children's attitudes toward the environment. *Science Education, 66,* 689–692.

Jenness, D. (1990). *Making sense of the social studies.* New York: Macmillan.

Johnson, D. W., & Johnson, R. T. (1981). Effects of cooperative and individualistic learning experience on interethnic interaction. *Journal of Educational Psychology, 73,* 444–449.

Johnson, D. W., & Johnson, R. T. (1982). Effects of cooperative, competitive, and individualistic learning experiences on cross-ethnic interaction and friendships. *Journal of Social Psychology, 118,* 47–58.

Johnson, D. W., Johnson, R. T., Tiffany, M., & Zaidman, B. (1984). Cross-ethnic relationships: The impact of intergroup cooperation and intergroup competition. *Journal of Educational Research, 78,* 75–79.

Johnson, J., & Ettema, J. S. (1982). *Positive images.* Beverly Hills, CA: Sage.

Johnston, L. D., O'Malley, P. M., & Bachman, J. G. (1987). *Drug use among American high school students, college students, and other young adults* (DHHS ADM 86-145D). Washington, DC: U.S. Government Printing Office.

Keats, B. E. (1990). The effects of an Afro-American history course on the racial attitudes of high school students. *Dissertation Abstracts International, 50,* 3200A.

Kehoe, J. (1980). An examination of alternative approaches to teaching the universal declaration of human rights. *International Journal of Political Education, 3,* 193–203.

Kinder, B. N., Pape, N. E., & Walfish, S. (1980). Drug and alcohol education programs: A critical review of outcome studies. *International Journal of Addiction, 15,* 1035–1054.

Kirby, D. (1980). The effects of school sex education programs: A review of the literature. *Journal of School Health, 50,* 559–563.

Kirby, D. (1984). *Sexuality education: An evaluation of programs and their effects.* Santa Cruz, CA: Network Publications.

Kirby, D. (1985). Sexuality education: A more realistic view of its effects. *Journal of School Health, 55,* 421–424.

Kliebard, H. M. (1987). *The struggle for the American curriculum: 1893–1958.* New York: Routledge.

Kobus, D. K. (1983). The developing field of global education: A review of the literature. *Educational Research Quarterly, 8,* 21–28.

Koeller, S. (1977). The effect of listening to excerpts from children's stories about Mexican-Americans on the attitudes of sixth graders. *Journal of Educational Research, 70,* 329–334.

Koop, C. E. (1986). *Surgeon general's report on AIDS*. Washington, DC: U.S. Public Health Service.

Kourilsky, M., & Campbell, M. (1984). Sex differences in a simulated classroom economy: Children's beliefs about entrepreneurship. *Sex Roles, 10*, 53–66.

La Farge, P. (1988). Nuclear teaching: Propaganda or problem solving? *Bulletin of the Atomic Scientists, 44*(6), 14–20.

Lessing, E. E., & Clarke, C. C. (1976). An attempt to reduce ethnic prejudice and assess its correlates in a junior high school sample. *Educational Research Quarterly, 1*, 3–16.

Leviton, D. (1977). The scope of death education. *Death Education, 1*, 41–56.

Lieberman, M. (1981, Summer). Facing history and ourselves: A project evaluation. *Moral Education Forum*, pp. 36–41.

Lieberman, M. G. (1991). *Facing history and ourselves: Evaluation report 1990*. Wellesley, MA: Responsive Methodology.

London, H. I. (1987). *Armageddon in the classroom: An examination of nuclear education*. Lanham, MD: University Press of America.

Marsiglio, W., & Mott, F. L. (1986). The impact of sex education on sexual activity, contraceptive use and premarital pregnancy among American teenagers. *Family Planning Perspectives, 18*, 151–161.

Massialas, B. G. (1991). Education for international understanding. In J. P. Shaver (Ed.), *Handbook of research on social studies teaching and learning* (pp. 448–469). New York: Macmillan.

McAlvin, D. W. (1989). The effect of alternative contents and instructional strategies on the global knowledge and attitudes of sixth- and ninth-grade students (Doctoral dissertation, Georgia State University, 1989). *Dissertation Abstracts International, 51*, 130A.

Miller, L., & Downer, A. (1988). AIDS: What you and your friends need to know—A lesson plan for adolescents. *Journal of School Health, 58*, 137–141.

Mitsakos, C. L. (1978). A global education program can make a difference. *Theory and Research in Social Education, 6*, 1–15.

Morgan, M. (1982). Television and adolescents' sex role stereotypes. *Journal of Personality and Social Psychology, 43*, 947–955.

Nagourney, P., Bolkosky, S., Ellias, R. R., & Harris, D. (1990). *Life unworthy of life: A holocaust curriculum* (Project submission to the Program Effectiveness Panel, U.S. Department of Education). Unpublished manuscript.

Noland, M., Richardson, G. E., & Bray, R. M. (1980). The systematic development and efficacy of a death education unit for ninth-grade girls. *Death Education, 4*, 43–59.

Oishi, S. S. (1983). Effects of team-assisted individualization in mathematics on cross-race interactions of elementary school children. *Dissertation Abstracts International, 44*, 3622A.

O'Neil, J. (1989, January). Global education: Controversy remains, but support growing. *ASCD Curriculum Update*.

Pentz, M. A., Dwyer, J. H., MacKinnon, D. P., Flay, B. R., Hansen, W. B., Yul, E., Wang, M. S., & Johnson, C. A. (1989). A multicommunity trial for primary prevention of adolescent drug abuse. *Journal of the American Medical Association, 261*, 3259–3266.

Piaget, J. (1965). *The moral judgment of the child*. New York: Free Press. (Original work published 1932)

Power, F. C., Higgins, A., & Kohlberg, L. (1989). *Lawrence Kohlberg's approach to moral education*. New York: Columbia University Press.

Ramsey, J. M., & Hungerford, H. (1988). The effects of issue investigation and action training on environmental behavior in seventh grade students. *Journal of Environmental Education, 20*, 29–34.

Ramsey, J., Hungerford, H. R., & Tomera, A. N. (1981). The effects of environmental action and environmental case study instruction on the overt environmental behavior of eighth-grade students. *Journal of Environmental Education, 13*, 24–30.

Reardon, B. A. (1988a). *Comprehensive peace education: Educating for global responsibility*. New York: Teachers College Press.

Reardon, B. A. (1988b). *Educating for global responsibility: Teacher designed curricula for peace education*. New York: Teachers College Press.

Reppucci, N. D., & Herman, J. (1991). Sexuality education and child sexual abuse prevention programs in schools. In G. Grant (Ed.), *Review of research in education* (Vol. 17, pp. 127–168). Washington, DC: American Educational Research Association.

Rose, S. D. (1988). The effects of participation in the global education project on children's attitudes toward foreign people (Doctoral dissertation, University of Connecticut, 1988). *Dissertation Abstracts International, 50*, 75A.

Rosenthal, N. R. (1980). Adolescent death anxiety: The effect of death education. *Education, 101*, 95–101.

Rundall, T. G., & Bruvold, W. H. (1988). A meta-analysis of school-based smoking and alcohol use prevention programs. *Health Education Quarterly, 15*, 317–334.

Ryerson, A. (1986). The scandal of "Peace Education." *Commentary, 81*, 37–46.

Sadker, M., Sadker, D., & Klein, S. (1991). The issue of gender in elementary and secondary education. In G. Grant (Ed.), *Review of research in education* (Vol. 17, pp. 269–334). Washington, DC: American Educational Research Association.

Schapps, E., Dibartalo, R., Moskowitz, J., Palley, C., & Churgin, S. (1981). A review of 127 drug abuse prevention program evaluations. *Journal of Drug Issues, 11*, 17–43.

Schau, C. G., & Scott, K. P. (1984). Impact of gender characteristics of instructional materials: An integration of the research literature. *Journal of Educational Psychology, 76*, 183–193.

Schloss, J. L. (1989). A comparison of selected instructional media and methods for teaching global studies. *Dissertation Abstracts International, 50*, 1621A. (University Microfilms No. 8921969)

Schultz, J. B. (1989). AHEA's survey of American teens. *Journal of Home Economics, 81*, 27–38.

Scott, K. P. (1984). Effects of an intervention on middle school pupils' decision making, achievement, and sex role flexibility. *Journal of Educational Research, 77*, 369–375.

Scott, K. P. (1986). Effects of sex-fair reading materials on pupils' attitudes, comprehension, and interest. *American Educational Research Journal, 23*, 105–116.

Selman, R. (1980). *The development of interpersonal understanding*. New York: Academic Press.

Sex education and sex-related behavior. (1986). *Family Planning Perspectives, 18*, 150, 192.

Shaffer, D., Garland, A., Underwood, M., & Whittle, B. (1987). *An evaluation of three suicide prevention programs in New Jersey.* Report prepared for the New Jersey State Department of Health and Human Services.

Shaffer, D., Vieland, V., Garland, A., Rojas, M., Underwood, M., & Busner, C. (1990). Adolescent suicide attempters: Response to suicide-prevention programs. *Journal of the American Medical Association, 264,* 3151–3155.

Sharan, S., Kussell, P., Hertz-Lazarowitz, R., Bejarano, Y., Raviv, S., & Sharan, Y. (1984). *Cooperative learning in the classroom: Research in desegregated schools.* Hillsdale, NJ: Erlbaum.

Sharan, S., & Shachar, C. (1988). *Language and learning in the cooperative classroom.* New York: Springer.

Shaver, J. P. (1989). Lesson from the past: The future of an issues-centered social studies curriculum. *Social Studies, 80,* 192–196.

Shedler, J., & Block, J. (1990). Adolescent drug use and psychological health. *American Psychologist, 45,* 612–630.

Shirley, O. L. (1988). The impact of multicultural education on the self-concept racial attitude, and student achievement of black and white fifth and sixth graders (Doctoral dissertation, University of Mississippi, 1988). *Dissertation Abstracts International, 49,* 1364A.

Six percent of U.S. teenagers report having attempted suicide. (1991, April 4). *New York Times,* p. A21.

Slavin, R. E. (1979). Effects of biracial learning teams on cross-racial friendships. *Journal of Educational Psychology, 71,* 381–387.

Slavin, R. E. (1990). *Cooperative learning: Theory, research and practice.* Englewood Cliffs, NJ: Prentice-Hall.

Sleeter, C. E., & Grant, C. A. (1987). An analysis of multicultural education in the United States. *Harvard Educational Review, 57,* 421–444.

Smith, V. A. (1977). The effect of a global studies course on the international attitudes of junior high school students (Doctoral dissertation, Indiana University, 1977). *Dissertation Abstracts International, 38,* 5388A.

Soley, M. E. (1982). The effects of a global studies curriculum on the perspective consciousness development of middle school students (Doctoral dissertation, Indiana University, 1982). *Dissertation Abstracts International, 43,* 2626A.

Sonenstein, F. L., & Pittman, K. J. (1984). The availability of sex education in large city school districts. *Family Planning Perspectives, 16,* 19–25.

Sprinto, A., Overholser, J., Ashworth, S., Morgan, J., & Benedict-Drew, C. (1988). Evaluation of a suicide awareness curriculum for high school students. *Journal of the American Academy of Child and Adolescent Psychiatry, 27,* 705–711.

Stapp, W. B., Cox, D. A., Zeph, P. T., & Zimbelman, K. S. (1978). The development, implementation, and evaluation of a transportation curriculum module for middle school-aged youth. *Journal of Environmental Education, 14,* 3–12.

Stout, J. W., & Rivara, F. P. (1989). Schools and sex education: Does it work? *Pediatrics, 83,* 375–379.

Strasburger, V. C. (1985). Sex, drugs, rock 'n roll: An introduction. *Pediatrics, 76*(suppl.), 659–663.

Strom, M. S., & Parsons, W. S. (1982). *Facing history and ourselves: Holocaust and human behavior.* Brookline, MA: Facing History and Ourselves Foundation.

Templer, D. I. (1970). The construction and validation of a death anxiety scale. *Journal of General Psychology, 82,* 165–177.

Tobler, N. S. (1986). Meta-analysis of 143 adolescent drug prevention programs:

Quantitative outcome results of program participants compared to a control or comparison group. *Journal of Drug Issues, 16*, 537–567.

Torney-Purta, J. (1982, May). *Research and evaluation in global education: The state of the art and priorities for the future.* Paper presented at the National Conference on Professional Priorities: Shaping the Future of Global Education, Easton, MD.

Torney-Purta, J. (1985). *Predictors of global awareness and concern among secondary school students.* Columbus: Mershon Center, Ohio State University.

Totten, S. (1991). [Review of *Facing history and ourselves; Elements of time*]. *Theory and Research in Social Education, 19*, 111–117.

Tye, K. A. (1991). *Global education from thought to action* (1991 ASCD yearbook). Alexandria, VA: Association for Supervision and Curriculum Development.

Valett, R. E. (1991). Teaching peace and conflict resolution. In J. S. Benniga (Ed.), *Moral character and civic education in the elementary school* (pp. 243–258). New York: Teachers College Press.

Vincent, M. L., Clearie, A. F., & Schluchter, M. D. (1987). Reducing adolescent pregnancy through school and community-based education. *Journal of the American Medical Association, 257*, 3382–3386.

Volk, T. L., & Hungerford, H. R. (1981). The effects of process education on problem identification skills in environmental education. *Journal of Environmental Education, 12*, 36–40.

Wass, H., Miller, M. D., & Thornton, G. (1990). Death education and grief/suicide intervention in the public schools. *Death Studies, 14*, 253–268.

Weigel, R. H., Wiser, P. L., & Cook, S. W. (1975). The impact of cooperative learning experiences on cross-ethnic relations and attitudes. *Journal of Social Issues, 31*, 219–244.

Wentzel, J. R. (1990). American families: 75 years of change. *Monthly Labor Review, 113*, 4–13.

Women's Educational Equity Act Program. (1983). *Fair play: Developing self-concept and decision-making skills in the middle school.* Newton, MA: Education Development Center.

Yocum, M. J. (1989). An investigation of the effects of global education on the attitudes of high school students (Doctoral dissertation, Michigan State University, 1989). *Dissertation Abstracts International, 53*, 620A.

Yulish, S. M. (1980). *The search for civic religion: A history of the character education movement in America, 1890–1935.* Lanham, MD: University Press of America.

Zeigler, S. (1981). The effectiveness of cooperative learning teams for increasing cross-ethnic friendship: Additional evidence. *Human Organization, 40*, 264–268.

Zelnik, M., & Kim, Y. J. (1982). Sex education and its association with teenage sexual activity, pregnancy and contraceptive use. *Family Planning Perspectives, 14*, 117–126.

Zolik, E. S., & Nair, D. (1987, May). *Evaluation of a nuclear war psychoeducational program for adolescents.* Paper presented at the International Physicians for the Prevention of Nuclear War conference, Moscow.

Chapter 4

Politics of Curriculum: Origins, Controversies, and Significance of Critical Perspectives

WILLIAM F. PINAR
Louisiana State University

C. A. BOWERS
University of Oregon

In one sense, most educational theorists who have envisioned using the educational process to support social reform could be said to exhibit a "critical perspective." John Dewey (1916), George Counts (1932), and Harold Rugg (1929–1932), as well as recent conservative critics such as Alan Bloom and William J. Bennett, have argued for educational reforms that presumably would foster students' capacity to think in more critical and informed ways about the issues of the day. While progressives, social reconstructionists, and conservatives would understand the social goals and consequences of such critical thinking from quite different ideological points of view, all could be said to advocate "critical perspectives." (Indeed, President Bush, in his January 1991 State of the Union address, used the term *empowerment*, long a favorite of left-wing theorists in education.) In the field of education, particularly in the field of curriculum, the concept of "critical perspectives" has been appropriated by a group whose intellectual roots can be traced, variously, to the Frankfurt School of Critical Theory and to Marxist and neo-Marxist theoreticians as varied as Antonio Gramsci, Raymond Williams, and, in education proper, Paulo Freire.

Within this group there can be said to be two subgroups. One has identified its agenda with the concept of "critical pedagogy," and the most visible of its spokesmen include Henry Giroux, Peter McLaren, and Ira Shor (1987a). Another group is more allied with concepts of critical scholarship and the politics of curriculum, the most visible being Michael W. Apple and his many students. Philip Wexler may represent the vanguard of a third group that both condemns the work of the first two and suggests

---

A shorter version of this essay, coauthored with William M. Reynolds, was presented at the annual meeting of the American Educational Research Association, Boston, April 1990.

an alternative. Collectively, this effort—increasingly diverse and acrimonious—to understand curriculum as a political text represents the largest body of contemporary scholarship in the field. Especially due to the efforts of Michael W. Apple and Henry A. Giroux, political scholarship functioned to reconceptualize the curriculum field from its moribund, atheoretical state Schwab decried in 1970 to its dynamic and complex configuration today.

The origins of this work will be outlined in the first section, focusing on its conceptual development (i.e., reproduction and resistance theory). In the second section, controversies both within this discourse and from outside will be reviewed. In a third section, we shall conclude with a discussion of our concerns with this body of work.

CONCEPTUAL DEVELOPMENT

Reproduction Theory

The first step in the effort to understand curriculum as a political text involved the concept of reproduction or correspondence. In their widely read *Schooling in Capitalist America*, Bowles and Gintis (1976) regarded schools as functioning in the stratum of superstructure, a stratum determined by society's economic base. Strike (1989, p. 26) portrays this relationship as follows:

Consciousness

Superstructure

Institutions

. . . .

Relations of production

Base

Material productive forces

Causality occurred unidirectionally, from base to superstructure. Elements in the base are used to account for elements in the superstructure (Strike, 1989, p. 26). In classic Marxian terms, the base determines the superstructure. Bowles and Gintis argued that schools prepare students to enter the current economic system via a correspondence between school structure and the structure of production.

The structure of social relations in education not only inures the student to the discipline of the workplace but develops the types of personal demeanor, modes of self-presentation, self-image, and social class identifications that are the crucial ingredients of job adequacy. Specifically, the social relationships of education—the relationships between administrators and teachers, teachers and students, students and students, and students and their work—replicate the hierarchical divisions of labor. Hierarchical relations are reflected in the vertical authority lines from administrators to teachers to students. Alienated labor is reflected in the student's lack of control over his or her education, the alienation of the student from the curriculum content, and the motivation of school work through a system of grades and other external rewards rather than the student's integration with the process (learning) or the outcome (knowledge) of the educational "production process" (Bowles & Gintis, 1976, p. 131).

Relying on this principle of correspondence, Apple (1979) and Giroux (1981a) argued that schools functioned to reproduce the class structure of the workplace (Liston, 1986). Although originating outside the curriculum field, the principle of correspondence was an important first step in understanding curriculum as a political text.

A second concept imported from other fields aided politically oriented curriculum scholars to advance their argument. Louis Althusser's (1971) understanding of ideology provided another major concept in curriculum scholarship. McLaren (1989) explains:

> Simply put ideology refers to the production of meaning. It can be described as a way of viewing the world, a complex of ideas, various types of social practices, rituals and representations that we tend to accept as natural and as common sense. It is the result of the intersection of meaning and power in the social world. Customs, rituals, beliefs and values often produce within individuals distorted conceptions of their place in the sociocultural order and thereby serve to reconcile them to that place and to disguise the inequitable relations of power and privilege; this is sometimes referred to as "ideological hegemony." (p. 176)

Wexler (1987) regarded ideology as the first key concept of the new sociology of education and curriculum. Rejecting what some characterized as more "vulgar" interpretations of the base/superstructure relationships in Marxian theory, Althusser argued that the relation of the economic base to the institutions of society cannot be reduced to any linear cause/effect determinism (Giroux, 1983a, p. 79). Institutions were termed "ideological state apparatuses" by Althusser (1971), who claimed that institutions functioned to subjugate the working class. Giroux (1983a) interpreted the Althusserian conception of ideology for curricularists:

> First, it [ideology] has a material existence: rituals, practices, and social processes that structure the day-to-day workings of schools. . . . Second, ideology neither produces consciousness nor a willing passive compliance. Instead it functions as a system of representations, carrying meanings and ideas that structure the unconsciousness of students. (p. 81)

The concept of ideology became central in understanding curriculum as political text. Curriculum itself became conceptualized as an ideological mystification (Apple, 1990a; Giroux, 1981a, 1981b, 1981c). Both Apple and Giroux described how the content and form of the curriculum were ideological in nature (Apple, 1990a; Giroux, 1981c). Generally, the ideas and culture associated with the dominant class were argued to be the ideas and content of schooling. Dominant culture was described as those "social practices and representations that affirm the central values, interests, and concerns of the social class in control of the material and symbolic wealth of society" (McLaren, 1989, p. 172).

By the early 1980s, the largely economic version of reproduction (correspondence) was being criticized by many of the same scholars who had embraced it in the 1970s. Now reproduction theory was characterized as deterministic and simplistic (Giroux, cited in Olson, 1981), as lacking a cultural analysis (Apple, 1979, 1980), as lacking an adequate theory of agency (Strike, 1989), and as basically mechanistic (Giroux, 1983a). In an essay titled "Contradiction and Reproduction in Educational Theory," Bowles and Gintis (1980) themselves criticized their earlier work:

The most critical [problem] is simply this: by standing in our approach as the *only* structural link between education and the economy and by its character as an inherently *harmonious* link between the two, the correspondence principle forced us to adopt a narrow and inadequate appreciation of the *contradictions* involved in the articulation of the educational system within the social totality. (p. 53)

Bowles and Gintis acknowledge that their earlier argument missed certain essential aspects of reproduction. As Liston notes (1986), they specify notions of sites and practices as crucial in understanding reproduction. They suggest that society be regarded as "an ensemble of structurally articulated sites of social practice," the primary three of which are state, family, and school (Bowles & Gintis, 1980, p. 55). Social practices represent "fundamental and irreducible elements of social dynamics" (Bowles & Gintis, 1980, p. 56).

Bowles and Gintis theorize four types of social practice, the point of which is social transformation. The first is the appropriative, the goal of which is the creation of useful projects. The second is the political, the goal of which is the transformation of social relations. The third is termed the cultural, which is said to transform the tools of discourse. The fourth is the distributive, which functions to alter the distribution of power and income (Bowles & Gintis, 1980; see also Liston, 1988). Sites and practices "add up" to what Bowles and Gintis term a "contradictory totality" (Bowles & Gintis, 1980, p. 56).

This mixing of sites and practices produces two dynamic tendencies that have distinct consequences: reproductive and "contradictory" (and undermining) effects (Liston, 1988, p. 54). Bowles and Gintis acknowledge that their earlier work (1976) failed to account for this fundamentally contradictory character of social relations. Schools are said to exhibit these contradictory tendencies. The stage is set for "resistance theory."

Wexler notes that "the Frankfurt School analysis of culture was also used to establish the view of education as a site for reproduction" (1987, p. 40). Among those associated with the Frankfurt School are Theodor Adorno, Walter Benjamin, Jurgen Habermas, Max Horkheimer, and Herbert Marcuse, all of whom are cited by politically oriented curriculum scholars, especially by Giroux and Wexler. What was the intent of the Frankfurt School? Broadly speaking, the Frankfurt School departed from the major Marxisms of the 1930s, namely Stalinist Marxism. Frankfurt School scholars deemphasized the economic foundations of Marxism and its theory of the historical inevitability of class conflict, including the close identification of the Communist Party with the "dictatorship of the proletariat." Frankfurt School scholars tended to emphasize culture rather than economics—culture in its anthropological as well as aesthetic sense. They might be said to typify the "early Marx" rather than the scientism characteristic of the "late Marx" (Miller, 1979). More specifically,

It was the hope of Horkheimer and the others that their work would help establish a critical social consciousness able to penetrate existing ideology, sustain independent judgment and be capable, as Adorno put it, of maintaining its freedom to think that things might be different. (Held, 1980, p. 38)

Giroux (1983a) regarded the Frankfurt School as fundamental to understanding curriculum as a political text (i.e., as expressing a political meaning):

I argued that the foundation for a radical theory of schooling can, in part, be developed from the work of the Frankfurt School and the more recent literature on the hidden curriculum. Whereas the Frankfurt School provides a discourse and mode of critique for deepening our understanding of the nature and the function of schooling, critiques of the hidden curriculum have provided modes of analysis that uncover the ideologies and interests embedded in the message systems, codes and routines that characterize daily classroom life. (p. 72)

The hidden curriculum, first popularized by Philip Jackson (1968), was another important conceptual tool for politically oriented curriculum scholars. The concept refers to those unintended but quite real outcomes and features of the schooling process (Apple, 1975, 1990a; Giroux & Purpel, 1983; McLaren, 1989). The hidden curriculum is to be distinguished from the "overt" curriculum, or the planned curriculum, including objectives. McLaren (1989) defines the concept:

The hidden curriculum deals with the tacit ways in which knowledge and behavior get constructed, outside the usual course materials and formally scheduled lessons. It is part of the bureaucratic and managerial "press" of the school—the combined forces by which students are induced to comply with the dominant ideologies and social practices related to authority, behavior and morality. (pp. 183–184)

Michael W. Apple, the first to reassert curriculum as a political text in the 1970s, defined the hidden curriculum in a way that pointed to the concept of hegemony, another important conceptual tool for politically oriented curriculum scholars.

The hidden curriculum in schools serves to reinforce basic rules surrounding the nature of conflict and its uses. It posits a network of assumptions that, when internalized by students, establishes the boundaries of legitimacy. This process is accomplished not so much by explicit instances showing the negative value of conflict, but by nearly the total absence of instances showing the importance of intellectual and normative conflict in subject areas. The fact is that these assumptions are obligatory for the students, since at no time are the assumptions articulated or questioned. (Apple, 1975, p. 99)

Owing to Apple's efforts, in part, the concept of the hidden curriculum became taken-for-granted curriculum knowledge, widely cited by those who insisted that the curriculum functioned to maintain social stratification as well as other stratifications, especially those of class, race, and gender (Apple, 1982b, 1990a; Aronowitz & Giroux, 1985; Beyer & Apple, 1988; Giroux, 1981a, 1983a, 1988b; Giroux, Penna, & Pinar, 1981; Giroux & Purpel, 1983; Oakes, 1985; Shapiro, 1981, 1983a; Sharp, 1980; Shor, 1986; Weis, 1988).

Another major concept employed in understanding curriculum as a political text was hegemony, borrowed from the Italian Marxist Antonio Gramsci (1971/1972), who borrowed the term from Marx and Engels (1974). Gramsci emphasized "the role of the superstructure in perpetuating class and preventing the development of class consciousness" (Carnoy, cited in Apple, 1982a, p. 86). He employed hegemony in two senses. First, hegemony referred to a process of domination whereby the ruling class is said to exercise political control through its intellectual and moral leadership over allied classes (Gramsci, 1985). (This is the sense in which Marx and Engels used the term.) Second, hegemony referred as well to the use of force and ideology in the reproduction of class relations (Aronowitz & Giroux, 1985, p. 88). Thus, hegemony is understood to occur via the use of force and via the shaping of human consciousness.

The concept of hegemony helped critical curriculum scholars refine the basic "base/superstructure" model of reproduction that had been accepted during the 1970s. Relying on Raymond Williams (1976), Apple declared that the concept of hegemony captures the complexity of processes of "saturation." In particular, Apple draws upon Williams's concept of "selective tradition" to point to the ways in which curriculum functions to privilege certain sets and orders of knowledge over others (Williams, cited in Apple, 1990a, pp. 5–6). Other politically oriented scholars rely on this concept in their analysis of cultural reproduction (Apple, 1982b, 1986, 1990a; Giroux, 1980, 1981a, 1983a, 1988b; McLaren, 1989; Sharp, 1980).

Henry Giroux worried that an overreliance on the concept of reproduction risks a "discourse of despair." If reproduction occurred as incontestably as Bowles and Gintis and the critical scholars of the 1970s insisted that it did, there was little hope for significant change aside from alterations in the economic base. The concept of ideology portrayed teachers and students as accomplices in the reproduction of the ruling class. Hegemony seemed to suggest that no escape was possible, as consciousness itself was saturated, "forged into the cognitive chains which bind the minds of the working class" (Strike, 1989, p. 137). In a word, reproduction theory lacked a concept of agency. Almost overnight, reproduction theory would give way to resistance theory.

From Reproduction to Resistance

In his widely read *Learning to Labour*, Paul Willis (1981) introduced the concept of resistance to an eager audience disenchanted with reproduction theory. Willis observed that the working class boys he studied resisted both the official and hidden curriculum of their English secondary school. The roots of this resistance, he wrote, "are in the shop-floor cultures occupied by their family members and other members of their class" (Giroux, 1983b, p. 283). Willis's concept of resistance allowed politically oriented scholars to view the process of reproduction as contestable, thereby correcting the nondialecticism of the Bowles and Gintis (1976) thesis.

The early 1980s saw considerable discussion of resistance theory. Particularly during the period 1980–1984, many scholars discussed and developed resistance in terms of curriculum (e.g., see Anyon, 1979, 1988; Apple, cited in Olson, 1981; Apple, 1982a, 1982b; Apple & Weis, 1983; Giroux, 1981a, 1983a, 1983b). Giroux cited resistance theory as important insofar as it corrected the failure of both conservative and radical curriculum theory. Conservatives, he alleged, tended to view oppositional behavior via psychological categories such as deviate, disruptive, and inferior. Radical theorists had overemphasized economic and cultural determinants. Put differently, in radical curriculum theory there had been an "underemphasis on how human agency accommodates, mediates and resists the logic of capital and its dominating social practices," including school curriculum and instruction (Giroux, 1983b, p. 282). As noted above, Giroux characterized the reproduction theory of the 1970s as a "discourse of despair," as it ignored the pedagogical possibilities of human thought and enlightened action.

In his introduction to Giroux's *Ideology, Culture and the Process of Schooling* (1981a), Stanley Aronowitz advocated resistance as a positive step for radical educators. Radical educators should begin to concentrate on the "cracks and disjunctions created by oppositional forces" (Aronowitz, cited in Giroux, 1981a, p. 31). Doing so would permit the contestation of power in the schools. Giroux asserted that struggles can be waged over administrative and curricular issues. Reproduction failed to inspire struggle; it was, in Giroux's words, a "myth of total domination" (Giroux, 1981a, p. 99).

A special issue of *Interchange*, edited by Paul Olson and published in 1981, illustrated the rapid shift from reproduction to resistance theory. A collection of papers from a conference held at the Ontario Institute for Studies in Education in Toronto, the issue was titled "Rethinking Social Reproduction." In his introduction, Olson (1981) noted that "social constructivists" wished to integrate knowledge of hegemony with strategies design to counter it. In his essay titled "Reproduction, Contestation, and Curriculum: An Essay in Self-Criticism," Michael Apple notes that his previous work lacked analysis that "focused on contradictions, conflicts, mediations and especially resistance—as well as reproduction" (Apple, 1981, p. 35). He cautions his audience, however, that it is not enough to conduct research into resistance; one must actually resist, in practice.

In Giroux's "Hegemony, Resistance, and the Paradox of Educational Reform," the outline of his scholarly agenda for the decade is evident. Discussing the assets and liabilities of reproduction theory, he praises resistance theories, which "perform a theoretical service" (Giroux, 1981b, p. 13). These theories demand analyses of those social practices that constitute the class-based experiences of day-to-day existence in schools. He calls for the development of a notion of radical pedagogy based on the pioneering work of Paulo Freire (1970/1971). "At the core of radical pedagogy," Giroux insists, "must be the aim of empowering people to work for change in the social, political, and economic structure that constitutes the ultimate source of class-based power and domination (Giroux, 1981b, p. 24). (Curiously, the concept of empowerment—used by relatively few radical curriculum theorists in 1981—would become educational cliche by the 1990s.) For Giroux, however, resistance theory quickly becomes a transitional concept to pedagogy.

In 1982, Michael Apple published two works examining issues of reproduction and resistance. In his introduction to *Cultural and Economic Reproduction in Education*, Apple distinguishes between two forms of reproduction theory, that which focuses on economic or macrostructural issues and that which concentrates on cultural or microstructural matters.

According to Apple, the school curriculum belongs to the latter category. Acknowledging difficulties with "pure" reproduction theory, Apple suggests that resistance and reproduction theory are intertwined, that studies inspired by this synthetical view would point to struggles in specific places. He alludes to issues of race, class, and gender, foreshadowing his own scholarly agenda for the decade.

In an essay titled "Curricular Form and the Logic of Technical Control," Apple (1982a) outlines the pervasiveness of resistance. He alludes to developments such as the so-called poststructuralism, which he depicts favorably, an attitude toward that work that would change. Concluding the essay is a discussion of resistances, especially those curricular sites of resistance. Despite increasing state control, Apple declares, there are moments of individual resistance. Teachers' resistance is said to be never "far from the surface" (Apple, 1982a, p. 269). The question becomes, What is the status of these resistances? Are they, in fact, counterhegemonic? Or do they function to reproduce the status quo? Willis notes, for instance, that the resistance of his "lads" functioned reproductively: Their resistance to mental labor functioned to reproduce their entrapment in the working class.

In his second 1982 work, Apple continues his examination of resistance theory and, in particular, the possible reproductive consequences of resistance. Apple worries that even the terrain of resistance can be viewed as determined by the interests of capital, not by those resisting (Apple, 1982b). Despite resistance, Apple concludes, reproduction proceeds. In fact, he continues, reproduction will continue "as long as the penetrations into the nature of work and control generated by working-class youths and their parents are unorganized and unpoliticized" (Apple, 1982b, p. 108). Only a few years after its introduction, then, resistance itself seemed to be in danger of being swallowed by reproduction.

Two important efforts to understand the curriculum as a political text appeared in 1983: Henry A. Giroux's *Theory and Resistance in Education: A Pedagogy for the Opposition* and Michael W. Apple and Lois Weis's *Ideology and Practice in Schooling*. In both books, one discerns movements away from resistance theory. For Giroux, resistance points to possibilities of oppositional pedagogy (1983a). He calls for a reformulation of the relations among ideology, culture, and hegemony, one that would "make clear the ways in which these categories can enhance our understanding of resistance as well as how such concepts can form the theoretical basis for a radical pedagogy that takes human agency seriously" (Giroux, 1983a, p. 111).

Apple and Weis (1983) discuss the movement beyond simple reproduction theory, stating that "hegemony is not and cannot be fully secure" (p. 28). Their view that the cultural sphere is relatively autonomous leads them to move beyond resistance to a belief in the possibility of meaningful intervention in the schools. However, they caution that this action must be a kind of praxis and that the connections between the schools and the larger society must be made.

The concept of resistance would continue to appear in political analyses of curriculum, but after 1985 it seemed to be a point of departure rather than arrival. For instance, Geoff Whitty calls for a movement away from reproduction theory and academic critique generally to radical intervention. He warns against romanticizing the resistances of the working class, even those that are reproductive. What is important now, he argues, is the elaboration of intervention strategies.

What the American work increasingly recognizes is that whether or not particular aspects of education are ultimately reproductive or transformative in their effects is essentially a political question concerning how they are to be worked upon pedagogically and politically, and how they become articulated with other struggles in and beyond the school. (Whitty, 1985, p. 90)

Pedagogy and Practice: Issues of Race, Class, and Gender

By 1985, scholarly efforts to understand curriculum politically had begun to turn away from reproduction and resistance theories to issues of political and pedagogical practice. This shift away from resistance theory is evident, for example, in the work of Henry A. Giroux, which, beginning in 1985, moved to questions of literacy, liberal arts, and transformative pedagogy. In his 1985 book *Education Under Siege* (coauthored with Stanley Aronowitz), Giroux discusses reproduction and resistance insofar as they lead to radical action. In the field of curriculum, what is necessary is a "language of possibility" (Aronowitz & Giroux, 1985). Educators must become transformative intellectuals rather than "skillful technicians" (Aronowitz & Giroux, 1985). What is now necessary is to "link emancipatory possibilities to critical forms of leadership by rethinking and restructuring the role of curriculum workers" (Aronowitz & Giroux, 1985, p. 142).

This shift toward action and practice is evident in Giroux's many journal articles (1985a, 1985b, 1987, 1988c, 1989), emphasizing always the importance of transformative struggle, both in school and in society generally. His most recent books, *Schooling and the Struggle for Public Life: Critical Pedagogy in the Modern Age* (1988a), *Teachers as Intellectuals: Toward a Critical Pedagogy of Learning* (1988b), and *Critical Pedagogy, the State and Cultural Struggle* (1989, coauthored with Peter McLaren), emphasize critical or transformative pedagogy.

Freire's work, which had been enormously influential in the late 1960s and early 1970s, reemerged as central to the effort to understand curriculum as a political text during the latter half of the 1980s. In *The Politics of Education* (1985), *A Pedagogy for Liberation: Dialogues on Transforming Education* (Freire & Shor, 1987), and *Freire for the Classroom: A Sourcebook for Liberatory Teaching* (Shor, 1987b), Freire elucidated the significance of critical pedagogy. Among those Freire influenced was George Wood, v hose interest in transformation (Wood, 1988) led to the establishment of the Institute for Democracy in Education at Ohio University. (By mid-decade Giroux had established the Center for Cultural Studies at Miami University in Oxford, Ohio.) Peter McLaren's *Schooling as a Ritual Performance* (1986) makes Freirian suggestions to teachers attempting to foster resistance in their own schools. In *Life in Schools* (1989), McLaren provides examples of critical educators, among whom are John Dewey, Michael Apple, and Henry Giroux.

Apple's work during the mid-1980s shifts also from reproduction and resistance to pedagogy and politics, especially as these are understood in terms of race, class, and gender. Apple's scholarship during this time emphasizes political and pedagogical struggle (Apple, 1987a, 1987b, 1988; Apple & Ladwig, 1989; Apple & Teitelbaum, 1986). In *Teachers and Texts: A Political Economy of Class and Gender Relations in Education*, Apple (1986) examines the textbook industry, particularly as it perpetuates the "selective tradition." He calls for political and pedagogical action by critical scholars, teachers, students, and parents. Apple argues that the effort to democratize the curriculum must be a collective one involving interested parties in addition to educational professionals.

[Critical scholars] need to be closely connected to feminist groups, people of color, unions, and to those teachers and curriculum workers who are now struggling so hard in very difficult circumstances to defend from rightist attacks the gains that have been made in democratizing education and to make certain that our schools and the curricular and teaching practices within them are responsive in race, gender and class terms. After all, teaching is a two-way street and academics can use some political education as well. (Apple, 1986, p. 204)

In essays published in 1988, Apple extends his analysis of curriculum as a political text. In "Race, Class and Gender in American Education: Toward a Nonsynchronous Parallelist Position," Cameron McCarthy and Apple call for theoretical work that demonstrates how race, class, and gender interconnect, and how economic, political, and cultural power expresses itself in education (McCarthy & Apple, 1988). (A description of scholarly efforts to understand curriculum as a racial text requires a separate essay.) As well, they point to a shift in strategies for fundamental change in curricular content, pedagogical practices, and social structures (McCarthy & Apple, 1988). Landon Beyer and Apple's "The Curriculum: Problems, Politics and Possibilities" (1988) concentrates on issues of political and pedagogical agency. Fundamental to these issues is the concept of praxis, which involves "not only a justifiable concern for reflective action, but thought and action combined and by a sense of power and politics. It involves both conscious understanding of and action in schools on solving our daily problems" (Beyer & Apple, 1988, p. 4).

Praxis implies "critical reflective practices that alter the material and ideological conditions that cause the problems we are facing as educators in the first place" (Beyer & Apple, 1988, p. 4). In "The Politics of Pedagogy and the Building of Community" (1990b), Apple continues this emphasis on collective and concrete action, narrating episodes from his "Friday Seminar." He reminds his students and himself that educational politics are inseparable from national politics: "I am constantly reminded of how important it is that we participate in those larger struggles as well" (Apple, 1990).

The effort to understand curriculum as a political text has shifted from an exclusive focus on reproduction of the status quo to resistance to it, then again to resistance/reproduction as a dialectical process, then again—in the mid-1980s—to a focus on daily educational practice, especially pedagogical and political issues of race, class, and gender. The major players in this effort have continued to be Apple and Giroux— Apple through his voluminous scholarship and that of his many students (Schubert, Schubert, Herzog, Posner, & Kridel, 1988, pp. 173–174) and Giroux through his prodigious scholarly production. Clearly, serious students of curriculum understand, as never before, the complexities and significance of curriculum's political dimensions.

This effort to understand the curriculum as a political text has not been without criticism, however. The "sociology of curriculum" has been critiqued from within its ranks as well as outside them. It is accurate to say that this work was generally ignored by the mainstream field during the 1970s. As it became too important to be ignored, it became criticized.

CONTROVERSIES

A major critique appeared in 1988. In *Capitalist Schools: Explanation and Ethics in Radical Studies of Schooling*, Daniel Liston alleges numerous weaknesses in Marxian analysis of schooling. Specifically, Liston criticizes the purposes and methods of explanation, justification, and empirical validation in radical scholarship. After analyzing the work of Bowles and Gintis (1976, 1980), Giroux (1981a, 1983b), Apple and Weis (1983), Apple (1980, 1982), and Wexler (cited in Barton, Meigham, & Walker, 1980), Liston criticizes what he terms the "facile functionalist" assertions found throughout these works. The fundamental base/superstructure model is assumed, not explained. Additionally, there is no empirical evidence presented in these works to support the model. Liston (1988) states: "Marxist explanations of public schools, while critical of functionalist approaches, nevertheless rely on functionalist assertions and ignore rigorous empirical assessments of these claims" (p. 101).

Liston is interested in ethical as well as methodological problems in the effort to understand curriculum politically. He notes that "freedom was and still remains the central ethical standard" in radical thought (Liston, 1988, p. 143). However, the major scholars seem to be unaware that Marx himself viewed concepts of justice and equality as problematical. Liston argues that "Marx viewed (and a consistent Marxist tradition would construe) justice as a deficient standard. Marx criticized capitalism morally but his standard was freedom (not justice), a standard embedded in the naturalist ethic" (Liston, 1988, p. 168). Marx's primary concern was freedom.

Yet, Liston observes, Marxist scholars often base their criticisms of the status quo on notions of justice. "Marxist educators claim that the structure of the larger socioeconomic system is unjust and that the schools contribute to the reproduction of the unjust system" (Liston, 1988, p. 123). If Marxist criticism of the status quo cannot be based on a concept of injustice, upon what can it be based? "What is the basis of condemnation? The basis exists, I believe, in Marx's notion of freedom" (Liston, 1988, p. 136). Liston asserts that those Marxists who currently employ standards of justice in their arguments "must at least recognize and appraise the merits of Marx's own critique of these standards" (Liston, 1988, p. 143). Furthermore, he advises radical scholars to "revise the basis of their critiques or argue against Marx's position" (Liston, 1988, p. 143). Liston makes clear that his critique is a "friendly" one: "Without enhanced explanatory claims or moral justification it does not seem likely the radical tradition will convince reasonable skeptics. I hope it does. Without convincing these skeptics it will inevitably fail. I hope it does not" (p. 174).

Liston's claim of "friendliness" appears to have been rejected, at least by one major theoretician. Peter McLaren portrays Liston's effort as having resurrected "an old and theoretically threadbare Marxian orthodoxy" (McLaren, 1990, p. 1). In his view, Liston's call for empirical validation represents a return to an epistemological position radical scholars long ago rejected, that is, a "stance of objectivity and . . . the scientific goal of Truth" (McLaren, 1990, p. 8). Liston's logic is characterized as "reductionist" (McLaren, 1990, p. 4). Recent debates regarding poststructuralism make it "difficult to remain sympathetic to Liston's penchant for causal mechanism and his reduction of ideology to flow charts and empirically based formulae" (McLaren, 1990, p. 5). Furthermore, McLaren characterizes Liston's treatment of Giroux's scholarship as "monumental in its speciousness" (McLaren, 1990, p. 7). Liston's analysis can be likened to "what Sartre called 'bad faith,' a kind of ritualistic blood-letting, an exaggerated self-flattery, and an inflated moral right-eousness that is damaging to the books' intent and purpose'' (McLaren, 1990, p. 7).

Like Liston, C. A. Bowers (1987) and Kenneth Strike (1989) allege that Marxist scholars have abandoned Marx. More specifically, Bowers and Strike regard radical scholars as moving away from Marxism toward liberalism. "Marxism has been decisively rejected. It appears that no alternative hard core is on the horizon, unless it is some variety of liberalism" (Strike, 1989, p. 166). Among those whose work Strike discusses include Apple, Bowles and Gintis, Giroux, Willis, Levin and Carnoy, McLaren, and Dale. Strike meticulously describes the movement of these efforts to understand the curriculum as a political text, a movement that increasingly exhibits "idealist and indeed liberal terms" (Strike, 1989, p. 160). The emphasis on cultural autonomy characteristic of radical scholarship after 1985 represents, for Strike, an "abandonment of the core of the Marxist program" (Strike, 1989, p. 155). Marx's labor to substitute materialism for idealism has been forgotten, apparently, by those who claim the heritage. The consequence is the absence of a coherent research program. Strike (1989) concludes:

Finally, the fact that many of those who gave us the Marxist critique of schooling seem to have lapsed into liberal construction is at least a piece of evidence that allows us to see Marxism as a degenerative research program. It suggests that the problems generated by a Marxist research program cannot be solved without abandoning the program's central assumption. (p. 167)

In *Elements of a Post-Liberal Theory of Education*, Bowers (1987) critiques the work of Paulo Freire, upon which much of the transformative or critical pedagogy literature rests. That work shares four basic assumptions with liberalism, according to Bowers: (a) Change is inherently progressive; (b) the individual is the basic social unit "within which we locate the source of freedom and rationality" (Bowers, 1987, p. 2); (c) human nature is basically good or at least changeable via environmental manipulation; and (d) rationality is "the real basis of authority for regulating the affairs of everyday life" (Bowers, 1987, p. 2). Bowers argues that Freire's version of Marxism is more appropriately characterized "as a form of democratic humanism, an ideology which would be scorned as a form of revisionist liberalism in those countries that rely upon a more scientific Marxism as a basis of their social organization" (Bowers, p. 1987, p. 37).

Perhaps the most compelling but caustic critic of political analyses of curriculum is Philip Wexler, himself an "insider" to debates regarding base/superstructure, ideology, hegemony, and so forth (Wexler, 1976). In *Social Analysis of Education: After the New Sociology* (1987), Wexler charts the rise and fall of the so-called "new sociology." Central to his analysis is the linkage of academic work to social movements outside the academy. He points out, for instance, that the "new sociology" arose in the aftermath of the radical student and civil rights movements of the 1960s. Wexler suggests that radical critics romanticized those movements. Politically oriented scholarship amounted to "a displaced imitation of it [the student movements of the 1960s], an attempt culturally to recapitulate the practical historical course of the movement, *in theory*" (Wexler, 1987, p. 26). Radical scholarship suggested a rediscovered but unfortunately idealized interest in educational change (Wexler, 1987). However, such change would occur within institutions and within professional roles. Wexler (1987) states:

In this view, new sociology of education is a rationalized cultural representation of identity politics: it is part of a *post*-"movement" effort to create a meaningful professionalism that is consonant with the ideal of a defeated social movement. (p. 27)

In this respect, politically oriented scholars are committed to their own professional advancement, which they mistake for political activism (see also Pinar, 1981b, p. 440). Their activism constitutes the "cultural formation of an identifiable social group which is engaged in sociocultural action on its own behalf" (p. 4). Consequently, "left professional middle class institutional intellectuals became a socially residual remnant, rather than the institutional vanguard of an ascendant social class segment" (Wexler, 1987, p. 123).

Furthermore, Wexler alleges, the primary concepts employed to understand curriculum politically—reproduction and resistance—represented a "combination of functionalist structuralism and romantic individualism" (Wexler, 1987, pp. 16, 42). In consequence, "the new sociology of education is historically backward-looking and ideologically reactionary, although its ideals combine the values of the New Left and traditional socialism" (Wexler, 1987, p. 127). The history of the effort to understand curriculum as a political text can be located in the social path of its producers (i.e., the so-called "new sociologists" of curriculum). That path leads from

ideology-critique to awareness of systematic reproduction through the accumulation of cultural capital; and then from idealized and socially displaced individual cultural resistance to the dissonant bifurcation between idealized social mobilization and an unconscious politics of internally exiled speech. (Wexler, 1987, p. 45)

The "new sociology" may have become trapped in the progressive-liberal paradigm, but Wexler allows for the possibility of escape. Certain strands of feminist thought and the so-called poststructuralism point to social movements and change that might inspire collective mobilizations. Particularly, literary "textualism" is "potentially transformative, . . . the starting point for a counter-practice" (Wexler, 1987, p. 180). Additionally, poststructuralism might free political scholars from "a socially inauthentic identification with 'the working class' or with the triadic oppressed groups of 'class, race and gender' " (Wexler, 1987, p. 181). Reaction to this claimed political potential of poststructuralism has begun to appear in print (McCarthy, 1988, p. 8). A feminist critique of critical pedagogy appeared late in 1989 (Ellsworth, 1989). (Understanding curriculum as a gender text intersects, of course, with understanding curriculum as a political text. The former, because of its significance, requires reporting in a separate essay, however.)

A revised version of a presentation made at the 1988 Bergamo Conference, Elizabeth Ellsworth's (1989) "Why Doesn't This Feel Empowering? Working Through the Repressive Myths of Critical Pedagogy" critiques both the conceptual structure and daily practice of critical pedagogy. Critical pedagogy should not be confused, Ellsworth insists, with feminist pedagogy, which "constitutes a separate body of literature with its goals and assumptions" (Ellsworth, 1989, p. 298). The key terms of critical pedagogy—"empowerment," "student voice," "dialogue"— represent "code words" and a "posture of invisibility" (p. 301). Relying upon a decontextualized and universalistic conception of reason, critical pedagogy leads to "repressive myths that perpetuate relations of domination" (pp. 298, 304). Critical pedagogy, she continues, leaves the structure of domination and authoritarianism in place. In Ellsworth's (1989) words, "[critical pedagogy] fails to challenge any identifiable social or political position, institution or group" (p. 307). The "utopian goals" of critical pedagogy are unattainable.

In working with participants in the course Curriculum and Instruction 607 (C & I 607), Ellsworth (1989) conceives of her role as interrupting institutionally imposed limits on "how much time and energy students . . . could spend on elaborating their positions and playing them out to the point where internal contradictions and effects on the positions of other social groups could become evident and subject to self-analysis" (p. 305).

This position is in contrast to that of critical pedagogy, which Ellsworth (1989) views as enforcing rational deliberation (p. 305). Furthermore, she alleges that critical pedagogy—here she quotes Freire, Shor, and Giroux—fails to question its own stance of superiority of teachers' understanding over students' (p. 307). Ellsworth is skeptical of critical pedagogy's claim of "emancipatory authority," judging that the goals for which such authority is used remain "ahistorical and depoliticized" (p. 307). "Empowerment," she insists, is defined so broadly that it fails to challenge specific social or political groups or positions. This failure represents critical educators' inability to confront the paternalism of traditional education (p. 307). Ellsworth goes on to criticize the concept of "student voice," as it is discussed in the critical pedagogy literature, relying on her experience in C & I 607 to suggest its flaws. From a gender perspective, for instance, she notes: "The desire by the mostly White, middle-class men who write the literature on critical pedagogy to elicit 'full expression' of student voices . . . becomes voyeuristic when the voice of the pedagogue himself goes unexamined" (p. 312). Concluding, Ellsworth notes that as long as critical pedagogy fails to understand issues of trust, risk, fear, and desire, especially as these are expressed through issues of identity and politics in the classroom, its "rationalistic tools will continue to fail to loosen deep-seated, self-interested investments in unjust relations of, for example, gender, ethnicity, and sexual orientation" (pp. 313, 314). After critiquing critical pedagogy's ahistorical use of "dialogue" and "democracy," Ellsworth suggests a "pedagogy of the unknowable" (p. 318), in which knowledge is understood as "contradictory, partial and irreducible" (p. 321). Reflecting on her own teaching, Ellsworth (1989) writes:

A preferable goal seemed to me to become capable of a sustained encounter with currently oppressive formations and power relations that refused to be theorized away or fully transcended in a utopian resolution—and to enter into encounter in a way that owned up to my own implication in those formations and was capable of changing my own relation to and investments in those formations. (p. 308)

Ellsworth (1989) advocates labor toward a position informed by feminism and poststructuralism (p. 304) in which none of us is permitted to be "off the hook." Ellsworth's criticism has provoked considerable controversy, as suggested by letters published in the August 1990 issue of the *Harvard Educational Review*. It is too early to judge to what extent this controversy will lead to a new stage of formulation of key terms in the effort to understand curriculum as a political text.

CONCERNS

A systematic effort to understand the curriculum as a political text asserted itself forcefully during the 1970s and 1980s. The notion that curriculum development, evaluation, and so forth could be conducted in a politically neutral fashion quickly became one conceptual casualty of this effort, as was the taken-for-granted assumption that schools functioned as avenues of upward social and economic mobility. The rejection of these mainstream and taken-for-granted ideas of the curriculum field was accompanied by the building of a relatively elaborate conceptual edifice, among the foundations of which were notions of reproduction, resistance, ideology, and hegemony. As the effort grew more voluminous and complex, so did the range of its interests, incorporating notions of critical pedagogy and literacy as well as issues of race, class, and gender. Tensions have developed within the movement, as have criticisms from outside it. These, we think, are testimonies to its power and importance as well as to its internal flaws. What has become indisputable is that to understand the curriculum, one must understand it, at least in part, as a political text.

The phrase "in part" leads to our first concern with this work. There are tendencies in critical scholarship to ignore those efforts that attempt to understand curriculum in other ways, as if curriculum were only a political text. It is, for instance, also a phenomenological text. There is a large literature on understanding curriculum and pedagogy as a phenomenological text (Pinar & Reynolds, 1992), a sector of the field generally ignored by the critical scholars. Likewise, literatures reflecting efforts to understand curriculum as autobiographic and biographic texts are passed over. The contextualization of political discourse in "place" seems largely ignored (Kincheloe & Pinar, 1991). There is a tendency here that goes beyond focusing on those subjects that interest one or are germane to one's specialization within a field. There is here an orthodoxy, a refusal to even acknowledge the existence of work outside the current boundaries of radical research. Politically oriented scholars are hardly alone in this, of course. During the 1970s, traditionalists refused to acknowledge their presence. It is also true that phenomenologists tend to look the other way when political issues are raised (there are exceptions; see Carson, in press). What we are worrying over here is a problem with the field at large (Pinar, 1990) that goes beyond the literature we have described herein. It is what we might loosely term conceptual isolationism, with attendant "nationalism" and "militarism." There are what we might term—tongue in cheek—high tariff barriers (a skeptical outsider might term them "jargon taxes") that work against "trade" (i.e., the exchange of ideas). Phenomenological curriculum theory could benefit, in our judgment, from judicious employment of the knowledge the politicists have constructed and collected, and vice versa. Clearly, the political work does engage other traditions, sometimes appreciatively, as in the case of feminist theory (although there are thorny issues here; cf. Ellsworth), sometimes ambivalently, as in the case of poststructuralism (cf. McCarthy, Wexler). At other times, it simply refuses to engage other segments of the field. One might expect such smug self-involvement from conservatives, but from those who espouse "dialogue," "voice," "democracy," and the like, this is a disheartening state of affairs. If the major players in this sector would engage other traditions of scholarship in an open and nonadversarial fashion, not only would their boundaries be expanded, perhaps their internal disputes would be less vitriolic. It is almost as if, locked up behind ideological walls, these individuals seem to become a bit stir crazy, and what should be amicable differences of opinion as well as serious efforts to educate each other have tended to become cold-war confrontations.

A second concern involves the import of this scholarship for the school community. That this work will exert significant influence on classroom practice in the United States is doubtful. This prognosis is based partially upon the failure of leading theorists such as Michael Apple, Henry Giroux, Ira Shor, and Peter McLaren to respond constructively to critics outside the critical tradition and, most recently, to critics who, like Elizabeth Ellsworth, address issues in the politics of curriculum but who do not wish to be identified with the more messianic tendencies in critical pedagogy and critical perspectives. More significant, our view that this work will fail to enlist widespread support among school personnel is based on the failure of critical perspectives to address major cultural shifts currently under way. First, this work has inadequately addressed issues of multiculturalism and racism. Second, it has failed to address the intensifying ecological crisis.

Cameron McCarthy (1990a, 1990b) has pointed to the failure of radical or critical curriculum scholarship to address adequately the concept of race. Critical scholarship was largely silent regarding race during its first and more significant phase—reproduction theory—and when mentioned, race tended to be subsumed under class or assumed to be of equal explanatory weight (McCarthy, 1990b). Relatedly, the very rhetoric of emancipatory or critical pedagogy resides in a European-American discourse (Ellsworth, 1989). Consider the following assumptions of critical pedagogy: (a) individual emancipation and empowerment are to be achieved through a process of critical reflection that demystifies historically and socially constituted forms of authority. (b) The goal of the educational process is emancipation; it must be viewed as a social process furthered through dialogue. For instance: "In the process of knowing the reality which we transform, we communicate and know socially even though the process of communicating, knowing, changing has an individual dimension" (Shor & Freire, 1987, p. 14). (c) The historical process of change (i.e., progress) is facilitated as individuals pursue social and individual emancipation. Put another way, the continual process of authorizing ideas and values through critical reflection (a process experimental in nature when viewed within the longer time frame of a cultural group) displaces other forms of knowing. (d) The educational process of transmitting culture is preeminently political. The goal of critical work involves "making the pedagogical more political and the political more pedagogical" (Giroux, 1988b). (e) A participatory democracy, wherein every member evidently experiences the same freedom and equality in all aspects of public life, represents the only legitimate form of political organization. (f) Finally, reality can be grasped in binary terms. For instance: "Liberatory dialogue is a democratic communication which disconfirms domination and illuminates while affirming the freedom of the participants to re-make their culture. Traditional discourse confirms the dominant mass culture and the inherited, official shape of knowledge" (Shor & Freire, 1987, p. 13).

As these assumptions suggest, critical pedagogy and critical scholarship generally incorporate the ideals of the Western Enlightenment. As well, they incorporate the silences, misconceptions, and hubris of this tradition. Applied to education, this tradition reproduces modes of reasoning Alvin Gouldner identified as distinctive to the "culture of critical discourse" (i.e., intellectuals whose own position is strengthened by forcing others to adhere to the rules of discourse they themselves have created and now represent as universally applicable to all culture). The highly abstract concepts characteristic of political scholarship (reproduction, resistance, emancipation, liberation, etc.) are, in principle, disconnected from the lived beliefs, values, and practices of specific cultural groups. Members of cultural groups who attempt to retain or recover their identities by repudiating the homogenization associated with mainstream consumer culture would take exception to many of these assumptions. Cameron McCarthy (1988) notes, in this regard:

I have begun to see contemporary Marxism as something of a classical realist text in which the subjective and omniscient speaking positions are reserved for white new middle class male intellectuals. Much of radical education theory is there part of the enabling linguistic competence of a peculiarly unreflexive community. In these frameworks, third world people are constituted as the objects of radical forms of intellectual tourism (Roman, 1987). (p. 8)

Probably, few cultural groups, including the advocates of critical perspectives themselves, would be willing to live continuously in the state of liminality that would accompany an ongoing and thorough process of demystification (which, in this work, requires politicizing) of "the historically and socially constructed forms by which they live" (Giroux, 1988c, p. 177). Such utopian language threatens to sweep away not only blocks to genuine progress, but sacred and life affirmative traditions as well. Additionally, the uncritical embrace of dialogue as the sole means of education discloses a lack of understanding of how group identity and traditions are maintained over time (e.g., Hopi and Jewish traditions, to name just two). For many cultural groups, it is doubtful that progress would be viewed as achievable through the critical pedagogy ideal of continual doubt: "to doubt everything, and to try to identify those forms of power and control that operate in their own lives" (McLaren, 1989, p. 233).

The goals of critical scholarship (empowerment, emancipation, and so on) and the limited suggestions offered by advocates to inspire classroom practice must be viewed as problematical because of the anthropological and linguistic poverty of their fundamental assumptions. As well, there are gender and racial blind spots that underline the fact that the leading advocates are European-American men. It may be that critical scholarship is self-referential scholarship and that, because these men insist on the primacy of their concepts and methods, their work functions in politically conservative, even reactionary ways. This divergence between rhetoric and political consequence has been noted at least twice (Pinar, 1981b; Wexler, 1987).

A third concern involves the relative silence of critical scholarship regarding the ecological crisis, a silence that could be linked with the cultural provincialism of this work. While the crisis merits book-length treatment, suffice it to say here that the accumulation of "greenhouse gases," coupled with massive disruptions in the ecosystem caused by deforestation, topsoil loss, and toxic waste dumping (to cite only a few aspects of the crisis), has led to an increasing awareness regarding the interdependence of human and natural systems. Evidence of the diverging trend line, with human population and consumer demands on the sustaining capacities of natural systems moving upward at a rapid rate while there is a corresponding decline in toxic-free ecosystems, has prompted many to question whether the most basic assumptions upon which modern (Western) consciousness rests may be exacerbating the crisis. Aldo Leopold was among those to first recognize that modern consciousness is based on an anthropocentric view of the universe and that long-term survival of the species depends on the emergence of new modes of cognition that represent the community of all life forms. Others, among them Gary Snyder, Wes Jackson, Wendell Berry, and Gregory Bateson, have begun to question the adequacy of Western contemporary cultural assumptions for long-term species survival. How we understand freedom and knowledge, among other fundamental constructs, must be linked with these considerations, not with a narrower—and historically discredited—Marxist agenda of class conflict and social change.

Although Giroux and McLaren (Giroux, 1991) have incorporated the discourses of postmodernism and feminism into their more recent expositions, critical scholarship as a whole remains located within a mode of reasoning that is unresponsive to the urgent, overwhelming character of the ecological crisis. Because space is limited, we limit this discussion to two essential issues: (a) how the ecological crisis leads to a different way of thinking about human freedom, and (b) the forms of knowledge that contribute to living in a sustainable relationship with the habitat. First, the recognition that humans are part of a biotic community leads to a view of human freedom different from that of the anthropocentric formulation of critical scholarship. Interdependence with the biotic community requires a concept of freedom as self-limitation for the sake of others, for the sake of species survival. According to Leopold, "a thing is right when it tends to preserve the integrity, stability, and beauty of the biotic community. It is wrong when it tends otherwise" (Leopold, 1966, p. 262). In providing an ecological model of cognitive processes that overcomes the human/nature dichotomy, Bateson challenges an anthropocentric concept of human freedom: "In no system which shows mental characteristics can any part have unilateral control over the whole. In other words, *the mental characteristics of the system are immanent, not in some part, but in the system as a whole*" (Bateson, 1972, p. 316). Bateson also challenges the Western bias that understands the human and natural worlds as separate because of the rational capabilities of human beings:

The total self-corrective unit which processes information, or, as I say, "thinks" and "acts" and "decides" is a *system* whose boundaries do not at all coincide with the boundaries either of the body or what is popularly called the "self" or "consciousness." (Bateson, 1972, p. 319)

He warns us to recognize that human beings, regardless of their ideas of freedom and progress, cannot survive independent of their habitat; that is, "*the unit of evolutionary survival turns out to be identical with the unit of mind*" (1972, p. 483).

The position advanced by Leopold, Bateson, and others leads to a profoundly different view of "human" freedom than that implied in critical scholarship. In that work, the past seems to represent domination and the future an ever-expanding realm of possibility, each succeeding generation demystifying the forms of domination presumably perpetuated by the previous generation. This sense of ever-expanding human freedom—material, political, intellectual—does not incorporate a vision of human freedom as contingent and interwoven with the larger biotic community of which we are part. Their anthropocentric view of freedom reproduces the Western myth of perpetual progress (Lasch, 1991), a myth that has led to the current crisis, and is silent upon this, the most fateful issue of the present time.

The ecological crisis implies that we must understand the very concept of knowledge differently. Recent and sympathetic studies of primal cultures that have evolved in ecologically sustainable ways (Hughes, 1983; Nelson, 1983) point to how these low status cultures (as viewed by mainstream Western intellectuals) function in species-affirmative ways. These ecologically wedded ways of knowing must not be interpreted as sentimentalized calls for earlier, premodern forms of social living, but rather as suggestive of knowledge forms unreflected in critical scholarship. Wes Jackson uses the rural farm to illustrate knowledge that is accumulated over generations, tested in the context of family and community survival—knowledge that has sustained, variously, the Koyukon hunter, the Hopi, and even certain European-American groups. In *Altars of Unhewn Stone*, Jackson writes:

> Species extinction and genetic narrowing of the major crops aside, the loss of cultural information due to the depopulation of our rural areas is far greater than all the information accumulated by science and technology in the same period. Farm families who practice the traditions associated with planting, tending, harvesting, and storing the produce of the agricultural landscape, gathered information, much of it unconsciously, from the time they were infants: in the farm household, in the farm community, and in the barns and fields. They heard and told stories about relatives and community members who did something funny or were caught in some kind of tragedy. From these stories they learned basic lessons of agronomy. But there was more. There was information carried by a farmer who looked to the sky and then to the blowing trees or grasses and made a quick decision as to whether or not to make two more rounds before quitting to do chores. Much of that information has already disappeared and continues to disappear as farmers leave the land. It is the kind of information that has been hard won over the millennia, from the time agriculture began. It is valuable because much of it is tuned to the harvest of contemporary sunlight, the kind of information we need now and in the future on the land. (Jackson, 1987, pp. 11–12)

We are not arguing, of course, that tacit, analogical forms of knowledge can or should be accepted uncritically as guides to social practice. Jackson's statement reminds us that the rationalistic and experimental forms of knowledge embedded in the work of the critical group, forms that can be traced back through Descartes to Plato, may not always represent the highest forms of knowledge, nor may they be suited for living in ecological balance.

The ecological crisis confronts educators with a new set of challenges unacknowledged by critical scholars. Required are forms of authentically radical thinking that enable us to articulate and provisionally answer fundamental questions hinted at by the following:

1. What cultural beliefs, values, and social practices currently transmitted via school curricula are ecologically affirmative, and which are rooted in assumptions formed in earlier historical periods when the natural environment was regarded as an exploitable resource?

2. How are ecologically unsustainable aspects of our culture encoded in those language patterns that are employed in classrooms and that teachers tend to take for granted?

3. How can we reframe in curricular forms the study of diverse his-

torical narratives in ways that enable students to recognize the consequences of anthropocentric biases, as well as prejudices against learning from ecologically sustainable cultures that have, in the main, been labeled pejoratively as "primitive"?

4. What forms of folk knowledge (technologies, social practices, etc.) that have evolved in harmony with the bioregion are pertinent to the crisis today, and how might these forms be incorporated into the curriculum?

5. How do we, as educators, begin to develop the languages of dance, painting, music, and narrative that primal peoples used as a means of encoding the moral templates for living in ecologically sustainable relationships, and for providing members of the community the experience of transforming the ordinary into an extraordinary, transformative sense of reality?

If, as a field, we become aware that the ecological crisis requires both cultural and technical responses, the scholarly forms of understanding curriculum as a political text will shift. Wexler's insistence that critical scholarship functions in politically reactionary ways—an allegation our summary of the ecological crisis supports and expands—will be acknowledged generally, and this judgment will then produce epistemological as well as thematic revisions in the effort to understand curriculum as a political text. The preoccupation with reproduction, resistance, and critical pedagogy will disappear or be revised dramatically. An ecologically informed agenda will occupy a central place in future political scholarship. The time for such change is past due.

Conclusion

Fundamental problems plague the contemporary effort to study the politics of curriculum. These are problems associated with gender, race, and culture; they are embedded in the modes of cognition as well as the themes and slogans of critical scholarship. Additionally, the political practices of critical scholars themselves seem self-involved, walled-in, and at times even militaristic. Finally, critical scholarship tends to omit what must be regarded as one of the political issues, if not the central issue, of the day—the ecological crisis. These are staggering problems, and, if unsolved, they threaten to undermine the most voluminous body of scholarship in the field.

Despite these significant problems, the effort to understand curriculum as a political text must be regarded as one of the great achievements of the curriculum field since the 1970s. At that time, the curriculum field, many readers will recall, was judged as moribund, dead, and arrested by Joseph Schwab (1970), Dwayne Huebner (1976), and William Pinar (1978), respectively. Concerns and questions notwithstanding, no commentator would regard the present field in these terms. Indeed, the volume and dynamism of critique testify to the vitality and complexity of the field. Without doubt, the apolitical blindspot of the traditional field has been corrected. Now we understand, to an extent few could have fathomed 20 years ago, the complex of ways in which curriculum is a political text.

REFERENCES

Althusser, L. (1971). *Lenin and philosophy and other essays* (B. Brewser, Trans.). New York: Monthly Review Press.

Anyon, J. (1979). Ideology and United States history textbooks. *Harvard Educational Review, 49*, 361–386.

Anyon, J. (1988). Schools as agencies of social legitimation. In W. F. Pinar (Ed.), *Contemporary curriculum discourses* (pp. 175–200). Scottsdale, AZ: Gorsuch, Scarisbrick.

Apple, M. W. (1975). The hidden curriculum and the nature of conflict. In W. F. Pinar (Ed.), *Curriculum theorizing: The reconceptualists* (pp. 95–119). Berkeley, CA: McCutchan.

Apple, M. W. (1979). Curriculum and reproduction. *Curriculum Inquiry, 9*(3), 231–252.

Apple, M. W. (1980). Analyzing determinations: Understanding and evaluating the production of social outcomes in the schools. *Curriculum Inquiry, 19*(1), 55–76.

Apple, M. W. (1981). Reproduction, contestation, and curriculum: An essay in self-criticism. *Interchange, 12*, 27–47.

Apple, M. W. (Ed.). (1982a). *Cultural and economic reproduction in education: Essays on class, ideology, and the state.* New York: Routledge & Kegan Paul.

Apple, M. W. (1982b). *Education and power.* New York: Routledge & Kegan Paul.

Apple, M. W. (1986). *Teachers and texts: A political economy of class and gender relations in education.* New York: Routledge & Kegan Paul.

Apple, M. W. (1987a). Producing inequality: Ideology and economy in the national reports on education. *Educational Studies, 18*, 195–220.

Apple, M. W. (1987b). Will the social context allow a tomorrow for tomorrow's teachers? *Teachers College Record, 88*, 330–337.

Apple, M. W. (1988). Social crisis and curriculum accords. *Educational Theory, 38*, 191–201.

Apple, M. W. (1990a). *Ideology and curriculum* (2nd ed.). New York: Routledge & Kegan Paul.

Apple, M. W. (1990b). The politics of pedagogy and the building of community. *Journal of Curriculum Theorizing, 8*(4), 7–22.

Apple, M. W., & Ladwig, J. (1989, December 20). Educators reel from decade of right-wing attacks. *The Guardian,* p. 9.

Apple, M. W., & Teitelbaum, K. (1986). Are teachers losing control of their skills and curriculum? *Journal of Curriculum Studies, 18*, 177–184.

Apple, M. W., & Weis, L. (Eds.). (1983). *Ideology and practice in schooling.* Philadelphia: Temple University Press.

Aronowitz, S., & Giroux, H. A. (1985). *Education under siege: The conservative, liberal and radical debate over schooling.* South Hadley, MA: Bergin & Garvey.

Barton, L., Meigham, R., & Walker, S. (Eds.). (1980). *Schooling, ideology and the curriculum.* Sussex, England: Falmer Press.

Bateson, G. (1972). *Steps to an ecology of mind.* New York: Ballantine.

Beyer, L. E., & Apple, M. W. (1988). *The curriculum: Problems, politics and possibilities.* Albany: State University of New York Press.

Bowers, C. A. (1987). *Toward a post-liberal theory of education.* New York: Teachers College Press.

Bowles, S., & Gintis, H. (1976). *Schooling in capitalist America: Educational reform and the contradictions of economic life.* New York: Basic Books.

Bowles, S., & Gintis, H. (1980). Contradiction and reproduction in educational theory. In L. Barton, R. Meigham, & S. Walker (Eds.), *Schooling, ideology, and the curriculum* (pp. 51–65). Sussex, England: Falmer Press.

Carson, T. (in press). Remembering forward. In W. F. Pinar & W. M. Reynolds (Eds.), *Understanding curriculum as phenomenological and deconstructed text.* New York: Teachers College Press.

Counts, G. S. (1932). *Dare the schools build a new social order?* New York: John Day Co.

Dewey, J. (1916). *Democracy and education.* New York: Macmillan.

Ellsworth, E. (1989). Why doesn't this feel empowering? Working through the repressive myths of critical pedagogy. *Harvard Educational Review, 59,* 297–324.

Freire, P. (1970/1971). *Pedagogy of the oppressed.* New York: Continuum.

Freire, P. (1985). *The politics of education: Culture, power and liberation* (D. Macedo, Trans.). South Hadley, MA: Bergin & Garvey.

Freire, P., & Shor, I. (1987). *A pedagogy for liberation: Dialogues for transforming education.* South Hadley, MA: Bergin & Garvey.

Giroux, H. A. (1980). Beyond the correspondence theory: Notes on the dynamics of educational reproduction and transformation. *Curriculum Inquiry, 10*(3), 225–248.

Giroux, H. A. (1981a). *Ideology, culture and the process of schooling.* Philadelphia: Temple University Press.

Giroux, H. A. (1981b). Hegemony, resistance, and the paradox of educational reform. *Interchange, 12,* 3–26.

Giroux, H. A. (1981c). Toward a new sociology of curriculum. In H. A. Giroux, A. N. Penna, & W. F. Pinar (Eds.), *Curriculum and instruction: Alternatives in education* (pp. 98–108). Berkeley, CA: McCutchan.

Giroux, H. A. (1983a). *Theory and resistance in education: A pedagogy for the opposition.* South Hadley, MA: Bergin & Garvey.

Giroux, H. A. (1983b). Theories of reproduction and resistance in the new sociology of education: A critical analysis. *Harvard Educational Review, 53,* 261–293.

Giroux, H. A. (1985a). Toward a critical theory of education: Beyond a Marxism with guarantees—A response to Daniel Liston. *Educational Theory, 35,* 313–319.

Giroux, H. A. (1985b). Thunder on the right: Education and the ideology of the quick-fix. *Curriculum Inquiry, 15*(1), 57–62.

Giroux, H. A. (1987). Liberal arts, public philosophy, and the politics of civic courage. *Curriculum Inquiry, 17*(3), 331–335.

Giroux, H. A. (1988a). *Schooling and the struggle for public life: Critical pedagogy in the modern age.* Minneapolis: University of Minnesota Press.

Giroux, H. A. (1988b). *Teachers as intellectuals: Toward a critical pedagogy of learning.* South Hadley, MA: Bergin & Garvey.

Giroux, H. A. (1988c). Border pedagogy in the age of postmodernism. *Journal of Education, 170,* 162–181.

Giroux, H. A. (1989). Rethinking education reform in the age of George Bush. *Phi Delta Kappan, 70,* 728–730.

Giroux, H. (Ed.). (1991). *Postmodern, feminism and cultural politics.* Albany: State University of New York Press.

Giroux, H. A., & McLaren, P. L. (1989). *Critical pedagogy, the state, and cultural struggle.* Albany: State University of New York Press.

Giroux, H. A., Penna, A. N., & Pinar, W. F. (Eds.). (1981). *Curriculum and instruction: Alternatives in education.* Berkeley, CA: McCutchan.

Giroux, H. A., & Purpel, D. (Eds.). (1983). *The hidden curriculum and moral education: Deception or discovery?* Berkeley, CA: McCutchan.

Gramsci, A. (1972.) *Selections from the prison notebooks of Antonio Gramsci* (1st ed.; Q. Hoare & G. N. Smith, Eds. and Trans.). New York: International Publishers. (Original work published 1971)

Gramsci, A. (1975). *Selections from cultural writings* (D. Forgacs & G. N. Smith, Eds.; W. Boelhoemer, Trans.). London: Lawrence and Wishart.

Held, D. (1980). *Introduction to critical theory: Horkheimer to Habermas.* Berkeley: University of California Press.

Huebner, D. (1976). The moribund curriculum field: Its wake and our work. *Curriculum Inquiry, 6*(2), 153–167.

Hughes, J. D. (1983). *American Indian ecology.* El Paso: Texas Western Press.

Jackson, P. (1968). *Life in classrooms.* New York: Holt, Rinehart and Winston.

Jackson, W. (1987). *Altars of unhewn stone.* San Francisco: North Point Press.

Kincheloe, J. L., & Pinar, W. F. (Eds.). (1991). *Curriculum as social psychoanalysis: Essays on the significance of place.* Albany: State University of New York Press.

Lasch, C. (1991). *The true and only heaven: Progress and its critics.* New York: Norton.

Leopold, A. (1966). *A Sand County almanac.* San Francisco and New York: A Sierra Club/Ballantine Book.

Liston, D. P. (1988). *Capitalist schools: Explanation and ethics in radical studies of schooling.* New York: Routledge & Kegan Paul.

Marx, K., & Engels, F. (1974). *Karl Marx and Frederick Engels on literature and art: A selection of writings* (L. Baxandall & S. Morawski, Eds.). New York: International General.

McCarthy, C. R. (1988). Slowly, slowly, slowly the dump speaks: Third world popular culture and the sociology of the third world. *Journal of Curriculum Theorizing, 8*(2), 7–23.

McCarthy, C. R. (1990a). Rethinking liberal and radical approaches to racial inequality in schools: Making the case for non-synchrony. *Harvard Educational Review* (reprint series 21), 265–279.

McCarthy, C. R. (1990b). *Race and curriculum.* London: Falmer Press.

McCarthy, C. R., & Apple, M. W. (1988). Race, class, and gender in American education. In L. Weis (Ed.), *Class, race, and gender in American education* (pp. 3–39). Albany: State University of New York Press.

McLaren, P. (1986). *Schooling as a ritual performance: Towards a political economy of educational symbols and gestures.* Boston: Routledge & Kegan Paul.

McLaren, P. (1989). *Life in schools: An introduction to critical pedagogy in the foundations of education.* New York: Longman.

McLaren, P. (1990). Capitalist schools: Explanation and ethics in radical studies of schooling, a review. Unpublished manuscript.

Miller, J. (1979). *History and human existence: From Marx to Merleau-Ponty.* Berkeley: University of California Press.

Nelson, R. K. (1983). *Make prayers to the raven: A Koyukon view of the northern forest.* Chicago: University of Chicago Press.

Oakes, J. (1985). *Keeping track: How schools structure inequality.* New Haven, CT: Yale University Press.

Olson, P. (1981). Rethinking social reproduction. *Interchange, 12,* 1–2.

Pinar, W. F. (1978). Notes on the curriculum field 1978. *Educational Researcher, 7*(8), 5–12.

Pinar, W. F. (1981). The abstract and the concrete in curriculum theorizing. In H. A. Giroux, A. N. Penna, & W. F. Pinar (Eds.), *Curriculum and instruction: Alternatives in education* (pp. 431–454). Berkeley, CA: McCutchan.

Pinar, W. F. (1990). Impartiality and comprehensiveness in teaching curriculum theory. In J. T. Sears & J. D. Marshall (Eds.), *Teaching and thinking about curriculum* (pp. 259–264). New York: Teachers College Press.

Pinar, W. F., & Reynolds, W. M. (Eds.). (1992). *Understanding curriculum as phenomenological and deconstructed text.* New York: Teachers College Press.

Rugg, H. (1929–1932). *Man and his changing society* (Rugg social science series of the elementary school course, Vols. 1–6). Boston: Ginn.

Schubert, W. H., Schubert, A. L. L., Herzog, L., Posner, G., & Kridel, C. (1988). A genealogy of curriculum researchers. *Journal of Curriculum Theorizing, 8*(1), 137–184.

Schwab, J. (1970). *The practical: A language for curriculum.* Washingon, DC: National Education Association.

Shapiro, H. S. (1981). Functionalism, the state, and education: Towards a new analysis. *Social Praxis, 98*(3/4), 5–24.

Shapiro, H. S. (1983). Class, ideology and the basic skills movement: A study in the sociology of educational reform. *Interchange, 14*(2), 14–24.

Sharp, R. (1980). *Knowledge, ideology and the politics of schooling: Towards a Marxist analysis of education.* London: Routledge & Kegan Paul.

Shor, I. (1986). *Culture wars: School and society in the conservative restoration 1969–1984.* New York: Routledge & Kegan Paul.

Shor, I. (1987a). *Critical teaching and everyday life* (3rd ed.). Chicago: University of Chicago Press.

Shor, I. (Ed.). (1987b). *Freire for the classroom: A sourcebook for liberatory teaching* (1st ed.). Portsmouth, NH: Boynton/Cook.

Shor, I., & Freire, P. (1987). What is the dialogical method of teaching? *Journal of Education, 169,* 11–31.

Strike, K. A. (1989). *Liberal justice and the Marxist critique of education.* New York: Routledge & Kegan Paul.

Weis, L. (Ed.). (1988). *Class, race and gender in American education.* Albany: State University of New York Press.

Wexler, P. (1976). *The sociology of education: Beyond equality.* Indianapolis, IN: Bobbs, Merrill.

Wexler, P. (1987). *Social analysis of education: After the new sociology.* Boston: Routledge & Kegan Paul.

Whitty, G. (1985). *Sociology and school knowledge: Curriculum theory, research, and politics.* London: Methuen.

Williams, R. (1976). *Keywords: A vocabulary of culture and society.* New York: Oxford University Press.

Willis, P. (1981). *Learning to labour.* Hampshire, England: Gower.

Wood, G. H. (1988). Democracy and the curriculum. In L. E. Beyer & M. W. Apple (Eds.), *The curriculum: Problems, politics and possibilities* (pp. 166–190). Albany: State University of New York Press.

Chapter 5

Computers and Education:
A Cultural Constructivist Perspective

TONY SCOTT, MICHAEL COLE, and MARTIN ENGEL
Laboratory of Comparative Human Cognition

The general topic of the use of computers in education has not been systematically dealt with in the *Review of Research in Education*, although some specific aspects of it have been touched on (e.g., Sherrie Gott's chapter on apprenticeship and intelligent tutoring systems in Volume 15). Since the field is so vast, no review could do justice to it; thus, we must circumscribe the research we are reviewing to a manageable portion.

Our major restriction is that we concentrate on Grades K–12 with an emphasis on late elementary school and early secondary school ages, including higher levels of the education system only in a few cases to illustrate a point. Within this still-vast field we select some research that has been widely disseminated, and therefore characterizes the field, and some research that, in our view, holds special promise for the future. Also, we come to this topic from a particular theoretical perspective, which acts as a further filter on the topics discussed.

We call that perspective *cultural constructivism*. The basic idea of this approach can be grasped most readily by contrasting it with Piagetian constructivism.

Piaget is justifiably famous for demonstrating the need to consider children to be constructors of their own development through their actions. By contrast, a *cultural constructivist* approach assumes not only an active child but an equally active and usually more powerful adult in interaction (we are speaking of educational settings). Moreover, cultural constructivism emphasizes that all human activity is mediated by cultural artifacts, which themselves have been constructed over the course of human history.

The general framework of this approach is derived from the axioms of the cultural-historical school of psychology, which asserts that the unique character of human activity is that it is mediated through socially constituted, and historically developing, systems of artifacts (see Wertsch, 1985, for a general treatment).

From this perspective, the historically conditioned forms of activity mediated through computers must be studied for the qualitatively distinctive forms of interaction that these artifacts afford and the social arrangements that they help to constitute. Moreover, one is encouraged to seek explanation of current uses of computers in terms of the history of the technology and the social practices that the technology mediates; one needs to consider the "effects" of interacting in this medium not only as they are refracted through transfer tests or in local activity systems (such as classroom lessons) but also in the entire system of social relations of which they are a part.

We begin to construct such a framework for computer use in education by sketching the historical context in which computer technology came to prominence and the perceived state of American education at the time it did so. We begin by sketching the 20th-century origins of computing in the military establishment. These origins are important to note not only for historical reasons but because the military remains the most important organization promoting research in computer-based education. We then look at various characterizations of the pedagogical use of computers in civilian education, concentrating primarily on the K–12 component.

Our account of these pedagogical factors is broadly cast and includes consideration of patterns of computer provision and of teacher education. We trace some of the leading trends in the use of computers for educational purposes over the past few years (those we feel to be most relevant to actual classroom practice).

After an all-too-brief excursion into the broader contemporary social context of computer use in education highlighted by recent research on gender and ethnicity, we discuss some projects that embed computer use in education in wider social contexts and conclude with a discussion of the evaluation issues in computer education.

THE CULTURAL-HISTORICAL CONTEXT OF COMPUTER USE IN EDUCATION

Computers and our conception of computers (if we may be allowed a commonsense dualism) were both constituted by and helped to constitute changes in the world between the mid-1940s and 1990. In this process, they were touted both as the positive agent of an optimistic vision of the future and excoriated as the negative agent of a grimmer version of that future (Evans, 1982; Stonier, 1983; Weiner, 1950/1989).

In the years before World War II, the word *computer* referred to a person who computed numbers. During the war, as the technology of ballistics developed and as encoding devices became more sophisticated, there was an urgent need to calculate enormously complex equations. As Winston (1986) put it, ENIAC (the first practical digital computer) was "still, in essence, a calculator designed to work out ballistic firing tables" (p. 137).

The great rapidity of the calculating machine soon displaced the (human) computer, whose work was restricted to programming input and using output. The spread of modern computers far beyond the confines of the military in recent decades somewhat masks the continued influence of its origins through military-sponsored research. Joseph Weizenbaum (1976, p. 568) has commented that "the computer in its modern form was born from the womb of the military." A good deal of historical research has supported his conclusion (P. Edwards, 1985, 1986, 1988; Noble, 1989; Slater, 1990; Winston, 1986).

Many of the developments in educational uses of computers we will discuss in this chapter have their origins in research on creating man-machine systems for military purposes and in improving military training, an origin that continues to shape the very structure of those educational practices. Whether one thinks this state of affairs is good or bad is a matter of personal values and estimation of effects. Our own view is that one needs to be suspicious of educational technology that embodies presupposed fixed tasks and goals and a restricted range of social arrangements of a top-down, authoritarian nature.

The development of computer technology also bears an interesting relationship to Americans' views of their educational system and to their views of the utility and importance of the "computer revolution" within and outside the education system. Although alarm has been expressed about the quality of education in American schools throughout this century, the period just following World War II found the United States a dominant world power whose technological achievements were a matter of pride and emulation. So secure did matters seem in the early 1950s that J. K. Galbraith could celebrate the prospect of a permanent freedom from want in his highly publicized book *The Affluent Society* (1958). The country seemed to support his optimistic view by responding with a series of educational reforms that implemented, more or less, the vision of progressive educationalists following in the tradition of John Dewey (see Cremin, 1976, for a review and discussion).

Activity-based curricula became the order of the day. Classrooms were reorganized to facilitate group and project work. A "checklist" focus on assessing achievement was displaced by a concentration on outcomes derived from integrative studies illustrating a range of curriculum achievements. However, while the evidence shows that these activity-based curricula (many of which were highly technological in character) were successful as first implemented, follow-up surveys indicate that once external funding was withdrawn, their use declined. At present, activity-centered curricula continue to be found in less than 10% of classrooms in the United States (see Kyle, 1984, for reviews and discussions of these efforts).

As America leapt ahead of the Soviet Union in the competition for dominance in space exploration during the 1960s, U.S. attention turned from concern to maintain high-powered education of the nation's middle classes to address those portions of the population not included among the affluent as part of a highly publicized "war against poverty."

Education was seen as the key to breaking the "cycle of poverty" so that all could partake of the affluent life. It is no secret that although many children benefited from the improved health care and nutritional services provided by Project Head Start, the "war on poverty" was most decidedly not won and the educational achievements of poor, minority-group children have continued to remain distressingly low (Muenshaw, 1980; Payne, 1973; Washington, 1987).

While this situation evoked and continues to evoke concern, the 1960s emphasis on eradicating economic injustice inside the United States was soon displaced by international concern evoked by our former World War II enemies, Japan and Germany. In recent decades, one has heard frequently the rhetoric so pithily captured by the title of the report of the National Commission on Excellence in Education (1983), *A Nation at Risk: The Imperative for Educational Reform.* When compared with the achievement of any number of economic competitors, American children continue to perform poorly in "basic" school subjects, especially those associated with the trinity of mathematics, science, and technology, upon which the United States' economic well-being is assumed to rest (see Stevenson, Lee, & Stigler, 1986; Walberg, Harnisch, & Tsai, 1984). Moreover, our achievements, such as they are, are very unevenly distributed.

At the same time that alarm was being expressed over comparative achievement levels in the schools, the American business community was distressed about the ability of school leavers to fill even entry-level jobs adequately. As various sectors of the American economy were eroded by foreign competition, industry began to get involved in education. Whereas the common wisdom in the 1970s had been that modern technology would simply *deskill* the work of lower and middle-level workers, by the 1980s it began to appear that highly educated workers were needed to run high-technology machinery and that such workers were in increasing demand (Sherman, 1985).

The issue of whether the amount of education was insufficient or the kind of education was inappropriate as preparation for "the average worker" still remains under dispute: Between 1976 and 1988 occupational groups that had above-average educational attainments grew by 51%, while those where low levels of educational achievement dominated grew by 19% (Bailey, 1989). In addition, skill levels demanded within occupational groups rose, especially on workers with low and average educational achievements.

There have been a variety of responses to this situation, including various forms of "back to basics" movements within the school curriculum and intensification of efforts by industry and business to upgrade the education and training of their employees. Recently, it has been estimated that U.S. firms spend $30 billion annually on formal training, a figure that is manifestly inadequate to the demand and is likely to grow (Commission on Workforce Quality and Labor Market Efficiency, 1989).

However important such job-related training may be, it is almost certainly not going to decrease the importance of education. In fact, some observers claim that

increasingly workers' positions in the labor market are determined prior to their entry into the labor market, in the course of their access to the vocational and higher educational systems. . . . The vocational and higher educational systems will need to undergo fundamental changes if they are to respond to these new pressures. (Noyelle, 1985)

When we combine these considerations with projected increases in the intensity of economic competition, it is easy to see why there is cause for concern. By the same token, it is easy to see why so many people are attracted to the use of computers as a means of extracting us from a difficult situation; computers seem to promise a technological "quick fix," a relatively cheap, clean, and unproblematic solution to what we believe to be long-term, expensive, and dirty problems (Kerr, 1991).

Given the record of past technology-driven reforms, we approach the question of computer use as a solution to educational problems with some skepticism. As Cuban (1986) has recently warned, the education system will absorb each successive quick fix offered by technology and restore the status quo. This view is supported by recent studies conducted by Rosenberg (1991) of the notion of "computer literacy" as both making irrelevant promises in terms of the economic situation and failing to deliver on those promises, and the reactions to that work by computer educators.

Cuban points out that in the early phases, at least, the introduction of computer technology in the schools is recapitulating patterns of adoption of film and radio, two media whose transformational potential for education was announced in almost precisely the same terms as computer-led transformations in schooling are announced today. Consider, for example, the following claim:

The central and dominant aim of education by [computers] is to bring the world to the classroom, to make universally available the services of the finest teachers. . . . The time may come when a [computer] will be as common in a classroom as a blackboard. [Computer] instruction will be integrated into school life as an accepted educational medium. (Cuban, 1986, p. 19)

Visions of this kind are encountered so often with respect to computers in school that one must pause and think about the fact that this particular claim was written in 1932: The medium in question was radio!

Cuban's work reminds us that one of the few firm laws concerning the effects of introducing a new technology is the tendency of the social system to retain current goals and social organizations (and seek to achieve old goals more efficiently). To be successful as an agent of change (*reform*), technologically based strategies should be based in a self-conscious effort to construct a social environment with a new morphology of interpersonal communication. It follows that there is a fundamental contradiction between the education system's conservative tendency to restore the status quo in response to each new technological innovation and the intended and unintended consequences of that innovation with respect to the goals of activities to be realized in the classroom (Cuban, 1986). We believe this to be especially significant in attempts to use computers as an agent of change in education, although the record thus far provides only meager justification for optimism.

THE PEDAGOGICAL CONTEXT OF EDUCATIONAL COMPUTING

Because our perspective highlights the importance of cultural and social factors as determinants of computer-mediated classroom practice, it also emphasizes the "close-in" components of the educational system. Curriculum design, building organization, teacher preparation, and their histories all profoundly affect the realization of the potential of the computer in the classroom.

First we look at the issues of computer literacy, computer competence, and computer programming, then we explore some applications of computing to education from a curriculum/subject perspective. The curriculum areas that we choose to highlight are science, English, and mathematics. (We bypass social studies because Thornburg and Pea [1991] have analyzed the synthesis of instructional technologies and educational culture in the context of social studies teaching from a perspective similar to our own.)

Our selection, though partial, serves to highlight the variety of approaches possible for computer use in the classroom. All but art are also presently the focus of debate on how to make a national assessment of progress toward nationally promulgated educational goals.

In examining this range of subjects, we draw on the characterization of various "modes" of human/computer interaction, developed for the purposes of comparative study by Makrakis (1988). Makrakis proposes, as a compromise between the various views of the computer as a device for individualized instruction and as a medium of interaction, that one should attend to no less than eight distinct "modes of delivery and interaction":

It would be more practical to consider eight modes of delivery and interaction: (1) drill and practice; (2) tutorial; (3) instructional games; (4) simulation; (5) problem solving; (6) spreadsheet; (7) word processing; and (8) database management-processing. These modes have been placed in a hierarchical order from low to high according to their levels of cognitive/mental thinking and degree of learner/computer interaction. *It is of particular importance to note that any computer program may explore more than one of these modes.* A tutorial mode, for example, formally includes drill-and-practice exercises. Likewise, instructional games and simulations may be incorporated into problem-solving activities. (pp. 12–13) [italics added]

In our examination of the application of computing to English and mathematics, we concentrate on drill-and-practice as a historically significant approach, giving somewhat less space to the other modes of interaction.

When considering art and science, we look at some approaches that can perhaps best be characterized as "modeling and simulation." But first we begin with a consideration of computing activities as curriculum content.

Computer Literacy and Computer Competence

With advances in computer technology and its spread into a variety of social spheres, educators began to focus on the need to train new generations of students to program such devices for a variety of purposes, and the concept of computer literacy was born. The earliest reference to computer literacy that we have been able to find occurred in an article by John Nevison about the ways in which involvement with computers was being integrated into the curriculum at Dartmouth College in the 1970s. Nevison noted that the ability to write computer programs was becoming part of the assumed foundation of a liberal education.

Because of the widespread use of elementary computing skill, there should be an appropriate term for this skill. It should suggest an acquaintance with the rudiments of computer programming, much as the term literacy connotes a familiarity with the fundamentals of reading and writing, and it should have a precise definition that all can agree on. It is reasonable to suggest that a person who has written a computer program should be called literate in computing. (Nevison, 1976, p. 401)

Adopting a rather narrow notion of literacy as mastery of the systems of symbol manipulation, one finds that during the late 1970s and early 1980s debates over computer literacy focused on the extent to which students need to be able to work with hexadecimal and binary number systems and to understand the principles of hardware construction. Perhaps the core conception of the pedagogical goals of computer use in schools at the time was provided by Arthur Leuhmann, who was quoted as asserting:

> One who is truly computer literate must be able to "do computing"—to conceptualize problems algorithmically, to represent them in the syntax of a computer language, to identify conceptual "bugs," and to express computational ideas clearly, concisely, and with a degree of organization and readability. (Douglas, 1980, p. 18)

During the 1980s one begins to see a shift in the terms of this discussion. With the advent of relatively inexpensive microprocessors, the dominant image of the computer as a machine driven by a card with rectangular holes punched in it is replaced by that of a microcomputer with a munched-upon rainbow apple on its screen.

It is not that the "traditional" emphasis on programming and learning to use quantifiable algorithms disappears. Rather, the advent of personal microcomputers, for which off-the-shelf applications programs were soon available, brought about a shift in emphasis among educators, researchers, and commentators. Instead of focusing on basic programming and engineering skills, computer literacy came to be seen as the ability to choose appropriate software applications and to modify them if necessary (but not at the level of the source program). Significantly, these applications began to extend beyond computing (understood as calculation) to word processing, that is, literacy activities of a more traditional kind, modified to take advantage of the microprocessor as medium.

The availability of the microprocessor also facilitated the development of computing as an entertainment medium, in the form of arcade-style games, which enjoyed extensive popularity and were later appropriated as the templates for "educational" games, which used the same style of human-computer interface for "less-trivial" pursuits.

With the addition of word processing to number manipulation, another significant element was brought to popular discussions of computer literacy—the need for an ability to use computers as communication devices through which one could interact with other people as well as with databases of a variety of kinds (Kinzer, Sherwood, & Bransford, 1986; Trainor, 1984).

By the beginning of the 1980s the broadening capacity and availability of microcomputers and communication networks made it clear that the initial focus on computers as, literally, devices for making computations had been supplemented by a general conception of the computer as (potentially) a general purpose tool for the manipulation of information, as a medium for pursuing educational goals that have nothing intrinsically to do with computer programming, and as a source of entertainment.

Concomitantly, one began to encounter the notion that what one should seek educationally is "functional computer literacy," which Longstreet and Sorant (1985, p. 119) suggested must encompass "the ability to be flexible and to modify existing procedures to new hardware and software" (Chandler & Marcus, 1985; Stonier & Conlin, 1985).

Alongside these changing views of computer literacy (see Pryczak, 1990, for a collation of views about "minimal" skills), people were also developing notions of the computer as a conveyor of pedagogy, either directly (the computer taking a tutorial role) or indirectly (the computer as a tool and as a resource; Levin & Souviney, 1983). This variety of uses was captured by Taylor (1980), who spoke of computers as "tutors, tools and tutees."

The 1983 National Commission on Excellence in Education report, *A Nation at Risk*, was the first formal, national document to include a consideration of computer literacy as a component of the national profile of educational progress, alongside a review of the traditional "three Rs." *Computer Competence: The First National Assessment* (Martinez & Mead, 1988), based on 1985–1986 data, followed this up with a more detailed review of computer competence.

The transition from "computer literacy" to "computer competence" is not without significance. Indeed, the major findings of Martinez and Mead, the authors of the latter, Educational Testing Service report, are so significant to our ensuing analysis that we quote them in full.

Several key findings emerge from this first national assessment of computer competence:

Students generally had difficulty answering questions on the assessment, especially questions about computer applications and programming.

The experiences of having used a computer, or studying computers in school, and of having a computer at home are positively related to computer competence.

Most students like using computers and want greater access to them.

Much learning of computers takes place outside of school and independent of formal instruction. Across demographic subgroups, the increased competence associated with having a computer at home is comparable to the advantage linked to studying and using computers at school. Students who study computers at school and have a computer at home are the most competent.

Computers are seldom used in subject areas such as reading, math or science. Rather, the use of computers in schools is largely confined to computing classes.

Males, in general, demonstrate a slightly higher level of computer competence than females.

There are clear racial/ethnic differences in computer competence, favoring White students over Black and Hispanic students. These differences are present even between students who have comparable levels of experience, but the differences are accentuated by greater experience with computers among White students.

Other subgroup comparisons show an advantage for:
—students whose parents are college graduates
—students who attend non-public schools
—students who live in high socioeconomic metropolitan areas
—students who live in the Northeastern United States
These subgroups are most likely to have used a computer, to be studying computers at school, and to have a computer at home.

Many computer coordinators have minimal training in computer studies and rate themselves mediocre in their ability to use computers. (Martinez & Mead, 1988, pp. 5–6)

Many of these issues will be taken up in ensuing sections. Here we wish to comment on the notion of "computer competence" and the persistence of "ability to program" as a component of that competence. The National Assessment of Educational Progress (NAEP, 1985) report specified three areas as constituting computer competence: knowledge of computer technology, understanding of computer applications, and understanding of computer programming.

This is a much wider specification than the view of functional computer literacy specified by Longstreet and Sorant (1985) and others who argued that the important quality is the ability to apply computers to changing circumstances without a deep knowledge of the component parts or internal algorithms. The view implicit in the NAEP document, by contrast, is that now one *should* be able to recognize the functions of the various component devices and peripherals (and that by the 11th grade over 90% of those asked were able to do so).

In addition to these "knowledge" questions, students should also show skills in applying software to a task. Martinez and Mead (1988) report results from testing directed at the application of word processing, graphics, databases, and spreadsheets over Grades 3, 7, and 11. They report levels of success at Grade 11 for these four domains of 72.2%, 60.7%, 53.4%, and 31%, respectively (p. 20). They point out that these differentials might arise from frequency (better *in*frequency) of use.

The pattern of use is, of course, a reflection of a complex range of issues: Teacher preparation as a significant element of this kind of patterning will be taken up later.

Students were also asked about three programming languages—LOGO, BASIC, and Pascal—all designed primarily for educational circumstances (although Pascal plays a more definite role in "real" programming, its function as a language for education and training was an important element in its design). Programming performances reported were very low, never rising over 40% (Martinez & Mead, 1988, pp. 26–27).

Combining the various results, NAEP concluded with an estimate of computer competence over Grades 3, 7, and 11 in which "most students appeared to have difficulty answering the assessment questions. No grade averaged even 50% correct, and third graders were able to answer only a third of the questions correctly'' (p. 28).

Martinez and Mead (1988), quite correctly, note that perhaps the most important problem to be addressed is the continuing disagreement among educators about what should be taught about computers in American schools. This debate concerns the teaching of computing per se and the integration of computing in the curriculum.

One can see, in the different categories of results, that the desire to include programming as an element of computer competence has the effect of pulling down overall performance. Whereas the categories of ''knowledge about'' and ''how to apply'' computing can be clearly motivated in the classroom, and teachers can understand how to teach the ''knowledge about computers'' category fairly well, relatively few teachers, and even computer coordinators, are adequately prepared to teach the more subtle aspects of programming or are themselves competent to make effective use of authoring languages.

Computer Programming

Much research into computer use in education in recent years has focused on those situations in which ''the child is controlling the computer.'' The ''computer as tutee'' is conceived of as a ''protean'' object with potential to be applied to a whole range of problem domains. In some sense, the attention paid to students' abilities to manipulate the ''computer as tutee'' is a rehearsal of the original drive in computer literacy for programming competence not for its own sake but because of beliefs about the cognitive consequences of programming.

Rather than programming the original, underlying computer, however, the ''computer as tutee'' in the precollege curriculum is inevitably a computer under the control of an ''educational'' language environment. Most commonly, in the range of education we are concerned with, that environment is either the LOGO or BASIC programming language.

Examples can, of course, be found of experiments with the ''computer as tutee'' using a wide range of programming languages, including comparatively early and less sophisticated languages and more recent and complex approaches such as Prolog (Ennals, 1985). We have chosen, however, to concentrate on LOGO (as the computer language of the middle years of schooling) rather than engage in the debates about appropriate educational languages because the educational rationale is best articulated for this language and its use has been the subject of most research (Dyck & Trent, 1990; Pea & Kurland, 1984; Pea & Sheingold, 1987; Weir, 1987).

Miniprogramming

There is a class of programs that is sometimes used as a precursor to fuller implementations of LOGO or LOGO-like languages or cast in the guise of discovering and developing mathematical concepts or programming routines. For example, Pond (Sunburst Corporation) provides an environment in which children discover repeating patterns of lily pads. A notational system of directional arrow keys and numbers is used to instruct a frog to move according to an algorithm created to reflect the recursive pattern of lily pads in a given pond. The goal of the game is to move the frog from the beginning of the pond to the designated "magic" lily pad at the end of the pond. Students are given an opportunity to rehearse the required patterns by trials in which they can move the frog across the pond jump by jump. To move on through the program and to "win the game," however, they must commit themselves to inputting a complete sequence of instructions. The frog will follow the instructions and either succeed or fail in its mission of crossing the pond. The frog's moves are constructed in terms of patterns, such as "two right and three down": The child must construct the appropriate number/direction: number/direction pairings (Griffin, Belyaeva, & Soldatova, in press; Lemons, 1990). Whereas such programs are cast in terms of the teaching of mathematical principles, they also inter alia require the development of certain programming skills.

Programming: Gaining Mastery of Microworlds

Perhaps the most influential line of research on the instructional uses of computers, and one that enables a sharp break with the tradition of teacher-led lessons followed by drill-and-practice, is that led by Seymour Papert. By his introduction of a graphics element and a steerable turtle into the LOGO programming environment, he enabled students to approach programming through the "mini-programming with visual feedback" route outlined above. In his enormously influential book *Mindstorms* (1980), Papert presented a constructivist theory of learning and development and showed how LOGO, considered as a "microworld," could be used by children to construct a variety of interesting objects in various knowledge domains: geometry, music, art, and so on. Papert suggested that in addition to whatever domain-specific knowledge children accumulated, they would also accumulate powerful ideas about their own knowledge and learning process (often given the generic label "metacognitive" skills.) As he described the core idea:

In Turtle geometry we create an environment in which the child's task is not to learn a set of formal rules but to develop sufficient insight into the way he moves in space to allow the transposition of this self-knowledge into programs that will cause a Turtle to move. (1980, p. 205)

Using Piaget as the major source of inspiration, Papert's claims about the intellectual consequences of creating objects through LOGO were grounded in notions about assimilation and accommodation that seemed to promise broad transfer, much in the character of a Piagetian stage. He suggested that the computers (in particular, as used by members of the Massachusetts Institute of Technology [MIT] Media Lab) permitted children to "concretize (and personalize) the formal" (p. 21).

Although he denied a technological determinist interpretation of "the effects of LOGO," his descriptions of the character of the social setting were sufficiently backgrounded that a number of tests of Papert's claims about LOGO were conducted using more or less controlled procedures, a conventional experimental group–control group experimental design, and various measures of transfer.

Some of these studies failed to produce evidence of transfer, whereas others were successful for reasons that are hotly debated (for access to this literature, see Salomon & Perkins, 1988). The failures of replication were interpreted by Papert as a failure by the researchers to realize that the "effects of LOGO" were not intended to be the result of programming per se. Rather, these effects should be seen in something akin to a cultural constructivist account, as emerging from the entire reconfiguration of educational interactions, a reconfiguration that constitutes a culture in which mediation of activity through LOGO (and not just programming per se) generates widely applicable "tools of thought" (Burns & Coon, 1990; Papert, 1987; Weir, 1987). Palumbo (1990), in an important review of the relationship between programming and problem solving, makes similar criticisms of research on other languages. Nevertheless, researchers using treatment conditions that fail to accord with Papert's expanded characterizations of the crucial processes involved continue to report "failure of transfer" of programming and other skills from the LOGO microworld to other domains (Swan, 1991).

Control Technology: Reaching Out From the Microworld

One approach that offers promise in overcoming this failure-to-transfer problem is the development of programming environments in which the student must control the technology's interaction with the real world. A study by Resnick and Ocko (1990), for example, investigates the coupling of the LOGO programming language with "technical LEGO" construction kits:

In using LEGO/LOGO, children start by building machines out of LEGO pieces, using not only the traditional LEGO building blocks, but newer pieces like gears, motors, and sensors. Then they connect their machines to a computer and write computer programs (using a modified version of the programming language LOGO) to control the machines. For example, a child might build a LEGO merry-go-round, then write a LOGO program that makes the merry-go-round turn three revolutions whenever a particular touch sensor is pressed. (p. 121)

Although Resnick and Ocko ground their findings in a growing appreciation of the design process, it seems important that the physicality of the environment enables constructive and authentic conversations to take place. In some way, the setting up of the problems of interaction between the designed device and the real world is more meaningful than the process of designing entirely based within the graphical world of the LOGO screen.

Computing Across the Curriculum

The 1991 (second) edition of a text oriented to teachers and teacher educators, *Classroom Applications of Microcomputers* (Bullough & Beatty, 1991), opens with a chapter concerned with the description of computer systems analogous to NAEP's set of knowledge skills. Only when this ground is covered do the authors turn, in Chapter 4, to a consideration of computers in the curriculum.

As Bullough and Beatty point out in their introduction, many more educators have embraced the idea of integrating computers into the existing curriculum since the previous (1987) edition of their text. They also point out that teaching about the computer has decreased somewhat except in computer science and some mathematics classes. The trend is toward the use of technology to enhance teaching in traditional subject areas. A significant component of the argument in favor of computing across the curriculum is that it does indeed enhance the character of traditional teachers; indeed, the computer may be a catalyst to promote positive changes in the teaching of such subjects. We find it significant and disappointing that, nevertheless, Bullough and Beatty feel the need to open with a view of computer literacy for teachers as "naming of parts," in contrast, for example, to Hunter (1984), who begins an assessment of computer literacy as taught throughout the K–8 curriculum in terms of the objectives of the entire curriculum.

Examining the computer literacy issue from a comparative perspective, Makrakis (1988) proposed a schema of the relation between interaction and cognition that provides a useful index of the various "modes" of computer-assisted teaching and learning in different parts of the curriculum. When we turn to an examination of selected curriculum/subject areas (for the English and mathematics areas), we present some examples of Makrakis's scheme of computer use (Figure 1). This is a device to achieve some economy on our part, and we wish to stress that examples of each mode of use can be found in all subject areas.

Perhaps this is a convenient point to mention that teacher-oriented pub-

FIGURE 1
Makrakis's Model of Computer Use in Schools

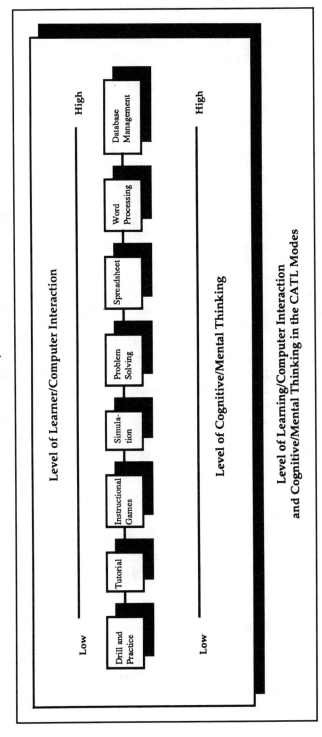

Level of Learner/Computer Interaction

Level of Cognitive/Mental Thinking

Level of Learning/Computer Interaction
and Cognitive/Mental Thinking in the CATL Modes

From "Education 2001: Learning Networks and Educational Reform," by J. Cummins and D. Sayers, 1990, *Computers in Schools*, 7(1/2), p. 24. Copyright 1990 by the Haworth Press, 10 Alice St., Binghamton, NY 13904. Used by permission.

lications, from interest groups, professional societies, and software developers, are all extremely important sources for tracing activity in the domain of computer use in the curriculum, and often contain information not available through conventional research sources. For example, Taylor is currently directing a project to develop preservice materials for computer use in art, science, music, social studies, the language arts, and mathematics (Taylor, 1991).

Complementing his previous formulation of the computer as tool, tutor, or tutee, Taylor indicates that (in each of these roles) three different specific functions of the computer need to be considered: *state resurrection*, *time compression*, and *graphical representation*.

By state resurrection Taylor means the ability of the computer to resurrect a particular set of prior conditions in the current computing situation. He points out the security that this ability provides to the user (e.g., to return to an earlier draft of a word-processed document), and the greater propensity to take risks and, therefore, to take an experimental approach to learning.

Time compression is the ability to compress into a short time activities that in everyday life would take much longer. Whether the activities are real, simulated analogues of the real world or fanciful, the compression factor again enables an experimental approach to learning.

Taylor also highlights the computer's ability to perform graphic representations, to represent and manipulate pictures easily. "Much of art and science has always been accessible to those who could visually represent what they were trying to understand" (p. 3). Of course, the state resurrection capability of the computer allows users to rehearse their attempts at visualization, just as the word processor allows rehearsals of text.

Modes of Computer Use

A wide variety of strategies for using computers as a tool of instruction have been developed over the past three decades. We will review these strategies, beginning with the most controlling and proceeding to those that afford treatment as "tools to think with."

Drill and Practice

When digital computers began to spread within scientific and engineering fields in the 1950s, they were initially seen as a way to implement drill-and-practice exercises whose utility derived from a behaviorist, association, and neural view of learning. This approach was viewed as especially appropriate to basic skills in literacy and mathematics education.

As computers became more sophisticated, drill-and-practice programs began to be described in terms of computer-assisted instruction (CAI)—a terminology expressing a view of teaching and learning grounded in the training and achievement of preset objectives. Many applications of, and research studies on, the use of computers in educational settings continue to be couched in this framework.

Even a good deal of the work on intelligent computer-assisted instruction (ICAI) is cast in these same terms. Although we do not object to drill-and-practice exercises per se, all too often drill-and-practice methods are used in the absence of the higher order concepts and activities for which the particular skills targeted in drill-and-practice are presumably components. In their most pernicious form, such methods are justified in terms of presumed differences in learning styles or abilities, such as Jensen's (1973) distinction between "Level 1" (rote learning) and "Level 2" (higher order learning") abilities.

As discussed in Cole and Griffin (1987) and Laboratory of Comparative Human Cognition (1989), this veiled ideology reveals itself in cases where wealthy and poor schools have equal numbers of computers but poor children spend their time on drill-and-practice exercises while better-off students spend their time in more meaningful activities (Center for the Social Organization of Schools [CSOS], 1984). This patterning makes existing reports of the effectiveness of computer-based drill-and-practice (and CAI) somewhat difficult to interpret.

Henry M. Levin (1986), for example, concluded that although CAI might improve the relative cost effectiveness of educational efforts,

evidence at the present time suggests, however, that educators should not assume blindly that CAI is a more cost-effective intervention than other alternatives. Clearly, the overall choice must depend on a school's instructional goals, available resources for reaching those goals, proficiency of using computers, and many other factors. (p. 173)

In fact, the subject of the Levin review was a specific intervention based on the Computer Curriculum Corporation's (CCC) drill-and-practice curriculum. The CCC curriculum itself was an outcome of The Stanford Project, begun in 1964 and one of the first large-scale attempts at developing computer-assisted support for learning in the K–12 age range. The reading support components were directed by Richard Atkinson; Patrick Suppes directed the mathematics components.

The CCC implementation and development of the Atkinson-Suppes material (further reviewed in Bork, 1985) is the software used by the largest number of students participating in CAI at the public school level. According to Suppes (1988), in the mid-1980s there were more than 400,000 students, most of them "culturally disadvantaged" or handicapped, using CCC materials on a daily basis.

Suppes points out that "the main effort at CCC has been in the development of drill-and-practice courses that supplement instruction in the basic skills, especially in reading and mathematics" (p. 108). His description is important, because there has been a tendency to see such programs as a substitute for instruction by the teacher, not a supplement.

Reacting to such overzealous interpretations, Balajthy (1989) introduces a survey of computer use in the reading curriculum with an admonishment: "Computer-based education's cart has been assigned a place in front of reading/language arts education's horse" (p. vii). This admonishment, unfortunately, also is true of much computer use in mathematics education. This situation arises, at least in part, because of the facility with which computing supports drill-and-practice approaches to teaching and learning and the view of the market that such approaches, especially in literacy and numeracy, constitute a profitable area of development. Balajthy concludes:

1. The lower the grade level or ability of the students the more effective Computer-based Instruction (CBI) is.
2. CBI is consistently more effective than traditional instruction, but the amount of improvement is low to moderate and cost-effectiveness is uncertain.
3. Structured CBI, with emphasis on direct instruction, is more effective in producing achievement gains than unstructured CBI.
4. CBI results in considerable savings of learning time.
5. CBI results in favorable attitudes towards computers. (Balajthy, 1989, p. 77)

As Balajthy points out, research into the impact of computer-based learning on reading has been carried out since the 1960s (pp. 69–81). He isolates a number of questions: Is computer-based instruction effective? Is it more effective than other methods? What is the best use of computer-based instructional technology? Is computer-based instructional technology "just another tool"? His answer is that computer-based instruction is indeed effective—often more effective than other methods—but one must bear in mind several limiting factors.

Foremost among these limitations is the particular nature of the programmed instruction model underlying much computer-based learning (which leads to a focus on the microskills of reading inappropriate to contemporary, holistic approaches), followed by concerns that the computer is often not well enough used by teachers ("exemplary" computer-mediated teaching is rare), that too much research focuses on older rather than younger readers, and that several experimental effects have not been sufficiently allowed for.

A very important dimension of the CAI characterization of the curriculum as developed at Stanford is that of "strands," where a strand represents one content area. So, for example, in the division strand, the decimals strand, and the equations strand included in the CCC mathematics curriculum, the student progresses through a string of related items, from easier to more difficult repertoires of questions. Performance records are checked against preset criteria, and the program determines whether more practice is required. The reading curriculum in the Stanford project was initially sequenced in six strands: reading readiness, letter recognition, sight words, phonics, spelling patterns, and word meaning and sentence completion (Balajthy, 1989, p. 79).

Suppes and other advocates of traditional CAI see great virtue in the potential of linking such stranding and criterion-referenced performance monitoring to district and school grade levels and in the ability of the computer to maintain detailed strand-by-strand ratings. The achievements possible in this tradition are exemplified in the report of Suppes and his colleagues' work in applying CAI to postsecondary education at Stanford (Suppes, 1981).

Integrated Learning Systems

One of the most critical characteristics of the deployment of information technologies into education has been its commercial dynamic; that is, the manufacturers of computers, the publishers of software, and the middlemen, resellers, dealers, and system integrators have had the most to gain by understanding U.S. school systems as a marketplace. Not to appreciate the significance of the economic motive in the sale and distribution of hardware, software, collateral print materials, computer courses, and the services of the cohort of experts who provide the training and guidance for the use of these technologies is to miss the central driving force behind the technology revolution in education.

Thus, it has not been the university- or school-based educators, researchers, developers, or administrators who have spearheaded the deployment of technological adjuncts to learning, so much as those who stand to profit by the sale of these electronic systems. Among the corporations who have been aggressive advocates for the deployment of these technologies in the schools are those who create "integrated learning systems" (ILS). These corporations (Jostens Learning Corporation, itself the combining of Prescription Learning and Educational Systems Corporation, is the largest; CCC, WICAT Education, Wasatch Education Systems, Ideal Learning Systems, Computer Systems Research, Innovative Technologies in Education, and the new Century Education Corporation are some of the vendors in the ILS business) are sometimes called value-added vendors or resellers (i.e., VARs).

These corporations provide the integration of hardware, software, and curriculum for instruction. Their heritage is grounded in programmed learning and drill-and-practice, the PLATO project, and the work of Patrick Suppes at Stanford (see above). The "product" in the sale of an ILS usually consists of 20 to 30 networked computer stations, one for each student. In addition, a file server contains a vast array of instructional material, most of it not unlike the workbooks one finds in nearly every school subject and grade level. These computer labs have sometimes been compared unfavorably with the language labs that once occupied a prominent position in foreign language instruction.

The ostensible advantage of these systems is that they can give immediate performance feedback to the teacher or the student. The systems, with considerable variation depending on the vendor, all provide comprehensive basic skills training in a computer-managed program (Kelman, 1989).

These programs are accountability driven, providing continuous performance feedback (e.g., how many questions answered correctly, how many wrong, how many completed). Thus, even though students may be working on a variety of topics, skills, or tasks, the system can give the instructor a moment-by-moment analysis of what work has been done at each student station. These ILS programs account for a large portion of computer and software sales in the nation's schools, and many of them use federal and state funding sources (Chapter I, Chapter II, etc.) that supplement the standard annual operating budgets of a particular school district.

Two issues arise: the intent to tap into available funding resources that are aimed at disadvantaged, at-risk school populations who perform well below preestablished norms and the implications of differential performance expectations based on socioeconomic status and ethnic classifications. Stated simply, federally funded underperforming minority populations are the target of ILS vendors.

Computer-managed instructional systems—ILS programs—have come under criticism for their pedestrian use of repetitive multiple-choice questions and problems, underuse of the creative power of computers, and targeting of inner-city, at-risk student populations with workbook-like practice problems (while more affluent, suburban communities use computers for tutorial and simulation activities as well as for more creative applications of spreadsheets, databases, and word-processing capabilities). It is for the schools' disadvantaged populations that state and federal sources provide supplementary funding.

In a lengthy review of ILS, Kelman (1989) criticizes the targeting of at-risk students with "drill-and-kill" software, which is the standard fare of ILS systems. He identifies the heritage of ILS systems dating back to the 1960s and early 1970s, the era of CAI supported by mainframes, terminals, behavioral objectives, repetitive isolated lessons, and single-correct-answer exercises.

While most ILS vendors claim that student performance and productivity increase—scores improve—Kelman finds such research wanting on several dimensions. He criticizes ILS on the philosophical and pedagogical grounds that these systems are implicitly intended to replace some of the teaching function; they control the student's behavior, not the other way around, and they are bureaucratically convenient, "teacher proof," and teacher free.

To highlight the perceived limitations of ILS-managed instruction, Kelman lists a number of instructional areas that could be improved by computer-assisted learning but are not supported by available ILS programs: higher order thinking skills, creative expression, personal and professional productivity, cooperative learning, multiple-modality learning, and individual empowerment. Although computers can support such programs within a local area network arrangement characteristic of ILS laboratories, so far the vendors have chosen not to target these meta-cognitive domains, presumably because there is no market for them among the present customer base of administrators of inner-city and large school districts.

Tutoring

Some recent developments provide the opportunity to embed a more holistic view of learning into the CAI process and to go beyond even the most sophisticated drill-and-practice to more sophisticated pedagogical practices (Mandl & Lesgold, 1988; Psotka, Massey, & Mutter, 1988).

Intelligent tutoring systems are made up of four components: an expert knowledge component, a learning modeling component, a tutorial planning component, and a communication component. Suppes (1990), reviewing the Mandl and Lesgold and Psotka et al. books, points out that although each of the four components is thoroughly covered, there is little systematic data about the achievements of the intelligent tutoring systems approach; the references given are almost all "soft" and "qualitative" in character, in contrast to the more quantitative research characteristic of traditional CAI and drill-and-practice studies.

By contrast, Hammill (1989), in his discussion of as-yet-unmet issues in the design of intelligent tutoring systems, calls for more soft or qualitative research. He comments:

Insights into social aspects of ITS's are also beginning to appear in the literature. For example, Schofield and her colleagues have used qualitative recording and analysis procedures to document effects on students, teachers, and even school administrators of the introduction of Anderson's Geometry Tutor into high school classrooms. Students using the tutor individually in geometry classes evidenced more effort and involvement related to geometry tasks and more competition with their peers in the classroom than did students in traditional geometry classes. Teachers using Tutor in their classes devoted more attention to slower students; assumed more of a collaborative role with individual students using the tutor and less of an authoritative expert classroom teacher role; and changed their grading practices in response to the relatively individualized pace fostered by the tutor. And an administrator found it difficult to evaluate teachers who used the Tutor because of the very different manner in which such classes were run and the different skills required of the teachers. (p. 179)

Hammill's concerns about as-yet-unmet objectives of intelligent CAI indicate an increasing awareness of the complexity of teacher-student interactions. Hammill calls for further work on how nonformal domains of knowledge and skill might be represented in intelligent tutors, exploitation of a greater variety of instructional strategies, and greater awareness of the means of interaction between the machine and the user.

This work is nicely complemented by Gott (1989), who provides a review of intelligent tutoring systems that focus on apprenticeship relations between "tutor" and student.

Instructional Gaming

As an example of the "instructional gaming" approach or "mode of interaction," to use Makrakis's terminology, we discuss a program that is designed to increase understanding of the concept of the "number line" and that also effectively demonstrates how "gaming" approaches can be used in the teaching of well-defined concepts or microskills.

"Sharks" is a family of educational games developed by James Levin and his colleagues (see Levin & Souviney, 1983, for a general description of Levin's group's approach to constructing educational games). The metaphor undergirding various versions of the game is that of shark hunters attempting to harpoon sharks.

Different versions of the game have the shark visible or invisible at the time the estimation is made, differ the size of the shark (and therefore the ease of "hitting" it), and have the shark as stationary or moving. In addition, the games have an authoring potential; children or teachers can vary the complexity of the scale (e.g., by selecting endpoints that vary from 0 to 10, -27 to 54, or even .001 to .01).

Levin and his colleagues' design of the shark games was motivated by evidence that Japanese children who were highly practiced in the use of the abacus displayed high levels of arithmetic skill, even in its absence, and contrasting evidence that children who experienced difficulty in learning long division also had a poor knowledge of the number line, such that they had difficulty rapidly estimating "remainders." Thus, the set of games was designed to encourage learning by requiring the child to make rapid and accurate estimations of the shark's position. This was seen as a means of providing children with practice in executing operations known to be important in a valued educational context. Unfortunately, in this case, as in many cases of game-embedded computer-based activities, there is as yet no real formal evaluation of its effectiveness.

Problem-Solving Tools

Judah Schwartz's Geometric Supposer is a good representative of problem-solving-tool approaches to mathematics education. Sunburst Corporation's catalog now offers a series of Geometric Supposer programs designed to support a change in the teaching/learning process in which teachers become facilitators of geometric inquiry and students become active learners of geometry.

In the Geometric Supposer series students begin with a triangle, quadrilateral, or circle and then use menu options to attempt various constructions. Students' conjectures of appropriate constructions are based on numerical and visual data they collect, and the program provides them with tools to carry out measuring operations. Ultimately, a yearlong geometry syllabus, covering a range of concepts and including textbooks as well as the program, is built as a context for the use of the computer-focused activities.

Viewing the essence of mathematical activity as the making and exploring of mathematical conjectures rather than the ability to manipulate the operations associated with mathematics made by other people, Schwartz (1989) places the Geometric Supposer into a genre of software that he describes as "intellectual mirrors," a genre

in which the users, be they students or teachers, can explore an intellectual domain. In these cases [i.e. the Geometric Supposers] the domain in question is Euclidean plane geometry. Because the software environment reduces the difficulties associated with the exploration of the domain and indeed provides rich tools for such exploration, those who have access to such an environment can, with the appropriate stimulation, use that access to explore the domain. I say appropriate stimulation because I believe that, for most of us, problem posing and problem solution are in large measure social activities. We need the stimulation of our peers, our students, and our teachers. (p. 58)

Other aspects that make the Geometric Supposer an intellectual mirror are that "it has no built-in pedagogic agenda" (p. 60) and that it provides a rich set of primitive operations in plane geometry—not so rich that they contain and embed answers to preset problems but rich enough that combining them in various ways allows for interesting insights about geometry.

Although the ability of the supposer to capture a construction and generalize it is important to us, and we can see how it "provides a special and supporting environment for allowing students to understand how the particularity of their efforts fits into a larger mathematical generality'' (p. 60), Schwartz's intimation of the importance of the social milieu within which the investigations are carried out strikes us as equally important and deserving of more explicit attention.

Word Processing

Much of the recent work in computer use in the language arts has concentrated on support for writing, both for its own sake and as a stimulus to literacy more generally (Daiute, 1985). There have been several initiatives to create software products that enable hesitant writers to approach writing through outlining approaches, at the precollege, college, and professional levels of process writing, or that approach writing as a collaborative process, one of computer-mediated partnership (Salomon, Perkins, & Globerson, 1991).

Perhaps the most significant computer-based initiative in the textual aspects of the language arts—if the degree to which it has been taken up by schools is taken as the criterion—is the IBM Writing to Read curriculum package. This package, in which the microcomputer is but one element among several media, was evaluated positively by the Educational Testing Service (1984). Slavin (1991) reviews several evaluations of the application of the Writing to Read program at the kindergarten and first-grade level and gives further access to the literature on this approach.

Computer Modeling and Simulation

In 1989 Robert Tinker and Seymour Papert made a number of recommendations about how computers might be used in science education, including that they should be used in science classrooms as tools for communication, for interfacing, for theory building, for creativity, for database access, and for programming. Tinker and Papert (1989) also recommend the use of computers as tools for programming in the science classroom. In their view,

programming languages continue to be important because they give the student the greatest control of computers, putting that resource at the student's disposal in the service of student-originated activities. It is important to note that programming languages are *just below the surface* in powerful applications like Hypercard, in many word processors, and most advanced databases. As a result programming concepts are *increasingly* important. (Tinker & Papert, 1989, p. 6) [italics added]

Abruscato (1986), in a primer for computer-using science educators, also sees an extensive area of computer literacy as a valid component of the science curriculum, and Ellis (1991) reports on the ENLIST Micros project, which defined, through extensive consultation with computer-using science teachers, a set of 22 competencies needed by the teacher to make effective use of the computer in the science classroom (see Table 1).

We see here a desire to employ each of Taylor's characteristics of time compression, state resurrection, and graphic representation and visualization.

Time compression in science education. Baird (1991) summarizes the current state of computer use in science education in the context of an analysis of preservice preparation for science teachers. He notes that the extent to which science teachers are using computers in their classrooms is disappointing—computers are used by very few science teachers, most of whom feel underprepared for their use.

Second, microcomputers and similar interactive devices offer unique opportunities for promoting multiple perspectives on science learning. Research findings point to enhanced learning outcomes and greater efficiency in classrooms where computers are used appropriately. Third, new teachers are most likely to use tools and techniques that they have been shown and used themselves. Such uses should be frequent, well-integrated into the curriculum, and involve field-based classroom settings. Finally, the rapid rate of change in educational software and hardware will require new approaches to preservice teacher education. Apprenticeships with master teachers who use computers effectively can develop confidence in newly-certified teachers and promote continuing growth. (Baird, 1991, p. 5)

Baird also reports that teachers at all levels employed roughly similar patterns, except teachers in Grades 7–12 made heavier use of the computer as a laboratory tool for simulations than those in Grades K–6. A particular instance of computer facilitation of complex problem spaces is that of the microcomputer-based laboratory, or MBL. Unfortunately, to find worthwhile instances of research in this domain would require us to go beyond our focus on K–12 education to higher education.

One stage in the transition from "drill instructor" to "object to think with" is the use of the computer to provide a framework of designed explanation, one in which a "dry lab" or simulation or "structured workbench" is run alongside the real world.

Such uses of computers have been available for a number of years; however, only recently, as scientists themselves, as part of their own practice, have adopted simulation as a research tool, have science teachers really awarded this approach to computer use much legitimacy.

State resurrection in art education. Art education has seen fundamental changes in recent years, and the role of the computer in the mediation of the new art education points to specially significant issues. State resurrection seems to be the key process in enabling computer use in art, with the availability of extremely powerful graphics programming facilities on even small microcomputers providing students opportunities to attempt,

TABLE 1

Essential Competencies and Factors

Awareness of Computers

Upon completion of *ENLIST Micros* the participant will be able to:

- Demonstrate an awareness of the major types of applications of the computer--such as information storage and retrieval, simulation and modeling, process control and decision making, computation, and data processing.
- Communicate effectively about computers by understanding and using appropriate terminology.
- Recognize that one aspect of problem solving involves a series of logical steps, and that programming is translating those steps into instructions for the computer.
- Understand thoroughly that a computer only does what the program instructs it to do.
- Demonstrate an awareness of computer usage and assistance in fields such as:

health	business and industry
science	transportation
engineering	communications
education	military,

- Respond appropriately to common error messages when using software.
- Load and run a variety of computer software packages.

Applications of Microcomputers in Science Teaching

Upon completion of *ENLIST Micros* the participant will be able to:

- Describe the ways the computer can be used to learn about computers, to learn through computers, and to learn with computers.
- Describe appropriate uses for computers in teaching science, such as:
  - computer-assisted instruction (simulation, tutorial, drill and practice)
  - computer-managed instruction
  - microcomputer-based laboratory
  - problem solving
  - word processing
  - equipment management
  - record keeping
- Apply and evaluate the general capabilities of the computer as a tool for instruction.
- Use the computer to individualize instruction and increase student learning.
- Demonstrate appropriate uses of computer technology for basic skills instruction.

(Continued)

TABLE 1 *(Continued)*

Implementation of Microcomputers in Science Teaching

Upon completion of *ENLIST Micros* the participant will be able to:

- Demonstrate ways to integrate the use of computer-related materials with non-computer materials, including textbooks.
- Plan appropriate scheduling of student computer activities.
- Respond appropriately to changes in curriculum and teaching methodology caused by new technological developments.
- Plan for effective pre- and post-computer interaction activities for students (for example, debriefing after a science simulation).

Identification, Evaluation, and Adoption of Software

Upon completion of *ENLIST Micros* the participant will be able to:

- Locate commercial and public domain software for a spécific topic and application.
- Locate and use at least one evaluative process to appraise and determine the instructional worth of a variety of computer software.

Resources for Educational Computing in the Sciences

Upon completion of *ENLIST Micros* the participant will be able to:

- Identify, evaluate, and use a variety of sources of current information regarding computer uses in education.

Attitudes About Using Computers in Science Education

Upon completion of *ENLIST Micros* the participant will be able to:

- Voluntarily choose to use the computer for educational purpose.
- Display satisfaction and confidence in computer usage.
- Value the benefits of computerization in education and society for contributions such as:
  - efficient and effective information processing,
  - automation of routine tasks,
  - increasing communication and availability of information,
  - improving student attitude and productivity, and
  - improving instructional opportunities.

From "The BSCS Perspective on Preparing Science Teachers in Educational Computing," by James D. Ellis, 1991, *SIGCUE Outlook, 21*(1), pp. 26–27. Reprinted by permission.

rehearse, and revise design processes in the search for solutions of graphic problems.

However, for the most part, art teachers have not yet realized the full potential of computers in art education contexts. Where the capabilities of computers to mediate art processes are recognized, the computer is often limited to art making. Although this computer-mediated art making may reach very high levels of practice and sophistication (e.g., three-dimensional computer animation), the more traditional roles of computer as information processor are not fully developed.

This situation obtains in a context where the Getty Trust is recording on videodisc all the major art collections of the world, collections that have themselves been indexed in computer databases since the 1960s, and in a climate where art education has broadened to a more organic inclusion of aesthetics, criticism, and history—with consequent requirements for a range of responses, including, especially, textual responses.

There have been two limitations, perhaps, on the development of a more complete approach to computer use in art education: cost and teacher training. Although costs are not yet (or ever in education) insignificant, they have sufficiently decreased in recent years to allow teachers the opportunity to take initiatives in computer-mediated art education if they wish to do so.

The limiting factor, therefore, is teacher training. This point is taken up by Hubbard (1991), who points to both the need for greater involvement in computing in teacher education and the potential leverage for instruction generally if art teachers are able to develop the potential of hypermedia and computer graphics in school contexts.

Hypermedia

An interesting dimension of the introduction of Apple's HyperCard was the difficulty that many expert computer users had in placing the new product in any existing category of computer software or pattern of computer/student interaction. Not only was HyperCard "content free" in the sense that word processors, databases, spreadsheets, and graphics programs are not limited to a particular domain, it also appeared to be "context free" and not mappable to any domain. Even its originators were unclear as to its purpose:

We didn't know what we were making, but we knew it was going to be great. (C. Espinosa, HyperCard product manager, cited in Jones & Myers, 1988, p. vii)

Even the program's creator, Bill Atkinson, was hard put to place the product in a niche large enough to house all its implications. (Daniels & Mara, 1988, p. 11)

HyperCard provides a mixture of text and graphics, text storage and retrieval, an object-oriented programming language (HyperTalk) and point-and-shoot programming, a flat-file database with the ability to input relational information, and a graphics package with sound facilities. The manufacturer's rhetoric states that "what makes Hypercard great is what makes personal computers great. They both put the power of information into the hands of ordinary people, and they both make that power usable" (Jones & Myers, 1988, p. viii). Daniels and Mara (1988) point out that system generalization of the kind evinced by HyperCard is "a two-sided coin": "If Hypercard is too general to fit all its functions into any existing category, it is also too unspecialized . . . to fully satisfy all of any given category's criteria" (p. 11).

Leaving aside the problematic issue of the affordability of the Macintosh computer by "ordinary" people, the important question to ask of HyperCard in education is not what it is, but what do people make of it? In a sense, HyperCard re-creates the flexibility of the underlying hardware substrate, and, to use Papert's phrase from *Mindstorms* (1980, p. viii), is "protean."

Padilla comments on its membership in a particular class of software programs little understood at the time of HyperCard's launching outside a limited development and research community (because powerful microcomputers are necessary):

Hypercard is one implementation of the concept of hypertext, [a means] to break the linearity of the traditional printed text . . . one should be able to explore ideas in a multidimensional space in which related ideas are linked. The learner can then jump from one to another in a free-flowing and self-directed manner. (Padilla, 1990, p. 211)

Given such sophistication, Padilla comments: "Despite all of these complex capabilities Hypercard and Hypertalk should be viewed simply as a potential tool for productivity enhancement in regular and bilingual classrooms" (p. 212). He then proceeds to discuss some of the ways in which teachers can construct HyperCard stacks in bilingual contexts (e.g., the ability to drive interactive video through interfaces written in alternative languages). Some of the more successful uses of HyperCard in the classroom have, in fact, been in terms of students designing individual cards that are then merged together to form the group's stack. In the Apple Global Education Network (Scott & Woodbridge, 1991), several schools produced "self-portraits" of themselves through class and group projects, the resulting stacks being exchanged through the Applelink network. (A few schools even went so far as to construct stacks cooperatively, assembling them across the network.) Also, in projects analogous to Papert and Harel's "instruction-as-design" experiment, students have pursued projects to produce stacks to teach other students (Allen, 1990).

As with LOGO, HyperCard can fulfill the various roles of tutor, tool, and tutee. Also as with LOGO, the medium constituted by the tool only becomes an effective interactive learning environment when it is used as a medium of social interaction as well.

THE RESOURCE CONTEXT

In the early 1980s it was possible to be optimistic about the diffusion of computer use in education because hardware costs were decreasing rapidly and many interesting ideas were in the air; today, we seem to have reached at least a temporary cost plateau that is very high indeed for all but the wealthiest school districts.

Perhaps the power of the Next machine, the Sun work station, and Macintosh-level technology will eventually approach the affordability of the Apple IIe, still the basic U.S. school computer in terms of numbers purchased, but we do not presently see any quick movement in this direction. (Certainly, Apple Computer Incorporated's current attempt to return to a high-volume, not-too-high-cost manufacturing strategy is to be welcomed by educationalists.)

Rather, we see a desperate scramble, coincident with the ending of the manufacture of the Apple II, to acquire software and hardware from secondary suppliers before the supply completely dries up and a somewhat jaundiced view, based on previous bitter experiences of similar promises from other manufacturers, of the promised downward compatibility of the new range of low-cost Macintosh computers.

We observe this to be happening as software developers, presented with increasingly powerful platforms by the manufacturers, produce ever-more sophisticated and memory-hungry microworld and simulation programs. The new System 7 operating system for the Macintosh, for example, requires a minimum of two megabytes of random-access memory, more than 250 times the size of the total memory available on the immediate postwar generation of computers, and this is not untypical of contemporary microcomputers.

Of course, Apple is not the only manufacturer to be engaged in supplying the U.S. school market. Since the advent of the personal computer, IBM has paid increasing attention to the potential returns from investing in the education market: There is much software available for the IBM microcomputer and its "clones." The Commodore Amiga has established a particular niche in graphics-oriented educational applications. All three of these manufacturers, and several others, supply microcomputers as components of sophisticated classroom-oriented packages: network and computer laboratory bundles, interactive video systems, and the like.

Patterns of Provision

Whatever the preparation of the teacher, of course, the controlling factor in the use of computers in education in a particular school is the organizational strategy adopted at that school, which relates in turn to density of provision. Although it would be pleasing to observe a process of rational design, whereby teachers work from curriculum specifications toward computer provision requirements, the reality is quite the reverse.

Decisions on the level of provision are often made at levels removed from the classroom: School districts or boards of education, computer advisers, access to commercial benefactors, current architectural fashions, and even local interpretations of safety regulations may have much more influence on determining the pattern of provision than any curriculum content requirement. Even so, one can discern some basic patterns of provision, ranging from one computer per classroom to a concentration of computers into a computer laboratory maintained by a specialist teacher.

At the midpoint of the decade, the Center for the Social Organization of Schools (CSOS, 1984) found that major issues in the diffusion of computer use in schools were the organization of student computer time, how to deal with time not spent on computers, and how to provide useful and adjustable access to few computers by many classes. The biases in this process have long been evident. CSOS reported:

Our data show that in schools where use is concentrated among above-average students, the primary computer-using teacher reports a more "individual use" pattern than in schools where the "average" students get a proportionate share of the time. Use by "average" students is instead associated with students using computers in pairs. (Cole & Griffin, 1987, p. 53)

Cole and Griffin (1987) report on a considerable body of evidence that suggests that one computer per child is not the optimum number and that considerable benefits are to be derived from students working in pairs or small groups. These findings have both theoretical and practical significance. Their practical meaning is, of course, that one does not have to worry about getting one computer per child but can work with higher (and cheaper) ratios. Shavelson et al. (1984) looked at the strategies of teachers who had been judged successful by their peers in providing microcomputer-based instruction in cases where there were several computers in the classroom. They found that teachers employed one of four strategies for organizing computer use: enrichment, adjunct instruction, drill-and-practice, or orchestration.

Only orchestration, which represented the widest variety of instructional applications and linked those applications to the regular curriculum, provided "the appropriate integration of microcomputer-based learning activities with teachers' instructional goals and with the ongoing curriculum, which changes and improves on the basis of feedback that indicates whether desired outcomes are achieved" (Shavelson et al., 1984, p. vi).

The orchestrating teachers used several types of software that they integrated into the curriculum, coordinated the activities with other instructional means, and stressed both cognitive and basic goals. The orchestrating style seems to arise naturally when higher order, "intelligent" programs, either as focused lessons or as mixed games and lessons, are the medium of instruction, but not when drill-and-practice is used. This result stems, perhaps, from the orchestrating teacher's pattern of inclusion. Orchestrated classrooms depend, for their success, on a considerable degree of student autonomy and responsibility: They provide a context in which students naturally develop responsibility for their own learning. It follows that each new technology is able to "find a seat" in the orchestra without wholly disrupting the pattern of the classroom.

An Experiment in Computer Saturation: ACOT

Although most schools still employ one or two computers per classroom (on average) for standard educational topics, some schools have invested far more in computers in the belief that the full value of computerization cannot be properly judged without some "utopian" experiments that provide one computer for each child and, in some cases, one to take home as well (see Kiesler & Sproull, 1987, and International Federation of Information Processing, 1987, for discussions of similar viewpoints in higher education). One project that has evaluated computer use in such circumstances is the Apple Classroom of Tomorrow (ACOT) experiment, initiated by Martin Engel.

The ACOT project was initiated on the premise that every person would one day possess continuous access to computation. It was the marketing premise of Apple Computer, Inc., that there would be "one person, one computer." Corporate slogans promoted "the computer for the rest of us," and the Apple IIe was deemed "the most personal computer." The ACOT project sought to embody these slogans and concepts, implementing them within a school setting. Even this context was an extension of the corporate marketing slogan "changing the world through education, one person at a time."

In the spirit of the General Electric "kitchens of tomorrow" and the Oldsmobile "car of tomorrow," the ACOT project was intended to become a demonstration of what was possible in the hypothetical future when everyone had continuous computer access. In this case, computer saturation was created within a school setting. Every student and teacher had his or her own computer during the school day and another computer at home. Using off-the-shelf technology and commercially available software, selected classrooms around the United States were equipped with sufficient hardware to place one on each desk. Six schools began the project, offering classrooms at different grade levels.

In addition to the hardware, which was provided by Apple, and the software, which was donated by various education software vendors, Apple provided a sum of money to employ an additional person for each of the classrooms—a computer coordinator/technical person to assist the teacher in curriculum development and hardware/software support. Apple and the participating school signed a letter of intent, Apple agreeing to provide hardware and cash and the school agreeing to provide a willing teacher and coordinator as well as consenting to the "rules" of keeping the equipment concentrated within the one chosen classroom. The school also agreed to permit evaluators identified by Apple to visit the classroom, and Apple agreed to any assessment programs required by the school.

A number of assumptions drove the initiation of this project. First, it was assumed that the computer/student ratio was far from optimal and that until and unless there was a "critical mass" of technology accessible to all students at all times cause-effect impact studies would be vitiated. Furthermore, it was assumed that in order for students to benefit sufficiently from computer access, they would have to have a proprietary relationship with that technology, just as they have with their desks and hallway lockers. This idea was associated with the notion of locus of control (i.e., the degree to which students have control over the events of their teaching/learning situation). The basic assumption is that increased internal locus of control leads to increased performance.

The catalyst effect was still another assumption behind the ACOT project. The school reform movement advocates dramatic transformations in the instructional process as well as the curriculum. It was assumed that the teaching role would change in the ACOT classroom, along with the curriculum content. It was hoped that the ubiquitous presence of computers would refocus the instructional process toward the development of higher order thinking skills, problem solving, and thematic- and project-oriented approaches to the study of various subjects. It was also expected that if all students had constant computer access, they would attain greater independence, and thus the controlling role of the didactic teacher would also change toward facilitation and support.

It was expected that different schools would implement their ACOT model differently, and that, over time, through trial and error, many changes would be taking place. The notion behind this emphasis on experimentation was twofold: (a) Since there is no "science" of education, change can only come from trial and error, and (b) the ACOT classrooms would be "living laboratories," environments in which schools and school districts could learn how to make computers work for them and enhance student as well as teacher productivity and performance.

Since the project (and its research) was sponsored by a for-profit computer vendor, research credibility would be suspect, and because there continues to be a raging debate about the validity and value of educational testing, standardized and norm-referenced criteria would not be used to measure computer impact. Informal and anecdotal data were systematically collected, with participant teachers providing regular audiotaped narrations and various university-based educational ethnographers conducting observational studies.

Observational and descriptive studies of the project found the following characteristics, after 2 to 4 years of ACOT classroom development (Baker, Gearhart, & Herman, 1989, 1990, 1991):

1. Students became "empowered." The locus of control shifted to the student. Students assumed responsibility for their own academic work.

2. Students wrote more, faster, and better. Their vocabulary, sentence, and compositional complexity and comprehension improved more and faster than that of students who did not "word-process" as much or at all.

3. Students became spontaneous peer teachers and cooperative learners if permitted by their teachers.

4. Teachers diminished their didactic roles and moved toward the orchestration model.

5. Students took more initiative for their own learning. They became much less passive receivers.

6. Students learned the basic skills at their own rate more efficiently, freeing teachers from those instructional tasks to conduct higher and more personal levels of pedagogic interaction.

Although the initial investment from Apple was considerable, there was an assumption of amortization; that is, as with all capital investments, the costs, when spread over time, would increase cost-effectiveness. Furthermore, in the original conception of ACOT, costs would be controlled such that other school districts could generate similar models of saturated classrooms, if only for purposes of localizing the laboratory and investigatory aspects. Even the coordinator costs would be highest in the first year, diminishing in proportion to the number of additional ACOT classrooms added in out-years.

Looking at the impact of ACOT on teachers, Sandholtz, Ringstaff, and Dwyer (1990) report that initially teachers' concerns with management predominate and that "instructional innovation begins to emerge" only "when teachers have achieved a significant level of mastery over management issues" (p. 7). The teachers' initial problem when faced with a "high access to technology classroom," as they see it, is simply to survive.

Then they need to gain some mastery over the technology themselves before they can use it to their advantage in managing the classroom. In addition to the problems of coping with a new and therefore rather unstable technology, teachers also experience problems of infrastructure (suppliers could not keep up with the intense demand for new software and replacement media), of unanticipated misbehaviors (illegal software copying, disruption of classroom network systems, students resisting transition to noncomputer activities), and of radically changed classroom dynamics.

Many teachers initially were troubled by the increase in noise level and the necessity for students to move freely around the classroom. Having become accustomed to students sitting in their seats and the teacher in front of the classroom, some teachers worried whether the students were on task and learning. . . . Since computers facilitated independent learning, some teachers felt that they were no longer teaching and suggested that classrooms had become "technology centered, not instruction centered." They wondered if they were accomplishing their main goal of "teaching students the content." (Sandholtz et al., 1990, p. 10)

Despite these problems, however, some teachers see positive virtues in the saturation provision of technology once they gain mastery of the new classroom order and of the technology it encompasses (Dwyer, Ringstaff, & Sandholtz, 1990, p. 11).

Preparation of Teachers to Use Computers

One conclusion to be drawn from the ACOT experience is the high degree to which teachers need support when integrating computer technology into K–12 classrooms with instructional change as the goal. A major problem involved in providing such support, through either pre-service or in-service training, results from the complexity and range of educational applications of computing. This is confirmed in a report on an IBM-funded saturation project (Cline et al., 1986) that includes no less than 10 recommendations related to staff development in its concluding statement (pp. 136–137).

Other research confirms that even in circumstances where there is a requirement for teachers to gain some education in the use of computers in the classroom themselves, the development of competencies in this area is limited both from lack of time (the number of required hours is often quite limited) and from approach (such tuition is often focused on a narrow range of programs). The importance of adequate preparation is highlighted by Sheingold, Martin, and Endreweit (1985).

Lampert and Ball (1990) are developing an interesting computer-mediated approach to improving the ability of math teachers to develop "authentic" situations of mathematics learning that could serve as an exemplar for teacher training concerning both the use of computers in education and for teaching in general. In an effort to try to communicate the complexities of teaching and learning authentic mathematics in school, Lampert and Ball have collected a large amount of information from two classrooms in which this kind of teaching and learning occurred over an entire school year: videos of most lessons from two vantage points, students' drawings and writing for each day, the teacher's journal of reflections and plans, interviews across the year with the teacher and the students, and observations of every lesson written in some sort of outline form.

Lampert and Ball have catalogued this material and are now designing hypermedia "terrains" for exploration by prospective and practicing teachers who want to learn about this "new" kind of teaching. The materials they are producing could be thought of as "tutorials" that are intended to support the construction of ideas about how teaching and learning might be structured to engage participants in authentic mathematical activity. It is their intention to have those ideas firmly grounded in actual practice rather than based on theoretical ideas about what should happen in classrooms. The work stations for users will be organized around a file server that enables access to the information they have collected about teaching and learning in video, audio, graphic, and print form. Users will also be able to keep notebooks and create collaborative annotations of the information.

This kind of approach may do much to reduce student-teacher anxiety about computer use, which seems to be widespread. Attitudes toward, and anxiety about, the use of computers in the classroom will naturally affect a teacher's propensity to attempt such teaching. Woodrow (1991a, 1991b, 1991c) has made an extensive survey of various attitude scales; although he found them to be reliable predictors, he has not as yet discovered significant correlations of the determinants of the achievement of computer literacy by student teachers.

Access to computers and tuition by sensitive computer-using teacher educators seems to be a necessary but not sufficient condition for the achievement of teacher computer literacy. Certainly, student teachers seem ill prepared on the whole to cope with the social dimensions of computer use, to which we now turn.

SOCIAL ISSUES IN EDUCATIONAL COMPUTING

Microcomputer use in education raises questions of inequalities of access in terms of gender, ethnicity, and class, as already indicated.

Gender

In this section we look at recent representative work on strategies to redress imbalances in access as far as educational computer use is concerned. In considering adult roles in computer education, Michael W. Apple and Susan Jungk (1990) point out that there is a gendering that arises because teaching has been seen historically as "women's paid work," a tendency they illuminate through a study of a computer literacy curriculum.

Starting with the labor process, they point out that among the consequences of the rationalization and standardization of jobs are the separation of conception from execution and deskilling. Examining how these processes work through the job of teaching, they note that even though teachers have acquired slowly increasing amounts of skill and power in most school systems, they have only limited rights of choice of what is to be taught and how.

The composition of the teaching labor force is also of importance:

Historically, teaching has been constructed as women's paid work. In most western industrialized nations, approximately two thirds of the teaching force are women, a figure that is higher the lower one goes in the educational system. Administrators are overwhelmingly male, a figure that increases significantly the higher one goes in the educational system. (Apple & Jungk, 1990, p. 232)

Assessing the effect of these pressures on the introduction of a computer literacy program in a school district, Apple and Jungk show that the "expertise" is situated in male teachers, two of whom are employed over the summer to write the curriculum package. Their perception of the abilities of the female teachers led them to deskill the curriculum, including attempts to encapsulate parts of it on tapes and worksheets. (There were significant limitations on hands-on access to the computer laboratory.) One outcome was reluctance of the female teachers with regard to the packaged units. Because of the time commitments (in Apple and Jungk's view, female teachers are already "doing two jobs"),

when a new curriculum such as computer literacy is required, women teachers may be more dependent on using the ready-made curriculum materials than most male teachers. Intensification here does lead to an increased reliance on "outside experts." An understanding of the larger structuring of patriarchal relations, then, is essential if we are to fully comprehend both why the curriculum was produced the way it was and what its [gendered] effects actually were. (Apple & Jungk, 1990, p. 249)

At the student level, many studies show considerable differences between the computing experiences of boys and girls. Boys habitually have more access, whether to school, home, or recreational (arcade game) computers. Where computer programming is offered, more boys take the subject than girls (the girls take word-processing courses). Parents are more likely to buy computers for their sons than their daughters, and boys are more likely to attend after-school computer club meetings than girls.

Although computer programming is often presented as a model of a career that affords girls with unlimited opportunities, or one that affords good coordination with the demands of family, commercial programmers and systems analysts are predominantly male. "Although computers are not restricted to the male domain, the inequality is controlled by the activities that computers are a part of and that continue to be divided along traditional lines" (Cole & Griffin, 1987, p. 55).

Computer use involves language and interactivity. These are "traditionally female 'domains' of expertise" (Cole & Griffin, 1987, p. 56): Given appropriate transformations of teaching contexts designed to redress gender imbalance (such as those in math, science, and technology surveyed by the American Association for the Advancement of Science, 1984), it follows that computers *can* be employed in the curriculum in such a way that girls find them accessible and inviting.

However, girls are not currently well provided for:

Boys seem to prefer game formats which include fantasy and violence. They do not require the same depth of instruction and understanding of a game prior to playing it. They like games and programs which are fast-paced. There is a profusion of such male oriented software on the market today. There is a dearth of software designed to attract girls. Girls, while they enjoy fantasy, are not as likely to become involved with it as boys. They do not like games which are violent or fast-paced. They prefer to have clear instructions and time to reflect on solutions. (Whooley, 1986, p. 15)

Whooley's study of gender differences in boys' and girls' use of computers concludes with an analysis of attitudes and outcomes of computer use in writing. She concludes that girls' "advantage in writing skills may provide an ideal inroad into the technology for girls to enter the field" (p. 121) and, more generally, that in collaborative work around the computer, boys working together will tend to be more impulsive, aggressive, and independent, whereas girls will tend to be less competitive and more cooperative. Of course, as she points out, none of these behaviors are of themselves negative or positive; rather, they need to be considered in context. Of cross-sex pairings, she concludes:

If girls working with boys learn to be more confident in their application of trial and error strategies, a strategy which relies heavily on intuition for success, and boys learn from girls to cooperate in their problem solving, then each party to the interaction will have gained new ways to meet the challenges of troubles on or off computers. (Whooley, 1986, p. 122)

Given an appropriate environment in which the computer-based activities are taking place (one that promotes and expects growth and provides an opportunity for students to learn from their differences), "males and females can and do experience success in their own way" (p. 125).

One cannot leave a discussion of computing and gender without mentioning Turkle's recent work, especially Turkle and Papert (1990), which both generalizes some of Turkle's earlier work on hard and soft mastery, as presented in Turkle (1985), and proposes that cognitive styles of computer use are associated with gender.

Turkle and Papert (1990) suggest that the computer is

an instrument for observing different styles of scientific thought and developing categories for analyzing them. [We] find that besides being a lens through which personal styles can be seen, it is also a privileged medium for the growth of alternative voices in dealing with the world of formal systems. (p. 346)

There are obvious links here to discussions on the cognitive consequences of learning to program. As a result of their investigations, Turkle and Papert wish to advance the notion of *ecological pluralism*, a notion that we see as a useful way of describing the range of interactional styles that computers allow, as well as the range of styles and attitudes brought to computer-based activities across the range of ethnicities.

Ethnicity

The use of computers in multicultural settings has a recent but rich history. Moll and Diaz (1987) showed that in a computer-mediated context where students were free to speak English or Spanish, facility with both languages improved. Bellman and Arias (1990) showed that it was feasible, through the use of telecommunications, to set up a cross-border project involving students from half a dozen institutions of higher education (three American, three Mexican) and in that context to pursue a common purpose and a common syllabus. DeVillar and Faltis (1991) have recently gathered some of the relevant literature together.

These efforts must contend with a number of underlying problems, stemming primarily from the consequences of minority ethnic group status, poverty, and problems of socially limited access to computers. The seriousness of these problems is reflected in the fact, mentioned earlier, that when minorities do have equitable access to computing resources, one observes "low-quality usage" in which drill-and-practice programs are used in place of enrichment activities, styles of classroom organization and management are adopted that reduce effectiveness of computer use, and telecommunications activities are pursued exclusively in English (CSOS, 1984; Mehan, Moll, & Riel, 1985; Shavelson et al., 1984).

In this connection there is the fact, not often enough acknowledged, that computer hardware carries cultural content: Computers can be adapted to work in Spanish, but they are designed in English. This Englishness, or more properly, Americanness, of computers is deeper than the keyboard layout and the screen driver, which can both be replaced to facilitate inverted exclamation marks and question marks. The menu structure, the design of icons, and the styles of problem decomposition and solution together construct the computer-human interface. So whatever communicative processes can be created within and between ethnic minorities, insofar as they are mediated by computer, they are also mediated by the Anglo culture.

In the light of these and other problems, DeVillar and Faltis (1991) advocate a "socioacademic achievement model" to facilitate learning in the multicultural setting—one that promotes a combination of social learning and independent practice. Although there are formidable barriers to the establishment of such models, the computer can be an effective facilitator:

The use of technology poses a threat to group socioacademic success only insofar as educational policy relegates its use toward divisive rather than integrative means or ends . . . educational equity, then, is intimately and irreversibly tied to computer integrated technology. Without equity in the use of technology for instructional purposes, the existing disparities in academic achievement can and will only widen. (Devillar & Faltis, 1991, p. 130)

Cummins and Sayers (1990) indicate some strategies for the creation of genuine joint interaction in cross-cultural contexts in a discussion of Project Orillas and the critical pedagogy of Celestin Freinet. The Modern School Movement, founded by Freinet, linked sister classes around the world in "interscholastic exchanges" for 60 years. These were class-to-class partnerships between teachers working on joint curricular projects and making extensive use of educational technology. The exchanges, being between cultures, served to promote intercultural understanding:

Through student learning networks, the MSM attempted to promote in young people a heightened awareness of all aspects of a community's life. Thus, technology-mediated learning networks encouraged students' development in many domains, including but not limited to academic achievement, by reestablishing the students' "psychic equilibrium" in an era increasingly dominated by mass media. (Cummins & Sayers, 1990, pp. 17–18)

Using computers as a medium of communication, rather than trying to program the machines to teach students or getting the students to program the machines, is a recent concept, Cummins and Sayers point out. They discuss as illustration of this point Project Orillas (from the Spanish De Orilla a Orilla, "From Shore to Shore"), one of several such projects

FIGURE 2

A. Educator Pedagogical Assumptions	
Transmission Orientation	Interactive/Experiential Orientation
Language:	
Decomposed	Whole
Learning:	
Hierarchical internalization from simple to complex	Joint interactive construction through critical inquiry within the zone of proximal development
B: Educator Social Assumptions	
Social Control Orientation	Social Transformation Orientation
Curricular topics:	
Neutralized with respect to societal power relations	Relevant to societal power relations
Student Outcomes:	
Compliant/Uncritical	Empowered/Critical

established or inspired by research taking place in San Diego in the early 1980s. Orillas brings together about 60 teachers in North and South America, in collaborative pairings.

As part of their agenda to elaborate the empowerment potential of computer-mediated learning networks for minority students, Cummins and Sayers point to an impasse of the pedagogical and social context for realizing such networks within the conservative educational agenda current in the United States. They present this in the form of a table, given here as Figure 2.

Cummins and Sayers hypothesize that when networking projects are implemented using their interactional/experiential orientation, "they do have *the potential* to act as a catalyst for critical analyses by students of societal issues that may pose a challenge to the status quo" (p. 25).

This potential arises because the collaborative input to the joint activity cannot be *pre*scripted to exclude joint critical inquiry on relevant social issues. Rather, it takes place against a background of a counterpotential in the form of the conservative approach to education, which combines a transmission-oriented pedagogy with a social-control-oriented curriculum. In order for computer-mediated learning networks to have the desired effect in heterogenous classrooms, a critical pedagogy combining interactive/experiential and social transformation orientations seems to be required.

EMBEDDING COMPUTER-MEDIATED ACTIVITY IN WIDER SOCIAL CONTEXTS

In this section we discuss three projects in which computer activities are socially structured to extend the learning activity beyond the confines of the single lesson/topic.

The Fifth Dimension Activity System

The Fifth Dimension is a deliberately constructed mixture of educational, play, and peer-oriented activities in which computed-based games and writing through telecommunications play a central role (Griffin & Cole, 1987; Laboratory of Comparative Human Cognition, 1982).

Its play elements are manifested in the culture of the Fifth Dimension, which is ruled over by a benevolent but somewhat unreliable and avuncular wizard. All of the adults, as well as all of the children, are hypothetically loyal citizens of this symbolic benefactor. Aside from collusion and playfulness in their joint "subordination" to the wizard, there is much play in the games themselves, which are a mixture of different kinds of commercially available software (including both arcade-style and drill-and-practice programs) and mixed genre history and geography games and clever mathematico-logical pretend worlds.

There is education, too, evoked not only by the content of the games but by the norms of the Fifth Dimension, which encourage mastery of the software and confront the children with various tasks set by the wizard at different levels of excellence. To obtain credit for their successes, the children routinely send written reports to the wizard, who arbitrates claims to expertise and privilege by getting the children to discuss such issues with other "citizens." These writing exercises are simultaneously functional, although effortful from the child's point of view, and, by the researcher's analysis, cardinal moments promoting cognitive development.

While introducing children to the world of computers, the Fifth Dimension is providing the occasion for the development of a community of writers, promoting writing through authentic (if somewhat playful) correspondence. There is also education in the fact that the children are learning to make telecommunications an integral part of what they understand about technologies of communication in modern life.

The Jasper Project

"Anchored instruction" is The Cognition and Technology Group of Vanderbilt University's (1990) recommendation for dealing with the inert knowledge problem, identified by Whitehead in the 1920s as a product of conventional classroom instruction and explored in recent years by Scardamalia and Bereiter (in press); Brown, Collins, and Duguid (1989); and others.

The major goal of anchored instruction is to overcome the inert knowledge problem. We attempt to do so by creating environments that permit sustained exploration by students and teachers and enable them to understand the kinds of problems and opportunities that experts in various areas encounter and the knowledge that these experts use as tools. We also attempt to help students experience the value of exploring the same setting from multiple perspectives (e.g., as scientist or historian). (Cognition and Technology Group, 1990, p. 3)

Note that the mediating object that the group is using is a set of interactive videodiscs, a technology within which the computer is embedded, yet accessible.

There are important ways in which interactive video and CD-Rom technology both embed and are embedded in computer technology that allow the manipulation (at the teacher level) and the exploration (at the student level) of macrocontexts. We should note also the important sense in which the group is attempting to use macrocontexts "to simulate the real world":

Brown et al. (1989) emphasized the importance of looking carefully at what we know about everyday cognition and of creating apprenticeships composed of authentic tasks. They noted that authentic activities are most simply defined as the "ordinary practices of the culture" (p. 34). Our anchored instruction projects simulate apprenticeships that comprise authentic tasks. . . . A focus on everyday cognition and authentic tasks also reminds us that novices who enter into a particular apprenticeship have a reasonable chance to develop expertise, in part because apprentices have the opportunity for sustained thinking about specific problems over long periods of time. (Cognition and Technology Group, 1990, p. 6)

The Jasper Series is a project to develop and evaluate the use of videodisc adventures that focus on mathematical problem formulation and problem solving. The project also has cross-curricular ambitions in science, history, and literature. The series will eventually comprise 6 to 10 adventures, primarily designed for fifth-grade students. The initial adventure, which is expected to be a template for those that follow, poses a very complex mathematical problem.

Students have to generate the problem to be solved and then find relevant information pertaining to the problem. All the data needed to solve the problem are embedded in the story.

By contrast, Vanderbilt's "Young Sherlock" project, also based on an initial stimulus of a videodisc-based story and associated database, encourages students to find material "off-line" as well as within the package. What are the basic properties of these teaching resources activities that the Vanderbilt group wishes to describe as macrocontexts? First, the problems to be derived and solved are anchored in the videodisc resource; more important, the problems derived are felt by the students to be authentic. Second, the work is sustained and challenging. The discs present whole problems rather than trimmed-down representations of problems: To cope with such complexity, students must work consistently over a period of time. Third, there is a rich motivational context, in this case the visual and dynamic qualities of the videodisc medium coupled with the narrative presentations of strong story lines.

Overall, our goals for anchored instruction include the establishment of semantically rich, shared environments that allow students and teachers to find and understand the kinds of problems that various concepts, principles and theories were designed to solve, and that allow them to experience the effects that new knowledge has on their perception and understanding of these environments. (Cognition and Technology Group, 1990, p. 9)

The Instructional Software Design Project

Harel and Papert (1990) provide one theoretically compelling model of the kinds of interactions that a fruitful computer-mediated environment would afford in their use of software design as a learning environment. This technique is essentially an embodiment of the common wisdom that one learns a great deal in the process of teaching.

In Harel and Papert's hands, the mechanism for implementing this common wisdom is a dual process where first the child must instruct the computer (e.g., write a program in LOGO to explain something) and then use that program to help teach another child, who also learns with LOGO as a mediating tool.

Harel and Papert (1990) present the Instructional Software Design Project (ISDP) as a paradigm of this constructionist vision. Fourth graders worked for 15 weeks, 4 hours a week, toward the goal of designing software to teach fractions to other children. This inversion had the benefit of making an area of learning that is usually passive into an active and exciting area of investigation. Papert (1990) describes "constructionism" as including but going beyond what Piaget would call "constructivism." The word with the V expresses the theory that knowledge is built by the learner, not supplied by the teacher. The word with the N expresses the idea that this happens especially felicitously when the learner is engaged in the construction of something external or at least shareable . . . a sand castle, a machine, a computer program, a book. This leads to a model using a cycle of internalization of what is outside, the externalization of what is inside, and so on. (p. 3)

Significant for us is the recognition of technology as a means of expression; the definition of the project to be tackled as one that is large-scale, meaningful, and whole; and the sustained effort required of the students toward the achievement of their goals. Also important is the shared construction of those goals. The children, the teacher, and the researcher jointly defined the meaning and purpose of the instructional software, and discussed software with which the students were familiar. Harel (the researcher) discussed her experiences as a programmer and those of other software designers, attempting to give the children a view of the process rather than of the finished product.

The binding of the researcher into the teaching/learning process, the sensitivity to atmosphere, and the replicability of the project are also significant: "The teacher and the researcher [Harel] collaborated and actively participated in all the children's software design and programming sessions . . . they looked at . . . programs, helped when asked, and discussed . . . designs, programming and problems" (Harel & Papert, 1990, p. 24).

We would say of this scenario that the designed inversion of role (students as designers and teachers) did indeed have considerable motivational power, but that the "expert" role of the researcher was also of great importance in relation to the novice designers. This was, in effect, a project about cognitive apprenticeship.

Papert attributes the fluency of the children's work with fractions to the fact that the knowledge is situated within "computational microworlds" and compares this with the situatedness of Jean Lave's (1987) respondents (weight watchers) in familiar territory (the kitchen). Harel and Papert recognize a consistency between the ISDP work and that of Lucy Suchman, Jean Lave, and John Seeley Brown: "Like these researchers we are strongly committed to the idea that no piece of knowledge stands or grows by itself. Its meaning and efficacy depend on its being situated in a relation to supporting structures" (Harel & Papert, 1990, p. 40).

But Harel and Papert also propose the situation of knowledge in "internalized supporting structures" and posit "mental environments as supporting and interacting with knowledge in much the same way as external, physical environments" (Harel & Papert, 1990, p. 40).

From our point of view, the interactive learning environment approach, which views either or both the internal and the external contexts of cognition as providing a range of props for knowledge building, misses the *mediational* essence of the "object of learning."

It is also significant to us that when Papert and Harel do discuss the importance of communication to learning, in largely Piagetian terms, the environmental situatedness of the communication is not addressed at that point.

NETWORKING

Following from our general approach to learning and development (LCHC, 1983), we consider the problem of using computers to promote educational objectives to have two sets of facets, each requiring development of different potentials in the use of computer technology.

The first, which we have been discussing for the majority of this paper, focuses on the organization of educational activity within the classroom. The second, to which we now turn, focuses on links between classroom-level activity and the broader context of which the classroom is a part.

In our view, computers always function as communications systems that mediate the interactions of their users. In this section, though, we wish to focus specifically on that category of educational computing research that most directly addresses the mediational aspects of computing: telecommunications.

As we have noted previously, "modern computer technology, when used as a component in a telecommunications system, offers a link between children, teachers, and the outside world in educationally powerful ways" (LCHC, 1989, p. 80).

The use of telecommunications affords opportunities for children to formulate and articulate new goals, to reflect on their own learning, and to use writing to create social contexts of joint activity. The joint activity does not come easily. Initial steps in children's use of networks are often difficult and uninspiring. Telecommunications use in education requires considerable planning and forethought to overcome the contradictions between the rhythms and goals of the interacting educational systems. Levin, Newman, and Crook have variously discussed aspects of the coordination issues arising in the use of computer networks to promote joint activity (Crook, 1987, 1991; Levin, Rogers, Waugh, & Smith, 1989; Newman, 1990; Newman, Brienne, Goldman, Jackson, & Magzamen, 1988).

A further issue is that of the timeliness of synchronous and asynchronous communication; computer-mediated communication provides organizers of joint activity at a distance with a crucial resource in that it occurs asynchronously, that is, in nonreal time (Black, Levin, Mehan, & Quinn, 1983; Scollon, 1983).

The fact that an answer is not normally expected for 24 hours or more means that recipients of messages can work on them "off-line," looking up information they are lacking, consulting with more expert speakers of a foreign language, getting a partner or teacher's reaction to a proposed answer, and so on. This reduced time pressure not only mitigates problems of translation but can convert them into useful learning experiences. (LCHC, 1989, p. 81)

The time-shifting properties of electronic mail also introduce, by contrast, a sense of immediacy where the corresponding school is on the "other" side of the globe: Students in one school may proceed with a next step in the joint activity "overnight" in the other school's terms.

Studies of the use of telecommunications as an integral part of overall educational activity consistently find that, when properly organized, telecommunications provides rich opportunities for children to articulate new goals. It enables them to reflect on their own learning, to use writing as a tool for both communication and thought, and to create social contexts that are not merely "passive backgrounds" for learning but are arenas for goal-oriented, reflective problem solving (LCHC, 1989; Levin, Rogers, Waugh, & Smith, 1989; Levin & Souviney, 1983; Riel, 1986).

At the same time, there has been a tendency for enthusiasts of telecommunications-mediated instruction to replicate the errors of those who focused on the technology of mediation within classrooms in the vain hope that telecommunications access to other people and contexts (classroom, databases, and so on) is sufficient to make a positive difference in the quality of classroom instruction. It is not (Cole & Griffin, 1987; Riel, 1986). Rather, as was true of within-classroom computer use, telecommunications activities have proven powerful when they encourage both collaboration among students and a new role for the teachers. In order to do so, telecommunications activities must provide rich opportunities for children to communicate in detail about jointly addressed problems (Katz, McSwiney, & Stroud, 1987; Riel, 1988).

Apple Global Education

Among the most ambitious global networking projects, although by no means the largest or most extensive, is Apple Global Education (AGE), a program sponsored by the Apple Computer Corporation. The intent was to connect schools all over the world on an easy-to-use computer network of electronic mail and bulletin board without requiring participation in a specific program or curriculum.

The assumption was that by virtue of shared and common interests in other cultures, teachers and students at great distances from each other would have much to discuss and would be given the opportunity to create a global learning community in the spirit of Marshall McLuhan's "Electronic Global Village." In other words, AGE was meant to be a learning environment, not a particular curriculum, and a virtual classroom without walls, not a set of learning objectives.

AGE was intended to be a distributed network, democratic and egalitarian in spirit and self-renewing. Teachers and students, when on-line, were recognizable by name and contribution, not rank or age. Curriculum projects were developed by and for the schools themselves rather than imposed from outside.

The project enjoyed 2 years of rapid growth and saw the emergence of leader teachers from different parts of the world inspire large numbers of students to produce a rich array of writing and graphic projects, including animated HyperCard graphics with sound, daily newsletters, and political discussions generated out of and within hours of world events. The question the network's promoters and participants now face is whether they can build on this beginning and incorporate new schools effectively (Scott & Woodbridge, 1991).

Earth Lab

The work of Denis Newman and his colleagues at Bolt, Beranek and Newman (BBN) on the Earth Lab project indicates that the promotion of communication between children, between subjects, and between teachers through local area networks *within* the schools can also have considerable benefits for learning.

For the last 4 years, the Earth Lab project has been designing, implementing, and observing the effects of a local area network system intended to facilitate collaborative work in elementary school earth science (Newman, 1990). Newman and his colleagues' plan was to create a prototype local area network system and demonstrate it in a New York City public school using an earth science curriculum. The pedagogical rationale was that students should use technology the way real scientists do: to communicate and share data (i.e., to collaborate). The demonstration school is a public elementary school (Grades 3 to 6) located in central Harlem, New York City. The school population of approximately 700 students is predominantly African American, with a minority of Hispanic and other groups. The school's achievement scores are about average for New York City but lower than the national averages. With a few exceptions, the staff took a traditional approach to teaching through whole-class lessons, textbook reading, and worksheet drills. Under normal circumstances, the school would have been a likely customer for an integrated learning system. However, in this case, the school's computer teacher, who had a different vision, was able to play a leadership role and make use of the technology provided by the project.

Earth Lab supports restructuring through a decompartmentalization of instruction. In designing the environment, the BBN team assumed that students would benefit from seeing the connections between such topics as math and science and science and writing. Projects that groups of students undertake could be made more authentic and perhaps more motivating if related to real-world concerns where the disciplinary boundaries do not necessarily hold. Students could also become more motivated if their schoolwork were, to a greater extent, under their own control rather than tightly controlled by the school schedule. Classroom tasks might have to extend beyond the single lesson period since once students begin working with some autonomy, the project may involve new goals that are discovered in the process. Teacher relationships, including distribution of expertise and collaboration among the teaching staff, might change as student projects begin to cross over the compartmentalized curriculum structure. Evaluation of students might also have to move from the typical short-answer tests of individuals to assessments of the group performance on the project itself.

A yearlong formative experiment began in the fall of 1986. In the initial setup, a local area network connected the 25 Apple IIe computers in the school to a hard drive that allowed for central storage of data, text, and programs. The Bank Street Writer word-processing program was enhanced with an electronic mail system (Newman, 1990). The Bank Street Filer was another basic tool that made it possible for students to create databases that could be accessed from any computer in the school. Along with the technology, BBN introduced a yearlong earth science curriculum designed in collaboration with the teachers (Brienne & Goldman, 1989). The formative experiment took as its goal an increase in the frequency of collaborative work among students. At least for the one year in which systematic research was funded, BBN was prepared to modify the design of the technology, introduce new software, develop curriculum materials, and conduct staff development workshops as needed. After the first year, the school obtained an additional 20 Apple IIGS computers through an award from Apple Computer, Inc., and over the last few years has added several other computers (including five Macintosh computers). Several other application programs are in use on the network, including "hypermedia" systems, LOGOWriter, telecommunication programs, and Macintosh programs including desktop publishing tools.

Databases are used extensively both within and outside the earth science curriculum. During the lunch hour, students are found inventing databases of their favorite action figures. In social studies, students research almanacs and other sources to fill in databases about countries of the world and figures from African-American history. In earth science, they examine databases of dinosaur fossils and earthquakes and create databases of the weather readings and indicators of seasonal change that small groups of students collect over a period of several months.

The primary means for supporting collaborative groups is the Earth Lab's network interface, which makes it easy for individuals or groups to store and retrieve data pertaining to their projects. The work of the project, in the form of text, database, graphics, and code files, is stored in work spaces, folders or directories on the network file server. These work spaces, available to any computer on the school local area network, give groups a location for their work together. Students and teachers can be assigned to any number of work spaces. For example, work spaces are set up for pairs of students to work on writing assignments together.

Other work spaces served schoolwide clubs or other projects. Each individual also had a personal work space. In the first year of the project, the science teacher, who had the students for two periods a week, had the class form groups of three or four for the purpose of conducting investigations in the science lab. The science groups gave themselves names that were used for group work spaces on the network. Students share different data with different students or groups in the school, for instance, a science group, a noon-hour club, and the whole class. The current Earth Lab network system is designed to present the same information when students are on either Macintosh or Apple II computers.

When the project began, BBN's explicit goal was to create a classroom environment in which students used technology the way scientists did: for collaborative work. Their analysis of what actually happened led them to a broader conception of how the local area network technology can function. While direct support of collaborative work groups is still important, they have increasingly become interested in the decompartmentalization of the school that can result from this kind of use of a local area network. Teachers are better able to collaborate, students are better able to carry their work from one context to another, and the computer lab was increasingly used in a heterogeneous manner, with several projects or groups from different classes working simultaneously. This restructuring supports both individual and group work and contributes to a sense of community in the school.

EVALUATION

There has been relatively little attention paid to the evaluation issues that our emphasis (and Papert's) on culture creation and reciprocal-transformative interaction raises. Salomon (in press) has proposed an "in-principle" way to evaluate the relative significance of different aspects of curriculum activities viewed as cultural systems, but as yet practical applications of his technique are lacking. There is also considerable uncertainty about how to parse joint activity with respect to individual contributions. Crook's (1991) position is nearest to our own; he stresses the need to address "common knowledge" (Edwards & Mercer, 1987) in any assessment of the acquisition of computer knowledge.

At present, some combination of experimental-control group designs at the level of the system and detailed clinical descriptions of individual minihistories of mediating activity seem the most appropriate form of evaluation, although they raise issues of interpretation for which there are as yet no widely accepted criteria.

A large question concerning socially framed research projects is their replicability. Harel is attempting further instantiations of ISDP, focusing on whether and how teachers are able to appropriate the ideas and incorporate them into their own teaching, addressing the social character of learning in the ISDP environment, and assessing the difficulties of integrating such an environment into the traditional school curriculum. Such projects also raise many methodological issues, especially about evaluation and experimental "rigor." Harel has reported that the ISDP students learned quantitatively measurable skills in programming and in standard school domains vis-a-vis the control groups.

Of much interest in the pursuit of a better understanding of children's computer-mediated learning is the "thick description" of "the microgenetic moment." ISDP students' work was preserved in the students' design notebooks, in computer files preserving the state of each project at the end of each working day, and in direct daily observations by researcher, by teacher, and by video camera (either a mobile camera following "interesting events" or a static camera, constantly focused on a specific workstation.) Of this thick description data, Papert says, "it is the richness of observation obtained from so many different sources that yielded a coherent sense of the development of individual subjects as well as that of shared developmental trends" (p. 31).

The research strategy employed by LCHC concerning the Fifth Dimension activity system falls within the theoretical tradition that agrees, with Papert, that in order to create powerful educational environments one must grow a culture to support that newly created form of activity. The LCHC group and the Media Lab group share a methodology that treats the formation of cultures, of systems of activity with all the theoretically necessary ingredients for maximal development, both as a central tool of analysis and as a utopian goal. This method is what Soviet psychologists refer to as a "formative experiment," defined by Davydov (1989a, 1989b, 1989c) as "an experiment in genesis-modelling," and what Werner (1948) and Vygotsky (1978) referred to as a "microgenetic" experiment (except that in this case the modeling is at a sort of "mesogenetic" level, being measured, as it is, in terms of months and years as well as seconds and minutes).

There is a growing perception by educational researchers of methodological dilemmas that arise from research focusing on the use of the microcomputer in the classroom. Foremost among these is the growth of *qualitative* research. As Harold Levine (1990) points out:

As interest in placing microcomputers in the classroom has increased, the interest by researchers/evaluators in assessing their potentially diverse effects has correspondingly grown. The assessment questions that arise are, in essence, no different from those posed for other educational innovations, technology-based or not: 1) What changes (and how permanent and illusory is any change)?, and 2) How do we make sense of the changes so as to justify, and properly conceptualize, their principled export to additional classroom environments? The answers to such questions are always difficult to provide, and investigators typically find themselves searching for new study designs and data collection strategies. (p. 461)

Levine then goes on to specify six models of qualitative data design and use: anecdotal, structured observations, case study, multisite case study, ethnography, and microethnography. Each of these has contributions to make to the qualitative evaluation of the impact of the microcomputer on the classroom; furthermore, each helps to focus on the embedding of the microcomputer in learning and teaching processes, in contrast to many quantitative experiments in which the microcomputer is evaluated as "an object of novelty." No doubt, the most effective educational research is that which contains a judicious mixture of quantitative and qualitative approaches and is embedded in formative, illuminative processes of evaluation rather than a search for "the one right answer." It is reassuring to note, however, the increasing emphasis on qualitative approaches as a counterbalance to the many experiments reported that are embedded in both the traditions of "computer-assisted instruction" and "controlled experiments."

The Cognition and Technology Group at Vanderbilt, the Epistemology and Learning Group at MIT, and LCHC, although differing in detail in their experimental approaches and informing ideologies of learning, share two commonalities: an interest in sustaining learning contexts over time and beyond the confines of their own experimentation and an interest in creating a research agenda (and therefore an educational practice) in contrast to the technicist tradition in educational computing.

Among the growing number of research centers with similar agendas is the Research Unit on Classroom Learning and Computer Use in Schools (RUCCUS) at the University of Western Ontario. RUCCUS's informing philosophy is drawn from the constructionist perspective in sociology, as framed originally by Berger and Luckman (1966). The initial work at RUCCUS is concentrating on the ethical and evaluative implications of methodologies derived from the constructionist perspective.

Their central concern is the ongoing examination of the social organization of classroom learning, and particularly the educational use of information technology as mediated by that social organization (and vice versa). The special interest that this program holds for us is, in part, the almost complete neglect of these issues in the general currency of educational computing research and the attempt to develop a

new theoretical synthesis [which would] account for school computer use in context, both the context of the particular use which recognizes the constraints of a particular application, and the larger context of the educational enterprise. What is needed is research which not only avoids easy generalizations, but which questions in each instance whether worthwhile pedagogical purposes are being served. Research which examines computing in context will ask whose interests are served by a given application, how it might impact the social organization of schooling, and what consequences might be anticipated for the process of knowledge production in general. (Goodson & Mangan, 1991, p. 4)

CONCLUSION

We have seen that there are a number of ways to characterize basic conceptions of how computers should be integrated into strategies of curriculum implementation. In considering these various positions, it also helps to locate them in terms of changes in the sophistication of the computers finding their way into classrooms.

First, the computer may be considered to be the content of and constitution of a specialized area of the curriculum. In this view, the computer is a device to be programmed and a machine to be technically understood. This focus has been expanded over time to include some considerations of the social consequences of computing and definitions of computer literacy as the ability to make appropriate use of content-free software (word processors, databases, spreadsheets, business graphics).

Second, the computer may be thought of and employed as a substitute for the teacher. This approach has also been elaborated over time, moving from a focus on drill-and-practice programs as remediation of reinforcement of basic skills and "difficult" topics to a focus on the use of computer-assisted instruction as "enrichment" for the more gifted student and to the development of intelligent computer-assisted instruction programs that use expert-systems techniques to "learn about the learner."

Third, the computer is considered as a tool that the teacher can use in a variety of ways to achieve traditional pedagogical goals. This focus has been elaborated, too, and now includes open-ended programming environments such as LOGO, curriculum applications of content-free software (word processing in English, databases in history, spreadsheets in home economics, graphics programs in art), educational games, computer networks, and HyperCard. These uses are not mutually exclusive, and one may encounter classrooms where a combination of uses is found.

An intriguing question, of course, is "What variety of new forms of activity can we create more easily with computers (conceived of, in a wide sense, as including all sorts of different computer support for activity)?" But this brings us around to the questions that motivated us at the beginning of this paper: How are we to think about the relation between computers and education, writ broad across the society? and Under what conditions are they productively used, and under what conditions are they merely a diversion?

REFERENCES

Abruscato, J. (1986). *Children, computers, and science teaching: Butterflies and bytes.* Englewood Cliffs, NJ: Prentice-Hall.

Allen, D. (1990). *Global learning guild.* Unpublished manuscript.

American Association for the Advancement of Science. (1984). *Equity and excellence: Compatible goals. An assessment of programs that facilitate increased achievement of females and minorities in K–12 mathematics and science education* (AAAS Publication 84-14). Washington, DC: Office of Opportunities and Science.

Apple, M., & Jungk, S. (1990). You don't have to be a teacher to teach this unit— Teaching, technology and gender in the classroom. *American Educational Research Journal, 27,* 227–251.

Bailey, T. (1989). *Changes in the nature and structure of work: Implications for skill requirements and skill formation* (Technical Paper No. 9). New York: Columbia University, Conservation of Human Resources.

Baird, W. (1991). Perspectives on computer science in pre-service science teacher education. *SIGCUE Outlook, 21*(1), 5–12.

Baker, E. L., Gearhart, M., & Herman J. L. (1989). *The Apple classrooms of tomorrow: 1988 UCLA Evaluation Study* (Report to Apple Computer, Inc.). Los Angeles: UCLA Center for the Study of Evaluation.

Baker, E. L., Gearhart, M., & Herman, J. L. (1990). *The Apple classrooms of tomorrow: 1990 UCLA Evaluation Study* (Report to Apple Computer, Inc.). Los Angeles: UCLA Center for the Study of Evaluation.

Baker, E. L., Gearhart, M., & Herman, J. L. (1991). *The Apple classrooms of tomorrow: 1988 UCLA Evaluation Study* (Report to Apple Computer, Inc.). Los Angeles: UCLA Center for the Study of Evaluation.

Balajthy, E. (1989). *Computers and reading: Lessons from the past and the technologies of the future.* Englewood Cliffs, NJ: Prentice-Hall.

Bellman, B., & Arias, A. (1990). Computer-mediated classrooms for culturally and linguistically diverse learners. *Computers and School, 7*(1/2), 227–241.

Berger, P. L., & Luckman, T. (1966). *The social construction of reality.* Garden City, NY: Doubleday.

Black, S., Levin, J., Mehan, H., & Quinn, C. (1983). Real and non-real time interaction: Unraveling multiple threads of discourse. *Discourse Processes, 6,* 59–75.

Bork, A. (1985). Children and interactive learning environments. In M. Chen & S. W. Paisley (Eds.), *Children and microcomputers* (pp. 267–275). Beverly Hills, CA: Sage.

Brienne, D., & Goldman, S. (1989, April). Networking: How it enhanced science classes in a New York school . . . And how it can enhance classes in your school. *Classroom Computer Learning,* pp. 44–53.

Brown, J. S., Collins, A., & Duguid, P. (1989). Situated cognitions and the culture of learning. *Educational Researcher, 18*(1), 32–42.

Bullough, R. V., & Beatty, L. F. (1991). *Classroom applications of microcomputers*. New York: Macmillan.

Burns, B., & Coon, H. (1990). LOGO programming and peer interactions—An analysis of process-oriented and product-oriented collaborations. *Journal of Educational Computing Research, 6*(4), 393–410.

Center for the Social Organization of Schools. (1984). *School uses of microcomputers: Reports from a national survey*. Baltimore: Johns Hopkins University.

Chandler, D., & Marcus, S. (1985). *Computers and literacy*. Philadelphia: Open University Press.

Cline, H. F., Bennett, R. E., Kershaw, R. C., Schneiderman, M. B., Stecher, B., & Wilson, S. (1986). *The electronic schoolhouse: The IBM secondary school computer education program*. Hillsdale, NJ: Erlbaum.

The Cognition and Technology Group at Vanderbilt University. (1990). Anchored instruction and its relationship to situated cognition. *Educational Researcher, 19*(6), 2–10.

Cole, M., & Griffin, P. (1987). *Contextual factors in education: Improving science and mathematics education for minorities and women*. Madison: Wisconsin Center for Education Research.

Commission on Workforce Quality and Labor Market Efficiency. (1989). *Investing in people*. Washington, DC: U.S. Department of Labor.

Cremin, L. (1976). *Public education*. New York: Basic Books.

Crook, C. (1987). Computers in the classroom: Defining a social context. In J. C. Rutkowska & C. Crook (Eds.), *Computers, cognition and development: Issues for psychology and education* (pp. 35–54). New York: Wiley.

Crook, C. (1991). Computers in the zone of proximal development: Implications for evaluation. *Education and Training Technology International, 28*(2), 81–91.

Cuban, L. (1986). *Teachers and machines: The classroom use of technology since 1920*. New York: Teachers College Press.

Cummins, J., & Sayers, D. (1990). Education 2001: Learning networks and educational reform. *Computers in Schools, 7*(1/2), 1–29.

Daiute, C. (1985). *Writing and computers*. Reading, MA: Addison-Wesley.

Daniels, J., & Mara, M. J. (1988). *Applied HyperCard*. New York: Prentice-Hall.

Davydov, V. V. (1988a). Problems of developmental teaching: The experience of theoretical and experimental psychological research. Part I. *Soviet Education, 30*(8).

Davydov, V. V. (1988b). Problems of developmental teaching: The experience of theoretical and experimental psychological research. Part II. *Soviet Education, 30*(9).

Davydov, V. V. (1988c). Problems of developmental teaching: The experience of theoretical and experimental psychological research. Part III. *Soviet Education, 30*(10).

DeVillar, R. A., & Faltis, C. J. (1991). *Computers and cultural diversity: Restructuring for school success*. New York: State University of New York.

Douglas, M. E. (1980). Computer literacy: What is it? *Business Education Forum, 34*(1), 18–22.

Dwyer, D. C., Ringstaff, C., & Sandholtz, J. H. (1990). *Teachers' beliefs and practices: Part 1—Patterns of change* (Apple Classrooms of Tomorrow Report No. 8). Cupertino, CA: Apple Computer, Inc.

Dyck, J. L., & Trent, A. P. (1990, April). *Learning to program in LOGO as a function of instructional method and cognitive ability.* Paper presented at the annual meeting of the American Educational Research Association, Boston.

Edwards, D., & Mercer, N. (1987). *Common knowledge.* London: Metheun.

Edwards, P. N. (1985). *Technologies of the mind: Computers, power, psychology, and World War III.* Santa Cruz, CA: Silicon Valley Research Group.

Edwards, P. N. (1986). *Artificial intelligence and high technology war: The perspective of the formal machine.* Santa Cruz, CA: Silicon Valley Research Group.

Edwards, P. N. (1988). *The closed world: Computers and the politics of discourse.* Unpublished doctoral dissertation, University of California, Santa Cruz.

Ellis, J. D. (1991). The BSCS perspective on preparing science teachers in educational computing. *SIGCUE Outlook, 21*(1), 17–32.

Ennals, R. (1985). *Artificial intelligence: Applications to logical reasoning and historical research.* Chichester, U.K.: Halsted.

Evans, C. R. (1982). *The mighty micro: The impact of the computer revolution.* London: Gollancz.

Galbraith, J. K. (1958). *The affluent society.* Boston: Houghton Mifflin.

Goodson, I. F., & Mangan, J. M. (1991). *Qualitative educational research studies: Methodologies in transition* (Occasional Paper No. 1). London, Canada: Research Unit on Classroom Learning and Computer Use in Schools, University of Western Ontario.

Gott, S. P. (1989). Apprenticeship instruction for real world tasks: The coordination of procedures, mental models, and strategies. *Review of Research in Education* (Vol. 15, pp. 97–169). Washington, DC: American Educational Research Association.

Griffin, P., Belyaeva, A. V., & Soldatova, G. (in press). Creating and reconstituting contexts for educational interactions including a computer program. In N. Minick, E. Foreman, & A. Stone (Eds.), *The institutional and social context of mind: New directions in Vygotskian theory and research.* Oxford, England: Oxford University Press.

Griffin, P., & Cole, M. (1989). New technologies, basic skills, and the underside of education: What's to be done? In J. A. Langer (Ed.), *Language, literature, and culture: Issue of society and schooling* (pp. 110–131). Norwood, NJ: Ablex.

Hammill, B. (1989). Three issues for intelligent tutoring. *Journal of Machine-Mediated Learning, 3,* 169–187.

Harel, I., & Papert, S. (1990). Software design as a learning environment. In I. Harel (Ed.), *Constructionist learning* (pp. 19–50). Cambridge, MA: MIT.

Hubbard, G. (1991). Computer applications in art education. *SIGCUE Outlook, 2,* 50–56.

Hunter, B. (1984). *My students use computers: Computer literacy in the K–8 curriculum.* Reston, VA: Reston Publishing.

International Federation of Information Processing. (1987). *Conference on a computer for each student and its impact on teaching and curriculum in the university.* New York: Elsevier.

Jensen, A. R. (1973). *Educability and group differences.* London: Methuen.

Jones, M., & Myers, D. (1988). *Hands-on HyperCard.* New York: Wiley.

Katz, M., McSwiney, E., & Stroud, K. (1987). *Facilitating collegial exchange among science teachers: An experiment in computer-based conferencing* (Technical Report of the Educational Technical Center). Cambridge, MA: Harvard School of Education.

Kelman, P. (1989, June). *Alternatives to integrated instructional systems.* Paper presented at the meeting of the National Educational Computing Conference, Nashville, TN.

Kerr, S. T. (1991). Educational reform and technological change: Computer literacy in the Soviet Union. *Computer Education Review, 35,* 222–254.

Kiesler, S., & Sproull, L. (1987). *Computing and change on campus.* New York: Cambridge University Press.

Kinzer, C. K., Sherwood, R. D., & Bransford, J. D. (1986). *Computer strategies for education: Foundations and content-area applications.* Columbus, OH: Merrill Publishing Company.

Kyle, W. C., Jr. (1984). What became of the curriculum development projects of the 1960's? How effective were they? What did we learn from them that will help teachers in today's classrooms? In D. Holdzkom & P.B. Lutz (Eds.), *Research within reach: Science education* (pp. 3–24). Charleston, WV: Appalachia Educational Laboratory.

Laboratory of Comparative Human Cognition. (1982). A model system for the study of learning difficulties. *The Quarterly Newsletter of the Laboratory of Comparative Human Cognition, 4*(3), 39–66.

Laboratory of Comparative Human Cognition. (1983). Culture and cognitive development. In P. H. Mussen (Ed.), W. Kessen (Vol. Ed.), *Handbook of child psychology* (Vol. 1, pp. 295–356). New York: Wiley.

Laboratory of Comparative Human Cognition. (1989). Kids and computers: A positive vision of the future. *Harvard Educational Review, 59,* 73–86.

Lampert, M., & Ball, D. H. (1990). *Using hypermedia technology to support a new pedagogy of teacher education* (Issue Paper 90-5). East Lansing: National Center for Research on Teacher Education, Michigan State University.

Lave, J. (1987). *Cognition in practice.* New York: Cambridge University Press.

Lemons, M. (1990, July). *Evaluation of a jointly created game.* Paper presented at the meeting of the Second Congress of the International Congress on Activity Theory, Lahti, Finland.

Levin, H. (1986). Cost and cost effectiveness of computer assisted instruction. In J. A. Culbertson & L. L. Cunningham (Eds.), *Microcomputers and education* (pp. 156–174). Chicago: National Society for the Study of Education.

Levin, J. A., Rogers, A., Waugh, M., & Smith, K. (1989, May). Observations on electronic networks: Appropriate activities for learning. *The Computing Teacher,* pp. 17–21.

Levin, J. A., & Souviney, R. (Eds.). (1983). Computers and literacy: A time for tools. *Quarterly Newsletter of the Laboratory of Comparative Human Cognition, 5*(3), 45–68.

Levine, H. (1990). Models of qualitative data use in the assessment of classroom-based microcomputer education programs. *Journal of Educational Computing Research, 6,* 461–477.

Longstreet, W. S., & Sorant, P. E. (1985, Spring). Computer literacy—Definition? *Educational Horizons,* pp. 117–120.

Makrakis, V. (1988). *Computers in school education: The cases of Sweden and Greece.* Stockholm, Sweden: Institute of International Education, University of Stockholm.

Mandl, H., & Lesgold, A. (1988). *Learning issues for intelligent tutoring systems.* New York: Springer-Verlag.

Martinez, M. E., & Mead, N. A. (1988). *Computer competence. The first national assessment* (ETS Report No. 17-CC-01). Princeton, NJ: Educational Testing Service.

Mehan, H., Moll, L. C., & Riel, M. (1985). *Computers in classrooms: A quasi-experiment in guided change* (NIE Final Report). La Jolla: University of California, San Diego.

Moll, L. C., & Diaz, S. (1987). Change as the goal of educational research. *Anthropology and Educational Quarterly, 18,* 300–311.

Muenshaw, S. (1980). *Head Start in the 1980's.* Washington, DC: U.S. Department of Health and Human Services.

National Assessment of Educational Progress. (1985). *The reading report card: Progress towards excellence in our schools: Trends in reading over four national assessments, 1971–1984.* Princeton, NJ: Educational Testing Service.

National Commission on Excellence in Education. (1983). *A nation at risk: The imperative of educational reform.* Washington, DC: U.S. Department of Education.

Nevison, J. M. (1976). Computing in the liberal arts college. *Science, 194,* 396–402.

Newman, D. (1990). Opportunities for research on the organizational impact of school computers. *Educational Researcher, 19*(3), 8–13.

Newman, D., Brienne, D., Goldman, S., Jackson, J., & Magzamen, S. (1988, April). *Peer collaboration in computer-mediated science investigations.* Paper presented at the annual meeting of the American Educational Research Association, New Orleans, LA.

Noble, D. D. (1989). Mental materiel. The militarization of learning and intelligence in U.S. education. In L. Levidow & K. Robins (Eds.), *Cyborg Worlds: The Military Information Society* (pp. 13–41). London: Free Association Press.

Noyelle, T. J. (1985, March). *The new technology and the new economy: Some implications for equal employment opportunity.* Paper presented to Panel on Technology and Women's Employment of the National Research Council.

Padilla, R. (1990). HyperCard: A tool for dual language instruction. *Computers in Schools, 7*(1/2), 211–226.

Palumbo, D. B. (1990). Programming language/problem solving research: A review of relevant issues. *Review of Educational Research, 60,* 65–89.

Papert, S. (1980). *Mindstorms, children, computers and powerful ideas.* New York: Basic Books.

Papert, S. (1987). Computer criticism vs. technocentric thinking. *Educational Researcher, 16*(1), 22–30.

Papert, S. (1990). Introduction. In I. Harel (Ed.), *Constructionist learning* (pp. 1–8). Cambridge, MA: MIT.

Payne, J. S. (1973). *Head Start: A tragicomedy with epilogue.* New York: Behaviorist Publications.

Pea, R., & Kurland, D. M. (1984). *On the cognitive effects of learning to program: A critical look* (Technical Report No. 7). New York: Bank Street College of Education, Center for Children and Technology.

Pea, R., & Sheingold, K. (Eds.). (1987). *Mirrors of mind: Patterns of experience in educational computing.* Norwood, NJ: Ablex.

Pryczak, F. (1990). Development of diagnostic tests for computer literacy. *Computers and Education, 14,* 213–216.

Psotka, J., Massey, L. D., & Mutter, S. A. (Eds.). (1988). *Intelligent tutoring systems: Lessons learned.* Hillsdale, NJ: Erlbaum.

Resnick, M., & Ocko, S. (1990). LEGO/LOGO: Learning through and about design. In I. Harel (Ed.), *Constructionist learning* (pp. 121–128). Cambridge, MA: MIT.

Riel, M. (1986, April). *The educational potential of computer networking.* Paper presented at the annual meeting of the American Educational Research Association, San Francisco.

Riel, M. (1988, September). *Telecommunications: Connections to the future.* Paper delivered to California State Educational Technology Committee.

Ringstaff, C., Sandholtz, J. H., & Dwyer, D. C. (1991, April). *Trading places: When teachers utilize student expertise in technology-intensive classrooms.* Paper presented at the annual meeting of the American Educational Research Association, Chicago.

Rosenberg, R. (1991). Debunking computer literacy. *Technology Review, 94*(1), 58–64.

Salomon, G. (in press). New information technologies for education. In M. C. Alkin (Ed.), *Encyclopedia of educational research* (6th ed.). New York: Macmillan.

Salomon, G., & Perkins, D. N. (1988). Transfer of cognitive skills from programming: When and how. *Journal of Educational Computing Research, 3,* 149–170.

Salomon, G., Perkins, D. N., & Globerson, T. (1991). Partners in cognition: Extending human intelligence with intelligent technologies. *Educational Researcher, 20*(3), 2–9.

Sandholtz, J. H., Ringstaff, C., & Dwyer, D. C. (1990). *Classroom management: Teaching in high-tech environments: Classroom management revisited. First–fourth year findings* (Apple Classrooms of Tomorrow Report No. 10). Cupertino, CA: Apple Classrooms of Tomorrow, Advanced Technology Group, Apple Computer Incorporated.

Scardamalia, M., & Bereiter, C. (in press). Schools as knowledge-building communities. In S. Strauss (Ed.), *Human development: The Tel-Aviv annual workshop: Vol. 7. Development and learning environments.* Norwood, NJ: Ablex.

Schwartz, J. L. (1989). Intellectual mirrors: A step in the direction of making schools knowledge-making places. *Harvard Educational Review, 59,* 51–61.

Scollon, R. (1983). Computer conferencing: A medium for appropriate time. *Quarterly Newsletter of the Laboratory of Comparative Human Cognition, 5*(3), 67–68.

Scott, A., & Woodbridge, S. (1991). *Transforming teachers through telecommunications: An evaluation of the Apple Global Education Network.* La Jolla: Laboratory of Comparative Human Cognition, University of California, San Diego.

Shavelson, R. J., Winkler, J. D., Stasz, C., Feibel, W., Robyn, A. E., & Shaha, S. (1984). *Successful teachers' patterns of microcomputer-based mathematics and science instruction* (Report No. N-2170-NIE/RC). Santa Monica, CA: Rand Corporation.

Sheingold, K., Martin, L. W. M., & Endreweit, M. E. (1985). *Preparing urban teachers for the technological future* (Technical Report No. 36). New York: Bank Street College of Education, Center for Children and Technology.

Sherman, B. (1985). *The new revolution: The impact of computing on society.* New York: Wiley.

Slater, R. (1990). *Portraits in silicon.* Cambridge, MA: MIT Press.

Slavin, R. (1991). Synthesis of research and cooperative learning. *Educational Leadership, 48*(5), 71–82.

Stevenson, H. W., Lee, S. Y., & Stigler, J. W. (1986). Mathematics achievement of Chinese, Japanese, and American children. *Science, 231,* 693–699.

Stonier, T. (1983). *The wealth of information: A profile of the post-industrial society*. London: Eyre Methuen.

Stonier, T., & Conlin, C. (1985). *The three C's: Children, computers and communication*. New York: Wiley.

Suppes, P. (Ed.). (1981). *University-level computer-assisted instruction at Stanford: 1968–1980*. Stanford, CA: Institute for Mathematical Studies in the Social Sciences, Stanford University.

Suppes, P. (1988). Computer-assisted instruction. In D. Unwin & R. McAleese (Eds.), *The encyclopedia of educational media communication and technology* (2nd ed., pp. 107–116). New York: Greenwood.

Suppes, P. (1990). Intelligent tutoring, but not intelligent enough. *Contemporary Psychology, 35*, 648–650.

Swan, J. C. (1991). Rehumanizing information: An alternative future. *Library Journal, 115*(1), 156.

Taylor, R. P. (Ed.). (1980). *The computer in the school: Tutor, tool, and tutee*. New York: Teachers College Press.

Taylor, R. (1991). Teacher training curriculum project: Integration of computing into the teaching of secondary school science and art. *SIGCUE Outlook, 21*(1), 2–5.

Thornburg, D. G., & Pea, R. D. (1991). Synthesizing instructional technologies and educational cultures: Exploring cognition and metacognition in the social studies. *Journal of Educational Computing Research, 7*, 121–164.

Tinker, R., & Papert, S. (1989). Tools for science education. In J. Ellis (Ed.), *1988 AETS yearbook: Information technology and science education* (pp. 1–23). Columbus: EIHIC/SMEAC Information Center, Ohio State University.

Trainor, T. N. (1984). *Computer literacy: Concepts and applications*. Santa Cruz, CA: Mitchell Publications.

Turkle, S. (1985). *The second self*. New York: Simon & Shuster.

Turkle, S., & Papert, S. (1990). Epistemological pluralism: Styles and voices within the computer culture. In I. Harel (Ed.), *Constructionist learning* (pp. 345–377). Cambridge, MA: The Media Laboratory.

Vygotsky, L. S. (1978). *Mind in society: The development of higher psychological processes* (M. Cole, V. John-Steiner, S. Scribner, & E. Souberman, Eds.). Cambridge, MA: Harvard University Press.

Walberg, H. J., Harnisch, J. L., & Tsai, S. L. (1984). *Mathematics productivity in Japan and Illinois*. Unpublished manuscript, University of Illinois at Chicago.

Washington, V. (1987). *Project Head Start: Past, present and future trends*. New York: Garland.

Weiner, N. (1989). *The human use of human beings: Cybernetics and society*. London: Free Association Press. (Original work published 1950)

Weir, S. (1987). *Cultivating minds: A LOGO casebook*. New York: Harper & Row.

Weizenbaum, J. (1976). *Computer power and human reason: From judgment to calculation*. San Francisco: Freeman.

Werner, H. (1948). *Comparative psychology of mental development*. New York: New York International University Press.

Wertsch, J. (Ed.). (1985). *Vygotsky and the social formation of mind*. Cambridge, MA: Harvard University Press.

Whooley, K. A. (1986). *Gender differences in computer use*. Unpublished master's thesis, University of California, San Diego.

Winston, B. (1986). *Misunderstanding media*. Cambridge, MA: Harvard University Press.

Woodrow, J. (1991a). A comparison of four attitude scales. *Journal of Educational Computing Research, 7*, 165–187.

Woodrow, J. (1991b). Locus of control and computer attitudes as determinants of the computer literacy of student teachers. *Computers in Education, 6*, 237–245.

Woodrow, J. (1991c). Determinants of student teacher literacy achievement. *Computers in Education, 6*, 247–256.

II.
HISTORY AND PHILOSOPHY

Chapter 6

Education Historians as Mythmakers

BARBARA FINKELSTEIN
University of Maryland

EDUCATION HISTORY AS A MESSAGE SYSTEM

In the contemporary context of everyday life in the United States, history has acquired an array of meanings. In popular parlance, it connotes a condition of irrelevance. "That boyfriend is history," announces a teenager. "Another word and you're history," suggests Telly Sevalas or Clint Eastwood. "Virulent racial discrimination is history," asserts a MacNeil-Lehrer guest scholar as she consigns affirmative action programs and the heritage of slavery to an historical junk heap. Scholars are more eloquent and grandiose in their understanding of how history functions in everyday life. Some, convinced apparently that history lives and historians lie, invoke Voltaire to suggest that history is "a pack of tricks we [historians] play on the dead." Still others pay homage to the power of historical discourse in conjuring Santayana's warning that "those who forget the past are bound to repeat it," as though history itself functions as a kind of bad example.

The simultaneous identification of history as utterly irrelevant on the one hand, and as totally powerful on the other, is hardly surprising. Implicit in these apparently contradictory views is an understanding, discovered in the last two decades, that historians function like other mythmakers in society. Through their analyses of the past, they reveal meaning and, when they do it well, make sense of reality. As they recover the past, they constitute vision, cultivate consciousness, consign to oblivion, and bring into collective memory (Bender, 1985; Bouwsma, 1981; Burstyn, 1987; LaCapra, 1983; Lasch, 1989; White, 1987).[1]

Few scholars have had the benefit of good colleagues like Michael Zuckerman and James B. Gilbert and friendly critics as astute and forthcoming as Joyce Antler and Joan Burstyn. Still fewer have had intelligent and dedicated help of the kind provided by Hilary Rosenfield, a rare and cherished student, or Carol Meyerson, a compassionate and supportive sister. To Joyce, Joan, Hilary, Michael, and Jim, and to Gerald Grant, who has functioned as an equally friendly critic and prod, I am grateful.

The discourse of historians, though focused on the recovery of the past, has another mythmaking function as well. It figures forth visions of future possibility and serves as a specialized kind of message system. Through the recovery of the past, historians can sow seeds of hope and/or despair, encourage and/or discourage particular courses of action, sound political or moral calls, or otherwise shape the forms that future imagining can take. It is not so much that historians intend to function in this way. Indeed, most would deny a role as mythmakers, dismissing such moral actions as pretentious, imperial, and/or inappropriate for scholars.

For historians of education especially, the role of mythmaker flows inexorably from the nature of the institutions they study and the special role schools have played in U.S. political and economic life. Both public and private schools have, from the time of their founding, served as ideological templates onto which generations of elite reformers, urban immigrant groups, women, and minorities have grafted aspirations, standards, and hopes for a better society (Anderson, 1988; Cremin, 1961, 1988; Finkelstein, 1985, 1988, 1991). Schools have served symbolically as emblems of political and economic reform and politically as objects of local, state, and national regulation (Cuban, 1989; Finkelstein, 1983). For two centuries, schools have been the subject of constant reform efforts and political pressure. They have functioned as social sponges, absorbing the shocks of social discontent and drawing heat from different classes of social actors seeking changes in their condition and status. School reform, unlike fundamental political or economic reform, is an ever-present reality in U.S. society.

Texts in Context

As chroniclers of institutions and processes perpetually in motion, historians of education occupy a unique historical vantage point. From their origin as university professionals almost a century ago, they have been situated within conflicting moral and professional worlds. They have been inhabitants of a contemplative universe of professional scholars, working at arm's length from the turbulent realities of life in classrooms, households, education policy-making agencies, and other social spaces. As participants in a world of professional education leaders, they have prepared new generations of teachers and school administrators, serving as consultants to education policymakers and decision makers (Ravitch, 1983). Their work has served sense-making functions for the turbulent world of education reform where various groups do battle for the minds and hearts of the nation's children, and often, in the process, forget the force of historical circumstance and the power of tradition.

Education historians have been the keepers of collective memory for the world of education practice and policy, revealing the evolution of the very realities that shape their consciousness, their questions, forms of inquiry, and definitions of their work. As disinterested intellectuals, they are fundamentally compromised. As chroniclers of their own enterprise, they dwell in a formative environment of compelling creativity for the recovery of diverse traditions and enduring visions with which to make sense (i.e., to formulate myths that reveal education purposes and practices).

This chapter explores how historians of education have situated their work in relation to the evolving contexts of education. It focuses on how they have dispatched their role as cultural authorities carrying messages from the past, informing education in the present, and charting possibilities and impossibilities for the future. Specifically, this chapter will explore how historians of education have functioned as mythmakers and social messengers. What aspects of the U.S. educational past have they considered important? What emphases have they revealed? What emphases are as yet unidentified?

Proceeding on an assumption that visions of future possibility are nested in their work, this chapter will analyze the role of historians as social messengers, paying close attention to the ways in which they have construed the evolving educational landscape, identified features that are fixed and/or subject to change, and wrested meaning from it. It will explore education history as an evolving series of narrative visions that, over time, have revealed ever more nuanced and complete aspects of the U.S. educational past and the limits and possibilities of education.

EVOLVING MYTHS: EVOLVING MESSAGES

The Myth of Benevolence: History as Professional Advocacy, 1900–1960

During the first 60 years of the development of education history as a professional field of study from 1899 to 1960, an array of historians including, among others, Ellwood P. Cubberley, R. Freeman Butts, Paul Munroe, Thomas Woody, and a young Lawrence A. Cremin recovered a professional past for an emerging new profession. Throughout the decades from 1900 to 1960, they transformed education history into professional advocacy, revealing intersections among the evolution of public commitments to education, the expansion of democratic forms and opportunities, and the cultivation of professional authority. As Ellwood Patterson Cubberley wrote in the preface of his influential text *Public Education in the United States* (1934), history learning and teacher pride were unvaryingly linked. Through the study of the American educational past, teachers could take "their political and social bearings, and see educational service in its proper setting as a great national institution evolved by democracy to help it solve its many perplexing problems" (Cubberley, 1919).

As chroniclers of a professional past, the "professional advocates" of education history acted as mythmakers and institution makers simultaneously, revealing education history as a progressive chapter in a history of benevolence, freedom, and egalitarian social arrangements; constituting and consolidating various bureaucracies; and fixing an historical canon that would draw fire from a new generation of education historians.

CRITICS OF THE TRADITIONAL CANON: INTELLECTUAL IMPERIALS AND JUSTICE WORKERS

Myths of an American Paideia

A second form of historical recovery exploded into view in the 1960s and 1970s and elaborated the mythmaking possibilities of education history in two distinct ways. One group, historians such as Bernard Bailyn and a self-reconstructed Lawrence A. Cremin, whom we will call "critics of the traditional canon," denigrated it for its evangelical qualities, characterizing it as the work of "powerful academic ecclesia" who viewed the past as the "present writ small" and, out of misplaced professional zeal, stripped history of its capacity to explain social change and public education from its origins in the 19th century (Bailyn, 1960). Their intention was to embed the history of education in a different professional and conceptual universe than was customary among historians in professional education schools (Axtell, 1974; Bailyn, 1960; Cremin, 1965b; Tyack, 1974).

Myths of Control and Oppression

A second group, those whom we will call "justice workers," turned traditional historical wisdom upside down. In pursuit of interpretive balance, historians such as Michael B. Katz and Jill K. Conway documented the existence of demagogic forms and processes in the evolution of schooling, linking aspects of the U.S. educational past to previously invisible historical processes—the hidden injuries of class, race, and gender and the evolution of wage labor, family structures, regulatory government, and new modes of social control (Bowles & Gintis, 1976; Conway, 1974; Katz, 1968, 1971, 1987; San Miguel, 1987; Schlossman, 1977). In the process, they projected a grim future for public schooling as an instrument of democracy, social reconstruction, and/or human justice.

Myths of Contested Terrains

In the late 1970s another group of education historians, inheritors of these contradictory interpretive traditions, aimed to synthesize the two traditions and found ways to link economic, political, social, and ideological dimensions into their historical narratives while reattaching educational history to the history of politics, ideas, and schooling. Skeptical about schooling as an instrument of social change, these historians nonetheless rediscovered democratic processes at work in the politics of education reform. For the last decade, several have reconnected the enterprise of history doing to contemporary policy concerns in education (Carnoy & Levin, 1985; Goodenow & Ravitch, 1983; Hogan, 1978; James, 1991; Katznelson & Weir, 1985; Labaree, 1988; Perlmann, 1988; Ravitch & Goodenow, 1981; Tyack & Hansot, 1982).

Myths of Community

A fourth group of historical vision makers have recently become impatient with intellectual traditions explaining the evolution of popular education strictly in relation to the centralizing tendencies of modern life. They are now in the process of retrieving history as lived and constructed in small social circles and among women, ethnic minorities, learners, and an array of the previously invisible (Anderson, 1988; Bender, 1978; Clifford, 1989; Finkelstein, 1974, 1989a, 1991; Katznelson & Weir, 1985; Sennett, 1980; Silver, 1983; Trennert, 1990; Wallace, 1970).

Each interpretive tradition integrates specific visions of education possibility and assumptions about the role of schools, school persons, and education professionals in the forming of new social possibilities. It is to this subject that I now turn.

EDUCATION HISTORY AS PROFESSIONAL ADVOCACY: THE MYTH OF BENEVOLENCE, 1900–1968

The founding generations of professional education historians entered into the history-making enterprise at an auspicious time. They were heir to a tradition of historical understanding forged by public school advocates working in policy-making environments at the national, state, and local level (Cremin, 1965b). As participants in the creation of research universities, they abetted the development of these specialized institutions as centers for the cultivation of professional authority and expertise, the advancement and production of knowledge, the pursuit of scientific understanding, the cultivation of technical expertise, and the preparation of an informed elite (Bledstein, 1976; Butts, 1939; Hall, 1983; Herbst, 1989). As university professionals in as yet unfixed settings, they helped to constitute a catalytic environment that served as a site for the transformation of fields of study into powerful professions, scholars and scientists into academic specialists trained in a discipline, and policymakers, school practitioners, and aspiring teachers into professional educators. Within an evolving context, they contributed to the formation of a specialized house of intellect, a kind of "watchtower" (Butts, 1939) within which they could integrate scholarship, professional preparation, public school advocacy, and social reconstruction (Monroe, 1940).

From their "watchtowers" inside universities, they participated in the shaping of professional identities for an expanding cadre of teachers planning to teach huge numbers of immigrant children in urban public schools. They organized campaigns to transform secondary and higher education curricula. They created model schools and mounted campaigns to privilege the importance of education knowledge and the authority of university professors (Ravitch, 1983). For historians of education, the emerging universities served as protected settings within which they could recover the past while participating in the making of professionals and the reconstruction of education in the present.

Not unexpectedly, within this context they plumbed documentary records in search of guiding traditions and, as university professors from then to now have done, identified universes of discourse within which to situate their work, hone their knowledge, contribute to understanding, elevate their status as professionals, and recover the past.

The founding generations of professional education historians latched on to interpretive traditions forged in policy-making settings where policy documents and prescriptive literatures served as important documents and where politics and the doing of education history were closely linked. It was a history that centered on institutions generally and schools specifically and that attached the evolution of public education to the evolution of political processes—nation building, public regulation at state and local levels, and the cultivation of democratic forms.

As architects and participants in the making of professionalism in education, professional education advocates shared and helped to develop what Michael Katz was later to describe as a "warm and comforting myth . . . portraying a rational enlightened working class, led by idealistic and humanitarian intellectuals, triumphantly wresting free public education from a selfish, wealthy elite and from the bigoted proponents of orthodox religion" (Katz, 1968, p. 1).

For historians charting the evolution of higher education, the story substituted a benevolent elite for an enlightened working class and constituted an equally comfortable narrative. A small cadre of enlightened intellectuals wrested intellectual freedom from intrusive and narrow-minded legislators and advocates of sectarian higher education. In the process, they helped to secure principles of corporate autonomy, intellectual freedom, control over the curriculum, and tenure rights (i.e., sufficient independence to make universities safe havens for the exercise of reason and the advancement of knowledge; Brubacher & Rudy, 1976; Butts, 1939; Hofstadter & Metzger, 1963).

First-Generation Messages

The history writers of these early generations revealed the evolution of education as a public good and public school advocates as heroes and heroines. It was a history that cast private school advocacy as a product of parochialism, conservatism, and ignorance and identified professionalism with expertise, benevolent advocacy, truth telling, and theological ecumenicism (Bremner, 1974; Butts & Cremin, 1953; Cubberley, 1919; Hofstadter & Metzger, 1963; Ravitch, 1974). It was a history that explained the emergence of public schools as a product of humanitarian concern and benevolent sentiment, a step forward in the realization of more humane institutions, and public opportunity structures for all classes of people. It was a history that indexed the evolution of public education to the emergence of regulatory legislation securing public support and government control for schooling, defining education as a right and school attendance as an obligation. It was a history that situated the evolution of education within an emerging and evolving array of policy environments (Butts & Cremin, 1953).

Focusing on the enabling character of literacy and the empowering qualities of public schools for individual mobility and civic nurture, first-generation professional advocates were at their most sophisticated when they documented connections among political philosophy, education policy, interest group politics, and the evolution of opportunity structures. Their myths were especially subtle and illuminating when they chronicled the ebb and flow of academic freedom in colleges and universities (Brubacher & Rudy, 1976; Butts, 1939; Cremin, 1961; Hofstadter & Metzger, 1963).

As mythmakers and message carriers, the first generations of professional education historians emphasized the promises rather than the realities, and the benefits rather than the costs, of educational expansion. Rooting their analyses in the rhetoric of social reform, the intentions of social planners, blueprints of institution building, analyses of civic philosophy, public administration, and local political battles, they privileged the role of ideas and words as compelling factors in social change. Mired in a particular narrative structure, they assumed that ethnic minorities, people of color, women, and Native Americans were automatic beneficiaries of new opportunity structures—public schools for children from the ages of 5 to 14, coeducational schools for women, tutorial environments for the children of slaves and Native Americans, and colleges serving more and more of the people.

Convinced that access to educational institutions constituted evidence of democratic commitments, professional advocates identified traditions that cut a definitive path to an improved future in education. An analysis of Thomas Woody's *History of Women's Education in the United States* illustrates the point. Writing in 1927, when women's right to vote was secure and coeducational institutions were everywhere in evidence, Woody announced that the struggle for equality in education was over, the rights of women in education secured. For Woody, the coeducational as well as the public school constituted the most sensitive index of democratic processes at work in education. For him, the merger over time of male and female institutions and the admission of women into previously all-male college bastions and professional schools constituted the most salient trend in the history of education for women. As he documented the evolution of coeducation, he proceeded on an assumption that the past and the future were dialectically linked, one leading the way for the other. If coeducation secured women's rights, then the future for women lay in the well-being of coeducational schooling. In a similar mode, historians of American education identified educational opportunity with the well-being of integrated schools for African Americans and nonsectarian public schools for religious minorities (Butts & Cremin, 1953).

As much could be said of all the founding-generation historians who, like Woody, identified public schooling with an array of enabling education reforms classified loosely under the label of "progressivism." Lawrence A. Cremin paid unwitting homage to the endurance of progressive visions in the conclusion of *Transformation of the School* (1961):

The transformation they [progressives] had wrought in the schools was in many ways as irreversible as the larger industrial transformation of which it had been a part. . . . The authentic progressive vision remained strangely pertinent to the problems of mid-century America. Perhaps it only awaited the reformulation and resuscitation that would ultimately derive from a larger resurgence of reform in American life and thought. (p. 353)

EDUCATION HISTORY AS DEMYSTIFICATION: THREE MYTHS FOR NEW REALITIES

When he wrote the sentences above as a relatively young scholar in the late 1950s and early 1960s, Cremin anticipated, but could not predict, the effects that new forms of scholarly thinking and social criticism would have on the nature of historical writing about education.

Together with other historians coming of academic age in the early 1960s and 1970s, Cremin approached the study of education history at an auspicious time—when the professional authority and legitimacy of the education professionals in universities were under attack from a variety of critics both inside and outside the university, and when the public school itself had become the focus of intense political debate and controversy.

The professional sensibilities of these historians would be nurtured in the 1960s and 1970s—a decade after the Brown decision had mandated the integration of schools, the civil rights movement had achieved new levels of success and militancy, poverty in the United States had been rediscovered, and women scholars and social activists had begun to reveal the hidden injuries of gender discrimination. It was a time when antiwar sentiment ran high and the authority of professional academics in universities and colleges was under attack. It was also a time when public schools acquired unprecedented new responsibilities for mending the social fabric.

It is hardly surprising that a newer generation of education historians would call the traditional enterprise into question and proceed from more skeptical assumptions and less celebratory dispositions than their predecessors. As apprentices situated in a universe of discourse that documented and celebrated the emergence of professional education expertise, public schooling, and the involvement of scholars in public life and education policy-making, they were awash in visions of the liberating possibilities of public education and the values that university professors attached to professional hegemony. But they also observed education and professional realities that contradicted these optimistic views of public schooling and benevolent expertise.

Situated in a peculiar professional situation, this newer generation of education historians was ready to generate new kinds of questions, integrate new conceptual dimensions, and rethink the nature of the field. And they did so in three uncommonly creative ways, attacking the traditional canon from three distinct perspectives and points of view.

A first group, the intellectual imperials, took leaves from the remarkable book of Bernard Bailyn titled *Education in the Forming of American Society* (1960) and approached a rethinking of education history with the intention of embedding it in a different professional universe—among historians who claimed a greater interest in historical reconstruction than in education reform, who defined education as more than public schooling, who disengaged history doing from moral and political advocacy, and who situated their analyses in the bedrock of culture and social structure rather than political processes and civic philosophy (Axtell, 1974; Bailyn, 1960; Cohen, 1974; Cremin, 1970, 1980, 1988).

A second group, the justice workers, made common cause with critical traditions of history writing constituted by historians such as Charles Beard, Merle Curti, and James Harvey Robinson. They attacked the traditional canon for transmitting what they believed to be retrograde and incorrect messages, and failing in their role as moral and social critics. As Michael B. Katz, the most original and creative of them, put it:

The historian's task is less to remind men of their obligation to the past than to force upon them an awareness of how the past could be used to effect an ethically responsible transition from present to future. . . . The old metaphor and its supporters invoke the past to nurture obligation, . . . rather than attempting to liberate them for a new educational future. (Katz, 1987, p. 5)

Historians such as Katz (1968, 1971), Karier, Violas, and Spring (1973), Karier (1975), Lazerson (1971), Tyack (1974), and Conway (1974), like Merle Curti before them (1959), were sensitive to the political aspirations of less-privileged groups—women, people of color, laboring groups—and approached the history of education in an attempt to understand the persistence of inequality, discrimination, racism, and sexism. As champions of the oppressed, they attacked the traditional discourse of education history for two of its most intransigent qualities: a myopic tendency to ignore the power of economic circumstance and narrow self-interest as a motive for reform and a failure to reveal the darker sides of educational history and recover some of the more oppressive and conserving qualities.

A third group, the sociological functionalists or modernization theorists, combined a concern with the history of less-privileged groups with an interest in the evolution of social as well as political institutions and attacked the traditional canon for its failure to reveal links between education and social structure (Kaestle, 1983; Kaestle & Vinovskis, 1978; Kett, 1977). They called attention to previously invisible facts—that for the first century and a half of education development, educational authority resided in families rather than in schools or legislatures, that concepts of public policy were as yet undeveloped, and that schooling and cultural transmission were not identical processes. Relying heavily on sociological, economic, and/or political theory, this third group accounted for the rise of schooling in relation to transformations in family structure and failures of family and church authority under the impact of wilderness conditions, urbanization, and commercial growth. As family authority faltered, schools emerged. However, they did not emerge as traditional education historians had suggested—as an expression of democratic commitments to mass education, humanitarian sentiments, dedication to political ideals, and the uses of reason—or necessarily as products of human greed and cunning. Rather, they emerged in response to the material realities of everyday life in the colonial period and a felt social need to shore up traditional authorities.

When they are taken together, these three attacks—of intellectual imperials, of justice workers, and of modernization theorists—on the traditional canon of education history challenged the interpretive and scholarly capacities of a new generation of education historians. Revisiting the American educational past, they left no explanatory framework or explanations of social change untouched. They integrated typically unidentified opposition to public schools and constituted new ways to reveal the past, capture previously invisible traditions, and constitute an array of contradictory messages and visions of future possibility.

Education History and the Myth of an American Paideia

One opening broadside on the original historiographical canon originated with Bernard Bailyn when he rejected the entire corpus of traditional education history, labeling it as anachronistic, parochial, ahistorical, and celebratory (1960). Among other things, Bailyn called for a redefinition of education that went beyond formal pedagogy and public schooling to include "the entire process by which a culture transmits itself across the generations" (1960, p. 14).

Picking up 8 years later on this particular line of reason, Lawrence A. Cremin developed a whole new vision of education history (1970, 1980, 1988). In a massive three-volume trilogy written over a 30-year period and titled *American Education*, Cremin integrated the study of schooling into a broader cultural context, analyzing its development in relation to other educative institutions (Church, Katz, Silver, & Cremin, 1989).

His histories privileged the recovery of education as it proceeded in all deliberative education settings: families, churches, community centers, and museums as well as schools, colleges, and universities. He was attentive to the educational possibilities of the mass media of communication: newspapers, the penny press, radio, and television as well as textbooks and reference books. He was as interested in the educative functions of farms and plantations as of schools and colleges.

By treating public schools as nothing more or less than one of a number of important educational institutions, Cremin was to embed it, as Daniel Boorstin had done with American history, in the entire cultural life of the nation, revealing hitherto unobserved resonances among educational institutions, reform ideals, and the uses to which people could put different educational opportunities (Axtell, 1974; Cohen, 1974; Cremin, 1970, 1980, 1988).

It is important to note, however, that Cremin's redefinition of education history to include multiple institutions did not alter his affinity with the interpretive traditions of first-generation education historians. Like the work of his predecessors, Cremin's interpretive lens brought the civic aspects of the American educational past into clear view. Cremin located the history of education in the history of ideas—albeit about a broader range of institutions than his predecessors, including the family, church, press, museum, and radio as well as the school and the college. Like the earlier historians, Cremin began each of his tomes with an analysis of grand and lesser traditions of educational thinking: Renaissance humanists, civic intellectuals, and theologians in the 16th and 17th centuries;

republican theorists, education utopians, and influential theologians in the 18th and 19th centuries; and civic philosophers, progressive reformers, cultural interpreters, and social critics in the late 19th and 20th centuries.

No less interested than his predecessors in understanding the origin and evolution of education in the forming of civic culture, Cremin (Church, Katz, Silver, & Cremin, 1989) also analyzed the evolution of institutions—families, churches, newspapers, museums, schools, colleges and universities—as specialized settings within which the "great common educative experiences of the American people" (p. 442) would be forged and their full diversities revealed.

Cremin was able to imbue education history with enormous narrative and visionary power while relieving schools of unrealizable social missions. His histories revealed the incapacities of public schools as legatee institutions absorbing a full array of reform agendas. By expanding the definition of education to go beyond public schooling, Cremin called attention to the overinflated agendas that he believed his historical predecessors had promoted (Cremin, 1988; Finkelstein, 1991; Tyack & Hansot, 1982).

As he was to say of *American Education: The Metropolitan Experience*, the last work of his trilogy: "The book is very much in the service of policy, insofar as I believe good history should be a lamp to light the present" (Church et al., 1989, p. 440). Its most important illumination was, for Cremin, the one that revealed the limitations of schools as instruments of fundamental social, political, and economic change.

As grand and elegant as Cremin's trilogy may have been, and as sensitive to the sources of consensus in U.S. history, his narrative revelations relegated power struggles and conflict to a kind of theoretical backseat. For him, conflict was neither the most important condition of U.S. education nor a reasonable explanation for change.

EDUCATION HISTORY AS DISCOURSE FOR THE OPPRESSED SECOND GENERATION: FIRST WAVE, 1968–1987

Justice Workers

A second group of historians, whom we have called justice workers, rejected all consensus-driven interpretations of educational evolution. Aiming to explain the persistence of inequality and the patent failures of public schools to equalize the social conditions of Americans, some justice workers integrated Marxist perspectives into their conceptual apparatus and revisited the origin of public schooling. Attacking the traditional canon of education history, they focused on four of its more intransigent tendencies: a disposition to ignore the power of economic circumstance and narrow self-interest as motives for reform, a failure to reveal the darker sides of educational history and recover some of its more oppressive and conserving qualities, a tendency to assume that public education had been an instrument of benevolent social transformation, and, finally, an indisposition to index education history to the evolution of the condition of women, children, youth, less-privileged minorities, and other dependent groups.

Michael Katz's influential *Class, Bureaucracy, and Schools* (1971) exemplified this particular face of revisionism. In it, Katz revealed a mosaic of four conflicting models of organization over which political battles were won and lost in the 19th century. Viewing organizational alternatives as "crystallizations of particular social values," he explored four contending models of school administration, focusing on issues of organization scale, control, professionalism, and finance to distinguish among and between them.

The first model, that of paternalistic voluntarism, was one espoused by elites aiming to sustain control over the developing political machinery of cities, to control the socialization of immigrant children whom they feared, and to wrest power from religious elites, establishing a single nonsectarian agency for the entire city. It was a vehicle, as Katz put it, for "one class to civilize another and thereby ensure that society would remain tolerable, orderly, and safe" (1971, p. 9). A second form, which he called democratic localism, was an alternative to corporate voluntarism leaving control over education to the "free and unrestricted action of the people themselves" (p. 15). The decentralized ward was the unit of control; a humanistic, broad conception of education, was, in Katz's view, its ideal. Within these two extremes of school administration are two related models. The model of corporate voluntarism involved "the conduct of single institutions as individual corporations operated by self-perpetuating boards of trustees and financed either wholly through endowment or through a combination of endowment and tuition" (Katz, 1971, p. 22). The last, and ultimately triumphant, model was that of incipient bureaucracy. The promoters of bureaucracy, including, in Katz's view, the architects of public education such as Horace Mann and Henry Barnard, concentrated on attacking democratic localism. Wearing blinders of rural nostalgia, frightened of diversity, and convinced that standardization and democracy proceeded simultaneously, their goal was to "uplift the quality of education by standardizing and systemizing its structure and content" (p. 33). Their goal was centralization; age grading, supervision, curriculum standardization, and a pan-Protestant moralism were the means of its realization. And they succeeded. As Katz put it, in an earlier work, *The Irony of Early School Reform* (1968), public schools were not a compound of democracy, rational planning, and humanitarian concern or

great democratic engines for identifying talent and matching it with opportunity. . . . They [were] rather, handmaidens of corporate capitalism—the product of self-interested, status seeking middle class elites seeking to conserve their advantages and extend them for their children. (p. 218)

What is perhaps most intriguing about Katz's transformation of vision about public education was an abstract empathy with the less privileged and an inexplicable indisposition to explore their voices. He was, at that time, seemingly more interested in exploding the traditional myth and exploring theories of social structure than he was in analyzing forms of human resistance, adaptability, or human lives. Katz inadvertently stripped working classes, less-privileged minorities, and women of human agency, defining them as unwitting victims of imperial elites.

By approaching education history in this manner, Katz and his early followers constituted a mythic transformation of American educational development that was as productive and insightful as it was blind to nuances in human motivation and aspiration. First-wave justice workers analyzed the expansion of publicly supported institutions—infant schools, Sunday schools, and public elementary schools in the first half of the 19th century; high schools, juvenile reformatories, and kindergartens in the second; and community colleges and youth organizations in the 20th century. They looked at these developments not as progressive and beneficial, but as part of a systematic strategy designed to create, train, and prepare a docile labor force for a newly emerging industrial order. Public institutions were the products of a coalition of elite reformers seeking to impose educational innovation upon reluctant working-class communities.

For this group of historians, the history of education constituted a chapter in the history of oppression and greed. Implicitly defining educational institutions as nothing more or less than structures of domination and control, they rooted the emergence of specialized institutions in the bedrock of economic structure, viewing ideas and institutions as weapons used by elites in an unending struggle for power.

Historians of this school described the emergence of tutorial complexes as driven by rational calculations of consciously recognized personal advantage. Some regarded nurture writers and other educational reformers as social class representatives, middle-class manipulators constructing social spaces or complexes through which they could transform and thereby control the socialization of working- or laboring-class children (Donzelot, 1979; Katz, 1968; Lasch, 1979b; Nasaw, 1980).

Others defined the array of educational institutions as attempts to undermine the capacity of laboring families to control the terms by which their children entered the labor force. There were no subtleties here. Defining institutionalization narrowly, strictly as a strategy of domination, they obscured the existence of motives other than greed and ambition as galvanizing mechanisms in the transformation of schooling (Bowles & Gintis, 1976; Cavallo, 1981; Lazerson, 1971).

Still others, less interested in the relationship between socioeconomic classes than in issues of gender discrimination and disinclined to mute the voices and experiences of their subjects, also upended a founding myth of progress, but on different grounds. Their approach is well illustrated by a pioneering article of Jill Kerr Conway (1974) documenting the persistence of inequality even in coeducational schools. In fact, she exploded two of Thomas Woody's reigning assumptions about the evolution of women's education—that coeducation was a liberating experience for women; that access to professional education naturally ensured equality with male professional peers.

Unlike others concerned with inequality, Conway was as interested in the lived lives of women as in the intentions of education reformers, especially those who had initiated coed higher education. Using women's capacity to develop their intellects or to cultivate new social roles for themselves as indices of equality, she analyzed the real lives of women at Oberlin as others were later to do (Lasser, 1987). She discovered, as we all now know, that gender inequalities did not disappear under the impact of coeducation. They simply took different forms. Within coeducational bastions, women continued to play compensatory roles, cooking, doing laundry, and otherwise sustaining the intellectual lives of men. They were valued and defined in relation to men, barred from higher status professions, and shunned when they used their minds to engage in economic and political rather than domestic and social responsibilities. Coeducational schools stripped women of intellectual capacity and expression, transforming and directing their aspirations into traditional service roles. As Conway (1974) puts it: "It [was] not access to educational facilities which is the significant variable in tracing the 'liberation' of women's minds. What really matters [was] whether women's consciousness of themselves as intellects [was] altered" (p. 9). The coeducational life of Oberlin did not encourage women to consider themselves the intellectual peers of men.

The real source of intellectual nurture lay in the establishment of women's colleges, where, as Conway argues, the development of women's intellect was unambivalently pursued, the curriculum was fundamentally challenged, and their most distinguished graduates, women like Jane Addams and Lillian Wald, could become the creators of new professions and social spaces for women. Later, other historians would claim as much for single-sex female liberal arts colleges founded after 1880 (Finkelstein, 1989b; Schwager, 1987; Solomon, 1985).

Messages From Justice Workers

The justice workers of education history had revealed a formidable and dark reality in the American educational past: that schools were not simply bastions of opportunity, democratic empowerment, and self-government but functioned as instruments of social control and social manipulation. For certain groups, they were instruments of cultural genocide as well (Adams, 1988).

Through the lenses of education history's justice workers, schools were transformed into institutions that reinforced and reproduced structures of inequality; constituted racism, sexism, ethnocentrism, and Anglo-Saxonism; and otherwise participated in the history of oppression.

Proceeding on an assumption that economic conditions were the most powerful engines of social change, one group of justice-seeking historians looked on education history as a manifestation of class struggle, a chapter in the evolution of capitalism and the power of its elites. They viewed education history as class warfare and defined it as an unequal contest with a predetermined outcome. Elites simply won and professional educators helped them do it, especially during the progressive era (Bowles & Gintis, 1976; Karier, 1975; Karier, Violas, & Spring, 1973; Nasaw, 1980).

As believers in the power of economic circumstance, justice workers were at their best when they accounted for and analyzed the emergence of vocational education, differentiated curricula, administrative divisions of labor, corporate ideology, and efficiency cults. They called attention to the myriad ways in which education constituted, sustained, and/or reproduced social inequalities; fostered gender, race, and ethnic discrimination; and constituted ideological hegemony for a relatively small group of status-seeking actors aiming to transform schools into publicly supported normative structures.

As mythmakers, this group of second-generation messengers turned traditional wisdom upside down and portrayed schools simply as instruments of oppression. As Katz was to put it, "by 1880, American education had acquired its fundamental structural characteristics, and . . . they have not altered since . . . it is, and it was, universal tax-supported, free, compulsory, bureaucratic, racist and class biased" (1971, pp. xviii–xx).

Historians such as Katz, Karier, Spring, and Violas had revealed schools as hegemonic institutions designed to suppress the aspirations of working and dependent classes and install particular forms of cultural authority (Katz, 1975a; Katz & Mattingly, 1975). Through their interpretive lenses, they had recovered traditions that structured and fixed inequalities—among and between men and women, races, ethnic groups, and social classes. They exposed public schools as places that served a hierarchical wage system and, through previously unidentified connections between work and school, reinforced economic inequalities.

Through the vision of most historical justice workers, the realization of a dignified future for all Americans required the overhaul, if not the dissolution, of organized public education and the construction of social agendas emphasizing economic and social transformation and a redistribution of material and cultural capital. For a few—historians of women, children, and families come to mind—ideological rather than institutional transformations were regarded as equally important (Aries, 1962; Burstyn, 1975; Conway, 1974; deMause, 1974).

Myths of Modernization

There was yet a third form of mythic reconstruction during the 1960s and 1970s, that of modernization theorists. Their approach to education history originated in a disposition to account for social and educational change differently—not as reflections of democratic trajectories or class struggles alone, but in relation to a series of "long revolutions" or modernizing tendencies (Butts, 1974; Foucault, 1965, 1970; Kaestle & Vinovskis, 1978; Kett, 1977; Lazerson, 1971, 1981; Lazerson & Grubb, 1974; Williams, 1961). As they explored the U.S. educational past, they called attention to the development of national and local government, the expansion of industrial networks, the transformation of family authority, and the emergence of specialized institutions for the transmission of culture and the socialization of the young. They emphasized the differentiation of public and private life, the popularization of high culture, the formation of national loyalties and civic sensibilities, and the triumph of rational planning and emergence of bureaucratic social forms.

With the exception of occasional historians such as Carl F. Kaestle and Maris Vinovskis (1978, 1980), R. Freeman Butts (1974), and Marvin Lazerson (1971), the work of modernization theorists did not directly address the historiographical debates raging within the history of education. Nonetheless, their work was to have formative effects on the creation of yet another sense-making myth in the history of education.

Modernization theorists called attention to an array of previously invisible facts. Through their transformative lenses, popular schools emerged as institutions designed to shore up or overhaul the declining authority of families, towns, and churches (Aries, 1962; Axtell, 1974; deMause, 1974). They portrayed concepts of public policy and the development of nation-states as relatively recent political inventions (Butts, 1974, 1978; Stanley, 1981). Some observed that group learning settings did not become popular until the 19th century (Finkelstein, 1974; Hamilton, 1980). Others observed that certain social phenomena were historically invented rather than universally determined. The discovery of childhood, adolescence, and the commitment to tutorial nurture constituted one cluster of modern phenomena (Aries, 1962; deMause, 1974; Finkelstein, 1985; Hawes & Hiner, 1985; Kett, 1977). So too did the emergence of the nuclear families, domestic women, and tutorial mothers and teachers (Badinter, 1980; Kerber, 1980; Welter, 1966).

Embedding education history in the bedrock of social structure, modernization historians explained the evolution and rise of schooling in relation to transformations of family structure; the effects of urbanization, immigration, and industrialization; and the deployment of public regulatory structures. They viewed the emergence of public schools as a complex response to material and ideological pressures—the separation of work and home, the emergence of wage labor, the bureaucratization of the labor force, the decline of self-sustaining domestic economies, and beliefs in progress and the value of state planning. They linked the emergence of schools to transformations in the nature of regulatory structures governing the young.

Through the transforming lenses of modernization historians, the history of education came into view as part of a long trajectory of efforts to constitute and regulate the course of social change by "establishing normative structures in the public world" (James, 1991, p. 170; Tyack & Hansot, 1982), by deploying organizational and pedagogical forms that would integrate meritocratic processes and thus attach schooling to the needs of the marketplace (Bowles & Gintis, 1976; Foucault, 1965, 1970; Graff, 1979; Hogan, 1985, 1989; Kaestle, 1973; Kantor & Tyack, 1982; Labaree, 1988; Lazerson, 1971), by constituting public authority structures and expert authority to manage social change (Finkelstein, 1985; Lasch, 1979a, 1979b), and by placing young people in settings that would effectively bind their impulses and imaginations, substitute the rule of reason for the rule of passion, bring emotion under the control of reason, and substitute the authority of the public for that of the home, church, and neighborhood (Aries, 1962; deMause, 1974; Donzelot, 1979; Finkelstein, 1974, 1985; Lasch, 1979b).

Messages From Revisionists of the 1960s and 1970s

Taken together, the three faces of revisionism constituted a form of history seeking that would reveal an array of previously invisible traditions within which to make sense of education and its relation to social life, social change, and future possibilities.

By the end of the 1970s, this group of education historians had constituted a broad, muscular, elite-centered vision of education history that went something like the following: The growth and character of public education represented a response to ideological and material transformations taking place over the course of the 19th and 20th centuries. The popularization and extension of schooling to ever-larger numbers of people over time represented the emergence of specialized institutions that could prepare people for labor in an industrial marketplace and for civic participation in emerging nation-states. Schools were places designed to complete a transition from agrarian to industrial modes of production and to prepare people for contractual rather than status-based relationships, impersonal rather than personal modes of association, and labor in factory and line rather than in field and craft. The popularization of schools, newspapers, the penny press, and magazines in the 19th century, and of radio and television in the 20th century, completed a transformation in modes of cultural transmission that elevated, if not substituted, the authority of public agencies, professional experts, and industrial leaders for the traditional authority of family, church, and neighborhood.

The second group of historical mythmakers, critics of the traditional canon, had constituted a sobered-up, problematic education history that planted education tradition in realities removed from human agency—the bedrock of social structure, large culture, capitalist enterprise, and national development. They suggested that schools were durable, "hard surfaces" (Geertz, 1973) and permanent, stable, fixed features of the social landscape—subject to tinkering but not transformation.

As historians that fixed the enterprise of schooling within traditions of capitalist development, republican ideology, and Protestant evangelicism (Kaestle, 1983), they created a message system that revealed educational history from the perspectives of elites even when they did so as justice workers revealing oppression. Although not all second-generation historians aimed to demystify the authority of professional education experts in universities, administrative offices, and classrooms, their histories did much to undermine it.

As large and elegant as their new portrayal was to be, and as conceptually more complex, it revealed education history from imperial lofts, much as the founders of American educational history had done. In fundamental respects, the myths and countermyths of the first two schools of education history professionals did little to uncover histories within which teachers, learners, women, members of less-privileged minorities, or local citizens groups had real traditions or voices—as promoters, resisters, participants, transformers, or simply learners. In the process, they constituted the history of education as the privileged domain of policymakers and intellectual elites.

Myths in Transition

The reigning myths of education history—benevolence, paideia, oppression, and modernization—had all favored the study of elites. They focused on the centralizing tendencies of modern urban life, linking education to the evolution of economic, political, and intellectual macrostructures. They revealed the evolution of high rather than folk culture, of large educational traditions rather than small ones. They recovered an educational past that revealed what was intended rather than what might have occurred, what was taught rather than learned. Whether they were critical or progressive in their vision, they accounted for the popularization of schooling as a by-product of the interests of a relatively small group of elite reformers and philosophers, social planners, and/or corporate leaders, and their histories reflected these dispositions.

EDUCATION HISTORY AS THE RECOVERY OF MULTIPLE PLURALISMS

Together with their predecessors, the second group of education historians had bequeathed a problematic legacy for a new generation of education historians who are approaching the doing of education history in the presence of overwhelming professional and political pressures. During the late 1970s and throughout the 1980s, assaults on the authority of professional educators were once again rampant (Ravitch, 1978). During this wave of reform, the voices of minority groups, women, and other champions of civil rights did not dominate the public discussion. They were overwhelmed, but not silenced, by other cacophonies: massive resistance to desegregation in the North and especially to school busing as a remedy, demographic and fiscal hardships, more powerful regulatory legislation, declining teacher morale, and diminishing public confidence in public schooling.

During the 1980s, public discourse about education reform revealed an omnipresent preoccupation with the meaning of academic achievement for the economic well-being of the United States. Over the course of the last decade and a half, hundreds of task forces and blue ribbon commissions, scholars, polemists, politicians, journalists, citizens groups, corporations, industrialists, and educators have lamented a 30-year decline in test scores and in the ability of American students to compete successfully on comparative international achievement tests.

These critics have issued a national call for new forms of professional accountability, private and public sector partnerships, stiffer graduation requirements, tougher tests, higher performance standards, and a reduction and diversion of federal education resources from national to state authorities (James, 1991). They have called for choice, vouchers, and an array of structural rearrangements to dismantle federal regulatory structures and restore total control over education to the states and localities. There are pressures to construct innovative certification mechanisms, to narrow the curriculum, and to constitute compulsory achievement examinations for high school students and teachers (Sedlak, 1989).

The work of a handful of education historians has resonated with these views by portraying a U.S. educational past dominated by an anti-intellectual professional education establishment that has diminished the power and possibility of public education, diluted the integrity of the curriculum, and otherwise fragmented intellectual coherence in schools (Grant & Riesman, 1978; Ravitch, 1983).

For most historians of education doing work in the context of the 1980s, however, the corpus of received wisdom, with rare exceptions, did not address the real world of schools or the effects of particular policies on the capacities of students to learn well. Education history had not typically revealed traditions of teaching style, learning, classroom practice, or local initiatives. Nor had it revealed the evolution of educational processes, community transformation, teacher education, professional practice, or the effects of policies aimed to secure equal education for all groups.

Traditions of history writing had stripped human agency from all but elite planners. Whether historians had functioned as professional advocates, imperials, or justice workers, they had revealed nonelites as powerless victims of unwanted educational arrangements and as acted-upon subjects of imperial tutelage.

EDUCATION HISTORY AS THE RECONCILIATION OF CONTRADICTION

Historians of the last 5 years have been sensitive to imperial traditions of history writing in education and uncomfortable with oversimplified explanations of social change. They have been attracted anew to the study of schooling and have inherited a sophisticated, if balkanized, scholarly canon. Like their predecessors, they have been plumbing education history to fill in yawning gaps and to conceptualize ways to analyze contradictions and dualisms—between the rhetoric of democracy and the realities of inequality, between the claims of opportunity and the structures of constraint, and between respect for pluralism and commitment to national unity and evolving tensions among federal, state, and local authorities.

Myths of Contradiction: African-American Dimensions

Historians seeking to recover the voices of the previously unheard have constituted a consistent theme with which to make sense of education history among laboring and dependent classes, people of color, and women. They have documented complex and contradictory realities within which to constitute new narratives and integrate previously unheard voices and group aspirations into education history. James Anderson (1988), in his introduction to *The Education of Blacks in the South*, suggests how contradiction has worked in the history of African-American education.

Both schooling for democratic citizenship and schooling for second-class citizenship have been basic traditions in American education. These opposing traditions were not . . . aberrations or isolated alternatives. *Rather, both were fundamental American conceptions of society and progress, occupied the same time and space, were fostered by the same governments and were usually embraced by the same leaders* [italics added]. (p. 1)

As Anderson tells the story, these two opposing traditions originated in the first half of the 19th century and clashed until well into the 20th century. Anderson analyzes the contradictions of popular education as reflections of long struggles between two ways of being and living. For African Americans, the long struggle was between their condition as slaves and peasants on the one hand and participants in a system of capitalism and wage labor on the other. For powerful elites, it was a struggle to suppress African-American aspiration and capacities.

According to Anderson, the origin and evolution of schooling constituted a history of ideological and hegemonic struggles between racist elites on the one side and African Americans and small numbers of friends with political aspirations, organizing power, and visions of education as a route to civic participation on the other. Anderson portrays the evolution of Afro-American educational efforts as the development of a distinct system of public and private education that was as much a product of Afro-American aspirations, ambition, organizing power, and commitment to education as it was of elite attempts to limit access, reproduce inequalities, and keep Blacks in their places.

By posing contradiction as the fundamental generative condition in the politics of educational development for African Americans, Anderson and others (Adams, 1988; San Miguel, 1987) have rescued cultural traditions from historical oblivion; integrated ethnicity, class, and culture into the recovery of the American educational past; and accounted for educational change in new and more complex ways.

Myths of Contradiction: Working-Class Dimensions

In a similar mode, there are historians seeking to integrate the voices and agency of working-class as well as racial minorities into their histories aiming to explore how immigrants and the urban poor shaped life in schools and the course of education history. An array of historians doing case studies of educational development in selected cities have paid close attention to the role of schools in the making of the working classes and their impact on the forming and development of American education (Hogan, 1978; Katznelson & Weir, 1985; Perlmann, 1988; Peterson, 1983; Ravitch, 1974; Reese, 1986).

For these historians, the experiences of immigrants in cities formed the contexts within which they framed their studies. Their histories reveal an unpredictable diversity of immigrant group motives and the relative importance of religious, ethnic, generational, and geographic, as well as economic, factors on the shaping of their commitments to education, their political alliances, and their sense of themselves as a class (Goodenow & Ravitch, 1983; Hogan, 1978; Katznelson & Weir, 1985; Ravitch & Goodenow, 1981; Reese, 1986).

Many late 1980s historians have documented a mosaic of working-class influences on urban educational development. Working-class groups emerge as actively engaged political partisans who constituted their sense of class from spiritual, sociocultural, and psychological identities as well as from their position within the labor force (Goodenow & Ravitch, 1981; Katznelson & Weir, 1985; Marsden, in press; Ravitch & Goodenow, 1981; Reese, 1986).

Not always at odds with urban elites about matters of school expansion and vocational preparation, working-class actors would, at times, reside their class consciousness in spaces outside the school and join in education battles only when schools provided access to new labor markets, as they did when certain vocational subjects were debated or when ethnic identities were fundamentally threatened. This, indeed, was the case during periods when English was defined as the only appropriate language of instruction (Katznelson & Weir, 1985).

Taken together, these histories tell a story of urban working classes entering into alliances of convenience, siding sometimes with organized labor and sometimes with corporate philanthropists, voting religion on some issues and economics on others, championing vocational curricula under some circumstances and classical curricula under others, and doing so in different combinations in different cities and at different times.

As revealed in an array of local histories of particular cities, working-class groups did battle over bond issues, the distribution of educational resources, the status of teachers, and the structure of schooling. They lost some battles. They won others. Some working-class representatives were teachers who entered into alliances with principals and failed to support organized labor (Wrigley, 1982). Others joined labor unions and did battle against the education powers that were. Working groups in the

South often conducted themselves differently from working groups in the North (Perlman, 1988).

Markets and Politics in Contest

Another group of historians, building on the diversities and pluralisms that historians of women and the working classes had revealed, has constituted contradiction as a fundamental reality in the evolution of education processes (Carnoy & Levin, 1985; Hogan, 1985; Kantor & Tyack, 1982; Katznelson & Weir, 1985; Labaree, 1988). Carnoy and Levin (1985), in a synthetic work exploring the origin and evolution of public schooling, have analyzed its development as a contest between markets and politics. Other historians have accounted for relentless diversities in urban educational history and described them as reflections of contradictory pulls on the schools.

During the period from 1890 to 1920, for example, schools emerged as social sites within which contradictory political, economic, and cultural purposes would be served and negotiated. A rich historical literature documents the emergence of schools as contested terrains, places within which larger conflicts in American society would be expressed and negotiated and their contradictions revealed and constituted. As Henry Levin and Martin Carnoy have suggested, schools would become sites for struggle for cultural authority and material resources, serving at least two taskmasters. The first were well-organized cultural minorities, seeking to transform the schools into instruments of democratic empowerment. The second were captains of capitalist enterprise and their allies, who, in search of profit and efficiency, turned to schools to sort workers and legitimize inequalities among and between categories of workers. Described variously by historians as struggles between markets and politics, person rights and property rights, cultural authority and economic necessity, equality and liberty, and localism and nationalism, the schools emerged in the 20th century with overloaded agendas and contradictory missions (Bowles & Gintis, 1986; Carnoy & Levin, 1985; Kantor & Lowe, 1987; Lie, 1987; Rogers & Tyack, 1982).

The interplay of democratic impulses and markets has become an interpretive focus for a whole new genre of studies exploring schools as problematic institutions. Not only have they failed in the attempt to fit students neatly into a hierarchical, unequal wage labor system, but they have miscarried in the effort to promote commitments to social democracy (Bowles & Gintis, 1986; Carnoy & Levin, 1985; Kantor & Lowe, 1987; Katznelson & Weir, 1985; Lie, 1987). The imperfections of schools as instruments of democratic empowerment on the one hand, and of economic planning on the other, emerge clearly when we look at the ways in which schools have intersected with the lives of women, where contradictions are metaphorically rearranged and emerge as complex associations between constraint and opportunity rather than simple oppositions between markets and politics.

Myths of Contradiction: Constraint and Opportunity for Women

A remarkably diverse group of studies exploring female educational experiences revealed the evolution of schools as contradictory message-conveying institutions functioning simultaneously to constrain and empower (Finkelstein, 1989b; Kaufman, 1984). Indeed, the capacity of women to transform discrimination-laden educational experience into visions of new social, moral, and intellectual possibilities for themselves and others became a major theme of historians exploring the evolution of education in a multiplicity of educational institutions. Whether they explored Republican girls' academies in the late 18th century (Burstyn, 1975; Jensen, 1984; Kerber, 1987; Schwager, 1987), single-sex liberal arts colleges and coeducational colleges and universities in the 19th and 20th centuries (Clifford, 1983; Lasser, 1987; Solomon, 1985; Tyack & Hansot, 1982), or normal schools, teacher training colleges, and university education faculties serving middle-class, immigrant, and minority women (Brenzel, 1983; Clifford, 1983, 1989; Finkelstein, 1985, 1988, 1989b; Hoffman, 1981; Jones, 1980; Kaufman, 1984; Tyack & Hansot, 1990), historians have concluded that institutions of higher education could be subversive. They had functioned as sources of feminism and intellectual cultivation, incubators of a new stage of female personality, and crucibles for political activism. Although originally intended to cultivate domestic commitments, they provided environments within which young women, as they were learning, were also acquiring time to experiment, immerse themselves in books, and discover alternative ways to live and be (Antler & Biklen, 1990; Clifford, 1989; Finkelstein, 1988; Tyack & Hansot, 1990).

Messages From Women's History

A striking array of historians, analyzing the empowering qualities of education for women in both coeducational and all-female schools, have revealed the existence of a peculiar dialectic between opportunity and constraint operating in almost all educational institutions throughout U.S. history.

Through the interpretive lenses of historians of the late 1970s and 1980s, women's struggles against constraint are privileged and revealed and their power to overcome the force of political, economic, and psychological circumstances documented. The future implicit in this past is one in which women continue to struggle for economic and political power and status, wrenching equality from an array of constraining environments and aiming to join men as equals in the workplace, the polling place, and the marketplace. Through these interpretive windows, women struggle to win and to overcome an array of marginal if not totally degraded statuses in the workplace. The past creates an invitation to future political and economic struggle.

Recently, historians of women have begun to excavate another aspect of the female past in education—that of traditional middle- or working-class women. They have done so with reference to two kinds of emphases: the relations of women to work as founders, volunteers, and teachers in the educating professions, and the relationship of institutions to the forming of women's consciousness and sensibility.

One such history, John Rury's (1991a) recently published *Education and Women's Work: Female Schooling and the Division of Labor in Urban America, 1870–1930*, is an exploration of intersections between education and the lives of women during the progressive era, the half-century from 1870 to 1930. Rury documents the same dialectic between opportunity and constraint that previous historians had revealed, but explores it in new social and ideological settings.

For women, the years between 1880 and 1930 were momentous. During this time, they campaigned for the right to vote, acquired some protection against labor exploitation and abusive husbands, and, perhaps most important of all, joined the wage labor force in slow but steady numbers (by 1930, 25% were working outside of the household). As new sectors of the economy became available to them, women took jobs as typists, clerks, teachers, and telephone operators, as well as traditional female jobs in manufacturing and domestic service.

The response of vocational education advocates to these new spaces in the economy was not, as historians have shown, to introduce commercial and/or professional studies for women but to create a gender-segregated curriculum. Even though they invoked relevance, opportunity, and practicality in their calls for the vocationalization of education, progressive reformers introduced industrial shop courses for boys and home economics courses for girls and narrowed the range of vocational opportunities being made available to students in schools. Invoking traditional visions of women as keepers of the domestic hearth, they organized curricula designed to transfer skills that were becoming obsolete in the evolving labor market (Kantor, 1988; Rury, 1991a; Tyack & Hansot, 1990).

The insensitivity of vocational education reformers to the structure of opportunities was coupled with an incapacity to respond to the vocational aspirations of parents and students, especially those of working-class boys and of women in general. Fearing social conflict and concerned, as Harvey Kantor (1988) has observed, that

girls outnumbered boys in high schools and . . . that rising rates of female participation in the labor force threatened women's traditional roles as wives, mothers, and housekeepers, much of the vocational movement in education was designed to hold more boys in school and to preserve traditional notions of women's sphere. (p. 188)

Academic education in high schools conferred credentials on middle-class women seeking entry into new white-collar sectors and newly evolving professions for women. For working-class and minority women, however, the commercial and/or home economics curricula offered no such route.

As Rury (1991a) has observed:

The vocational education movement which swept through American high schools created a gender-segregated curriculum. While boys studied "industrial" subjects and principles of business, young women studied home economics and stenography. By the third decade of the twentieth century, young women and men in high schools pursued different courses of study designed to prepare them for separate responsibilities in the social order. Women's work had entered the high school, and the result was a new gender-specific curriculum. The new occupations became sex typed and a highly developed array of institutional constraints emerged in women's education.

Without quite saying so, Rury has revealed a moment in the educational past of women when schools obstructed their entry into marketplace opportunities. Nonetheless, they moved into new economic spaces while education reformers tried, through the home economics curriculum, to remind them of their traditional duties. This particular dialectic—between the economic ambitions of women and the traditional domestic fears of school reformers—did not, as Rury has shown, contribute fundamentally to the nurture of radical feminist thinking. Rather, it contributed to the tensions between opportunity and constraint that previous historians had identified and to the struggle of middle- and working-class women to reconcile home and family life.

This theme permeates the work of historians such as Joyce Antler (1987), Linda Perkins (1989), and Roberta Wollons (in press) who combine biography, organizational and institutional history, and sophisticated psychological analyses as they explore women's lives and careers (Antler & Biklen, 1990). Joyce Antler's (1987) study of progressive reformer Lucy Sprague Mitchell, founder and director of the Bank Street College, focuses on Mitchell's struggle to mute conflict between professional work and engagement as a wife and mother. As Antler tells her story, she documents the inseparability of personal and professional concerns, family relationships, and educational work for women like Lucy Sprague Mitchell. These interactions, as Antler was to comment,

> had consequences beyond the boundaries of her own life: her contributions to the theory and practice of early childhood education helped redefine the public dimensions of women's private sphere, while her private life gave expression to the new patterns of sex-role behavior beginning to emerge in the twentieth century. (Antler, 1987, p. xvii)

The fusing of public and private life, a powerful theme in the life of individual women, was also apparent in the forms that organizational life was to take in the first half of the 20th century. In a study titled *Educating Mothers: A Study in Volunteerism*, Roberta Wollons (in press) explores the political styles, organizational behavior, biographies, and perceptions of opportunity and constraint among the group of women volunteers who created the Child Study Association of America.

Like Antler, Wollons has placed an analytic probe in the center of the lives of well-educated women and documented a certain level of acceptance with marginal status in the work force, and work commitments that comported with more established views that women's place was fundamentally in the home.

Ultimately, Wollons reveals the existence of women who were not so stuck on political and economic advancement if its cost was the quality of their family life, the well-being of their children, or the authenticity of their own values. She documents an almost tragic irony for women of the progressive era who sought intellectual legitimacy for the work of child study. The cost of their involvement with the real world of intellectual power was the public disappearance of their own voices, modes of mutual support, and cultural authority over mothering that they lost to the bevy of consultants who distrusted mothers and tendered truths for them to learn (Finkelstein, 1988; Lasch, 1979b). Needless to say, there are teachers, teacher researchers, and social workers who have experienced this pattern over and over again (Perkins, 1989).

Through the interpretive lenses of historians such as Antler, Wollons, and Perkins, the history of education reveals the evolution of a segmented society in which the separations between home and workplace, public and private life, the older and younger generations, and men and women created excruciating conflicts—in the women themselves and in the educational institutions that shaped and were shaped by their lives.

Second Messages From Women's History

From this historical perspective, women acquire a future that privileges an array of moral if not economic possibilities—as nurturers and care givers in home, school, and society and as seekers after power and wielders of economic, political, and cultural power. It is a future of contradiction and dilemma—one inviting them to balance commitments between work and home, ambition and compassion, and independence and companionship.

Myths of Contradiction: Policy Dimensions

The existence of tension and contradiction is also a theme in studies that focus specifically on the evolution of public authority in education. Over the last two decades, a collection of policy studies has documented the effects of public regulation on the forms and processes of schooling and the evolution of professional authority in education (Hurn, 1985; James, 1991; Johnson, 1989; Kantor & Tyack, 1982; Kliebard, 1987; Lazerson & Grubb, 1982; Ravitch, 1983; Sedlak, Wheeler, Pullin, & Cusick, 1986; Tyack, Lowe, & Hansot, 1984). Unlike the policy studies of the founding generation of historians of education, contemporary studies take conflict among and between central and local authorities, and professional and lay authorities, as analytic starting points. These kinds of studies have had an uncommon capacity to reveal intersections between policy and practice and have called attention to structural tensions created by regulatory pressures and the traditional cultures of schools and teachers (Cusick, 1983; Grant, 1988; Hurn, 1985; James, 1991; Powell, Farrar, & Cohen, 1985; Sedlak et al., 1986; Willis, 1976). As Thomas James (1991, p. 169) has observed, many scholars

have sought to clarify the assorted forms of control and resistance that collide in school settings as teachers act to protect their autonomy . . . (e.g., McNeil, 1983, 1988; Muir, 1986), and as the schools, despite a plethora of reforms . . . routinely fail to alter their hierarchical and unequal treatment of students from different groups in society (Everhart, 1983; Gamoran, 1987; Oakes, 1985; Willis, 1976).

Third-Generation Messages

When it is viewed as a whole, the work of late 1980s and early 1990s historians has advantaged the study of institutions, situated analyses in local contexts, and revealed schools as sites of conflict and reflections of the politics of pluralism. Through the interpretive lenses of these historians, schools have emerged as politically constructed realities, originating and changing as a result of conflict. Like their first-generation predecessors, "conflict historians" reveal public education as a form of public contest where an array of small interests, special interests, private interests, and particular interests did battle to secure control over the nature of schooling and the course of social change.

Beyond revealing conflict as a fundamental reality in education history and documenting the existence of what Silver (1990) has called "multiple pluralisms," recent historians have sustained a traditional piety—that the centralizing tendencies of modern life had overwhelmed traditional forms of communal solidarity and patterns of life. As Thomas Bender (1978) has observed, "history from [this] perspective unfolds a powerful logic: that modernization processes shatter social unity and communal solidarity, and substitute instead, associations based on interest" (pp. 42–43).

With this vision in mind, it is hardly surprising that historians of education rarely indexed the history of education to the evolution of private as well as public processes or situated their analyses in the bedrock of community life (Calhoun, 1973; Finkelstein, 1974, 1985). Focusing on the analysis of structure rather than process, prescription rather than practice, and ideology rather than consciousness, historians of education had unknowingly concealed private processes from view (e.g., the formation and evolution of community, the acquisition of identity, the cultivation of intellect, sensibility, and aspirations).

What is more, as Michael Sedlak (1990) has observed, they did not "go beyond exploring issues of simple access to schools to include . . . questions of access to learning, distinctions in the value of different kinds of learning, and the central place of pedagogy in equalizing learning outcomes" (p. 126).

Myths of Community: Some Beginnings

One interpretive reconstruction is being undertaken by a group of historians revisiting traditional pieties by systematically exploring cities and schools as something more than sites for competing interests or filters for the cultivation and transmission of knowledge and culture.

Over the last 25 years, a relatively small number of historians have been analyzing cities and urban schools as "teaching environments" (Calhoun, 1973), "learning settings" (Finkelstein, in press), and consciousness-forming environments within which individuals and groups learn to distinguish themselves from other people and in which social bonds are cemented and forged (Bender, 1978; Finkelstein, 1974, 1986; Lagerman, 1979; Weibe, 1975, preface).

These community historians have privileged the study of small social units, primary circles of identity, values, association, and goals. Thus, they have found ways to link education history to hitherto unrecognized traditions and concerns. As a result of explorations of the role of community in educational history, schools acquire new dimensions and emerge as powerful structures of persuasion mediating between the increasingly differentiated worlds of men and women, young and old, parent and child, and household and work site. They appear as protracted group-learning settings within which teacher culture, youth culture, work culture, and the culture of the book form important new forms of human association, ways of seeing, feeling, believing, and knowing (Aries, 1962; Biklen, 1990; Calhoun, 1973; deMause, 1974; Finkelstein, 1986, 1990, in press; Hamilton, 1980). Schools become institutions that cultivate identity, creating boundaries between groups, forging social bonds, evoking meaning, compelling allegiance, and exacting commitment.

Through the narrative vision of community historians, economic and political realities are reconstituted as forms of human association that generate new possibilities for schooling and human communities. Some of these historians, studying the relationships between capitalism and classroom pedagogy, define capitalism as Marx had originally done, as a form of human association as well as production. They analyze classroom culture, with its emphases on meritocracy, hierarchy, competition, and individual achievement, as both reflective and constitutive of new moralities and human relationships (Hogan, 1989; Labaree, 1988; Finkelstein, in press).

Other historians, exploring teachers and teaching, have revealed the emergence of schools as specialized kinds of message-sending institutions where children, women, workers, African Americans, Anglos, and others learn their place, affirm or reorganize their affiliative loyalties, reconstruct their networks of association, form social bonds, and create intercultural and international connections that were previously unimaginable in human history (Aries, 1962; Aries & Duby, 1987; Calhoun, 1973; Finkelstein, 1989b; Herbst, 1989; Schwager, 1987; Solomon, 1985; Warren, 1989).

Defining community as bonding and commitment, a handful of historians are beginning to study the way in which educational processes help to create communities of people who share in a universe of meaning. This is an especially fruitful approach for comparative study in education history and for an analysis of pluralism in education forms and practice. Through the study of the role of education in the forming of cultural habits, thinking styles, communication forms, affiliative processes, and the cultivation of human potential, historians have begun to constitute historical indices that can make education and cultural comparison meaningful and collaborative (Bender, 1978; Fass, 1989; Finkelstein, in press; Seller, 1991).

Studies in the history of literacy, pluralism, childhood, the cultures of teaching and learning, the practice of policy, and civic learning, like studies of community, are already revealing new sense-making possibilities in educational history and uncovering traditions within which to make sense of new social realities—a world of relentless pluralisms, interdependence, new communication forms, and continuing battles between nations as well as cultures for control over the hearts and minds of the next generation (Butts, 1989; Cuban, 1991; Franklin, 1981; Franklin, Gordon, Schwartz, & Fass, 1991; Graff, 1979, 1987; Grubb & Lazerson, 1982; Kaestle, 1991; Silver, 1990).

MYTHS IN THE MAKING: NEW DIMENSIONS

If education historians are to sustain their sense-making capacities and critical traditions, they must, in my view, cultivate whole new intellectual genres centering on three new domains: the history of human relationships, including the history of learners and learning and the history of culture; the history of communication; and the comparative study of education history and international educational relations.

Human Relationships

Historians of education have only rarely explored education as it might form or reflect the nature of human relationships, sensibilities, mentalities, or forms of connection. Indeed, they have stripped education history of human agency, assuming, somehow, that the study of educational intentions will tell us something about educational outcomes. They have not focused on the processes by which people might have created, incorporated, reproduced, or transformed the force of economic, political, or philosophical circumstances into new educational forms and visions. With rare exceptions, they have not explicitly explored the motives, qualities of intellect, or creative sensibilities that the architects of education brought to their task. As a result, we have little systematically documented understanding of how economic, political, or ideological realities became transformed through human agency. Nor have we the means to understand the conditions under which school achievement, human dignity, and commitments to freedom and pluralism flourish or to discover educational arrangements favoring the cultivation of human sensibility, intelligence, and diversity. To explore these kinds of questions, we must find ways to integrate learners and learning directly into our explorations of education history.

Learners and Learning

With the exception of a few historians of childhood and youth and psychohistory (deMause, 1974; Finkelstein, 1990; Fishman, 1979; Hawes & Hiner, 1985), a handful of biographers, and a few context-sensitive historians such as Daniel Calhoun (1973), Larry Cuban (1984, 1986), and David Hogan (1989), historians of education have not as yet availed themselves of conceptual or technical approaches that will reveal the data from which traditions of learning can emerge. The few who have done so proceed on an assumption that processes of cultural transmission will be incompletely understood and unnecessarily narrowed by a single-minded preoccupation with teaching, and they have called for the systematic study of two of education history's most neglected aspects—learners and learning itself.

As historians approach the task, they have introduced an intergenerational dimension deeply into their studies (Finkelstein, 1974; Hawes & Hiner, 1985, 1991). Some reflect this tendency in studies examining adult-child and teaching-learning relationships. Others explore how political and educational leaders convert what they have learned in one generation into new social forms, cultural constructions, and visions of educational, political, and economic possibility in the next (deMause, 1974; Finkelstein, 1990; Greven, 1977; Vine, 1979).

Implicit in a learning-centered approach to education history is a disposition not only to focus on relationships between adults and children, but to study education as a complete cycle, focusing on the results as well as the possibilities of educational arrangements and practices and modes of cultural transmission.

The integration of learners and learning into education history is no easy task. Children themselves leave few written records for us to study. Nor have explorations of educational fare—the formal and informal curriculum and the content of textbooks, catalogues, syllabuses, newspapers, televisions, computer programs, museum exhibitions, musical performances, and so forth—typically revealed what has been learned as well as what had been intended or taught.

Furthermore, education historians have relied less on the revelations of psychology, theology, literary theory, rhetorical analysis, and anthropology than on politics, economics, and sociology to explore education history, thus relying on a universe of understanding that has little to reveal about the content and processes of learning or the links between cognitive processes and the cultivation of political, economic, and/or social affiliations.

What is needed is systematic analysis, case study by case study, of the uses of literacy, the processes of schooling, and the consequences of deliberate education as they might be revealed in particular settings, communities and nations over time, and the lives of learners. Historians will also need a range of methodological skills—those of thick description, ethnography, psychohistory, speech act and rhetorical theory, material culture, iconography, and collective biography, among many others.

Culture

Another way to integrate a concern with human relationships directly into the study of education history is to privilege the study of culture as well as politics, economics, and ideology. Historians of education have traditionally focused on what Clifford Geertz (1973) calls the "hard surfaces of life"—political, economic, stratificatory realities within which people are born and in which they lead their lives. No matter what approach, they tend to reveal educational structure rather than behavior, drawing conclusions from an analysis of structural variables: density, size, heterogeneity, or economic or political systems. As a consequence, we learn much less than we need to know about the shaping of modern consciousness and the uses to which people might put education, not only in the pursuit of power and status but in the pursuit of dignity and meaning as well.

To bring these dimensions into view, historians will have to go beyond the study of structure, macropolitics, economics, and the lives of elites to include the study of ordinary people, face-to-face contexts, and the evolution of human consciousness. They will need to conceptualize schools, cities, politics, and markets as cultural and psychological as well as material and intellectual environments. They will need to analyze the work of educational elites in psychosocial as well as structural terms and analyze education as something experienced as well as planned—as reflections of interactions between teachers and learners in all kinds of settings.

The History of Communication

The communication revolution that is upon us also requires much more attention from historians of education. Some, like Cremin (1970, 1977) and Ong (1977), have explored "interfaces of the word" (Ong, 1977) and described transformations in modes of cultural transmission—from oral to written to electronic forms. A handful of others have investigated the evolution of education technologies and the material culture of families, schools, theaters, and so forth. None, to my knowledge, have integrated communication as a fundamental category in the definition of education and exploration of classroom practices and the culture of the home.

A systematic emphasis on communication would reveal traditions of cultural communication as mediated in classrooms under conditions of wartime occupation and contexts of international exchange, intercultural encounters, and the like. It would require the study of educational interaction between teachers and learners, between education policy and teacher culture, between conquering and vanquished peoples, and between nations and tribes within and across national boundaries. An emphasis on communication would root the history of education in the history of long-term cultural, political, economic, and technological transformations, forcing historians to account for the effects of education on the forming of political, economic, and intergroup relations under diverse conditions.

Comparative History, Culture, and International Relations

The realities of diversity and the emergence of new forms of communication are affecting transformations in the intensity and scale of communication among and between peoples of different nations and tribes. The sense-making possibilities of education history would benefit immensely from systematic comparative study and scholarly collaboration between teams of historians from different nations. As alien as traditions of scholarly collaboration are for historians who have generally worked alone, the effort promises to generate new indexes of change, categories of analysis, and transformations of historical consciousness. As we are now learning, the scholarly possibilities of international exchange are exquisite in their possibilities as consciousness-transforming, conceptually generative, and interculturally challenging modes of revisiting education history.

As yet, most historians of education have integrated neither intercultural nor comparative dimensions into their work, although the content of professional journals such as the *History of Education Quarterly* and *Educational Policy* and presses like Taylor and Francis reflect an increasing interest in education in other nations. Recently, books such as Robert J. Lifton's *The Nazi Doctors* (1986) and Steven Selden's forthcoming work, *The Capturing of Science: Eugenics, Race Betterment, and Education, 1883–1940*, suggest the existence of hitherto unappreciated traditions of affiliation among scientists with communities of shared knowledge and interest beyond nation, neighborhood, or even ethnic affiliation. Indeed, there is reason to believe that professional culture might, under certain circumstances, overwhelm the force of nationalism and local culture, thereby constituting higher education as a formative institution in the shaping of world relations.

CONCLUSION

Generations of education historians have been chroniclers of the institutions they serve and help to create. They have entered into a world of contemporary education debate, if not directly as advocates of one or another political interest in education, then as specialized scholarly emissaries conveying usable messages to an array of policymakers and citizens seeking to reform education. The best of them have distinguished themselves as revealers of an educational past, actors in contemporary debates, and crafters of future possibilities, and, hopefully, will continue to do so.

NOTE

[1] The concept of historical myths used in this essay is not the traditional concept of myth as lie or mystification. Rather, it is an idea of myth as revealing narrative,

linking chains of events, ideas, or circumstances compounded from documentable material and lived realities. For an exploration of the dilemmas of narrative in education history, see Joan Burstyn's "Narrative Versus Theoretical Approaches" (1990) and Hayden White's exploration of narrative form and historical discourse (1987).

REFERENCES

Adams, D. W. (1988). Fundamental considerations: The deep meaning of Native American schooling, 1880–1900. *Harvard Educational Review, 58,* 1–28.

Anderson, J. D. (1988). *The education of Blacks in the South, 1860–1935.* Chapel Hill and London: University of North Carolina Press.

Antler, J. (1987). *Lucy Sprague Mitchell: The making of a modern woman.* New Haven, CT: Yale University Press.

Antler, J., & Biklen, S. K. (Eds.). (1990). *Changing education: Women as radicals and conservators.* Albany: State University of New York Press.

Aries, P. (1962). *Centuries of childhood: A social history of family life* (R. Baldick, Trans.). New York: Knopf.

Aries, P., & Duby, G. (Eds.). (1987). *A history of private life* (Vol. 1). Cambridge, MA: Harvard University Press.

Axtell, J. (1974). *The school upon a hill: Education and society in colonial New England and New York.* New Haven, CT: Yale University Press.

Badinter, E. (1980). *L'Amour en plus: Histoire de l'amour maternel: 17ème-20ème siècles* [More than love: The history of maternal instinct: 17th to 19th centuries]. Paris: Flammarion.

Bailyn, B. (1960). *Education in the forming of American society: Needs and opportunities for study.* Chapel Hill: University of North Carolina Press.

Bender, T. (1978). *Community and social change in America.* New Brunswick, NJ: Rutgers University Press.

Bender, T. (1985, October 6). Making history whole again. *New York Times Book Review,* pp. 42–43.

Biklen, S. K. (1990). Confiding woman: A nineteenth-century teacher's diary. *History of Education Review, 19*(2), 19–35.

Bledstein, B. J. (1976). *The culture of professionalism: The middle class and the development of higher education.* New York: Norton.

Bouwsma, W. J. (1981). Intellectual history in the 1980's: From history of ideas to history of meaning. *Journal of Interdisciplinary History, 12,* 280–293.

Bowles, S., & Gintis, H. (1976). *Schooling in capitalist America: Educational reform and the contradictions of economic life.* New York: Basic Books.

Bowles, S., & Gintis, H. (1986). *Democracy and capitalism: Property, community, and the contradictions of modern social thought.* New York: Basic Books.

Bremner, R. (1974). *Children and youth in America: A documentary history* (5 vols.). Cambridge, MA: Harvard University Press.

Brenzel, B. (1983). *History of nineteenth-century women: A plea for inclusion of class, race, ethnicity.* Unpublished manuscript.

Brubacher, J. S., & Rudy, J. W. (1976). *Higher education in transition* (3rd ed.). New York: Harper & Row.

Burstyn, J. N. (1975). Catherine Beecher and the education of American women. *New England Quarterly, 48,* 386–403.

Burstyn, J. N. (1987). History as image: Changing the lens. *History of Education Quarterly, 27,* 167–179.

Burstyn, J. N. (1990). Narrative versus theoretical approaches: A dilemma for women. *History of Education Review, 19*(2), 1–8.

Butts, R. F. (1939). *The college charts its course.* New York: McGraw-Hill.

Butts, R. F. (1974). Public education and political community. *History of Education Quarterly, 14,* 165–183.

Butts, R. F. (1978). *Public education in the U.S.: From revolution to reform.* New York: Holt, Rinehart, & Winston.

Butts, R. F. (1989). *The civic mission of educational reform: Perspectives for the public and the profession.* Stanford, CA: Hoover Institution Press.

Butts, R. F., & Cremin, L. A. (1953). *History of education in American culture.* New York: Henry Holt Company.

Calhoun, D. (1973). *The intelligence of a people.* Princeton, NJ: Princeton University Press.

Carnoy, M., & Levin, H. M. (1985). *Schooling and work in the democratic state.* Stanford, CA: Stanford University Press.

Cavallo, D. (1981). *Muscles and morals: Organized playgrounds and urban reform, 1880–1920.* Philadelphia: University of Pennsylvania Press.

Church, R. L., Katz, M. B., Silver, H., & Cremin, L. A. (1989). The metropolitan experience in American education. *History of Education Quarterly, 29,* 419–447.

Clifford, G. J. (1983). "Shaking dangerous questions from the crease": Gender and American higher education. *Feminist Issues, 3*(2), 3–62.

Clifford, G. J. (1989). Man, woman, teacher: Gender, family and career in American educational history. In D. R. Warren (Ed.), *American teachers: Histories of a profession at work* (pp. 293–344). New York: Macmillan.

Cohen, S. (1974). *Education in the U.S.: A documentary history* (5 vols.). New York: Random House.

Conway, J. K. (1974). Perspectives on the history of women's education in the United States. *History of Education Quarterly, 16,* 1–13.

Cremin, L. A. (1961). *The transformation of the school: Progressivism in American education, 1876–1957.* New York: Random House.

Cremin, L. A. (1965a). *The genius of American education.* Pittsburgh: University of Pittsburgh Press.

Cremin, L. A. (1965b). *The wonderful world of Ellwood Patterson Cubberley: An essay on the historiography of American education.* New York: Teachers College Press.

Cremin, L. A. (1970). *American education: The colonial experience, 1607–1783.* New York: Harper & Row.

Cremin, L. A. (1977). *Traditions of American education.* New York: Basic Books.

Cremin, L. A. (1980). *American education: The national experience, 1783–1876.* New York: Harper & Row.

Cremin, L. A. (1988). *American education: The metropolitan experience, 1876–1980.* New York: Harper & Row.

Cremin, L. A. (1990). *Popular education and its discontents.* New York: Harper & Row.

Cuban, L. (1984). *How teachers taught: Constancy and change in American classrooms: 1890–1980.* New York: Longman.

Cuban, L. (1986). *Teachers and machines: The classroom use of technology since 1920.* New York: Teachers College Press.

Cuban, L. (1989). The persistance of reform in American schools. In D. R. Warren (Ed.), *American teachers: Histories of a profession at work* (pp. 370–393). New York: Macmillan.

Cuban, L. (1991). Policy making and the uses of history. *Educational Policy, 5,* 200–205.

Cubberley, E. P. (1919). *Public education in the United States: A study and interpretation of American educational history.* Boston: Houghton Mifflin.

Curti, M. (1959). *The social ideas of American educators with a new chapter on the last twenty-five years.* Patterson, NJ: Littlefield, Adams and Company. (Original work published 1935)

Cusick, P. A. (1983). *The egalitarian ideal and the American high school: Studies of three schools.* New York: Longman.

deMause, L. (1974). The evolution of childhood. In L. deMause (Ed.), *The history of childhood* (pp. 1–65). New York: The Psychohistory Press.

Donzelot, J. (1979). *The policing of the family* (1st English ed.). New York: Pantheon Books.

Everhart, R. B. (1983). *Reading, writing and resistance: Adolescence and labor in a junior high school.* Boston: Routledge & Kegan Paul.

Fass, P. S. (1989). *Outside in: Minorities and the transformation of American education.* New York: Oxford University Press.

Finkelstein, B. (1974). *Regulated children/liberated children: Education in psychohistorical perspective.* New York: The Psychohistory Press.

Finkelstein, B. (1985). Casting networks of good influence: The reconstruction of childhood in nineteenth century America. In N. R. Hiner & J. Hawes (Eds.), *American childhood: A research guide and historical handbook* (pp. 111–153). Westport, CT: Greenwood Press.

Finkelstein, B. (1986). Exploring community in urban educational history. In R. K. Goodenow & D. Ravitch (Eds.), *Schools in cities: Conflict and consensus in American educational history* (pp. 305–319). New York: Holmes and Meier.

Finkelstein, B. (1988). The revolt against selfishness: Women and the dilemmas of professionalism in early childhood education. In B. Spodek, O. N. Saracho, & D. L. Peters (Eds.), *Professionalism and the early childhood practitioner* (pp. 10–28). New York: Teachers College Press.

Finkelstein, B. (1989a). *Governing the young: Teacher behavior in popular primary schools in 19th century United States.* London: Falmer Press.

Finkelstein, B. (1989b). Conveying messages to women: Higher education and the teaching profession in historical perspective. *American Behavioral Scientist, 32,* 680–698.

Finkelstein, B. (1990). Perfecting childhood: Horace Mann and the origins of public education in the United States. *Biography, 13*(1), 6–21.

Finkelstein, B. (1991). *Dollars and dreams: Classrooms as fictitious message systems: 1790–1930* (American Studies Seminar Proceedings). Sapporo, Japan: University of Hokkaido Press.

Finkelstein, B. (in press). Redoing urban educational history. In W. Marsden & R. K. Goodenow (Eds.), *Education in four nations.* Cambridge, England: Cambridge University Press.

Fishman, S. (1979). The double vision of education in the nineteenth century: The romantic and the grotesque. In B. Finkelstein (Ed.), *Regulated children, liberated children: Education in psychohistorical perspective* (pp. 96–114). New York: The Psychohistory Press.

Foucault, M. (1965). *Madness and civilization: A history of insanity in the age of reason.* New York: Random House.

Foucault, M. (1970). *The order of things: An archaeology of the human sciences.* New York: Vintage Books.

Franklin, V. P. (1981). Continuity and discontinuity in Black and immigrant minority education. In D. Ravitch & R. K. Goodenow (Eds.), *Educating an urban people: The New York City experience* (pp. 44–66). New York: Teachers College Press.

Franklin, V. P., Gordon, L. D., Schwartz, M. S., & Fass, P. S. (1991). Understanding American education in the twentieth century. *History of Education Quarterly, 31*, 47–67.

Gamoran, A. (1987). The stratification of high school learning opportunities. *Sociology of Education, 60*, 135–155.

Geertz, C. (1973). *The interpretation of cultures.* New York: Basic Books.

Goodenow, R. K., & Ravitch, D. (1983). *Schools in cities: Consensus and conflict in American educational history.* New York: Holmes and Meier.

Graff, H. J. (1979). *The literacy myth: Literacy and social structure in the nineteenth-century city.* New York: Academic Press.

Graff, H. J. (1987). *The legacies of literacy: Continuities and contradictions in Western culture and society.* Bloomington: Indiana University Press.

Grant, G. (1988). *The world we created at Hamilton High.* Cambridge, MA: Harvard University Press.

Grant, G., & Riesman, D. (1978). *The perpetual dream: Reform and experiment in the American college.* Chicago: University of Chicago Press.

Greven, P. (1977). *The Protestant temperament: Patterns of child-rearing, religious experience and self in early America.* New York: Knopf.

Grubb, W. N., & Lazerson, M. (1982). *Broken promises: How Americans fail their children.* Chicago: University of Chicago Press.

Hall, P. D. (1983). *The organization of American culture, 1700–1900: Private institutions, elites, and the origins of American nationality.* New York: New York University Press.

Hamilton, D. (1980). Adam Smith and the moral economy of the classroom system. *Curriculum Studies, 12*, 281–298.

Hawes, J. M., & Hiner, N. R. (Eds.). (1985). *American childhood: A research guide and historical handbook.* Westport, CT: Greenwood Press.

Hawes, J. M., & Hiner, N. R. (Eds.). (1991). *Children in historical and comparative perspective: An international handbook and research guide.* Westport, CT: Greenwood Press.

Herbst, J. (1989). Teacher preparation in the nineteenth century: Institutions and purposes. In D. R. Warren (Ed.), *American teachers: Histories of a profession at work* (pp. 213–236). New York: Macmillan.

Hoffman, N. (1981). *Woman's "true" profession: Voices from the history of teaching.* Old Westbury, NY: Feminist Press/McGraw-Hill.

Hofstadter, R., & Metzger, W. (1963). *The development of academic freedom in the United States.* New York: Knopf.

Hogan, D. J. (1978). Education and the making of the Chicago working class, 1880–1930. *History of Education Quarterly, 18*, 227–271.

Hogan, D. J. (1985). *Class and reform: School and society in Chicago, 1880–1930.* Philadelphia: University of Pennsylvania Press.

Hogan, D. J. (1989). The market revolution and disciplinary power: Joseph Lancaster and the psychology of the early classroom system. *History of Education Quarterly, 29*, 381–419.

Hurn, C. (1985). Changes in authority relationships in school: 1960–1980. *Research in Sociology of Education and Socialization, 5*, 31–57.

James, T. (1991). State authority and the politics of educational change. In G.

Grant (Ed.), *Review of research in education* (Vol. 17, pp. 169–224). Washington, DC: American Educational Research Association.

Jensen, J. (1984). Not only ours, but others: The quaking-teaching-daughters of the mid-Atlantic, 1790–1850. *History of Education Quarterly, 24*, 3–58.

Johnson, W. R. (1989). Teachers and teacher training in the 20th century. In D. R. Warren (Ed.), *American teachers: Histories of a profession at work* (pp. 237–257). New York: Macmillan.

Jones, J. (1980). *Solders of light and love: Northern teachers and Georgia Blacks, 1865–1973.* Chapel Hill: University of North Carolina Press.

Kaestle, C. F. (1973). *The evolution of an urban school system.* Cambridge, MA: Harvard University Press.

Kaestle, C. F. (1983). *Pillars of the republic: Common schools and American society, 1780–1860.* New York: Hill and Wang.

Kaestle, C. F. (1991). *Literacy in the United States: Readers and reading since 1800.* New Haven, CT: Yale University Press.

Kaestle, C. F., & Vinovskis, M. A. (1978). From apron strings to ABCs: Parents, children and schooling in nineteenth-century America. In J. Demos & S. S. Boocock (Eds.), *Turning points: Historical and sociological essays on the family* (pp. 39–80). Chicago: University of Chicago Press.

Kaestle, C. F., & Vinovskis, M. A. (1980). *Education and social change in nineteenth-century Massachusetts.* New York: Basic Books.

Kantor, H. (1988). *Learning to earn: School, work and vocational reform in California, 1880–1930.* Madison: University of Wisconsin Press.

Kantor, H., & Lowe, L. (1987). Empty promises. *Harvard Educational Review, 57*, 68–76.

Kantor, H., & Tyack, D. B. (1982). *Work, youth, and schooling: Historical perspectives on vocationalism in American education.* Stanford, CA: Stanford University Press.

Karier, C. (1975). *The shaping of the American educational state.* New York: Free Press.

Karier, C. J., Violas, P., & Spring, J. (1973). *Roots of crisis: American education in the twentieth century.* New York: Rand McNally.

Katz, M. B. (1968). *The irony of early school reform: Educational innovation in mid-nineteenth-century Massachusetts.* Cambridge, MA: Harvard University Press.

Katz, M. B. (1971). *Class, bureaucracy, and schools: The illusion of educational change in America.* New York: Praeger.

Katz, M. B. (1975). *The people of Hamilton, Canada West: Family and class in a mid-nineteenth-century city.* Cambridge, MA: Harvard University Press.

Katz, M. B. (1987). *Reconstructing American education.* Cambridge, MA: Harvard University Press.

Katz, M. B., & Mattingly, P. H. (1975). *Education and social change: Themes from Ontario's past.* New York: New York University Press.

Katznelson, I., & Weir, M. (1985). *Schooling for all: Class, race, and the decline of the democratic ideal.* Berkeley: University of California Press.

Kaufman, P. W. (1984). *Women teachers on the frontier.* New Haven, CT: Yale University Press.

Kerber, L. K. (1980). *Women of the republic: Intellect and ideology in revolutionary America.* Chapel Hill: University of North Carolina Press.

Kerber, L. K. (1987). Nothing useless or absurd or fantastical: The education of women in the early republic. In C. Lasser (Ed.), *Educating men and women together: Coeducation in a changing world* (pp. 37–49). Urbana: University of Illinois Press.

Kett, J. F. (1977). *Rites of passage: Adolescence in America, 1790–1970*. New York: Basic Books.

Kliebard, H. M. (1987). *The struggle for the American curriculum 1893–1958*. New York: Routledge & Kegan Paul.

Labaree, D. F. (1988). *The making of an American high school: The credentials market and the central high school of Philadelphia, 1838–1939*. New Haven, CT: Yale University Press.

LaCapra, D. (1983). Rethinking intellectual history and reading texts. In *Rethinking intellectual history: Texts, contexts, language* (pp. 25–39). Ithaca, NY: Cornell University Press.

Lagerman, E. C. (1979). *A generation of women: Education in the lives of progressive reformers*. Cambridge, MA: Harvard University Press.

Lasch, C. (1979a). *The culture of narcissism: American life in an age of diminishing expectations*. New York: Norton.

Lasch, C. (1979b). *Haven in a heartless world: The family besieged*. New York: Basic Books.

Lasch, C. (1989). Politics and culture. *Salmagundi*, pp. 51–60.

Lasser, C. (Ed.). (1987). *Educating men and women together: Coeducation in a changing world*. Chicago: University of Chicago Press.

Lazerson, M. (1971). *Origins of the urban school: Public education in Massachusetts, 1870–1915*. Cambridge, MA: Harvard University Press.

Lazerson, M. (1981). Writing an educational history of New York City: Some observations. In D. Ravitch & R. K. Goodenow (Eds.), *Educating an urban people: The New York City experience* (pp. 240–250). New York: Teachers College Press.

Lazerson, M., & Grubb, W. N. (Eds.). (1974). *American education and vocationalism: Documents in vocational education: 1870–1979*. New York: Teachers College Press.

Lie, J. (1987). The road to nowhere? *Harvard Educational Review, 57*, 331–340.

Lifton, R. J. (1986). *The Nazi doctors: Medical killing and the psychology of genocide*. New York: Basic Books.

Marsden, W. (Ed.). (in press). *Education in four nations*. Cambridge, England: Cambridge University Press.

Monroe, P. (1940). *Founding of the American public school system: A history of education in the United States from the early settlements to the close of the Civil War period*. New York: Macmillan.

Nasaw, D. (1980). *Schooled to order: A social history of public schooling in the United States*. New York: Oxford University Press.

Oakes, J. (1985). *Keeping track: How schools structure inequality*. New Haven, CT: Yale University Press.

Ong, W. J. (1977). *Interfaces of the word*. Ithaca, NY: Cornell University Press.

Perkins, L. M. (1989). The history of Blacks in teaching. In D. R. Warren (Ed.), *American teachers: Histories of a profession at work* (pp. 344–369). New York: Macmillan.

Perlmann, J. (1988). *Ethnic differences: Schooling and social structure among the Irish, Italians, Jews, and Blacks in an American city, 1880–1935*. Cambridge, England: Cambridge University Press.

Peterson, P. (1983). Urban politics and changing schools: A competitive view. In R. K. Goodenow & D. Ravitch (Eds.), *Schools in cities: Consensus and conflict in American educational history* (pp. 223–249). New York: Holmes & Meier.

Powell, A. G., Farrar, E., & Cohen, D. K. (1985). *The shopping mall high school: Winners and losers in the educational marketplace.* Boston: Houghton Mifflin.

Ravitch, D. (1974). *The great school wars, New York City, 1805–1973: A history of the public schools as battlefield of social change.* New York: Basic Books.

Ravitch, D. (1978). *The revisionists revised: A critique of the radical attack on the schools.* New York: Basic Books.

Ravitch, D. (1983). *The troubled crusade: American education, 1945–1980.* New York: Basic Books.

Ravitch, D., & Goodenow, R. K. (Eds.). (1981). *Educating an urban people: The New York City experience.* New York: Teachers College Press.

Reese, W. J. (1986). *Power and the promise of school reform: Grassroots movements during the progressive era.* Boston: Routledge & Kegan Paul.

Rogers, D. T., & Tyack, D. B. (1982). Work, youth, and schooling: Mapping critical research areas. In H. Kantor & D. B. Tyack (Eds.), *Work, youth, and schooling* (pp. 167–184). Stanford, CA: Stanford University Press.

Rury, J. L. (1991a). *Education and women's work: Female schooling and the division of labor in urban America, 1870–1930.* Albany: State University of New York Press.

Rury, J. L. (1991b). Transformation in perspective: Lawrence Cremin's *Transformation of the school. History of Education Quarterly, 31,* 66–80.

San Miguel, G., Jr. (1987). *"Let them all take heed": Mexican Americans and the campaign for educational equality in Texas, 1910–1981.* Austin: University of Texas Press.

Schlossman, S. I. (1977). *The American delinquent: The theory and practice of progressive juvenile justice, 1825–1920.* Chicago: University of Chicago Press.

Schwager, S. (1987). Educating women in America. *Signs, 12,* 333–372.

Sedlak, M. W. (1989). "Let us go and buy a school master": Historical perspectives on the hiring of teachers in the United States, 1750–1980. In D. R. Warren (Ed.), *American teachers: Histories of a profession at work* (pp. 257–290). New York: Macmillan.

Sedlak, M. W. (1990). Review of *Governing the young. Teachers College Record, 91,* 126.

Sedlak, M. W., Wheeler, C. W., Pullin, D. C., & Cusick, P. A. (1986). *Selling students short: Classroom bargains and academic reform in the American high school.* New York: Teachers College Press.

Selden, S. (in press). *The capturing of science: Eugenics, race betterment, and education, 1883–1940.* New York: Routledge & Kegan Paul.

Seller, M. S. (1991). Boundaries, bridges, and the history of education. *History of Education Quarterly, 31,* 195–206.

Sennett, R. (1980). *Authority.* New York: Knopf.

Silver, H. (1983). *Education as history.* New York: Methuen.

Silver, H. (1990). *Education, change, and the policy process.* London: Falmer Press.

Solomon, B. M. (1985). *In the company of educated women: A history of women and higher education in America.* New Haven, CT: Yale University Press.

Stanley, M. (1981). *The technological conscience.* Chicago: University of Chicago Press.

Trennert, R. A. (1990). *Educating Indian girls at non-reservation boarding schools, 1878–1920.* In E. DuBois & V. Ruiz (Eds.), *Unequal sisters: A multicultural reader in U.S. women's history* (pp. 224–237). New York: Routledge & Kegan Paul.

Tyack, D. B. (1974). *The one best system: A history of American urban education.* Cambridge, MA: Harvard University Press.

Tyack, D., & Hansot, E. (1982). *Managers of virtue: Public school leadership in America, 1820–1980.* New York: Basic Books.

Tyack, D., & Hansot, E. (1990). *Learning together.* New Haven, CT: Yale University Press.

Tyack, D., Lowe, R., & Hansot, E. (1984). *Public schools in hard times: The Great Depression and recent years.* Cambridge, MA: Harvard University Press.

Vine, P. (1979). Preparation for republicanism: Honor and shame in the eighteenth-century college. In B. Finkelstein (Ed.), *Regulated children/liberated children: Education in psychohistorical perspective* (pp. 44–63). New York: The Psychohistory Press.

Wallace, A. F. C. (1970). *The death and rebirth of the Seneca.* New York: Knopf.

Warren, D. R. (Ed.). (1989). *American teachers: Histories of a profession at work.* New York: Macmillan.

Weibe, R. (1975). *The segmented society.* London: Oxford University Press.

Welter, B. (1966). The cult of true womanhood. *American Quarterly, 18,* 151–174.

White, H. (1987). *The content of the form: Narrative discourse and historical representation.* Baltimore: Johns Hopkins University Press.

Williams, R. (1961). *The long revolution.* New York: Columbia University Press.

Willis, P. (1976). The class significance of school counter-culture. In M. Hammersley & P. Woods (Eds.), *The process of schooling* (pp. 188–200). London: Routledge & Kegan Paul.

Wollons, R. (in press). *Educating mothers: A study in volunteerism.* New Brunswick, NJ: Rutgers University Press.

Woody, T. (1966). *A history of women's education in the United States* (2 vols.). New York: Octagon Books. (Original work published 1927)

Wrigley, J. (1982). *Class politics and public schools: Chicago, 1900–1950.* New Brunswick, NJ: Rutgers University Press.

Chapter 7

Interdisciplinary Education: Are We Faced With Insurmountable Opportunities?

HUGH G. PETRIE
State University of New York at Buffalo

Over 15 years ago, I wrote a paper titled "Do You See What I See? The Epistemology of Interdisciplinary Inquiry" (1976). In that article I discussed the relationship of interdisciplinarity to the disciplines and some of the features of the burgeoning field of interdisciplinary inquiry and education. Of course, a concern with a narrow focus on the disciplines predates my 1976 article. In the first part of the century, Dewey (1916, 1933, 1938) implicitly attacked a narrow formulation of the disciplines as the basis for education in his elaborate theory of the role of experience in learning. Similarly, the National Association for Core Curriculum, in its *Core Teacher* series, has been plumping for interdisciplinary education for over 40 years.

In the last 15 years, however, interest in interdisciplinary matters has increased significantly. The interest has been especially noteworthy in higher education. For example, Jane Roland Martin (1982) has questioned what she calls the dogma of God-given subjects in order to try to build a more adequate curriculum that does not rely solely on the traditional disciplines and that finds a place for such paradigm examples of inter-disciplinary education as African-American and women's studies. Ernest Boyer (1987) has analyzed the undergraduate experience, criticizing the overemphasis on the disciplinary major to the detriment of general education. Most recently, the Association of American Colleges has been active in trying to bring coherence to the undergraduate college curriculum and has just completed a 3-year project on the academic major (1991). One of the working groups participating in the project focused on interdisciplinary studies, and its results have been jointly published by the Association of American Colleges and the Society for Values in Higher Education (1990).

Turning to K–12 education, John Goodlad (1983) has decried the sameness of schools and the deadening formality of instruction that is closely

299

tied to the disciplines, especially in secondary schools. More recently, Ted Sizer's Coalition of Essential Schools (1984, 1988; Sizer, 1984) has had a significant impact on the educational scene with its slogan that "less is more" and the push for teachers as generalist coaches and students as workers. From a slightly different direction, Lauren Resnick, in her American Educational Research Association (AERA) presidential address (1987), contrasted the many differences between in-school and out-of-school learning to the disparagement of much of the abstract, decontextualized, disciplinary nature of in-school learning.

In addition to these, as well as many other educational developments that have sharpened our interest in interdisciplinarity, two volumes have recently appeared that provide excellent summaries and discussions of the growing literature in the field. In 1986 Daryl Chubin, Alan Porter, Frederick Rossini, and Terry Connolly edited *Interdisciplinary Analysis and Research: Theory and Practice of Problem-Focused Research and Development*. This book contains an excellent collection of the most important articles in the field. *Interdisciplinarity: History, Theory and Practice* (1990a), by Julie Thompson Klein, is a superb comprehensive study of the concept of interdisciplinarity. Klein explores the definition of interdisciplinarity, examines the relationship of interdisciplinary studies to disciplinary work, and surveys the state of the art of interdisciplinarity in areas such as research, education, and health care. The book also contains an extensive bibliography on interdisciplinarity, as does the volume edited by Chubin, Porter, Rossini, and Connolly. Any serious study of the present-day status of interdisciplinary education should start with these two outstanding books.

Because of this recent summary work in the area, I am going to approach this review with, perhaps, a bit more focus on specific areas and questions than is customary in a *Review of Research in Education* chapter. It seems to me that the somewhat unbridled enthusiasm within education for interdisciplinarity could profit from a serious analysis of the concept and origins of interdisciplinary thought and from a critical study of the implications for interdisciplinarity of some of the major theses of current cognitive science. In this way, I will suggest that an enthusiastic but naive view of interdisciplinary education could indeed lead us to "insurmountable opportunities."

There are three main conceptual strands I wish to trace throughout this discussion of interdisciplinary education. These are, first, Aristotle's (1941) distinction between theoretical and practical wisdom; second, the distinction between the locus of interdisciplinary thought residing primarily in the individual and primarily in groups or other social arrangements; and, third, the modern constructivist view of learning, foreshadowed by Dewey's (1938) theory of experience.

Let me say a little more about each of these strands of thought. The Aristotelian distinction between theoretical and practical wisdom is the distinction between answering the question "What is the case?" and the question "What ought one to do?" The former is concerned with understanding the world as it is, the latter with acting in the world. Theoretical wisdom pursues the truth; practical wisdom pursues the good. Indeed, one way of viewing the task of any modern professional school such as law or medicine or education is to ask how the understandings provided by the basic disciplines presumably undergirding that profession are connected to the practice of the profession. How does jurisprudence or legal theory improve the practice of law? What is the relationship between biology and chemistry on the one hand and health on the other hand? How can a knowledge of psychology or sociology or history inform education?

The standard notion that such "theoretical foundations" are simply "applied" to the practical question of what ought to be done has come under increasing attack (Kennedy, 1987). The problem is that the concept of some more or less recipe-like following of rules derived from theoretical understanding simply does not work in any reasonably complicated activity such as teaching and learning. Although still popular in some policy quarters, the idea that we can directly apply the knowledge of the disciplines to develop specifications of what teachers should routinely do to get students to learn has pretty much been abandoned in the face of our increasing understanding of the complexities of the work of teaching and learning (e.g., see Schön, 1983, 1987; L. Shulman, 1986, 1987b).

The second strand of thought I wish to pursue is the distinction between the knowledge of an individual and the knowledge of a group. I am here referring to a wide range of ideas. There is the relatively simple notion that in many instances, simply summing the individual knowledge of the members of a group results in more knowledge than is held by any single individual in the group. There is the more sophisticated notion that knowledge, theoretical or practical, depends essentially on a community of inquirers (e.g., Benson, 1989; Hamlyn, 1978). There is also the observation that we sometimes actually build knowledge into a social system of some sort or other. Resnick (1987) cites the example of modern navigation on U.S. Navy ships. Additional striking examples for those who participate in them are electronic mail networks and modern electronic library searching techniques. There simply seems to be more and a different kind of knowledge residing in the group or system than resides in any given individual.

The third area is that of a constructivist view of learning. Under this conception, knowledge is not just handed over from teacher to learner. Rather, the idea is that learners construct meanings that enable them to make sense of the situations in which they find themselves. This view has its roots in Kant's (1961) notion of the synthetic a priori, Dewey's (1938) concept of experience, Kuhn's (1970, 1974) theory of scientific revolutions, Toulmin's (1972, 1977) analysis of the development of conceptual understanding, my own examination of the dilemma of learning and understanding (1981), and more recent accounts of psychological development (e.g., see Anderson & Pearson, 1984; Brown, Collins, & Duguid, 1989; Gardner, 1983; Phillips & Soltis, 1985; Sternberg, 1985). The core idea is that a variety of ways of dealing with the world and its problems may be appropriate, especially within the sometimes limited ecology of an individual, and individuals construct their own ways of making sense of their environments in accordance with past history and the details of current situations, both physical and social. Teaching and learning then become ways of trying out alternative sense-making strategies in terms of their increasing adequacy.

The idea of interdisciplinary teaching and learning is a powerful and appealing one. We must, however, be careful to examine its limitations as well as its promises. The three strands of thought just sketched will help us in thinking about the limits of interdisciplinary education. What, for example, is the difference between interdisciplinarity as a means of promoting theoretical knowledge and its use in practical problem solving? How are the disciplines used in these two areas? Can interdisciplinary knowledge be located in the heads of individual teachers and students, or must we think harder about the use of social arrangements to pursue interdisciplinarity? What implications does the idea of the construction of knowledge have for interdisciplinary studies? Does it mean that "anything goes"? Or are there some kinds of limits to the constructions? These are the major questions to be addressed in what follows.

A LEXICON

Perhaps the seminal work in scholarship in the field of interdisciplinarity is the report of the First International Conference on Interdisciplinarity sponsored by the Organization for Economic Cooperation and Development. It is titled *Interdisciplinarity: Problems of Teaching and Research in Universities* (1972). The basic terminological distinctions were drawn in this work, and they enjoy a reasonably broad acceptance in the literature; however, given the state of development of the field, these distinctions are by no means universally accepted, and a number of other categorizations are also evident (Klein, 1990a).

Disciplinarity

Interdisciplinarity cannot be understood apart from the concept of disciplinarity. Roughly, the idea of a discipline today connotes a number of things, including:

- A specialization of knowledge within some sort of overriding unity of cognitive endeavor, such as the natural sciences (Toulmin, 1972)
- The fact that the unity of a discipline seems to come from (Hirst, 1974) a common set of core metaphors and concepts defining the field of inquiry, a particular set of observational categories for structuring experience in the field, specialized methods for investigation, a specification of the means for determining the truth or justification for claims made within the field, and, perhaps most important of all, an idea of the purposes to be served in investigating the field (e.g., in physics, the desire to understand the nature of the physical world in which we find ourselves)
- An organized grouping of people who study the discipline, train other practitioners, and form the social mechanism for arbitrating among varying truth claims within the discipline; this often involves university departments and degrees, national societies and conferences, and peer-reviewed scholarly journals and publications (Kuhn, 1974)

Clearly, there can be a variety of things that are called disciplines and a variety of arguments about whether something is or is not a discipline. A good recent analysis (Becher, 1989, 1990) describes the cultures that develop in and around the disciplines and points to the ways in which the disciplines are coming under increased scrutiny from a number of perspectives.

Multidisciplinarity

Given the above characterization of a discipline, the notion of multidisciplinarity is simply the idea of a number of disciplines working together on a problem, an educational program, or a research study. The effect is additive rather than integrative. The project is usually short-lived, and there is seldom any long-term change in the ways in which the disciplinary participants in a multidisciplinary project view their own work. Someone, perhaps a project manager, needs to glue the disciplinary pieces together, but that is all that happens by way of integration. Traditional distribution requirements in high school or college curricula are typically of this nature. Any integration is simply assumed to take place in the heads of individual students rather than there being a carefully thought-out system of general education. Health care and special education teams often operate in this mode. Klein (1990a, pp. 59–60) cites the middle phase of the Philadelphia Social History Project, where social scientists and demographers were added to the project's original historians, as an example of multidisciplinary work. It is *group* work rather than *team* work.

Interdisciplinarity

Interdisciplinary research or education typically refers to those situations in which the integration of the work goes beyond the mere concatenation of disciplinary contributions. Some key elements of disciplinarians' use of their concepts and tools change. There is a level of integration. Interdisciplinary subjects in university curricula such as physical chemistry or social psychology, which by now have, perhaps, themselves become disciplines, are good examples. A newer one might be the field of immunopharmocology, which combines the work of bacteriology, chemistry, physiology, and immunology. Another instance of interdisciplinarity might be the emerging notion of a core curriculum that goes considerably beyond simple distribution requirements in undergraduate programs of general education (Gaff, 1989; Newell, 1986, 1988). In many ways, the integrative thrust of interdisciplinary thinking is often a central feature of efforts to reform general education rather than a frill.

Turning to the schools, there are a number of national efforts, to be discussed in more detail below, to turn the "layer cake" (first biology, then topped by chemistry, then topped by physics) approach to American science education on its side. These efforts would require an interdisciplinary approach to teaching science since, at any given time, a combination of biology, chemistry, and physics would be studied. The various interconnections among these traditional disciplines would then need to be emphasized and fundamental principles, including mathematics, could be taught and learned more efficiently and effectively. Truly interdisciplinary special education teams might approach the special education student as a person with multiple needs rather than as a case to which different perspectives—all duly laid out in their hierarchical order of medical expert, psychologist, teacher specialist, and classroom teacher—can be brought. Klein (1990a, pp. 60–63) describes the later intention of the Philadelphia project noted above as interdisciplinary in that the results of the multidisciplinary research were to provide the basis for a holistic framework for studying cities. The fact that no such integrative framework ultimately emerged illustrates both the integrative idea of interdisciplinarity and the disciplinary paradox to be discussed below.

Transdisciplinarity

The notion of transdisciplinarity exemplifies one of the historically important driving forces in the area of interdisciplinarity, namely, the idea of the desirability of the integration of knowledge into some meaningful whole (e.g., see Kockelmans, 1979; Toulmin, 1982). The best example, perhaps, of the drive to transdisciplinarity might be the early discussions of general systems theory (Bateson, 1979; Boulding, 1956) when it was being held forward as a grand synthesis of knowledge. Marxism, structuralism, and feminist theory are sometimes cited as examples of a transdisciplinary approach (Klein, 1990b). Essentially, this kind of interdisciplinarity represents the impetus to integrate knowledge, and, hence, is often characterized by a denigration and repudiation of the disciplines and disciplinary work as essentially fragmented and incomplete.

If we now look at these rough and ready distinctions through the lenses of the three conceptual strands I noted above, some interesting results emerge. First, consider the theoretical-practical wisdom distinction. Strictly disciplinary activities tend primarily to be concerned with theoretical understanding, while multidisciplinary activities, and perhaps even some interdisciplinary projects, are more concerned with practical results. Transdisciplinary activities, to be sure, tend toward addressing questions of theoretical understanding, especially those of the unity of knowledge, but the distinction between theoretical concerns and practical questions in interdisciplinary work seems worth making.

"The Disciplinary Paradox"

Consider, for example, Klein's (1990a, chap. 7) analysis of what she calls the disciplinary paradox. The paradox is essentially that, on the one hand, the fragmentation of knowledge into the disciplines leads to the necessity for interdisciplinary approaches, yet, on the other hand, interdisciplinary approaches to knowledge can only receive an epistemic justification from the established disciplines. This may, indeed, be a paradox if one restricts one's attention to theoretical knowledge. In such a case, the only solution would seem to be to try to construct some transdisciplinary notion of knowledge that encompasses all of the disciplines and their specific methodologies and provides an overall epistemic justification for knowledge claims.

However, if one also remembers that interdisciplinary approaches are frequently interested in practical wisdom and the solution of practical problems, then it is not clear that in such cases there is a paradox at all. As one of the early pioneers of interdisciplinary thought, Rustum Roy, put it, there is the "inexorable logic that the real problems of society do not come in discipline shaped blocks" (1979, p. 163). Thus, if the disciplines are sometimes irrelevant to our practical concerns, it is not at all clear that they can provide the only justification for good interdisciplinary work. At least in those cases where the problem is a practical one, the success in solving the problem can be taken as a justification.

This has important implications for such movements in education as the increasing emphasis on teachers as researchers (e.g., Cochran-Smith & Lytle, 1990) and the importance of teachers' craft knowledge (e.g., J. Shulman & Colbert, 1987, 1988). Insofar as these endeavors are taken as interdisciplinary ways of addressing practical problems in the classroom, it is not clear that they will require a disciplinary justification above and beyond their success in the classroom.

Second, if we are interested primarily in interdisciplinarity as it applies to individuals, we typically begin with the disciplines and how they are learned by individuals. The subjects of secondary school and the disciplinary majors of the college curriculum are our reference points. We yearn after some, perhaps unattainable, vision of transdisciplinarity that we wish all children could acquire in the name of general education. However, the cooperative and collaborative problem solving of groups of people working in multidisciplinary or interdisciplinary teams may provide a more adequate characterization of what is possible by way of general education. It may even be that current movements toward a kind of multicultural diversity as a curriculum goal are more realistic than imposing a single consensus where none is available. Possibly, the only unity to be found in a plurality of cultures may be that they all, in their own ways, solve some of the very situation-specific practical problems with which each culture is faced. Furthermore, the coherence desired of general education would then become a coherence of seeing the diverse ways in which human beings can solve their practical problems of what to do.

Consider in this regard the implication of a constructivist theory of learning for interdisciplinarity. This implication is nicely drawn out by Jane Roland Martin (1982) in her discussion of two dogmas of curriculum (also Birnbaum, 1969). One of these dogmas is what she calls the dogma of God-given subjects—that the subjects and disciplines are, somehow, natural and God-given. It is Martin's contention, reinforced by the notion that meaning is constructed, that these subjects are chosen and constructed rather than handed over. They are chosen for some human purpose and can be more or less useful in serving that purpose.

There is, of course, probably an historical justification for most of the subjects and disciplines studied in school. These disciplines probably do codify reasonably widespread and useful human activities in that they have, at least in the past, given us the solution to a number of human problems. However, it is important to realize that they can be challenged and that they can be replaced with other subjects and disciplines. Thus, discussions of and arguments about women's studies, African-American studies, or Western civilization are to be expected and welcomed. If our disciplines and subjects are created rather than discovered and their usefulness lies in their ability to help us deal with the world in which we find ourselves, then curriculum will always be an essentially contested domain. Interdisciplinarity, in this sense, begins to look more and more like a way of trying to see and deal with the world in new and different ways that may be more adequate than the traditional ways.

The preceding discussion contrasting problem-solving approaches with epistemological issues of warranted belief echoes a related set of distinctions that runs through the literature on interdisciplinarity. This is the distinction between "bridge-building" between relatively firm, independent disciplines and "restructuring" by changing parts of the interacting disciplines (Group for Research and Innovation, 1975). It is the difference between "instrumental borrowing" to solve problems and "integration" or "synopsis" to achieve a new conceptual unity (Klein, 1985; Landau, Proshansky, & Ittelson, 1962; Taylor, 1969). It is the disparity between a "vacant" interdisciplinarity that simply looks for commonalities among existing frameworks and a "critical" one that rethinks the nature of knowledge (Kroker, 1980; Robbins, 1987). Finally, there is the distinction, on the one hand, between traditional disciplinarity and its associated interdisciplinarity, which pretty much accepts the disciplines as given, and, on the other hand, the postdisciplinary and deconstructionist critique of disciplinarity, in which a radical interdisciplinarity or cross-disciplinarity becomes a new kind of theoretical imperative (Brantlinger, 1990; Elam, 1990; Fish, 1989). Of course, from the radical deconstructionist perspective a truly critical interdisciplinarity is, in the end, impossible because no approach to knowledge has any justification. Such an argument, however, seems plausible only if one ignores practical wisdom and the justification derived from successfully pursuing our practical problems.

In my 1976 article I argued that the minimal necessary conditions for successful interdisciplinary work were that the participants understand the observational categories and meanings of the key terms in each other's disciplines—pointing out that different disciplines often see the same things differently and mean different things by the same words served, then, to illustrate reasons why interdisciplinary work is sometimes so difficult. We seldom take account of these differences.

What has happened is that the deconstructionists (e.g., Fish, 1989) have been impressed with the fact that different people see the same thing differently and mean different things by the same words and have concluded that warranted knowledge is, in principle, impossible. What they seem to ignore, however, is the centrality of the practical in interdisciplinary research and education. We do solve problems, of both a research and an educational nature, more or less well. Thus, pointing out that a given theoretical perspective with its associated observational categories is only one of several ways of looking at the world does not imply that we cannot make judgments of better and worse in comparing those ways of looking at the world.

Many scholars appear to have accepted the tentative nature of the disciplines and the need for interdisciplinary activity. However, as the disciplinary paradox suggests, we have only begun to look at ways of warranting knowledge beyond those contained in the disciplines (e.g., Giroux & McLaren, 1986). Interdisciplinary work, with its emphasis on the practical, may be a resource for correcting this deficiency.

Both the problem-solving and unity-of-knowledge perspectives have historically driven movements toward interdisciplinarity (Geertz, 1980; Klein, 1985). However, the motivations and goals have been quite dissimilar (e.g., see Turner, 1990). The former is aimed at practical problem solving in terms of current ways of understanding problems, while the latter focuses on the warrants for knowledge (i.e., on transdisciplinarity). The deconstructionists emphasize the lack of any absolute grounding for traditional disciplinary knowledge claims and, still captured by the vision of transdisciplinarity, seem driven to conclude that knowledge is impossible. There is no absolutely privileged place to stand from which to evaluate knowledge claims.

But all of this discussion actually depends on our continuing to draw a sharp distinction between theoretical and practical knowledge and assigning the disciplines solely to the realm of theoretical knowledge. If, however, we think of thought and action as much more closely connected, it can be argued that the disciplines themselves really are, or should be considered primarily as, means of solving practical problems of what ought to be done. In this sense, the disciplinary paradox can be solved not only for those cases in which the interdisciplinary work is focused on practical concerns, but also for the cases in which the justification for integrated, interdisciplinary systems of knowledge is being sought.

Throughout the history of ideas, the traditional disciplines have generated any number of theoretical problems of understanding that by now often have little connection to the ordinary problems of human existence. What is required is not some absolutely privileged standpoint from which to articulate a warranted notion of transdisciplinarity, but a return to the roots of the disciplines themselves as organized means of dealing with the world in which we find ourselves (e.g., see Casey, 1986; Geertz, 1980, Simons, 1989, 1990). As such, the worth of disciplines or interdisciplines can be judged on roughly pragmatic grounds. Both the traditional disciplines as well as new interdisciplinary and transdisciplinary ways of thinking are to be judged on the basis of how well any of these constructed systems of thought enable us to solve the historically conditioned, yet constantly evolving, practical problems of living (Petrie, 1981).

Having provided a lexicon, discussed the disciplinary paradox, and argued that the two imperatives for interdisciplinary work—the practical and the theoretical—can be understood as related, I want, in turn, to investigate interdisciplinary research and the growth of knowledge, the idea of interdisciplinary problem solving, and, finally, interdisciplinary approaches to education, both in higher education and in the schools. I hope to make clear that these areas are not separate and that preconceptions in one area often affect what happens in another.

INTERDISCIPLINARY RESEARCH AND THE GROWTH OF KNOWLEDGE

Donald Campbell (1969) has provided us with a compelling metaphor for interdisciplinary research and the growth of knowledge. In his paper, "Ethnocentrism of Disciplines and the Fish-Scale Model of Omniscience," Campbell argues persuasively for a social view of knowledge. Indeed, one of his main contributions in the paper is to note that even within the disciplines themselves, the integration and comprehensiveness we find is a collective product and not the accomplishment of any one scholar. Every scholar within any discipline has slightly different experiences and slightly different expertise. This fact actually follows from the notion of the construction of meaning by individuals from their unique experiences.

How, then, do we achieve the notion of a single discipline? Campbell's (1969) argument is that it is through the overlap among disciplinarians that a collective communication, competence, and breadth is attained. We have enough common experience to allow for the more or less common observational categories, more or less common methodological approaches, and more or less common core metaphors involved in defining a discipline (Petrie, 1976, 1981). In turn, these commonalities are roughly accounted for by the fact that as human beings living in the same world, we have similar basic problems to solve in terms of coping with that world. Thus, some commonality within individual disciplines is to be expected.

However, if one conceives of different subsections of the world as more or less well known by different individuals, much as a fish is covered by overlapping fish scales, then the problem with the disciplines is that the disciplinarians within a single discipline cluster much closer together than do the scales on a fish. It is as if small portions of the fish were covered by a large number of overlapping fish scales, but between these clusters, significant portions of the fish are uncovered. Campbell (1969) remarks how the social inventions of university departments with journals and tenure and the whole set of academic rituals actually work toward keeping the individual clusters closer together and against any individual's covering the spaces between the disciplinary clusters.

Interestingly, Campbell (1969) uses his social model to argue that rather than trying to create interdisciplinarians who truly know both disciplines (they would have to be equivalent to very large fish scales), it is more plausible to encourage individuals to become quite knowledgeable about small parts of the world that are not yet well known but that can overlap, at least to some degree, with a discipline that is already well known. In this way interdisciplinary work can move forward. Indeed, as these new areas are explored, it may even be that the "center of gravity" of a discipline may shift and different parts of the world may come to be more densely covered by the individual researchers working in the new area. Thus, geography can be seen as a movement away from drawing maps of the world toward a study of how we use space.

Of course, devising social structures that will allow for this kind of communication is not easy, especially in educational institutions. In higher education (e.g., see Caldwell, 1983; Rich & Warren, 1980), the traditional departmental structure tends to keep the disciplinarians quite isolated, although professional schools and area studies break down the gaps to some degree. In the schools, the problems of isolated classrooms and their inhibiting effects on communication among the professionals are well known. Indeed, this feature lies behind much of the reform-oriented analysis of the Holmes Group (see *Tomorrow's Teachers*, 1986, and especially *Tomorrow's Schools*, 1990). Fortunately, much general institutional and organizational analysis today (Peters & Waterman, 1982, continues to be a key reference) focuses on the need to build organizations in which there is a good deal less specialization (read "disciplinarity") and a good deal more integration (read "interdisciplinarity").

Viewing the disciplines and the movement toward interdisciplinary work as the result of social processes has a certain liberating effect on thinking about the growth of knowledge. On the one hand, the idea of a social grouping with more or less similar ways of seeing the world accounts for the strength of the notion of a community of inquiry in thinking about the growth of knowledge. People can communicate with each other, even given their more or less idiosyncratic experiences, if they are all working in roughly the same area of knowledge. Such a relatively stable social grouping explains why disciplinary work is thought to be so important. At the same time, it shows that some problems do not get addressed if these problems lie outside the interests of the dominant disciplinary group.

Campbell (1969) uses this idea to explain why university departments are so powerful and why they must sometimes be resisted. The fish-scale model would also explain the development of area studies in the social sciences and the more recent emergence of African-American studies, women's studies, and the like (e.g., Miller, 1982). Individuals are trying to cover new parts of the intellectual landscape, yet to be understood at all, they must have some ties to the traditional disciplines. However, as their studies progress, they may be able to develop new groupings of scholars in these fields themselves, rather than relying on the traditional disciplines (Clifford & Marcus, 1986).

This is reminiscent of Klein's (1990a) disciplinary paradox and Martin's (1982) dogma of the God-given subjects. In each case one can escape the straitjacket of the disciplines by looking for new and emerging human problems that need to be solved outside of the disciplinary ways of looking at things. At the same time, it explains the inherent difficulty of interdisciplinary research and the innate conservatism of the disciplines. People do tend to experience the world in the ways in which they were taught, and it requires major efforts to break out of the mold.

The theme of theory and practice is relevant here as well. The logic of interdisciplinarity, as laid out by Campbell (1969), suggests that sterile theory without the anchor of problem solving will keep the disciplines fragmented and incomplete. Thus, as Klein (1990a) has also noted, interdisciplinary thought often emphasizes the reunification of the theoretical and the practical. This movement toward unification has been exemplified recently for those of us in education in powerful critiques of the nature of schools and colleges of education (Clifford & Guthrie, 1988; Judge, 1982). The crux of the issue is the extent to which schools and colleges of education at the university level can conceive of themselves as true professional schools, committed to the practical needs of the profession as opposed to succumbing to the pull of the traditional disciplines that are enshrined in the departments of the university.

As a 10-year veteran dean of a graduate school of education and one of the founding members of the Holmes Group, I can personally testify to the importance of this issue. Our ability as schools of education to come to grips with the practical question of our place in the university will determine whether or not we will survive. A metaphor for the change I am advocating is the nearly yearlong debate we had at the State University of New York at Buffalo about whether or not to change our name from the "Faculty of Educational Studies" to the "Graduate School of Education." In the end, the Graduate School of Education was accepted, but not without considerable discussion of the nature of scholarship and the relationship of the theoretical and the practical. In addition, the 1991 national conference of the Holmes Group was devoted to the question of the nature of inquiry in the new, more collaborative relationships being proposed between schools of education and educational practitioners (Holmes Group, 1990).

The constructivist theory of learning and knowledge development is also quite congenial to interdisciplinary ideas of knowledge growth and development. Knowledge structures (i.e., disciplines) are constructed in response to certain general human needs and problems. The structures are built with certain purposes in mind and can be judged as more or less adequate depending on their long-range ability to provide solutions for the characteristic problems they were built to address. But as new problems arise (e.g., the growing awareness by minorities and women that their place in society has been largely defined for them), new, interdisciplinary areas of study may arise (Kolodny, 1984; Stimpson, 1988; Stimpson & Cobb, 1986). An excellent article by Brian Turner (1990) analyzes the sometimes conflicting reasons for pursuing interdisciplinary research and education in the medical field. The question is whether or not to accept the world as described currently and undertake interdisciplinary work to solve problems as currently defined or to use interdisciplinary investigation to challenge the fundamental conception of the traditional ways of defining our disciplines and problems. In such cases, however, there is, as Klein (1990a) puts it, always a "burden of comprehension" placed on the interdisciplinary advocate. The interdisciplinarian must continue to be able to communicate with the disciplinarians while standards of rigor and trustworthiness are being developed in the new area.

Ethnography is a good recent example in education of a field of study that had to face the criticisms of the disciplinarians as it struggled to find its own methods (Eisner, 1981; Firestone, 1987; Guba & Lincoln, 1988; Howe, 1988; Howe & Eisenhart, 1990; Jacob, 1988; Peshkin, 1988; L. Shulman, 1988). It was attacked in education as being both eclectic and lacking in rigor (Phillips, 1983, 1987a, 1987b), but, of course, those charges are themselves dependent on a particular, generally positivist view of science and disciplined inquiry, a view that is precisely under question. The danger is to fall back on the accepted disciplinary standards of evaluation rather than to develop the new standards appropriate to the new questions being asked. However, keeping the constructivist view clearly in mind helps one to avoid the temptation to dismiss new interdisciplinary proposals out of hand. All of knowledge is constructed, including the current disciplines, and the ultimate test of their worthwhileness is whether or not they allow us to deal more adequately with the world, physical and social, in which we find ourselves (Petrie, 1981).

INTERDISCIPLINARY PROBLEM SOLVING

The area of interdisciplinary problem solving has already been referred to in Roy's (1979) famous quotation that societal problems do not, at least nowadays, come in discipline-shaped blocks. This idea is exemplified in any number of other examples—from the interdisciplinary development of the atomic bomb during World War II to the most recent war on drugs and the debate over the adequacy of our nation's concern for children. In each instance, the problem that has been identified seems to call for an interdisciplinary solution. In most cases the solution tends to be what I have called "multidisciplinary." That is, a mission-oriented team of experts from a variety of disciplines is assembled and given the charge of solving the problem. A substantial literature has grown up around the problems of setting up, organizing, and running interdisciplinary problem-solving groups (Chubin et al., 1986; Klein, 1990a, chap. 8).

As might be expected, there are a number of difficulties with such undertakings. Key among these have to do with different modes of structuring their experience by different members of the team and the blind spots this causes. After all, if a given discipline has as part of its essential makeup a certain way of seeing things (Petrie, 1976), then that way of seeing things will automatically also be a way of *not* seeing things. This perceptual feature of interdisciplinary inquiry still seems to me to characterize one of the necessary conditions for successful interdisciplinary work. One must be able to see the world in the ways in which the other members of the interdisciplinary team do. It takes a certain kind of broad-gauge scholar (e.g., one who is not worried about obtaining tenure or who already has it) to be able to experiment with new ways of looking at and conceiving of things. The problem also needs to be clearly stated in relatively nontechnical language, and the group needs to have leadership sensitive to the many different ways of thinking about the issue.

The construal and definition of the problem to be solved is one of the key features of interdisciplinary problem solving (Chubin, Porter, & Rossini, 1986). How should our experience be structured so as to permit the greatest chance of coming up with a solution? Should we view the drug problem as one of interdicting supply? As one of decreasing demand? As one of social anomie? As one of building a character sufficiently strong so that one can "just say no"? All problem solving in Dewey's sense (1933) is interdisciplinary. How can we structure our experience so that inquiry can lead to solutions of our problems along with continued growth and education? The justification of both the traditional disciplines and any new interdisciplinary study must be that they ultimately allow us to pursue our purposes in an ever-changing but broadly stable world (Petrie, 1981).

Porter and Rossini (1984) have offered the very useful STRAP framework for thinking about interdisciplinary problem solving. The framework consists of analyzing the *S*ubstantive knowledge required to solve the problem, the *T*echniques needed, the *R*ange of substantive knowledge and techniques to be used, the *A*dministrative or organizational complexity required, and the *P*ersonnel needed. This approach offers a more systematic updating of the features I mentioned in 1976. The framework also throws into clear relief the salience of the theoretical-practical distinction, the question of individual or group expertise, and the construction of knowledge issues that I am pursuing.

The theoretical disciplines are pressed into service only insofar as they appear to cast some light on the problem to be addressed. In this sense, interdisciplinary problem solving is a paradigm case of exploring, on the one hand, the relationships between theoretical understandings of why something is the case derived from codifications of substantive knowledge, and, on the other hand, technique and practical questions of what ought to be done. The justification for a given discipline in interdisciplinary problem solving is how well it assists in solving the problem. Notice, too, that the notion of the "application" of theory to practice in interdisciplinary settings has a hollow ring to it. Only if a given problem falls wholly within a discipline does the notion of applying theory to practice even appear to make sense. The more typical case in mission-oriented interdisciplinary research is that the various disciplines offer different scenarios and point to different, possibly salient features of the problem situation, but the solution is not a disciplinary matter.

For example, the recent calls for establishing the school as a center for integrated social services for at-risk children (e.g., Kirst, 1991) can be conceived of in at least two ways. First, it might be that the school simply happens to be where the child is for a good portion of the day, so that on logistical grounds alone, that is where he or she should be treated medically, counseled psychologically, and remediated socially and educationally. An interdisciplinary approach to the problem, however, might look at the situation not from the point of view of the various disciplinary experts, but rather from the point of view of the overall welfare of the child. In the latter case, while the doctor, nurse, psychologist, social worker, and remedial teacher may all have something to say about the situation, none of their individual "solutions" will be primary. Nor is this simply a bland sort of eclecticism. It is, instead, a structuring of the problem in a way that gives hope of a real solution.

Interdisciplinary problem solving also illustrates very clearly the difference between conceiving of interdisciplinarity as located in the individual and conceiving of it as located in the group. Almost all interdisciplinary problem solving occurs in groups, groups that must be organized and administered. With respect to personnel, we still have the occasional individual who can master several disciplines, but as Campbell (1969) pointed out, such persons are rare indeed. It seems much more promising to look at the problem as one of blending the correct social and organizational means of interaction and integration. Collaboration, rather than competition, must be the watchword.

The implications of interdisciplinary problem solving for education are several. First, insofar as the calls for curricular revision (discussed more fully below) emphasize problem solving and "higher order thinking skills," as many of them do, the STRAP framework provides a good way of thinking about these suggestions. What is the appropriate substantive knowledge? The techniques? How much is required? How should we organize the delivery of the new curriculum? What personnel should be involved? Single teachers? Groups of teachers? Others?

Second, the literature on teacher thinking and problem solving (e.g., Berliner, 1986, 1989; Kennedy, 1987; Schön, 1983, 1987; L. Shulman, 1986, 1987a, 1987b) can benefit from the work on interdisciplinary problem solving. After all, the work of the typical teacher, especially the elementary school teacher, is fundamentally interdisciplinary. One of the more interesting implications is that team teaching, in both elementary and secondary schools, may be a most useful way to conceive of approaching complex curricula. Instead of the nearly impossible task of trying to devise elementary teacher education programs to put all of the necessary knowledge in the head of a single teacher (Petrie, 1987), one can perhaps organize elementary teaching as if it were an interdisciplinary problem-solving group.

INTERDISCIPLINARY EDUCATION IN COLLEGES AND UNIVERSITIES

Interdisciplinary education takes a number of different forms in higher education (Klein, 1990a; Levin & Lind, 1985; Organization for Economic Cooperation and Development, 1972). According to Klein (1990a, chap. 10), interdisciplinary education is found primarily in a few revolutionary institutions in which the entire program is conceived of as interdisciplinary, within so-called "area" studies at traditional universities, in liberal or general education programs, as a minor within a disciplinary structure, and within at least some professional programs.

Klein (1990a, pp. 157–163) mentions several institutions, primarily abroad, that have attempted to base the whole of their educational programming on interdisciplinary principles. All attempted to find a transdisciplinary definition of knowledge and knowledge acquisition. It is important to note that none of the institutions remains the same today as when it was founded. All have had to accommodate to a version of the disciplinary paradox in which the prevailing disciplinary mold puts enormous pressure on them to conform, from recruiting faculty to finding jobs for their students (Trow, 1984/1985). All, however, retain some important vision of interdisciplinarity within their mission.

Area studies, on the other hand, have a seemingly firm foothold in the modern university. Early geographical area studies, such as African, East Asian, and the like, blazed the way for more recent additions such as African-American studies, women's studies, American studies, and programs in science, technology, and society. Sometimes these programs borrow faculty from traditional disciplines; in other instances they have their own departments and degrees. Very often they are heavily associated with the institution's general education program. Professors who work in such programs frequently have to deal with a particularly virulent form of the disciplinary paradox in that they are usually held suspect by their disciplinary colleagues, yet must not succumb to the temptation simply to fall back into disciplinary ruts and modes of thinking.

Liberal or general education provides another important site for interdisciplinary education in the colleges and universities (Boyer, 1981, 1987; Clark & Wawrytko, 1990; Gaff, 1989; Newell, 1986, 1988). Typically, the impetus toward general education is not initially problem based, but rather arises from the drive toward integrative knowledge. The traditional distribution program is at best multidisciplinary, with any integration simply assumed to take place within the heads of individual students. Core curricula, on the other hand, usually involve at least some integration in the heads of the faculty who plan them and very often also include specially designed core courses that sometimes cross disciplines and nearly always aim at some sort of integrating function (Undergraduate College, 1990). Often there are also special projects and experiences and "capstone" seminars to pull it all together. The key feature of a truly interdisciplinary general education program is, ultimately, the extent to which the program itself attempts to synthesize the elements of the curriculum instead of simply leaving it to the students.

A particularly useful interdisciplinary model for implementing general education is provided by Hursh, Haas, and Moore (1983). They use the developmental theories of Dewey (1916, 1933, 1938), Perry (1968, 1981), and Piaget and Inhelder (1969) to argue for a "skills-based" general education curriculum—one that focuses on problem solving. They stress Perry's notion of acting on beliefs, another theme of interdisciplinarity. The model also emphasizes multiple perspectives on the problem and the salient concepts and the necessity for analyzing the strengths and limitations of various disciplinary approaches.

Hursh et al.'s answer (1983) to those who hold that an emphasis on skills alone ignores the critical role of substantive knowledge in attaining intellectual skills (e.g., McPeck, 1981) constitutes a powerful argument in favor of interdisciplinary skills. Essentially, they grant that cognitive development depends on the acquisition of substantive knowledge but point out that this begs the question of *what* substantive knowledge. Insofar as substantive knowledge depends on ever more specialized investigation, we become wholly dependent on the (disciplinary) experts in these specializations. What we need, however, is an ability to critically evaluate the claims of the "experts." In short, we need general intellectual skills, obtained through interdisciplinary education, to know when to trust the experts. Although they do not address the issue, I believe that Hursh, Haas, and Moore still feel that this capacity can be generated within each individual student. If, however, the locus of interdisciplinary wisdom might be better conceived of in the group, there still would be reason to follow a skills-oriented, problem-solving general education program, only with significant emphasis on learning group problem-solving skills as well.

Interdisciplinary minors are another feature of the modern university. They represent a relatively innocuous concession to interdisciplinarity on the part of the strong disciplinarians who dominate most universities. Such minors are frequently problem focused, with an attempt to show how the disciplinary majors can be used in these areas.

Another important site for interdisciplinary education in the university is the modern professional school (Schön, 1983). Many graduate and undergraduate programs in the professions of medicine, law, social work, engineering, and education attempt to provide integrated experiences for their students in order to help them situate themselves within the larger problematic of their professions. My earliest experience with interdisciplinarity was in a school of engineering with a program devoted to humanizing the education of engineers (Petrie, 1976). I have also mentioned the debate within education (Clifford & Guthrie, 1988; Judge, 1982) regarding the extent to which graduate schools of education need to broaden their programs beyond simply conceiving of themselves as populated by applied disciplinarians, without at the same time succumbing to the temptation to become modern versions of the old normal schools (i.e., without succumbing to the disciplinary paradox).

My own background over the past 10 years as dean of a graduate school of education in a major public research university is instructive. Klein's (1990a, p. 131) description of the qualities needed to lead interdisciplinary efforts struck a responsive chord with me. The need for previous interdisciplinary experience, sensitivity toward different paradigms, commitment to problem solving, group interaction skills, and enormous energy and patience conforms extremely well to my experience. Among the most difficult tasks a dean of education has is bringing together the various educational disciplinarians around common problems in ways that allow them to continue to value their cultural and disciplinary backgrounds. The interesting thing is that the "disciplinarians" needed are not only the educational psychologists, sociologists, philosophers, historians, and curricular specialists, but also the academics from the rest of the university and the very practically oriented professionals from the field. As the literature predicts, our most successful efforts have occurred when there is a clear problem, or problems, to be solved and when the participants take the time to understand each other's backgrounds and ways of looking at and dealing with the world (Chubin et al., 1986; Petrie, 1976).

The distinction between theoretical knowledge and practical knowledge finds expression in a number of places in college- and university-based programs of interdisciplinary education. The revolutionary institutions and at least some programs of general education tend toward an integration of theoretical knowledge, although there is often a strong, if not well-articulated, sense of the primacy of the practical in current discussions of liberal education. Area studies, interdisciplinary minors, and professional programs, on the other hand, tend to recognize the essentially practical focus of their activities and, thus, seem to be more problem oriented. The truly interesting questions cluster around the relationships between the traditional core of the academy, with its emphasis on theoretical understanding, and the increasing social and professional pressures to be relevant to society's problems. One critical question for future interdisciplinary research might well be the relationship between theoretical understanding and practical wisdom.

Ernest Boyer (1990) has recently begun to address this problem with his attempt to articulate a broadened definition of scholarship and its different relationships to knowledge. He and his colleagues have begun to speak of the discovery of knowledge, the synthesis and integration of knowledge, the application of knowledge, and the presentation and representation (teaching) of knowledge. Scholarship and research might well focus on any one of these relationships. We in the Graduate School of Education at Buffalo have begun to address a broadened notion of scholarship in promotion and tenure guidelines. Using the concept of "professional service" articulated by Elman and Smock (1985), we are attempting to fashion a category of equal prestige with traditional concepts of research and teaching. The category of professional service is a blending of the applied research and professional interpretation of theoretical scholarship for the field with the kind of service on a national or state commission or study group that may result in new and influential standards for the field. Such attempts within the university to articulate the relationships between theoretical and practical knowledge will only increase in the future.

The question of whether the locus of interdisciplinarity is to be found in the individual or the group also finds significant expression in higher education. Current attempts to redefine a core curriculum appear to assume that we really can create interdisciplinary individuals, if only we get the ideas and integrating concepts right (e.g., see Clark & Wawrytko, 1990; "General Education," 1989; Interdisciplinary Studies, 1978). On the other hand, the critique offered by multiculturalists raises some significant questions regarding our ability to define a core curriculum. Implicit in the multiculturalist argument is the claim that there are a number of different situation-specific ways of dealing with the problems of life and that none is necessarily better than any other (Kolodny, 1984; Stimpson, 1988). The multicultural curriculum would acquaint students with that fact and some cultural examples and then celebrate the diversity.

Others find this approach to be hopelessly relativistic and incoherent (e.g., D'Souza, 1991; Hirsch, 1987).

It is not at all surprising, given the constructivist approach to learning, that there now rages a significant debate regarding the nature of general education. There are those who believe that traditional subjects are simply "given" and any retreat from them would represent a retreat from standards (D'Souza, 1991; Hirsch, 1987). Even some who accept the constructivist point of view argue for the superiority of classical conceptions of knowledge (Ravitch & Finn, 1987). Others point out that great chunks of the experience of certain people have not only been omitted from the classics, but the classics have been used to oppress those people (Stimpson & Cobb, 1986). The constructivist point of view clearly indicates that such debates are to be expected and even welcomed as we try to achieve increasingly better ways of dealing with the world in which we live. It is not, however, the case that accepting a constructivist position on knowledge acquisition necessarily leads to a kind of relativism. As noted above, notions of better and worse ways of dealing with the world still seem to have application.

INTERDISCIPLINARY EDUCATION IN THE SCHOOLS

Educational programs in the schools, especially the secondary schools, often take their substance from programs in the colleges and universities. The influence of the traditional disciplines tends to extend with an iron hand into the secondary school curriculum, so it is not surprising to find similar debates within the schools. One good example is the recent furor caused by the publication in New York State of a report titled *A Curriculum of Inclusion* (Commissioner's Task Force on Minorities: Equity and Excellence, 1989). This report to the state education department argued that the state social studies curriculum had omitted and denigrated the contributions of many of the minority cultures represented in the state. As might be expected, people lined up on both sides of the issue. Like similar debates over the canon in higher education, this debate raises a number of issues of interdisciplinarity regarding theoretical versus practical knowledge, the construction of meaning, and the disciplinary paradox. The original report was followed by another, *One Nation, Many Peoples: A Declaration of Cultural Independence* (Social Studies Syllabus Review and Development Committee, 1991), produced by a distinguished panel of scholars, teachers, and laypersons. This second report, although still quite controversial, appears to be much more in line with the view I have been arguing that there are many ways of solving the practical problems with which human beings are faced and it is important for students to recognize that fact and approach all cultures with a sympathetic but critical eye.

Aside from the influence of the university, however, there are a number of developments in K–12 education that raise significant issues of interdisciplinarity and that deserve to be addressed independently. In order to do this, I have chosen three areas to discuss in some detail—the recent call for new standards in mathematics and science education, Ted Sizer's Coalition of Essential Schools, and Lauren Resnick's (1987) critique of school learning contained in her AERA presidential address.

New Standards for Mathematics and Science Education

Let me turn first to the proposed reforms in mathematics and science education (American Association for the Advancement of Science, 1989; National Council of Teachers of Mathematics, 1989). These reforms call for radically new ways of conceiving of the teaching and learning of mathematics and science. They are relevant to my discussion of interdisciplinary education primarily because of the suggestions they make regarding the relations of these disciplines to problem solving and everyday experience. In short, these proposals stem from a general acceptance of the constructivist theory of learning and a growing unease with the sharp traditional distinction between theoretical understanding in the disciplines and practical activity. As Toulmin (1977) suggests, even the traditional disciplines, conceived as growing, changing social institutions attempting to solve persistent human problems, often oscillate back and forth between discipline-oriented phases and problem-oriented phases. Mathematics and science education, with the new proposed standards, may now be entering one of the problem-oriented phases, and in this respect will begin to appear more interdisciplinary.

The new standards proposed by the National Council of Teachers of Mathematics (NCTM) (1989) presuppose a radically different conception of mathematics than the traditional, algebra, geometry, trigonometry, calculus way of proceeding. Goals for students are much more process oriented. They include the ideas that students should learn to value mathematics, reason mathematically, communicate mathematically, become confident of their mathematical abilities, and become mathematical problem solvers. In addition to more traditional concepts such as whole numbers, algebra, and geometry, the standards involve problem solving; communication; reasoning; seeing connections; understanding measurement, statistics, and probability; seeing patterns and relations; and attaining mathematical power. The sample problems illustrating the standards also move beyond the artificial "story problems" so familiar to most of us to encompass the multistep solution of real problems found in society. In this way, mathematics itself becomes an interdisciplinary approach to solving certain kinds of social problems involving quantity, space, measurement, and statistics.

The American Association for the Advancement of Science (AAAS) takes a similar tack in its Project 2061 (the name is derived from the next return of Halley's Comet) report, *Science for All Americans* (1989). The very title of the report indicates its focus not simply on those who will become the next generation of scientists, but on a scientific understanding for all. AAAS argues that our increasingly complex world requires a certain level of scientific literacy of all of us. This is to be accomplished by softening the boundaries between the traditional areas, turning the "layer cake" of current science education on its side and, in so doing, emphasizing the connections among the traditional areas of biology, chemistry, physics, and earth science.

AAAS (1989) proposes to lessen the amount of detailed knowledge that students are expected to retain and to increase the understanding of the essential processes and presuppositions of science, including the place of science in the overall history of ideas. The social implications of science, mathematics, and technology are stressed, along with inculcating scientific habits of mind. These habits of mind include the internalization of the scientific values of the respect for and use of evidence and reasoning; informed beliefs about the social costs and benefits of science; a positive attitude toward being able to understand science and mathematics; computational skills, especially with regard to estimation of reasonable answers; manipulation and observational skills, including the use of a computer; communication skills, including the use of graphs, tables, and diagrams; and critical skills that will enable the student to evaluate arguments and claims that invoke the mantle of science.

As with the NCTM standards (1989), *Science for All Americans* (1989) focuses more on process and underlying conceptions and connections than on traditional disciplinary content. There is a clear commitment to demystifying science and making it accessible to all Americans. Turning the layer cake on its side in curriculum development will necessitate at least an initial level of interdisciplinary work, as teachers of biology, chemistry, and physics will have to study the other sciences to learn the connections and underlying principles that they will need to teach their students. The focus on science, technology, and society also provides a clear recognition of the places at which science as a set of disciplines must connect with other disciplines.

Both the NCTM (1989) and the AAAS (1989) projects are involved in exploring the ways in which teacher education must be changed in order to accommodate these new conceptions of mathematics and science. Another report has recently been issued on this topic, *A Call for Change: Recommendations for the Mathematical Preparation of Teachers of Mathematics* (Mathematical Association of America, 1991). Almost all of the discussion calls for a much more problem-focused and contextualized approach to the preparation of teachers. This is fully consonant with the problem-solving theme found in the interdisciplinary literature, while the calls to integrate mathematical and scientific concepts partake of the transdisciplinary approach. Hursh et al.'s (1983) discussion of the implementation of a skills-based, problem-solving general education curriculum discussed above is also relevant here. Their emphasis on the tentative nature of knowledge, intellectual skills, problem solving, the methodologies of problem solving, reflection on the material, and one's approach to it are all very consonant with the NCTM and AAAS rhetoric.

Both the NCTM (1989) and the AAAS (1989) reports stress the connections that must be made between theoretical understanding and practical activity, especially for those who may not go on to make a career of mathematics or science. Both reports are also clearly informed by the constructivist approach to learning with their emphasis on student activities as opposed to the simple delivery of content. Neither has much to say, however, about whether or not these bold new conceptions of mathematics and science actually are teachable to individuals or whether we may not have to approach mathematical and scientific literacy from a more social perspective. Of all the disciplines, mathematics and science have perhaps the best-developed and most rigorous standards for competence in the field. Whether one can, by focusing on processes, skills, problem solving, and the social and historical connections of science, escape the disciplinary paradox remains to be seen. It is likely that many "real" scientists will find such interdisciplinary approaches inadequate. At the same time, there is also afoot in the land a serious reexamination, discussed above, of the dominance of the traditional disciplines. In any event, such efforts have much to gain from close collaboration with those who have been toiling for some time in the interdisciplinary vineyards.

The Coalition of Essential Schools

I turn now to Ted Sizer's Coalition of Essential Schools and its associated program with the Education Commission of the States, Re:Learning (Sizer, 1988). Sizer's Coalition of Essential Schools arises out of his Study of High Schools, documented in three volumes (Hampel, 1986; Powell, Farrar, & Cohen, 1985; Sizer, 1984). The Coalition is an ever-increasing group of schools committed to restructuring education around the common principles that emerged from the Study of High Schools. Re:Learning is an effort to link the restructuring effort going on at individual schools with district- and statewide educational reform. It is supported primarily by the Education Commission of the States, a nonprofit, nationwide, interstate compact formed to help governors, state legislatures, state education officials, and others develop policies to improve education.

The so-called Common Principles of the Coalition (Coalition of Essential Schools, 1988) include the following:

- The school should focus on helping adolescents learn to use their minds well.
- Each student should master a limited number of essential skills and areas of knowledge. "Less is more."
- The school's goals should apply to all students.
- Teaching and learning should be personalized.
- The governing practical metaphor of the school should be student-as-worker and teacher-as-coach.
- The diploma should be awarded upon a successful final demonstration of mastery—an exhibition.
- The tone of the school should stress values of unanxious expectation.
- The principals and teachers should perceive of themselves as generalists first and disciplinary specialists second.
- Student loads should not exceed 80 per teacher and costs should not exceed 10% more than traditional schools.

Sizer and his colleagues believe that these principles, conscientiously implemented in schools, will have a revolutionary impact on schooling. The Coalition is continuing its efforts, although with a number of problems analogous to those of instituting interdisciplinary education efforts (Chion-Kenney, 1987; see the newsletters of the Coalition of Essential Schools, 1984–1991).

Several features of Coalition schools reflect the interdisciplinary emphasis inherent in the principles. First, the notions that less is more and that teachers should be generalists first and specialists second clearly reflect a movement toward interdisciplinarity. Students are supposed to master the material and make significant intellectual achievements rather than just grasp traditional content. The requirement for an "exhibition" of mastery (i.e., a kind of practical "doing") rather than the accumulation of a list of courses taken is another hallmark of interdisciplinary work. It is also clear that what is intended here is a synthesis and integration rather than simply a multidisciplinary gathering together of traditional disciplines, although it does not appear that the Coalition is hostile to the traditional disciplines or that it attempts to foster some new notion of transdisciplinarity.

The distinction between theoretical and practical knowledge seems, in Coalition schools, to tilt toward the primacy of the practical. This is seen in the deemphasis of traditional subjects and the central place of an exhibition as a demonstration of mastery. The exhibition is described as a demonstration that the student can do important things using the mind.

Typical exhibitions appear to rely heavily on real-life projects such as environmental impact studies, artistic performances, and the like. They are not typically discipline-sized chunks, but rather real social problems.

The rhetoric for Coalition schools also seems to emphasize teamwork, although perhaps more so for faculty and staff than for students. It is clear that Sizer believes that the professionalization of teaching must result in much more interaction and sharing of knowledge among school professionals than is now the case. As generalists, teachers and other staff are also expected to take on multiple obligations such as teacher-counselor-manager and must have a sense of commitment to the whole school.

Clearly, Coalition schools subscribe heavily to a version of the constructivist theory of learning. The student-as-worker exemplifies the spirit of the constructivist position, as does the emphasis on students using their minds well to accomplish important things, although there could be some concern about who decides what the "important things" are and on what basis. It may be that these important things are still to be taken largely from the received wisdom, and, therefore, they might not obviously represent a fresh look at what the real problems of society are.

Although the rhetoric of interdisciplinary education is somewhat muted in the Coalition of Essential Schools, it is, nevertheless, clearly present, at least as a means to the end of restructuring schooling. As such, it is also predictable that many of the typical problems of interdisciplinary education also bedevil the work of the Coalition. In particular, several versions of the disciplinary paradox seem to attend the development of Coalition schools (Sizer, 1989; see the newsletters of the Coalition of Essential Schools, 1984–1991). As might be expected, teachers in Coalition schools have a good deal of trouble seeing themselves as generalists. Their bases in the disciplines provide a sense of security, and venturing forth from those secure bases is problematic. Similarly, there is a good deal of skepticism regarding the actual worthwhileness of work in a Coalition school. Parents, administrators, and policymakers want to know how the students will do on standardized tests, yet if the emphasis is on depth instead of coverage and on essential skills instead of the traditional academic disciplines, it is not at all clear how well the students will do on traditional, discipline-based assessment procedures.

It is at this point that the emphasis on the exhibition becomes particularly important. Coalition proponents realize that as one attempts to move away from the disciplines, new standards and forms of evaluation must be developed, or else the perspectives of the disciplines will devalue the new work. Coalition schools are beginning to join forces with those who are advocating the development of new forms of authentic assessments (Gardner, in press; Lave, 1988; Resnick & Resnick, in press; Stiggins, 1988; Wiggins, 1989; Wolf, Bixby, Glenn, & Gardner, 1991), and will need to pay attention to the extent to which the exhibitions really do appear to address real problems and issues of society. Recall that one way out of the disciplinary paradox was to judge interdisciplinary work on its success in dealing with practical problems. On the positive side the communities of inquiry, composed of both students and faculty, in Coalition schools will almost surely help each of the participants to experience the extent to which knowledge is a social construction. Work in the Coalition would certainly benefit from a more explicit acknowledgment that much of what is being advocated really is interdisciplinary education, with all of the attendant problems and challenges.

"Learning in School and Out"

Last, let me turn to Lauren Resnick's AERA presidential address, "Learning in School and Out" (1987). In this address Resnick draws four sharp contrasts between learning in schools and learning outside of schools, and, although her purpose was not to address directly the issue of interdisciplinary education, the distinctions she draws are clearly relevant.

Resnick (1987) first contrasts the kind of individual cognition we value in school with the shared cognition valued outside of schools. It seems that our social systems embody the realization that we cannot, in very many instances, solve the problems we must solve if we rely solely on individual performances. Yet, in school, we insist that the performances be individual. This insistence may be compatible with the function of the school as a social and economic sorting mechanism that assigns people to scarce desirable positions. However, it is not clear that it is helpful educationally in promoting student learning. Why do we not use socially cooperative modes of instruction and learning instead of individually competitive modes? Real work, even in the traditional disciplines, more often than not takes place collaboratively.

The second contrast cited by Resnick (1987) is the difference between pure mentation in school subjects and tool manipulation outside (see also Lave, 1988). Why, for example, do we continue to insist that students learn the multiplication tables instead of allowing them to use calculators as tools to obtain a deeper knowledge of mathematics? This contrast echoes the tendency for the traditional disciplines to focus on theoretical understanding and accepted ways of doing things, while interdisciplinary problems require practical reasoning and solutions.

The difference between symbol manipulation in school and highly contextualized reasoning outside of school forms Resnick's (1987) third contrast (see also Lave, 1988). She cites an interesting example of the difference between symbol manipulation in solving a mathematics problem and contextualized reasoning. The problem involves deciding how much more money would be needed to buy an ice cream cone costing 60 cents if one had in hand a quarter, a dime, and two pennies. The standard school solution involves the calculation that 23 cents more is needed. The typical situated real-world solution involves looking for some additional coins, perhaps another quarter. As interdisciplinarians point out, real problems seldom come in discipline-shaped chunks (Roy, 1979), and this helps explain why doing well in school often has so little effect on how well one does in the real world. The reasoning is simply different in the different places, and at least one impetus for interdisciplinary work, the solution of real problems, tends to focus on the situated reasoning needed in the real world.

Resnick's (1987) fourth contrast is that between generalized learning in school and situation-specific competencies outside. A version of the disciplinary paradox can be constructed here as well. On the one hand, generalized theoretical learning does not always apply directly to or translate well into specific situations. Thus, we often need some sort of situation-specific problem solving. On the other hand, if one only learns to solve specific problems, then the adaptability to new situations is suspect. As I have been arguing, interdisciplinary education provides a way of beginning to see how to relate theoretical knowledge to practical knowledge.

Resnick's (1987) discussion of learning in school and out raises quite forcefully several of the conceptual themes I have used to discuss interdisciplinary education. First, the distinction between theoretical and practical wisdom is implicit in all of Resnick's contrasts. Schools, especially secondary schools, in their massive reliance on traditional disciplines, clearly emphasize theoretical wisdom. The real world, however, often demands practical competence. Second, Resnick's powerful descriptions of the social nature of learning and performance outside of schools call into serious question whether or not interdisciplinarity can be reasonably located within individuals rather than in social systems. Finally, the idea of situation-specific learning implies the need for much more work on just how we do, indeed, seem to construct meaning from our individual experiences and yet, nevertheless, are able to deal with new situations that do not differ too radically from the ones in which we have learned the skills and competencies. The traditional answer is that we learn general principles and then apply them to specific situations. The interdisciplinary answer seems to be that we must bring different perspectives to bear in learning how to see specific situations in new and useful ways.

CONCLUSION

The impetus to interdisciplinary education comes from a variety of different directions. One impetus is integrative in the epistemological sense. The disciplines appear to have fragmented knowledge, and it would be better to have a unified system of knowledge. This impetus results in attempts to develop transdisciplinary approaches to knowledge, in many of the efforts to devise some kind of core curriculum, and perhaps in the new standards proposed for mathematics and science teaching. The underlying idea seems to be that somehow we should be able to unify knowledge in the head of the individual. Typically, this impetus is directed primarily at theoretical understanding, and is particularly subject to the disciplinary paradox. How can we develop standards of evaluation of a unified core curriculum separate from the standards of the disciplines?

Another impetus is the need to solve the practical problems of society. These problems seldom come in discipline-shaped chunks. At issue here is the extent to which these problems are defined by society as it exists as opposed to being the problems that a restructured society ought to address if the deficiencies of the disciplines were corrected. The press for interdisciplinarity can be conceived both as inherently conservative of how we currently understand our human milieu or as progressive in moving us forward and out of the straitjackets of the disciplines. Interdisciplinary research on weapons delivery systems may be an example of the former, while interdisciplinary research in women's studies may exemplify the latter.

Another press that seems to combine elements of both the theoretical and practical comes from the imperatives of civic and cultural education. Our educational system has always had as one of its chief functions the transmission of the culture. This culture has both theoretical and practical aspects. Indeed, as I have noted, one of the more interesting current debates has to do with the extent to which we do have or ought to have a common culture in the United States or whether we have moved or ought to move to a more multicultural perspective. One of the strengths of our liberal democratic heritage, in its emphasis on tolerance and the marketplace of ideas, is that the possibility of embracing a multicultural perspective is not an incoherent idea. In any case, interdisciplinary education is central to the debate in that it provides a major stimulus for the dialogue.

Indeed, the social nature of much interdisciplinary thought and education may even suggest a practical solution to the problem of possible social fragmentation that may follow upon an overly enthusiastic embracing of multiculturalism. We need only recall Campbell's (1969) fish-scale model and seek out those individuals who, though still specialized at the fringes of an existing culture, overlap with other cultures. We must then expand our mainstream cultures to incorporate these individuals into social arrangements in which we can take advantage of their bridging skills. Perhaps we can, through new social arrangements, celebrate our diversity and see it as a strength.

Finally, the impetus for interdisciplinary thought probably also comes from the disciplines themselves, at least if they are vibrant and active. As long as the disciplines remember their fundamental roles as systematized ways of helping human beings deal with the major problems of being human, they will never for too long be abstract and disconnected from practical affairs. Interdisciplinary problem solving and theoretical reflection will serve to remind the disciplines that they must constantly anchor their work in the real world of human thought and activity.

The idea of interdisciplinary teaching and learning is a powerful and appealing one. We must, however, think hard about its limitations. Can we overcome the disciplinary paradox in its various manifestations? Can interdisciplinary knowledge be integrated in the heads of individuals, or must it be located in social groupings? Does multiculturalism constitute a kind of locating of interdisciplinary knowledge in groups? What are the relations between theoretical understanding and practical wisdom? If we address these issues carefully and systematically, perhaps the opportunities promised by interdisciplinary education will not be insurmountable after all.

REFERENCES

American Association for the Advancement of Science. (1989). *Science for all Americans: Project 2061*. Washington, DC: Author.

Anderson, R. C., & Pearson, P. D. (1984). A schema-theoretic view of basic processes in reading comprehension. In P. D. Pearson (Ed.), *Handbook of reading research* (pp. 255–292). New York: Longman.

Aristotle. (1941). Nicomachean ethics, Book VI. In R. McKeon (Ed.), *The basic works of Aristotle*. New York: Random House.

Association of American Colleges. (1991). *Liberal learning and arts and sciences majors: The challenge of connecting learning, Vol. 1*. Washington, DC: Author.

Association of American Colleges and Society for Values in Higher Education. (1990). Interdisciplinary resources. *Issues in Integrative Studies, 8*.

Bateson, G. (1979). *Mind and nature: A necessary unity*. New York: Dutton.

Becher, A. (1989). *Academic tribes and territories: Intellectual enquiry and cultures of the disciplines*. Milton Keynes, Australia: Open University.

Becher, A. (1990). The counter-culture of specialisation. *European Journal of Education, 25*, 333–346.

Benson, G. D. (1989). Epistemology and science curriculum. *Journal of Curriculum Studies, 21*, 329–344.

Berliner, D. (1986). In pursuit of the expert pedagogue. *Educational Researcher, 15*(7), 5–13.

Berliner, D. (1989). Implications of studies of expertise in pedagogy for teacher education and evaluation. In *New directions for teacher assessment: Proceedings of the 1988 ETS International Conference* (pp. 39–78). Princeton, NJ: Educational Testing Service.

Birnbaum, N. (1969, July–August). The arbitrary disciplines. *Change, the Magazine of Higher Learning*, pp. 10–21.

Boulding, K. (1956). *The image: Knowledge in life and society.* Ann Arbor: University of Michigan.

Boyer, E. (Ed.). (1981). *Common learning: A Carnegie colloquium on general education.* Washington, DC: Carnegie Foundation for the Advancement of Teaching.

Boyer, E. (1987). *The undergraduate experience in America.* New York: Harper & Row.

Boyer, E. (1990). *Scholarship reconsidered: Priorities of the professoriate.* Washington, DC: Carnegie Foundation for the Advancement of Teaching.

Brantlinger, P. (1990). *Crusoe's footprints: Cultural studies in Britain and America.* New York: Routledge.

Brown, J., Collins, D., & Duguid, P. (1989). Situated cognition and the culture of learning. *Educational Researcher, 18*(1), 32–42.

Caldwell, L. (1983). Environmental studies: Discipline or metadiscipline? *Environmental Professional, 5,* 247–258.

Campbell, D. (1969). Ethnocentrism of disciplines and the fish-scale model of omniscience. In M. Sherif & C. Sherif (Ed.), *Interdisciplinary relationships in the social sciences* (pp. 328–348). Chicago: Aldine.

Casey, B. (1986). The quiet revolution: The transformation and reintegration of the humanities. *Issues in Integrative Studies, 4,* 71–92.

Chion-Kenney, L. (1987, Winter). A report from the field: The Coalition of Essential Schools. *American Educator,* pp. 18–28.

Chubin, D., Porter, A., & Rossini, F. (1986). Interdisciplinary research: The why and the how. In D. Chubin, A. Porter, F. Rossini, & T. Connolly (Eds.), *Interdisciplinary analysis and research: Theory and practice of problem-focused research and development* (pp. 3–10). Mt. Airy, MD: Lomond.

Chubin, D., Porter, A., Rossini, F., & Connolly, T. (Eds.). (1986). *Interdisciplinary analysis and research: Theory and practice of problem-focused research and development.* Mt. Airy, MD: Lomond.

Clark, M., & Wawrytko, S. (Eds.). (1990). *Rethinking the curriculum: Toward an integrated, interdisciplinary college education.* New York: Greenwood.

Clifford, G., & Guthrie, J. (1988). *Ed school.* Chicago: University of Chicago Press.

Clifford, J., & Marcus, G. (Eds.). (1986). *Writing culture: The poetics and politics of ethnography.* Berkeley: University of California.

Coalition of Essential Schools. (1984). *Prospectus.* Providence, RI: Coalition of Essential Schools, Brown University.

Coalition of Essential Schools. (1988). *The common principles of the Coalition of Essential Schools.* Providence, RI: Coalition of Essential Schools, Brown University.

Cochran-Smith, M., & Lytle, S. (1990). Research on teaching and teacher research: Issues that divide. *Educational Researcher, 19*(2), 2–11.

Commissioner's Task Force on Minorities: Equity and Excellence. (1989). *A curriculum of inclusion.* Albany, NY: State Education Department.

D'Souza, D. (1991). *Illiberal education: The politics of race and sex on campus.* New York: Free Press.

Dewey, J. (1916). *Democracy and education.* New York: Macmillan.

Dewey, J. (1933). *How we think.* New York: D. C. Heath.

Dewey, J. (1938). *Experience and education.* New York: Macmillan.

Eisner, E. (1981). On the differences between scientific and artistic approaches to qualitative research. *Educational Researcher, 10*(4), 5–9.

Elam, D. (1990). Ms. en abyme [Feminism in feminist scholarship]. *Social Epistemology, 4*, 293–308.

Elman, S., & Smock, S. (1985). *Professional service and faculty rewards: Toward an integrated structure.* Washington, DC: National Association of State Universities and Land-Grant Colleges.

Firestone, W. (1987). Meaning in method: The rhetoric of qualitative and quantitative research. *Educational Researcher, 16*(7), 16–21.

Fish, S. (1989). Being interdisciplinary is so very hard to do. *Profession 89* (pp. 15–22). New York: Modern Languages Association.

Gaff, J. (1989). The resurgence of interdisciplinary studies. *National Forum, 69*(2), 4–5.

Gardner, H. (1983). *Frames of mind.* Cambridge, England: Cambridge University Press.

Gardner, H. (in press). Assessment in context: The alternative to standardized testing. In B. Gifford, M. O'Connor, & M. Catherine (Eds.), *Rethinking aptitude, achievement, and assessment.* Boston: Kluwer Academic Publishers.

Geertz, C. (1980). Blurred genres. *American Scholar, 42*, 165–179.

General education. (1989, July/August). *Change, The Magazine of Higher Learning.*

Giroux, H., & McLaren, P. (1986). Teacher education and the politics of engagement: The case for democratic schooling. *Harvard Educational Review, 56*, 213–238.

Goodlad, J. (1983). *A place called school.* New York: McGraw-Hill.

Group for Research and Innovation. (1975). *Interdisciplinarity: A report by the Group for Research and Innovation.* Regents Park, England: Group for Research and Innovation, The Nuffield Foundation.

Guba, E., & Lincoln, Y. (1988). Do inquiry paradigms imply inquiry methodologies? In D. Fetterman (Ed.), *Qualitative approaches to evaluation in education: The silent scientific revolution* (pp. 89–115). New York: Praeger.

Hamlyn, D. W. (1978). *Experience and the growth of understanding.* London: Routledge & Kegan Paul.

Hampel, R. (1986). *The last little citadel.* Boston: Houghton Mifflin.

Hirsch, E. D. (1987). *Cultural literacy: What every American needs to know.* Boston: Houghton Mifflin.

Hirst, P. H. (1974). *Knowledge and the curriculum: A collection of philosophical papers.* London: Routledge & Kegan Paul.

Holmes Group. (1986). *Tomorrow's teachers.* East Lansing, MI: Author.

Holmes Group. (1990). *Tomorrow's schools.* East Lansing, MI: Author.

Howe, K. (1988). Against the quantitative-qualitative incompatibility thesis (or, dogmas die hard). *Educational Researcher, 17*(8), 10–16.

Howe, K., & Eisenhart, M. (1990). Standards for qualitative and quantitative research: A prolegomenon. *Educational Researcher, 19*(4), 2–9.

Hursh, B., Haas, P., & Moore, M. (1983). An interdisciplinary model to implement general education. *Journal of Higher Education, 54*, 42–59.

Interdisciplinary studies. (1978, August). *Change, The Magazine of Higher Learning*, pp. 6–48.

Jacob, E. (1988). Clarifying qualitative research: A focus on traditions. *Educational Researcher, 17*(1), 16–24.

Judge, H. (1982). *American graduate schools of education: A view from abroad.* New York: Ford Foundation.

Kant, I. (1961). *Critique of pure reason, translated by Norman Kemp Smith.* London: Macmillan.

Kennedy, M. (1987). *Inexact sciences: Professional education and the development of expertise* (Issue Paper 87-2). East Lansing: National Center for Research on Teacher Education, Michigan State University.

Kirst, M. (1991). Improving children's services: Overcoming barriers, creating new opportunities. *Phi Delta Kappan, 72,* 615-618.

Klein, J. T. (1985). The interdisciplinary concept: Past, present, and future. In L. Levin & I. Lind (Eds.), *Interdisciplinarity revisited: Re-assessing the concept in the light of institutional experience* (pp. 104-136). Stockholm: OECD/CERI, Swedish National Board of Universities and Colleges, Linköping University.

Klein, J. T. (1990a). *Interdisciplinarity: History, theory, and practice.* Detroit: Wayne State University Press.

Klein, J. T. (1990b). Interdisciplinary resources: A bibliographical reflection. *Issues in Integrative Studies: Interdisciplinary Resources, 8,* 35-67.

Kockelmans, J. (Ed.). (1979). *Interdisciplinarity and higher education.* University Park: Pennsylvania State University Press.

Kolodny, A. (1984). *The land before her: Fantasy and experience of the American frontier, 1630-1860.* Chapel Hill: University of North Carolina.

Kroker, A. (1980). Migration across the disciplines. *Journal of Canadian Studies, 15,* 3-10.

Kuhn, T. (1970). *The structure of scientific revolutions* (enlarged ed.). Chicago: University of Chicago Press.

Kuhn, T. (1974). Second thoughts on paradigms. In F. Suppe (Ed.), *The structure of scientific theories* (pp. 459-482). Urbana: University of Illinois.

Landau, M., Proshansky, H., & Ittelson, W. (1962). The interdisciplinary approach and the concept of the behavioral sciences. In N. Washburne (Ed.), *Decision, values and groups, II* (pp. 7-25). New York: Pergamon.

Lave, J. (1988). *Cognition in practice: Mind, mathematics, and culture in everyday life.* New York: Cambridge University Press.

Levin, L., & Lind, I. (Eds.). (1985). *Interdisciplinarity revisited: Re-assessing the concept in the light of institutional experience.* Stockholm: OECD, Swedish National Board of Universities and Colleges, Linköping University.

Martin, J. (1982). Two dogmas of curriculum. *Synthese, 51,* 5-20.

Mathematical Association of America. (1991). *A call for change: Recommendations for the mathematical preparation of teachers of mathematics.* Washington, DC: Author.

McPeck, J. (1981). *Critical thinking and education.* New York: St. Martin's.

Miller, R. (1982). Varieties of interdisciplinary approaches in the social sciences. *Issues in Integrative Studies, 1,* 1-37.

National Council of Teachers of Mathematics. (1989). *Curriculum and evaluation standards for school mathematics.* Washington, DC: Author.

Newell, W. (Ed.). (1986). *Interdisciplinary undergraduate programs: A directory.* Oxford, OH: Association for Integrative Studies.

Newell, W. (1988). Interdisciplinary studies are alive and well. *Association for Integrative Studies Newsletter, 10*(1), 6-8.

Organization for Economic Cooperation and Development. (1972). *Interdisciplinarity: Problems of teaching and research in universities.* Paris: Author.

Perry, W. (1968). *Forms of intellectual and ethical development in the college years.* New York: Holt, Rinehart, and Winston.

Perry, W. (1981). Cognitive and ethical growth: The making of meaning. In A. Chickering (Ed.), *The modern American college* (pp. 76-116). San Francisco: Jossey-Bass.

Peshkin, A. (1988). In search of subjectivity—One's own. *Educational Researcher, 17*(7), 17–22.

Peters, T., & Waterman, R. (1982). *In search of excellence: Lessons from America's best-run companies.* New York: Warner Books.

Petrie, H. G. (1976). Do you see what I see? The epistemology of interdisciplinary inquiry. *Educational Researcher, 5*(2), 9–15.

Petrie, H. (1981). *The dilemma of enquiry and learning.* Chicago: University of Chicago Press.

Petrie, H. (1987). Teacher education, the liberal arts, and extended preparation programs. *Educational Policy, 1*(1), 29–42.

Phillips, D. C. (1983). After the wake: Postpositivistic educational thought. *Educational Researcher, 12*(5), 4–12.

Phillips, D. C. (1987a). *Philosophy, science, and social inquiry.* New York: Pergamon.

Phillips, D. C. (1987b). Validity in qualitative research: Why the worry with warrant will not wane. *Education and Urban Society, 20,* 9–24.

Phillips, D. C., & Soltis, J. F. (1985). *Perspectives on learning.* New York: Teachers College Press.

Piaget, J., & Inhelder, B. (1969). *The psychology of the child.* New York: Basic Books.

Porter, A., & Rossini, F. (1984). Interdisciplinary research redefined: Multi-skill, problem-focused research in the STRAP framework. *R & D Management, 14,* 105–111.

Powell, A., Farrar, E., & Cohen, D. (1985). *The shopping mall high school.* Boston: Houghton Mifflin.

Ravitch, D., & Finn, C. (1987). *What do our 17-year-olds know? A report on the first national assessment of history and literature.* New York: Harper & Row.

Resnick, L. (1987). Learning in school and out. *Educational Researcher, 16*(9), 13–20.

Resnick, L., & Resnick, D. (in press). Assessing the thinking curriculum. In B. Gifford, M. O'Connor, & M. Catherine (Eds.), *Rethinking aptitude, achievement, and assessment in testing.* Boston: Kluwer Academic Publishers.

Rich, D., & Warren, R. (1980). The intellectual future of urban affairs: Theoretical, normative, and organizational options. *Social Science Journal, 17*(2), 53–66.

Robbins, B. (1987). Poaching off the disciplines. *Raritan, 6*(4), 81–96.

Roy, R. (1979). Interdisciplinary science on campus: The elusive dream. In J. Kockelmans (Ed.), *Interdisciplinarity and higher education* (pp. 161–196). University Park: Pennsylvania State University Press.

Schön, D. (1983). *The reflective practitioner.* New York: Basic Books.

Schön, D. (1987). *Educating the reflective practitioner.* San Francisco: Jossey-Bass.

Shulman, J., & Colbert, J. (1987). *The mentor teacher casebook.* San Francisco: Far West Laboratory for Educational Research and Development.

Shulman, J., & Colbert, J. (1988). *The intern teacher casebook.* San Francisco: Far West Laboratory for Educational Research and Development.

Shulman, L. (1986). Those who understand: Knowledge growth in teaching. *Educational Researcher, 15*(2), 4–14.

Shulman, L. (1987a). Assessment for teaching: An initiative for the profession. *Phi Delta Kappan, 69,* 38–44.

Shulman, L. (1987b). Knowledge and teaching: Foundations of the new reform. *Harvard Educational Review, 57,* 1–22.

Shulman, L. (1988). Disciplines of inquiry in education: An overview. In R. Jaeger (Ed.), *Complementary methods for research in education* (pp. 3–17). Washington, DC: American Educational Research Association.

Simons, H. (Ed.). (1989). *Rhetoric in the human sciences*. London: Sage.

Simons, H. (Ed.). (1990). *The rhetorical turn: Invention and persuasion in the conduct of inquiry*. Chicago: University of Chicago Press.

Sizer, T. (1984). *Horace's compromise: The dilemma of the American high school*. Boston: Houghton Mifflin.

Sizer, T. (1988, August). Creating a society that thinks: Re:Learning. *State Government News*, pp. 20–21.

Sizer, T. (1989). Diverse practice, shared ideas: The essential school. In H. Walberg & J. Lane (Eds.), *Organizing for learning: Toward the 21st century* (pp. 1–8). Reston, VA: National Association of Secondary School Principals.

Social Studies Syllabus Review and Development Committee. (1991). *One nation, many peoples: A declaration of cultural independence*. Albany, NY: State Education Department.

Sternberg, R. (1985). *Beyond IQ: A triarchic theory of human intelligence*. Cambridge, England: Cambridge University Press.

Stiggins, R. (1988). Revitalizing classroom assessment. *Phi Delta Kappan, 69*, 363–368.

Stimpson, C. (1988). *Where the meanings are*. New York: Methuen.

Stimpson, C., & Cobb, N. (1986). *Women's studies in the United States*. New York: Ford Foundation.

Taylor, A. M. (1969). Integrative principles and the educational process. *Main Currents in Modern Thought, 25*, 126–133.

Toulmin, S. (1972). *Human understanding: Vol. 1*. Princeton, NJ: Princeton University Press.

Toulmin, S. (1977). From form to function: Philosophy and history of science in the 1950s and now. *Daedalus, 1*, 143–162.

Toulmin, S. (1982). *The return to cosmology: Postmodern science and the theology of nature*. Berkeley: University of California Press.

Trow, M. (1984/1985). Interdisciplinary studies as a counterculture: Problems of birth, growth, and survival. *Issues in Integrative Studies, 4*, 1–15.

Turner, B. (1990). The interdisciplinary curriculum: From social medicine to postmodernism. *Sociology of Health and Illness, 12*(1), 1–23.

Undergraduate College. (1990). *A new general education curriculum for arts and sciences students at UB: A proposal from the undergraduate college to the university (revised)*. Buffalo: State University of New York at Buffalo.

Wiggins, G. (1989). A true test: Toward more authentic and equitable assessment. *Phi Delta Kappan, 70*, 703–713.

Wolf, D., Bixby, J., Glenn, J. III, & Gardner, H. (1991). To use their minds well: Investigating new forms of student assessment. In G. Grant (Ed.), *Review of research in education* (Vol. 17, pp. 31–74). Washington, DC: American Educational Research Association.

Chapter 8

Is Dewey's Educational Vision Still Viable?

EMILY ROBERTSON
Syracuse University

The Center for Dewey Studies' 50-page working bibliography of Dewey scholarship over the last 10 years provides ample evidence of continuing interest in Dewey's work. In philosophy, religion, ecology, feminist theory, economics, and law, as well as education, Dewey's pragmatic philosophy and theory of progressive education are being "revisited" and "revived," "applied" and "reconstructed." Within the field of education itself, Dewey's ideas have been used by theorists and practitioners in vocational education, teacher education, curriculum development (including social studies, art, music, foreign languages, science, and mathematics), Jewish studies, experiential education, career education, moral education, peace education, community education, higher education, and outdoor education.

Nor is this attention confined to the United States. More than 150 items from Dewey's corpus have been translated into languages other than English, with the result that Dewey's thought is at least partially accessible in 35 different languages. *Democracy and Education* alone has received 25 translations into languages including Arabic, Bulgarian, Chinese, Czech, German, Gujarati, Italian, Japanese, Korean, Marathi, Polish, Portuguese, Serbo-Croatian, Spanish, and Turkish (Boydston & Andresen, 1966; see also Passow, 1982).

Yet it would be premature to conclude that Dewey's enduring influence is as great as this brief survey may suggest. Alongside the articles just referred to are others that ask "Whatever happened to pragmatism? Why were its lessons not learned?" (Schwartz, 1988). Even Bernstein and Rorty, two leading advocates of Dewey's work, acknowledge that Dewey

I wish to acknowledge the assistance of Barbara Levine at the Center for Dewey Studies, Southern Illinois University, for sharing with me relevant portions from her draft of the *Checklist of Writings About John Dewey*, Third Edition, Enlarged, 1887–1990. I also am indebted to Jo Ann Boydston, Gerald Grant, and Maxine Greene for helpful comments on an earlier draft of this paper.

now is neglected within professional philosophy (Bernstein, 1987b; Rorty, 1991). In his recent book on Dewey's philosophy, Tiles comments that "what flowed through his [Dewey's] life and work and inspired his contemporaries seems, for the time being at least, to have run into the sand" (Tiles, 1988, p. 3). For example, while *Art as Experience* was referred to in 1966 by a leading figure in aesthetics as "the most valuable work on aesthetics written in English (and perhaps in any language) so far in our century," Fisher notes that Dewey's work is now missing from many leading anthologies in aesthetics. There are few or no "Deweyans" in aesthetics, and few articles are published on Dewey's aesthetics (Beardsley, 1966, p. 332; Fisher, 1989). Even Rorty's almost single-handed efforts to revive interest in Dewey's work have a paradoxical quality, for several commentators agree that Rorty's interpretation of Dewey omits or rejects central Deweyan themes (Bernstein, 1980, 1987a; Campbell, 1984b; Kolenda, 1986; Sleeper, 1986; West, 1989; Westbrook, 1991).

Turning from philosophy to education, while Tanner believes that "Dewey's educational reforms hold on," Kliebard's view is that Dewey's brand of progressive education failed to have much lasting impact on public schools. He doubts that "there is a public elementary or secondary school anywhere that self-consciously or conspicuously follows even the most elemental curricular principles that Dewey set forth" (Kliebard, 1987, p. 140; Tanner, 1987, p. 138; see also Kliebard, 1986). Greene notes that, ironically, progressive theories had more influence in private education, where social intelligence and commitment were nurtured through attention to problems stemming from ordinary life. But in mass public education, the emphasis was on preparation for future life and work and adjustment to what is given (Greene, 1986). Lazerson suggests that the ideal of the Dewey teacher, "flexible, experimental, knowledgeable of child development and of subject matter," was in sharp conflict "with the realities of schools and classrooms." Thus, teachers either rejected the ideal as "impractical" or felt guilty for failing to live up to it (Lazerson, 1984, p. 176). And Lagemann, speaking of traditions of educational research, claims that "one cannot understand the history of education in the United States during the twentieth century unless one realizes that Edward L. Thorndike won and John Dewey lost" (Lagemann, 1989, p. 185).

Yet there are also predictions of a Dewey renascence and claims that the time has come for a reappropriation of the pragmatic spirit (Bernstein, 1987b; Rorty, 1981, 1982, 1988, 1989, 1991; West, 1989). Greene argues that those who are searching for a critical pedagogy appropriate for schools in the United States would do well to turn to our own cultural heritage rather than to European theorists. She finds, in Dewey, some of the resources for such a pedagogy (Greene, 1986; see also Shusterman, 1989). Some socialist critics of schooling hold that elements of the Deweyan tradition, such as an emphasis on critical thinking, can be combined with the socialist tradition to develop a worthwhile philosophy of teaching that gives an active role to the teacher (Proefriedt, 1980). And Radest (1980) argues that progressivism provides a "usable politics" for schooling.

Is the time right for a Dewey revival? Can Dewey's vision provide guidance for our current educational efforts? Obviously, a major step in answering these questions is determining the appropriate criteria of assessment. Surely, a "viable" perspective has to be of more than historical interest. It must have some promise of being helpful in solving current problems. But, of course, whether or not a particular map proves helpful in reaching one's destination depends crucially on where one wants to go. Thus, the viability of Dewey's vision will be judged differently by those who share his goals than by those who do not. And some who support his goals may reject the means he proposed for reaching them. Furthermore, there are different interpretations of what is essential to Dewey's philosophy, of how best to understand his educational vision. Finally, there is a difference between concluding that Dewey was, in a general way, pointing in the right direction and believing that serious study of his work will prove fruitful in helping us deal with our current dilemmas.

In light of the many possibilities these issues present, I should make plain my own choices. This review focuses on Dewey's theory of radical democracy and his commitment to democratic education. Although there are other significant themes in his work, Dewey's advocacy of participatory democracy "goes to the heart of his philosophy" (Westbrook, 1991, p. xi). I agree with Burnett's claim that Dewey's ideal of progressive education is "but a part of something much larger and something to be sought on all fronts; viz., militant liberalism or socialized democracy" (Burnett, 1988, p. 211). From this perspective, whether it is reasonable to adopt Dewey's vision as our own depends fundamentally on whether we share these commitments. Dewey's ideal of the excellent teacher will continue to inspire individuals. And we might yet again turn to focusing on the interests of the child in one of our periodic rebellions against "mindless" learning, as Wirth (1981) predicts. But from a Deweyan perspective, this would be an incomplete victory without a commitment to the development of the radically democratic culture classrooms were to both model and help produce. "The reconstruction of philosophy, of education, and of social ideals and methods go hand in hand" (Dewey, 1916, p. 331). Accordingly, this review explores current interpretations and criticisms of the social ends Dewey sought through education, as well as the means he proposed for achieving them. More specifically, I survey interpretations and critiques of two central and intertwined strands of Dewey's social and political philosophy: experimentalism and radical democracy. The review begins with brief accounts of each of these themes to provide some orientation for the reader before turning to further elaboration and critical commentaries in the second part.

DEWEY'S SOCIAL AND POLITICAL IDEAL

Experimentalism and the Method of Intelligence

It is well known that Dewey was an advocate of the use of "intelligence" in human affairs and that he drew analogies between social intelligence and scientific method. But it is perhaps less widely understood what Dewey meant by this recommendation and what his reasons were for suggesting it.

The high value Dewey accords to science lies in its being a self-correcting enterprise and in the "moral relations" he sees it as embodying. It is a commonplace that what are regarded as scientific truths at one point in time may be rejected later in light of new discoveries. More important, from Dewey's point of view, the methods themselves are correctable. One can, even while employing faulty methods, discover their faults and make improvements. Experience can supply its own ideals and standards. Thus, while for Dewey there are neither foundations for knowledge nor absolute ends, this does not entail either skepticism or relativism. Our inherited beliefs can be tested and modified through a communal process of inquiry that makes use of intelligence. By discovering the causes and consequences of our actions, we can make judgments about the best course to pursue. By "intelligence," Dewey means the capacity for projecting into the future, through our knowledge of causal relations, the outcomes of a contemplated course of action, and for making foresight of what is desirable and undesirable in future possibilities the guiding factor in deciding what to do. The method of intelligence, then, frees us from dependence on tradition as our sole guide to conduct, and it replaces alleged a priori guides, whether arrived at through revelation or by rational philosophical argument. Openness to experience, that is, to discovering empirically what our actions lead to and reconstruction of our practices in light of these discoveries, is what makes possible more satisfactory human interactions with both the human and nonhuman environments. This enhanced quality of experience is what constitutes individual and social growth (Bernstein, 1987b; Dewey, 1916, 1933, 1938b).

The moral relations of science are found in the model the scientific community provides for the relationship between the individual and society, between freedom and authority. It is individuals who have new ideas and who are the source of hypotheses, but within the scientific community these hypotheses must be submitted for public testing and evaluation. Individuals must have the freedom to develop and put forward new ideas, which are then judged by collective intelligence. Thus, unlike the older individualistic philosophy that pitted freedom and authority against one another, Dewey's theory held them to be connected. Thus far, according to Dewey, organized intelligence has operated in a relatively small and technical area. But the application to human affairs of "the control of organized intelligence, operating through the release of individual powers and capabilities" offers, Dewey thought, our only real hope for the solution of the problem of freedom and authority (Dewey, 1946/1958, p. 109).

The application of the scientific outlook to human life was appropriate for Dewey, because his reading of evolutionary theory led him to regard human beings as part of nature rather than as standing over and against it (Dewey, 1929). Mind or intelligence, from this perspective, is an adaptation that allows for conscious control of experience. For these reasons, Dewey opposed what he called "the spectator theory of knowledge, . . . the idea that knowledge is intrinsically a mere beholding or viewing of reality" (Dewey, 1920/1948/1957, p. 112). To the metaphor of spectator, Dewey opposed the experimental agent who "proceeds to *do* something, to bring some energy to bear upon the substance to see how it reacts" (Dewey, 1920/1948/1957, p. 113; see also Dewey, 1929/1988).

The mind is within the world as a part of the latter's own on-going process. It is marked off as mind by the fact that wherever it is found, changes take place in a *directed* way, so that a movement in a definite one-way sense—from the doubtful and confused to the clear, resolved and settled—takes place. From knowing as an outside beholding to knowing as an active participant in the drama of an on-moving world is the historical transition whose record we have been following. (Dewey, 1929/1988, p. 232)

Dewey did not argue that "the scientific method" as used in the natural sciences should be applied directly to human affairs, but that "intelligence" or the "scientific attitude" should become part of human life. The scientific attitude involves a willingness to suspend action in the face of a problematic situation and an inclination to engage in inquiry in trying to decide how to resolve the problem. The general description of inquiry Dewey gave varies somewhat in different texts, but always included a puzzling or problematic situation as the impetus to inquiry, a clarification of the problem, the development of hypotheses about how to solve the problem, and their evaluation and testing either imaginatively or overtly through action. If the problem-solving episode was successful, the situation was unified or clarified and no longer, for the moment, problematic (Dewey, 1916, 1933). Scientific method, as thus far developed in the natural sciences, is simply the best means human beings have devised for coping with the environment. Specific methods for studying the human world have not yet been fully worked out, Dewey thought (Dewey, 1920/1948/1957). The methods of the sciences are specialist methods. But the scientific attitude is available to everyone, not merely an elite.

Dewey's "experimentalism," then, was his belief that we should use the method of intelligence in reconstructing unsatisfactory social and political norms. The crucial attitude is a willingness to assess our practices by their consequences. In general, his view was that as we lead our lives, we discover that some of the values we have previously held no longer work for us, in the sense that we cannot live satisfactory lives while continuing to adhere to those values. Our response to such problematic situations should be to initiate an inquiry in which, by examination of the causes and consequences of our present valuing, we come to adopt new values, which are then experimentally tested to see whether life led according to them is more satisfactory. We are able to make judgments about our changed circumstances because not all of our values are being tested at the same time. While some are being questioned, others continue to be accepted and provide the context for evaluation (Brodsky, 1988; Dewey, 1939b).

Science, Dewey thought, helps us to achieve particular aims and interests, especially interests in directing the course of experience. But there are other ways, Dewey acknowledged, for thinking about experience, the aesthetic, for example. And while science deals with the realm of "true-and-false meanings," Dewey held that the

realm of meanings is wider than that of true-and-false meanings. . . . Poetic meanings, moral meanings, a large part of the goods of life are matters of richness and freedom of meanings, rather than of truth; a large part of our life is carried on in a realm of meanings to which truth and falsity are irrelevant. (Dewey, 1929/1988, p. 332; see also West, 1989)

The educational implications of Dewey's experimentalism are straightforward: Children need to acquire the complex of dispositions that constitute the scientific attitude. And for that to happen, schools must become communities in which intelligence is freed for inquiry, places where knowledge is not offered ready made and where the record of knowledge is not mistaken for knowledge (Dewey, 1916). If children are to become genuine inquirers, they must have real questions they want to answer; they must encounter problematic situations other than wondering what the teacher wants. They must be active agents, doing things in order to discover the consequences, not mere passive recipients of facts discovered by others. And teachers, too, must be similarly freed, not mere enactors of methods handed down from above. By Dewey's criteria, efforts at reforming education that involved dictation of subject matter and methods to teachers by scientific experts were doomed to failure. Denying teachers the ability to be self-directing results in mechanical following of rules without interest and in the failure to recruit the best minds to teaching, since no self-respecting intelligence would put up with such conditions. Educational reform requires the freeing of intelligence, not its sub-jugation (Dewey, 1899/1980, 1916, 1938a, 1946/1958; Wirth, 1981).

Radical Democracy

Democracy for Dewey was social, not merely political. "It is primarily a mode of associated living, of conjoint communicated experience" (Dewey, 1916, p. 87). In *Democracy and Education*, Dewey argued that only in a democracy in this social sense could the criteria for the best form of social life be met. His criteria were that the "consciously shared" interests within the group were "numerous and varied" and that inter-action with other groups was "full and free" (Dewey, 1916, p. 83). This meant that there could be no barriers to communication either within the group or between the group and other groups. These criteria, Dewey said, "point to democracy," because sharing common interests means that they will be recognized as "a factor in social control," and free communication with other groups requires the continuous reconstruction of social habits in order to cope with new situations created by such communications. And these criteria are what characterize a democratic society. In other words, a "society which makes provision for participation in its good of all its members on equal terms and which secures flexible readjustment of its institutions through interaction of the different forms of associated life is . . . democratic" (Dewey, 1916, p. 99). Dewey held that democracy was not really an alternative form of community life but was, rather, the completion or perfection of community life itself:

Wherever there is conjoint activity whose consequences are appreciated as good by all singular persons who take part in it, and where the realization of the good is such as to effect an energetic desire and effort to sustain it in being just because it is a good shared by all, there is in so far a community. The clear consciousness of a communal life, in all its implications, constitutes the idea of democracy. (Dewey, 1927/1954, p. 149)

These definitions highlight the importance of communication to Dewey in the creation and maintenance of community life. It is through com-munication that people come to have common interests and beliefs and similar mental and emotional dispositions. And it is "in virtue of the things which they have in common" that people can be said to live in a com-munity. Neither physical proximity nor working toward a common end constitute a community without common interest in, and recognition of, the common end and deliberate cooperation in achieving it. But this, Dewey noted, would require communication. Because such communi-cation is educative, living in a community educates. Both the giving and receiving of genuine communication enrich and enlarge experience.

Experience has to be formulated in order to be communicated. To formulate it requires getting outside of it, seeing it as another would see it, considering what points of contact it has with the life of another so that it may be got into such form that he can appreciate its meaning. (Dewey, 1916, pp. 5–6)

It follows, then, that social arrangements that restrict communication (e.g., through barriers of class or race) diminish the ability of both individuals and groups to grow and to develop an enhanced quality of life. Such groups are not genuinely democratic. And wherever individuals are used without their consent to secure the interests of others, outcomes in which they do not share and had no role in shaping, the human relationships involved are not truly social and remain "upon the machine-like plane" (Dewey, 1916, p. 5). The giving and taking of orders between "parent and child, teacher and pupil, employer and employee, governor and governed . . . modifies action and results, but does not of itself effect a sharing of purposes, a communication of interest" (Dewey, 1916, p. 5).

Dewey saw important analogies between the democratic community and the scientific community in the former's openness to communication, its willingness to reconstruct social practices in light of new experiences, its commitment to the freedom of all to participate in the shaping of social interests and the communal evaluation of social experiments. Like science, democracy in this ideal sense is self-correcting and makes freedom and authority compatible.

The ideal of community described above can be fully embodied, Dewey thought, only in face-to-face associations, units small enough for direct communication among the members. Dewey held that the existence of face-to-face associations is important, not only for practical experience in self-government but, more important, for the creation of "stable loyal attachments" (Dewey, 1939a, pp. 160–161). Dewey noted the destruction of traditional face-to-face communities through the growth of an interconnected industrial economy, rapid transportation, and extended networks of communication. Social ties were loosened and customary loyalties disrupted by a growing regional, national, and international interdependence that produced causal interconnections not understood by ordinary citizens. Unable to appreciate the indirect consequences of their own actions or the way the actions of others affect them, citizens are unable to intelligently control or channel actions in the public interest. But awareness of interdependence, sustained by communication, and actions regulated by knowledge of the actions of others in efforts to accomplish common goals characterize the democratic community. Thus, the majority are unable to play an effective role in political life and cannot prevent private interests from seizing the economic and political apparatus to serve their own private ends. As a consequence, citizens retreat into their private worlds, leaving the public to take care of itself in the sort of individualism predicted by Tocqueville. Under these conditions, the society is an aggregate of interdependent human beings, but it is not a community. "The machine age in developing the Great Society has invaded and partially disintegrated the small communities of former times without generating a Great Community" (Dewey, 1927/1954, pp. 126–127).

Thus, two problems must be faced: the erosion of local communities of loyalty and attachment and the inability of the public to identify itself and act to secure its interests. In Dewey's account of the nature and office of the state, he noted that the indirect consequences of actions of private individuals often have a significant effect on others who are not themselves party to the actions. When this occurs, a public is created composed of those affected by the actions. If they organize to secure their interests and appoint officials to act in their behalf, a state exists. Thus, members of the public will be able to play their part in determining policies only if they are aware of the common interests they share (i.e., only if they are aware of themselves as a public). But this condition does not obtain under the conditions of modern life:

> The machine age has so enormously expanded, multiplied, intensified and complicated the scope of indirect consequences, has formed such immense and consolidated unions in action, on an impersonal rather than a community basis, that the resultant public cannot identify and distinguish itself. . . . The public is so confused and eclipsed that it cannot even use the organs through which it is supposed to mediate political action and polity. (Dewey, 1927/1954, p. 126, p. 121)

In *The Public and Its Problems*, Dewey united the problems of the lost community and the eclipsed public: "Unless local communal life can be restored, the public cannot adequately resolve its most urgent problem: to find and identify itself" (Dewey, 1927/1954, p. 216). In *Freedom and Culture*, he acknowledged that social agencies for dealing with consequences of extensive interdependence could not be "confined to localities." But he thought there needed to be "harmonious adjustment between extensive activities, precluding direct contacts, and the intensive activities of community intercourse" (Dewey, 1939a, p. 160).

What are the conditions for the revitalization of face-to-face communities and the transformation of the Great Society into the Great Community, the emergence of the public from its eclipse? In *The Public and Its Problems*, Dewey placed his faith in the formation of public opinion through the communication of the results of social inquiry into the complex network of interdependencies and consequences. He granted that adequate knowledge of this kind did not yet exist, but he regarded free inquiry and full distribution of results through the public press as an appropriate preliminary step. In *Freedom and Culture*, he acknowledged the power of the press "to create pseudo-public opinion and to undermine democracy from within" (Dewey, 1939a, p. 148). And he noted the belief of some (in his opinion, either thoughtless people or snobs) that the majority of people are not capable of adopting the scientific attitude. But he still believed that the future of democracy depended on "democratic extension of the scientific morale till it is part of the ordinary equipment of the ordinary individual" (Dewey, 1939a, p. 151).

It is hard to characterize Dewey in more specifically political terms, but he appeared to advocate a form of socialism (which he called "public socialism") that honored the values of democracy and inquiry (Dewey, 1930, pp. 119–120). Although his critique of capitalism at times sounded very Marxist, he did not think revolution was required. In place of what he understood as Marxist doctrines of the inevitability of economic progress and the dictatorship of the proletariat, he advocated experimentalism and cooperative political action based on rational persuasion. Social problems were indeed the result of conflicts of interests, but the question was how these conflicts were to be resolved in the interests of all. Dewey insisted that the competing claims could be democratically resolved when the conflicts were made public so that the claims could be discussed and organized intelligence brought to bear. This was possible, despite the Marxists' contrary claims, because there were overlapping interests, and this meant common ground. American society did embody moral understandings that provided common criteria of judgment. It was not ownership of capital goods per se that was at issue, but rather public control of the economy so that it was run in the interest of human values that led to a rich life for all rather than "pecuniary profit" (Dewey, 1930, pp. 118–119). Among these values were a commitment to full employment, to socially responsible businesses, and to meaningful work democratically managed, which would be educative and aesthetically fulfilling as well as socially useful. Campbell argues that democracy took priority over social justice for Dewey. He wanted to end economic exploitation, but more important was the goal of the individual participation of citizens in the economic and political decisions that determined their fate (Campbell, 1988).

Dewey did not think that bringing about a radically democratic culture would be easy. Initially, he called upon schools to foster the scientific attitude in students and to engage students in "joint activity, where one person's use of material and tools is consciously referred to the use other persons are making of their capacities and appliances" in order that "a social direction of disposition would be attained" (Dewey, 1916, p. 39). Individuals, he said, must have a type of education that gives them "a personal interest in social relationships and control, and the habits of mind which secure social changes without introducing disorder" (Dewey, 1916, p. 99). Also, he wrote:

It is not enough to see to it that education is not actively used as an instrument to make easier the exploitation of one class by another. School facilities must be secured of such amplitude and efficiency as will in fact and not simply in name discount the effects of economic inequalities, and secure to all the wards of the nation equality of equipment for their future careers. Accomplishment of this end demands not only adequate administrative provision of school facilities, and such supplementation of family resources as will enable youth to take advantage of them, but also such modification of methods of teaching and discipline as will retain all the youth under educational influences until they are equipped to be masters of their own economic and social careers. The ideal may seem remote of execution, but the democratic ideal of education is a farcical yet tragic delusion except as the ideal more and more dominates our public system of education. (Dewey, 1916, p. 98)

In remarks such as these, Dewey contributed to the progressive legacy of making schools an "essential institution of social policy," as Lazerson (1984, p. 177) argues. But in a later work, *Individualism Old and New*, Dewey feared that the schools, though providing opportunity to some who would otherwise be without it, were merely turning out "efficient industrial fodder and citizenship fodder in a state controlled by pecuniary industry" (Dewey, 1930, p. 127). Given the constraints on schools, economic and political institutions are the most powerful educators, the shapers of fundamental attitudes and beliefs, he thought. They, too, must be changed, not only the schools.

EXPERIMENTALISM AND RADICAL DEMOCRACY: ARE THEY VIABLE IDEALS?

For some, Dewey's ideas of experimentalism and radical democracy seem "simplistic, almost quaint, . . . a melody from the time when the world was young" (Frankel, 1977, pp. 10–11). Dewey's love affair with science and his aspirations for a revitalization of citizen participation in political and economic life seem naive and overly optimistic in light of our experience. And for some critics, Dewey's proposals are not merely naive but are positively pernicious, leading to social engineering or to liberal, incrementalist policies of social change when radical reform is called for. Yet these ideals are not easily abandoned, either, for they lie at the heart of deeply held American aspirations. Our schools still struggle with developing the critical capacities of students and with providing equal opportunity for "all the wards of the nation," a goal still "remote of execution."

This section surveys contemporary efforts to come to terms with Dewey's vision, whether by way of appropriation and reinterpretation or through rejection. The first strand of criticism explored concerns whether Dewey's views about the relationship between science and democratic politics commit him to social engineering and government by elites, despite his intentions to the contrary. Investigation of this charge leads to an exploration of Dewey's model of social inquiry as a means of education and empowerment rather than manipulation. Although Dewey was not an advocate of social engineering, he did stress similarities between scientific communities and democratic forms of social life. Some critics question the appropriateness of this analogy. Conflicts of interests and values must be resolved through political processes, they argue, not by social inquiry. A crucial issue here concerns whether Dewey was overly sanguine about the prospects for rational resolution of conflict. These criticisms are assessed in the second section. The third section pursues further critiques of Dewey's optimism about the role of social intelligence in establishing democratic communities and meliorating social problems. The fourth section considers whether Dewey's call for schools to create a "unity of outlook" among "youth of different races, differing religions, and unlike customs" is compatible with the current emphasis on multiculturalism and diversity (Dewey, 1916, p. 21). The final section presents Dewey's reasons for rejecting Marxism as he understood it and Marxist criticisms of Dewey's brand of liberalism. Contemporary efforts to combine the best in Dewey and in Marx are reviewed.

Scientific Expertise and Politics: Social Engineering?

Does Dewey's conception of the relationship between science and political practice embody a morally and politically unacceptable form of social engineering? That it does has been alleged by many critics. In a recent historical review of critics who represent Dewey as advocating social engineering, Kaufman-Osborn cites Bourne, White, Mills, Hartz, Crick, Santayana, Lasch, and Murphy (1985, p. 831). Featherstone, for example, writes that "Dewey thought of himself as a Populist radical, but from the outside he too often looked like a social engineer, preaching adjustment" (Featherstone, 1972, p. 29; see also Borrowman, 1980). In an extensive recent study of whether the social engineering tenor of contemporary policy sciences is consistent with Dewey's account of the role and nature of social inquiry, Kaufman-Osborn argues persuasively that it is not. To the contrary, he urges that "fidelity to pragmatism requires a fundamental reconceptualization of the practice of social science and a reconsideration of the organization of knowledge and power in a democratic society" (Kaufman-Osborn, 1985, p. 827).

As we have seen, for Dewey, knowledge is to be understood as an adaptive response to a specific problematic situation. For social science, this means that inquiry should grow out of the perception of a specific problem of public life and that the hypothesis discovered should, when acted upon, resolve the initiating problem. While, contrary to what has been claimed by some of his critics, Dewey did not deny that science proceeds by means of theory development and conceptualization remote from commonsense experience, he did insist that science both grows out of the conditions of ordinary life and is responsible to it (Dewey, 1938b; Gordon, 1984). A Deweyan social science should prove its worth, not merely by the causal explanation of social phenomena but by its power to reform social life for the better. Dewey believed that social science knowledge is called for when the public needs to formulate an effective plan of action to control the indirect consequences of private actions in the public interest.

Dewey's account of the role that social science plays within public life may seem compatible with a social engineering interpretation. Kaufman-Osborn finds two (interestingly enough, conflicting) arguments that it is. First, because science represents a knowledge of the relationship between means and ends, not a knowledge of which ends to pursue, social science will in fact serve the ends of those who employ it. In particular, it will serve the interests of those in power. When the ends are furnished not by the people, but by a centralized state, social engineering is the result. Bourne held this view when he criticized Dewey and his followers for their support of World War I. He accused them of "making themselves efficient instruments of the war technique, accepting with little question the ends as announced from above" (Bourne, cited in Kaufman-Osborn, 1985, p. 833). The second formulation is opposed to the first in that it accuses pragmatic social science of supplying the ends for public action, based on its assessment of social "needs." These ends are then implemented through the power of the state.

Whichever of these views is taken, whether social scientists are regarded as masters of technique in the service of state-imposed ends or as suppliers of the ends themselves, there is an illegitimate transfer of power to an elite that has the relevant social science expertise. This transfer is said to be inevitable once political problems are taken to be problems of the rational manipulation of the environment in the interest of human ends rather than the adjustment of conflicting interests through political institutions. This conclusion reinforces a standing complaint against Dewey that he exaggerated the similarities between a scientific and a democratic community (Bernstein, 1985; Frankel, 1977; Greene, 1986; Smiley, 1990). And it points toward what Lazerson (1984) describes as a progressive legacy of pitting democratic politics against the promise of knowledge, to the detriment of both.

The crucial question for assessing this line of criticism, argues Kaufman-Osborn, is, Who will judge the success or failure of a particular effort at social reconstruction? According to social engineering conceptions, it would have to be the experts, because only they have the knowledge required to pass judgment. Ordinary citizens, then, would be held incapable of judging their own interests. But this view is incompatible with

Dewey's democratic commitments, as Kaufman-Osborn notes. First, Dewey argued that the political structures of democracy must be subordinate to democracy considered as a form of social life. The healthy vitality of democracy as community required flexible institutions that could be reconstructed according to the shifting interests of variously emerging and disappearing publics, Dewey thought. Second, Dewey believed that unless the methods of inquiry and the knowledge they produce are made accessible to the citizens who experience the problems applied intelligence is to resolve, the methods and knowledge will be appropriated by a class of experts to serve private interests.

> The new science was for a long time to be worked in the interest of old ends of human exploitation. . . . It put at the disposal of a class the means to secure their old ends of aggrandizement at the expense of another class. The industrial revolution followed upon a revolution in scientific method. . . . But it is taking the revolution many centuries to produce a new mind. Feudalism was doomed by the application of the new science. But capitalism rather than a social humanism took its place. (Dewey, 1916, p. 283)

Thus, Kaufman-Osborn concludes, "the craft of a Deweyan social science [must] be rooted in the social practices that comprise its subject matter" (1985, p. 839).

Kaufman-Osborn locates this grounding in a novel conception of verification that, he argues, was developed by Dewey. As already noted, Dewey held that the testing of the hypotheses of social science must include solving the problems of social life that initiated the inquiry. But this requirement does not, in itself, eliminate the possibility of social engineering. However, Dewey further held that all individuals should be involved in testing theoretical findings and that these findings become knowledge only when distributed throughout the community so as to become an instrument of common understanding and communication (Dewey, 1927/1954, 1938b). Kaufman-Osborn (1985) interprets these requirements as meaning that

> a social scientific hypothesis can be warranted only after the members of the inchoate public to whom it is directed have first accepted it as a plausible account of the causes of their current distress as well as its possible remedy and, second, demonstrated that recognition through collective action which overcomes the difficulties that occasioned this distress. (p. 841)

This interpretation means that if a social science hypothesis fails to make a situation more understandable to its participants, then the hypothesis is falsified. Thus, the logic of social inquiry itself requires a revitalized public. The language in which the hypotheses are offered must be intelligible to the public. The aim of Deweyan science, therefore, "is not to facilitate manipulation but to educate and to empower" (Kaufman-Osborn, 1985, p. 842). It accomplishes this objective by revealing to a public the causes of its current distress and possibilities for action by means of which the distress might be overcome. In this process of collective inquiry, values held by the public, as well as ends to be achieved, undergo scrutiny, not merely the means of achieving preexisting values and ends. Thus, Kaufman-Osborn (1985) concludes that Dewey joined "a hermeneutic conception of social inquiry, i.e., one which takes as its end the clarification of discourse, to a pragmatic deliberation as the method through which a community of agents accumulates the power necessary to social and political reconstruction" (p. 843).

Kaufman-Osborn considers current educational policy proposals in light of this analysis. He holds that federal uniform curricular standards represent a social engineering approach, while another proposal, vouchers, represents an approach that appears to offer more consumer control. But, Kaufman-Osborn argues, neither approach would be a Deweyan one, for neither takes place in the context of a public inquiry into the causes of the present dissatisfaction with education through which the public can become educated to act collectively in changing their situation. Vouchers only introduce a "softer" form of state imposition.

Deweyan representation requires not the aggregation of individual interests through the medium of a public institution like that of the market, but rather the formation of a shared apprehension of a common good—in this case, improved educational services for an entire community—whose achievement may require a partial subordination of the immediate quest for private advantage. The cultivation of civic virtue, not enlightened self-interest, is the fruit of a reconstructed pragmatism. (Kaufman-Osborn, 1985, p. 847)

This interpretation of Dewey's conception of the relationship between scientific expertise and social practices is supported by Garrison's (1988) Deweyan analysis of teacher empowerment. Garrison argues that an "inquiry-oriented paradigm" is the best model of teacher education for those who believe both that teaching needs to have a firm scientific knowledge base and that teachers should be democratically empowered. Garrison borrows from Bertrand Russell the distinction between "scientific temper" and "scientific technique" to explain his position. The scientific temper, like Dewey's scientific attitude, involves the virtues required for scientific inquiry. The scientific temper is

cautious, tentative, and piecemeal; it does not imagine that it knows the whole truth, or even at its best knowledge is wholly true. It knows that every doctrine needs emendation sooner or later, and that the necessary emendation requires freedom of investigation and freedom of discussion. (Russell, cited in Garrison, 1988, p. 490)

"Scientific technique," on the other hand, represents the attitudes of social engineers and the institutions that employ them—"a temper full of a sense of limitless power, of arrogant certainty, and of pleasure in manipulation of even human material" (Russell, cited in Garrison, 1988, p. 490). The use of the products of science (technology and knowledge) by dogmatic authorities in government, industry, and schools denies the development and spread of the scientific temper, of the freedom of thought and communication essential to democracy. It represents a devotion to the products, but not the process, of science, Garrison concludes.

Teachers, Garrison argues, have too frequently been victims of scientific technique, talked down to by authorities who regard them as passive consumers of knowledge. Inquiry-oriented teacher education, in all the current models (action research, scholar-teacher, and critical inquirer), would focus on developing the Deweyan scientific attitude among teachers, thus reconstituting the social relations within schools as democratic by Dewey's definition.

The Political Community and the Scientific Community

Recent reappraisals of the progressive political movement have placed greater stress on the diversity of progressive social and political thought (Crunden, 1982; Tilman, 1984). Although some other strands of progressivism no doubt supported a social engineering conception of the relationship between science and social life, it seems reasonable to absolve Dewey of the charge of social engineering. But his conception of the relationship between science and political practice may be flawed nonetheless. Social inquiry has not proved as helpful in understanding and resolving social problems as Dewey hoped. As Greene (1986) notes, the news media present us with large amounts of information on matters of public interest, from nuclear energy to homelessness. Yet "we seldom hear of intentionally organized collaborative action to repair what is felt to be missing, or known to be wrong" (Greene, 1986, p. 549).

Some have doubts about the ability or interest of citizens in becoming participants in the sort of democratic culture Dewey described. In a recent *New York Times* op-ed piece, for example, Leslie Gelb rejects the efforts of Bill Moyers, the Kettering Foundation, and others to engage "the American people in the conversation of democracy." Instead, he places his hope in the return to public service of community elites:

An inert and uninformed public is a basic problem of democracy that can't be solved. . . . The fate of our democracy rests ultimately on the will and good sense of the people. But it rests daily and practically on the active and responsible involvement in politics of community leaders, people with the backgrounds and time to hold elected officials accountable. That involvement is what has diminished in American democracy. And it is the problem that can and must be solved if our nation is to forge a more effective democracy. (Gelb, 1991, p. A19)

Dewey himself persisted in thinking that the scientific attitude could become widely shared, that the willingness and interest required for contributing to the analysis and solution of social problems could become a commonplace part of the character of citizens. (Whether or not Dewey had an overly optimistic view of human nature will be considered in the next section.)

These considerations raise doubts about our ability to act as Dewey recommends, a serious question by Dewey's own lights, since for him the crucial test of ideas and ideals lies in the consequences that follow from trying to act on them, in particular in their ability to resolve our problematic situations. But from a more theoretical perspective, a foundational question is what explains these practical failures. A substantial group of critics holds that Dewey's analysis fails because it is not radical enough. This point will be discussed below. A second suggestion, as previously noted, is that practical failure results from Dewey's overestimation of the similarities between scientific and political communities. Did Dewey inflate the role that knowledge of the causes and consequences of social practices can play in resolving social problems and downplay the significance of conflicts of interests and of values?

Frankel argues that Dewey perhaps did, at least if he meant something more by his comparisons between science and democracy than just that democracy, like science, met certain basic standards of rationality such as freedom of speech and inquiry, openness of traditions to public criticism and correction, and commitment to collective judgment. "Democracy," Frankel says,

is a procedure for melding and balancing human interests. The process need not be conducted, and at its best is not conducted, without some regard to truth and facts. But a democratic polity is not a university, a scientific discipline, or a debating club. Its controlling purpose is collective action, not the accreditation of propositions as true. . . . A democratic consensus can be a consensus in illusion. . . . The validation of this consensus comes not from its correspondence with putatively independent facts but from its fidelity to state constitutional procedures. . . . Nor does the justification of the validating constitutional procedures involve showing that they are analogous to scientific models of sound procedure. The justification, if there is one, lies in pointing to their efficiency in producing an operative consensus and to their moral and intellectual consequences in the lives of citizens. (1977, pp. 18–20)

I think Dewey would agree with most of the points Frankel makes here. Certainly, he would approve of the view that democratic procedures are to be justified by their consequences for the lives of citizens. And he would concur that the democratic community does not aim at the production of truth, but rather at consensus about what course of action to take in specific circumstances of importance to the public. Dewey does not suppose that citizens will become scientific inquirers in the narrow sense. Nevertheless, he does believe that, given widespread dissemination of the scientific attitude, citizens will be able to make use of the findings of science in their deliberations. But reaching consensus, Dewey held, also will depend on a common set of meanings and values, which are themselves the product of the free communication democracy encourages and are also subject to reconstruction through the application of intelligence. What course of action to take depends not only on our projection into the future of the probable consequences of our actions, but also our evaluation of these consequences as desirable or undesirable.

While Dewey did not assimilate the democratic community to the scientific community, there are two remaining issues implied by Frankel's objections. First, one can wonder whether meanings and values are capable of rational evaluation and reconstruction and what role facts play in such evaluations. The crucial question, Lawler (1980, p. 335) asserts, is whether science can create ends. Dewey (1929) grants that "philosophers have denied that common experience is capable of developing within itself methods which will secure direction for itself and will create inherent standards of judgment and value" (p. 35). Gouinlock argues that when Dewey talked about making moral discourse more scientific, he meant exercising intelligence in determining the causes and conditions of valued events. He was not trying to derive evaluative statements from descriptive ones. Furthermore, he meant that moral inquiry should be public and collaborative, that is, democratic in ways that mirrored investigation in the scientific community. And, Gouinlock (1989) suggests, to intelligence and democratic procedures, Dewey added that the deliberating parties should possess the social dispositions and virtues of the democratic individual. Campbell (1988) suggests that our reluctance to accept Dewey's theory of valuation arises from our preference for believing that the justification for our actions has "come from 'on high' " rather than having "arisen in the give-and-take of political action, situated in a particular context and simplified in the process of debate" (p. 163). (For a discussion that fits this hypothesis, see Bloom, 1987; Bloom, who does believe that values come from "on high," accuses Dewey of preparing the way for "cultural relativism, the fact-value distinction and the neglect of civic culture" [pp. 29–30].) Recent interest within philosophy in nonfoundationalist ethics may make Dewey's theories of greater interest. As Frankel notes, "the new and fertile ground Dewey broke [in ethics and the theory of value] has not yet been fully cultivated" (Frankel, 1977, p. 38; for recent accounts of Dewey's ethics, see Caspary, 1990; Gouinlock, 1989; Noble, 1978).

Second, one might still follow Dewey in believing that values can be rationally evaluated in light of experience and yet wonder whether consensus is to be expected even among those who inquire with the best of will and with proper concern for the common good. In *The Public and Its Problems*, Dewey described democracy, "from the standpoint of . . . groups," as demanding "liberation of the potentialities of members of a group in harmony with the interest and goods which are common" (p. 147). According to Dewey, meeting this requirement means that there must be interaction among groups, since "every individual is a member of many groups" (p. 147). A "robber band" fails to meet the interaction condition, and this, presumably, explains the conflict between its ends and those of other groups. By contrast,

a good citizen finds his conduct as a member of a political group enriching and enriched by his participation in family life, industry, scientific and artistic associations. There is a free give-and-take: fullness of integrated personality is therefore possible of achievement, since the pulls and responses of different groups reinforce one another and their values accord. (Dewey, 1927/1954, pp. 148)

Dewey did recognize that values do not always "accord," that social problems are created by conflicting interests. But he thought such problems could be adjudicated through public debate in light of common interests (Dewey, 1927/1954). He believed that the cause of group conflict is that the interests of some groups are advanced at the expense of others. But when one group is harmed, he thought, so are all the others because of their ultimate interconnections. Hence, we cannot really separate our good from that of others (Westbrook, 1991). An end to conflicts is not guaranteed, but, Gouinlock (1989) writes, "Dewey's procedures . . . might reduce them more than any of the other methods hitherto undertaken" (p. 321).

One difficulty in interpreting and evaluating Dewey's views about the role of intelligence in democratic political practices concerns the various levels and types of communities Dewey refers to as at least potentially "democratic." There are small, face-to-face associations, which are the locus for personality development and the attainment of stable loyalties and commitments, as well as training grounds for democratic participation. These communities, Dewey said, have the greatest prospect for being democratic in the most ideal sense. But there are also various "publics," consisting of "all those who are affected by the indirect consequences of transactions [to which they are not party] to such an extent that it is deemed necessary to have those consequences systematically cared for" (Dewey, 1927/1954, pp. 15–16). Under conditions of modern life, the members of a public in this sense are not likely to be members of the same face-to-face association. And there is "the Great Community," which is, at present, only an aspiration for the "Great Society." In the Great Community "the ever expanding and intricately ramifying consequences of associated activities shall be known in the full sense of that word, so that an organized, articulate Public comes into being'' (Dewey, 1927/1954, p. 184). What is the potential relationship of scientific social inquiry and organized intelligence to each of these?

Kaufman-Osborn (1984) holds that Dewey conceives of the local community as ''the site upon which the knowledge of modern science is to fuse with the virtues that can only be cultivated within intimate surroundings'' (p. 1154). In the context of local communities, scientific findings will be made concrete, while the parochialism and lack of understanding of local community life will be enlarged and enlightened by scientific findings. Thus, the Great Community will find its ultimate justification in the revitalization and enrichment of local associations (Dewey, 1927/1954).

Kaufman-Osborn argues that this is not a coherent vision, because the Great Community, when structured as Dewey recommends, will undermine the local community's ability to perform the moral and political functions Dewey assigns it. First, there is the problem of how the language and findings of science will be made vital in small communities that have been ''invaded and partially disintegrated'' by the Great Society (Dewey, 1927/1954, p. 127). That is, Dewey's ''solution'' seems to require the vitality of the very communities whose disintegration poses the problem science is supposed to help cure. Second, social science findings will reveal to individuals the insufficiency of their commonsense experience for generating the knowledge their own political action in translocal publics requires. Thus, they must recognize their dependence on social science experts. And because political action will be located outside localities, the localities themselves will be drained of political significance and can no longer fulfill their function of teaching meaningful participation. Finally, Dewey's conception of publics requires the ability of citizens to organize and reorganize as social science points to new constituencies of common interest. Consequently, all bonds that restrict the ability to reorganize must be weakened. Citizens must be willing to shed ''the more concrete bonds of family, religion, neighborhood, custom, and so forth whenever these intrude upon their commitment to the communities of interest which they share with others whom they know but cannot see'' (Kaufman-Osborn, 1984, pp. 1158–1159). Thus, the commitments of face-to-face communities must be relatively weak if the political democracy of the Great Community is to be sustained.

While Kaufman-Osborn's arguments are compelling, part of their cogency does depend on supposing that the face-to-face associations Dewey sought to revitalize were local geographic and political communities. That does seem to be what Dewey thought in *The Public and Its Problems*, first published in 1927. But in *Freedom and Culture*, published in 1939, Dewey followed a quotation from *The Public and Its Problems* that ''de-

mocracy must begin at home, and its home is the neighborly community,''
with this commentary:

On account of the vast extension of the field of association, produced by elimination of
distance and lengthening of temporal spans, it is obvious that social agencies, political and
non-political, cannot be confined to localities. But the problem of harmonious adjustment
between extensive activities, precluding direct contacts, and the intensive activities of community intercourse is a pressing one for democracy. It involves even more than apprenticeship in the practical processes of self-government. . . . It involves development of local
agencies of communication and cooperation, creating stable loyal attachments, to militate
against the centrifugal forces of present culture, while at the same time they are of a kind
to respond flexibly to the demands of the larger unseen and indefinite public. To a very
considerable extent, groups having a functional basis will probably have to replace those
based on physical contiguity. In the family both factors combine. (Dewey, 1939a, pp. 160–161)

These comments suggest that Dewey no longer expected geographic communities to be the primary locus of communal association, although he
did not offer examples of what he meant by groups having a functional
basis beyond the family. Furthermore, it is unclear what functional groups
would meet all of Dewey's criteria for community (e.g., holding many
interests in common). (And Kaufman-Osborn's point that the loyalties
formed in intensive communities must be flexible enough to be compatible
with the demands of the larger public was accepted by Dewey.)

The research surveyed here suggests that Dewey's thinking about forms
of human association had a richness that has not yet been fully tapped.
Nor is further work in this area simply an exercise in Dewey scholarship.
For the issues with which Dewey was struggling are still pressing issues:
the relationship between political practice and expertise, the forms of
association that provide meaning and stable attachments in life and their
interaction with wider political groupings generated by common interests,
and the revitalization of a democratic political life.

How should we think about the contribution of schooling to the solution
of these problems? Are schools one of the "functional" groups Dewey
referred to that can take on the role localities once played in creating
"stable loyal attachments"? Or is the primary function of schools to form
the dispositions, the scientific attitude and social concern, that help children become participants in the public world? And if the answer is, as it
so often is in the case of schools, "all of the above," we have to ask how
it is possible for the same association to play both roles.

Optimism, Social Progress, and Human Nature

Dewey believed that experimental intelligence made it possible for
human beings to control their own social evolution, unlike lower life forms
that merely enact hereditary scripts. Caught up in problematic situations,

human beings could clarify the problems, project solutions, and experimentally test them. These potentials could be realized, however, only if the method of intelligence was nurtured in children, supported by the community, and communally regulated. "The distinctive optimism of the reformers in their various fields in the Progressive Era was linked to their faith that embodied 'social intelligence' could even in the midst of momentous economic changes, bring into existence such a community" (Greene, 1988, p. 43).

For some, the experience of the 20th century showed that these beliefs were not viable. Optimism withered with the horror of modern warfare, Hitler's concentration camps where Jews were "scientifically" exterminated, the precariousness of our efforts to control nuclear weapons, the recognition that saving the environment may entail trade-offs with other goods such as jobs, our seeming inability to end problems of poverty and social injustice, and the recognition that a government strong enough to solve social problems can develop goals of its own (Campbell, 1987; Greene, 1988; Radest, 1980). Intelligence no longer appeared adequate when confronted with mysterious and overwhelming military, economic, and political forces that seemed beyond our control. "The claims of human intelligence and of the possibilities of rational morality and coherent policy seemed irredeemably optimistic" (Radest, 1980, p. 320).

Later writers in the pragmatist tradition, including Hook, Mills, Du Bois, Neibuhr, and Trilling, no longer viewed the world as necessarily "hospitable to human aspiration. . . . Pervasive in their writings are a sense of the tragic, a need for irony, a recognition of limits and constraints, and a stress on paradox, ambiguity, and difficulty" (West, 1989, pp. 114, 124).

After World War I, Dewey himself grew more critical of American society and of education. He fully recognized that critical intelligence faced formidable foes:

The reactionaries are in possession of force, in not only the army and politics, but in the press and the schools. The only reason they do not advocate the use of force is the fact that they are already in possession of it, so their policy is to cover up its existence with idealistic phrases of which their present use of individual initiative and liberty is a striking example. (Dewey, 1946/1958, p. 139)

Nevertheless, he still made it clear that the method of reform indicated is "maximum reliance upon intelligence," because when force is used the ends become compromised. However, he granted that

when the forces in possession are so blind and stubborn as to throw all their weight against the use of liberty of inquiry and of communication, of organization to effect social change, they not only encourage the use of force in those who want social change, but they give the latter the most justification they ever have. (Dewey, 1946/1958, p. 139)

Were Dewey's continuing commitments to the method of intelligence and his hopes for a democratic community unwarranted? Some arguments that they were focus on Dewey's assumptions about human possibilities. Some see human beings as fundamentally selfish. Neibuhr said that "there are definite limits in the capacity of ordinary mortals which make it impossible for them to grant to others what they claim for themselves" (Neibuhr, cited in Campbell, 1987, p. 99). Another challenge concerns whether human beings are intelligent enough to make Dewey's scheme work. Krutch thought that humans had developed conditions beyond their capacities for control (Campbell, 1987). Social science is, on this view, unlikely ever to possess the capabilities for informing social action that Dewey's theory demands. And even if it could, some doubt the ability of the people to adopt the scientific attitude, that is, to be willing to order their lives by experimentally derived standards and values. Dewey assumed, says Lawler, that "the severe rational self-control of the scientist, who eschews all dogmatism and overcomes the metaphysical desire for certainty and security, is what every man ought to imitate in pursuing his most intense and most natural satisfaction" (Lawler, 1980, p. 334). Lawler, an existentialist, finds this a deficient conception of human beings, who are "irrational . . . spontaneous, impulsive, and vital" as well as rational. According to Lawler, Dewey's theories ignore or depreciate the value of solitude, contemplation unrelated to ordinary, everyday concerns or technical control, the inevitability of conflict, the tension between the individual and the social whole, and the significance of death (which may be the source of the need for certainty and security). These views are echoed by Peters (1977), who holds that human beings have a nature that must be simply accepted, in wonder and awe, rather than always transformed.

In a general defense of Dewey against such arguments, Campbell (1987) responds that Dewey never believed in the inevitability of progress or that nature is on our side, but rather that success in our efforts to overcome our social ills is possible. The world, Dewey granted, was a precarious place, and any specific intervention may produce new problems, or old problems thought solved may recur in the future in new forms or on a different plane. Dewey did not suppose that intelligence will necessarily save us from destruction, but only that it has a higher probability of success than the alternatives of unexamined social habits or caprice. But Dewey did have a faith, Campbell grants, "in human intelligence and in the power of pooled and cooperative experience. . . . If given a show, they will grow and be able to generate progressively the knowledge and wisdom needed to guide collective action" (Dewey, 1946/1958, p. 59). This faith stemmed from his faith in democracy, since Dewey thought the alternative to the authority of collective intelligence was rule by a "superior" few.

Campbell (1987) calls Dewey's commitments here a "faith" because he holds that there are no circumstances under which Dewey would have given up these commitments or regarded them as unwarranted. This conclusion does not seem to square well with Dewey's experimentalism. Could we conclude, after giving it a trial, that the commitment to intelligence (as Dewey understood it) is not warranted by our experience? Is pessimism more "realistic" or experimentally warranted? Radest reaches similar conclusions to Campbell's. He says that, for as long as we try to improve our situation, we "must assume the continuing possibility of the better in experience. From that point of view, progress is a political *stance*, . . . not a cognitive claim" (Radest, 1980, p. 336). McDermott argues that the proper way of articulating the pragmatist's commitments is that they espoused the virtue of hope, rather than faith. Neither optimism, nor pessimism, he suggests, is the proper response. The attitude of hope enjoins us to strive for improvement without believing in its inevitability or in the possibility of perfection (McDermott, 1984).

When Dewey's vision is viewed broadly enough, his claims about the value of social intelligence have considerable plausibility. As Featherstone (1972) says, "What else do we have to go by but science, broadly understood—the appeal to evidence, to logic, to inquiry validated by open public processes of criticism and debate?" (p. 32). Yet, even this much has been doubted by social groups ordered by traditional understandings of how life should be led, traditions that are not thought to be open to experimental modification. In this sense, Dewey's ideal is modernist, as he knew full well. One of the issues raised by this aspect of experimentalism will be considered in the next section on cultural pluralism. For now, it seems to me correct to suppose that Dewey did believe that experience warranted the experimentalist outlook in that a better quality of life could be led in societies that adopted it. Furthermore, he argued that absolute standards of conduct put forward as eternal truths always, in fact, embody the prejudices and interests of particular individuals and groups (Gouinlock, 1989, p. 312).

Several of the specific positions that have been referred to as constituting part of Dewey's "optimism" do seem worthy of further thought and study. How extensive a commitment to the common good can reasonably be expected of people when the common good appears to conflict with local loyalties? Does the theory need to take account of wider dimensions of human personality? Is extensive civic participation experienced as a good by many even under optimum conditions? Is living in harmony with nature an alternative to progressive "improvements"?

Some recent work has attempted to use Dewey's perspectives on religious and aesthetic experience and on nature to address some of these questions. Burnett (1988) suggests that Dewey's educational philosophy can be enriched by his work on art and religion, perspectives that he worked out after most of his major educational writings had been completed and that therefore were never fully integrated with his educational thought. Burnett takes the function of both art and religion in Dewey's thought to be the elucidation of what Dewey called the "enveloping whole." In *Art as Experience*, Dewey (1934b) argued that every experience has an "undefined pervasive quality . . . which binds together all the defined elements . . . of which we are focally aware, making them a whole" (p. 194). Art elicits a special form of experience (aesthetic experience), the

> quality of being a whole and of belonging to the larger, all-inclusive, whole which is the universe in which we live. This fact . . . is the explanation of that feeling of exquisite intelligibility and clarity we have in the presence of an object that is experienced with aesthetic intensity. (Dewey, 1934b, p. 195; see also Dewey, 1934a)

Burnett suggests that the Deweyan concept of the child as an "embryonic inquirer/scientist" be supplemented with an aesthetic/religious dimension that would develop the child's capacity for "aesthetic experience," which conveys a sense of the "enveloping whole."

In what may initially seem a surprising move, some environmentalists have argued that Dewey's philosophy is a potential source for the development of an environmental ethic (Chaloupka, 1987; Colwell, 1983, 1985; French, 1980). Chaloupka acknowledges that some may think that pragmatism is just what an environmental ethics is against. An emphasis on "progress" and the application of science to human affairs may appear to be the problem rather than the solution. But Dewey's naturalistic metaphysics has been appropriated by some environmentalists to further their philosophical project of overcoming the ontological dualism between human beings and nature, in order to situate the human species within nature while still maintaining full awareness of human dignity.

Colwell (1985) holds that Dewey's naturalism expresses a biocentric ecological perspective that sees "human life as part of a shared and mutually sharing earth household" (p. 255). In *The Quest for Certainty*, Dewey (1929/1988) wrote: "Mind is no longer a spectator beholding the world from without. . . . The mind is within the world as a part of the latter's own on-going process" (p. 232). Colwell (1985) finds, in Dewey's account of learning as awareness of organism-environment interactions through observation, reflection, and manipulation, an account of science that keeps "science close to human life and the rest of nature" (p. 262). Colwell suggests that the ecology movement is an example of Dewey's ideas of "social and moral self-correction" and, hence, believes that these ideas should not be dismissed as utopian. As a result of the ecological movement, industrial practices have been questioned and modified, recycling efforts increased, and attitudes altered.

Cultural Pluralism and Diversity

Is Columbus Day an occasion for *celebration*? Should it be renamed "Native American Day"? What should children be taught about Columbus in our history classes? What should every "culturally literate" American know about Columbus? Debates about the social studies curriculum are part of a larger debate taking place at all levels of our educational system, in the struggle to acknowledge previously suppressed voices of oppressed peoples. Should a multicultural curriculum have as its primary focus promoting awareness of, and tolerance for, cultural differences within the context of an assumed commonality? Or should its goal be empowerment, defined as forging "a coalition among various oppressed groups as well as members of dominant groups, teaching directly about political and economic oppression and discrimination, and preparing young people directly in social action skills?" (Sleeter, 1991, p. 12). Or is the proper focus of schooling on *unum* rather than *pluribus*, on a core curriculum and common values?

Obviously, Dewey did not address this controversy in its present form. In fact, Eisele (1983) points out, it is remarkable that he commented as much as he did on relevant issues, given that "Dewey was only four years shy of sixty when Horace Kallen first used the term 'cultural pluralism' " (p. 154). Nevertheless, there appears to be much in Dewey's philosophy that is relevant to the questions of diversity. As a good evolutionist, Dewey had reason to value at least some forms of diversity, for without variety a population has a greatly diminished chance of successfully adapting to a changed environment. The social analogue to genetic diversity for Dewey was variety in ideas, since "diversity of stimulation means novelty, and novelty means challenge to thought" (Dewey, 1916, p. 85). That is, encountering others who think differently can be the stimulus to critical reflection on one's own beliefs and practices. The question, therefore, is not whether Dewey valued diversity, but rather what kinds of diversity are compatible with his social and political outlook.

In a 1983 article, Eisele reviewed previous interpretations of Dewey's concept of pluralism. He found that Dewey had been held by some to be a supporter of cultural pluralism, while others thought him an Americanizer or supporter of the melting pot view. Eisele examines Dewey's writings on immigrants as a way of gaining access to his views on pluralism. In a 1915 letter to Horace Kallen, Dewey rejected assimilation to Anglo-Saxon culture as the proper goal for immigrants:

I want to see this country American and that means the English tradition reduced to a strain along with others. . . . I quite agree with your orchestra idea, but upon the condition we really get a symphony and not a lot of different instruments playing simultaneously. I never did care for the melting pot metaphor, but genuine assimilation *to one another*—not to Anglo-Saxondom—seems to be essential to an American. (Dewey, cited in Eisele, 1983, p. 151)

Eisele (1983) concludes: "This letter shows Dewey's pro-cultural pluralism stand (a variety of instruments), as well as his anti-melting pot and anti-Americanization feelings, and it reflects Dewey's opinion on . . . the value of immigrant culture" (pp. 151–152).

According to the sources Eisele assembles, Dewey affirmed cultural variation in literary and artistic traditions, regarded respect for cultural diversity as analogous to respect for the individuality of persons (which he certainly affirmed), regarded cultural diversity as necessary for "give and take" among people, and treated "nationality" as social rather than political and, hence, as separable from citizenship in the way in which church is separate from state. But it is unity more than diversity that is the problem Dewey addressed and the function he assigned the schools. He expressed concern that the "hyphens" in expressions such as "Polish-American" must connect rather than separate, for example. This reinforces his concern in the Kallen letter quoted above that the collection of different cultural instruments in our national "orchestra" result in a "symphony" and that "assimilation to one another" is "essential to an American." The schools are charged with developing a sense of the common problems Americans face, problems that, to be solved, require the cooperation of all. Nevertheless, it seems fair to say that he sought unity, not *uniformity*. (For a balanced discussion of Dewey's study of the Polish community in Philadelphia and its bearings on his conception of pluralism, see Westbrook, 1991.)

Responding to Eisele, Maxcy (1984) describes Dewey's ideal as "cultural diversity with intergroup harmony" (p. 302). Maxcy emphasizes that Dewey's pluralism is not an ethnic group pluralism that, in Maxcy's interpretation, Dewey would have thought limiting to the possibilities of new group formations and individual freedom of association. According to Maxcy, Dewey's support of comprehensive public high schools and his rejection of ethnic schools and vocational schools indicates that the school was, for Dewey, an agency for developing the Great Community. Thus, Maxcy rejects Eisele's claim that Dewey was a cultural pluralist if "culture" means "ethnic group." Maxcy believes that, for Dewey, pluralism had to be compatible with, and contained within, a democratic form of life that places a premium on free interaction between social groups and breaking down barriers to communication.

While Eisele notes that Dewey's pluralism was set within a democratic context, Appleton (1983) argues that there is a conflict between the ideals of democracy and cultural pluralism. Citing Dewey's claim that "democracy . . . means voluntary choice, based on an intelligence that is the outcome of free association and communication with others," Appleton argues that there is a "potential for groups to restrict an individual's opportunities and autonomy," despite the enriching role a diversity of groups can play in a democratic culture (p. 153). The more effectively a group succeeds in transmitting its culture to a new generation, the lower mobility between groups is likely to be: Traditional affiliation works against voluntary association. Appleton argues that a democratic pluralism of groups formed around common interests is compatible with democracy and individual choice, but not a *cultural* pluralism.

A second problem Appleton introduces is that acquiring the dispositions Dewey favored (experimentalism, for example) will tend to destroy traditional subcultures such as the Hopi of northeastern Arizona and the Amish, Mennonite, and Hutterite religious groups when these dispositions run counter to the beliefs on which the groups are based. Traditionally, such groups have been protected when they pose no threat to the larger society, Appleton says. But requiring all children to adopt the democratic ideal would violate this agreement. (Holton, 1981, adds "Christian fundamentalism" to the list of views challenged by Dewey's experimentalist ethics.)

The conflict between Dewey's experimentalism and traditional bases for social practices emerges as a substantial problem in the literature on Dewey and cultural pluralism. McDermott (1984) argues that "cultural imperialism is an abomination" (p. 671) and looks to pragmatism as a way of articulating a viable cultural pluralism. Yet this is a pluralism that requires a substantial common commitment:

The task, of course, is for each culture and subset thereof to be open to the experience, originality, and value of other cultures. This is possible only if individuals are taught from the outset to be open to experience and to avoid judgments based on conceptual a priori assumptions, inherited truths legitimated only by longevity or by authority. (McDermott, 1984, p. 671)

Commenting on McDermott's remarks, Pruitt holds that the attitudes favored by Dewey and the pragmatists are not "clearly universally adoptable options." He identifies pragmatic themes of pluralist meliorism and tolerance with American Protestantism and concludes that "the gift of pragmatism to non-Protestant nations, then, becomes an invitation to conversion, or disintegration" (Pruitt, 1984, p. 677; see also Ichimura, 1984). This view is shared in some measure by Sarlos, who rejects Dewey's "narrow and dogmatic insistence on scientific method as sole arbiter of social change" (Sarlos, 1983, p. 161; see also Itzkoff, 1969). She suggests that there are "existential and intuitive avenues toward attaining and retaining values" that can also serve as means to the development and maintenance of culturally diverse values (p. 161).

In a similar vein, Bowers (1987) argues that Dewey failed to fully recognize the effects of cultural traditions on his own thinking, traditions that led to his privileging the "ideological" assumptions of social progress, a belief in the ability of rational thought to control social life, faith in the scientific method, and a belief in the ability of society to cope with an increasing rate of social change. In particular, Bowers argues, Dewey failed to realize the positive benefits of tacit forms of knowledge as an authority in human life. Tacit forms of cultural practices and beliefs provide the foundation of social life, Bowers asserts. There are, he suggests, limits to basing community on "explicit and politically negotiable understandings" (p. 119). Bowers concludes that Dewey failed to recognize that "intelligence, as a form of consciously directed activity, is always limited and never entirely free from the implicit assumptions of the culture" (p. 120).

Furthermore, Dewey's assumption that all the background tacit assumptions of a form of social life should be open to experimentation, to being reassessed in light of the consequences of these beliefs and arrangements, is politically problematic, Bowers says. Dewey simply assumed that a new consensus can be generated on the basis of the evidence uncovered by experimental inquiry, that people can be rationally persuaded to change their beliefs and to reach agreement. Bowers holds that there are widely divergent belief systems in the world, systems not necessarily inferior to the forms of knowledge generated by scientific inquiry. Dewey's commitment to experience and intelligence as the basis of authority can be reasonably challenged by those who hold to other bases of social authority. Because Dewey had little appreciation of tradition as a basis of social life and authority, he presumed that all aspects of inherited traditions can be put to the test of the method of intelligence, which then

enables individuals to escape from their own historical embeddedness. . . . Dewey's stress on reconstructing experience causes students to misread their powers of rational self-direction, as well as the degree their lives represent the "end-state of a sequence of transmissions and modifications" that connects them to the cultural achievements of the past. (Bowers, 1987, pp. 123–124)

Bowers draws two educational implications from this critique. First, while all aspects of culture are theoretically open to experimental test, Dewey held that inquiry begins with a problematic situation. Thus, Bowers claims, those tacit aspects of culture that seem too self-evident to be discussed will largely remain unexplored. But it may be these very beliefs that are, in fact, responsible for the social problems we experience, such as ecological damage. Thus, students should be taught how to decode culture and to reveal background assumptions, not simply the techniques of problem solving (see Miranda, 1984, for a similar argument). Second, because the scientific techniques Dewey championed can be, and are, used by experts to advance special interests, education must conserve traditions that promote resistance to the technological manipulation of the political process. Bowers believes that "knowledge of the traditions that are a source of meaning, humane social relationships, and material and spiritual well-being, is not acquired through Dewey's problem-solving approach to education" (1987, p. 125).

Was Dewey a pluralist and, if so, of what sort? More important, does Dewey's vision offer any guidance in our present deliberations about diversity? Dewey, as Eisele (1983) notes, rejected the dualism of unity *or* pluralism in favor of "unity with pluralism" (p. 155). Ethnic group pluralism, or other forms of diversity based on tradition, seem compatible with the unity Dewey sought insofar as they are consistent with the conditions of democratic life, with a willingness to interact and communicate with other groups about common interests and to abide by the judgments of organized intelligence in deciding how to solve common problems. Contrary to Bowers's claims, Dewey certainly understood that much of social life is patterned by inherited traditions. He had no desire to encourage individuals to "escape from their historical embeddedness," since such escape would not make sense within the context of his theory. For him, the question was whether we can reconstruct our traditions in light of our experience in living our lives according to them. Inquiry is initiated by problematic features of experience, not for its own sake. Bowers himself seems to favor such reconstructions in his remarks about how deep cultural assumptions might need to be examined in order to solve the problem of ecological damage. It is true that Dewey's advocacy of standards of judgment derived from experience conflicts with the demand for continued adherence to standards based on other sources of authority, as Dewey himself made plain. He put forward experimentalism as an alternative to such bases of authority. Whether he was correct—that intelligence is a better guide than the available alternatives—is for us to judge in the give and take of communication. To silence or coerce those who think otherwise would surely be contrary to Dewey's claims that democratic means must be used if we are to achieve democratic ends. Bernstein says that the practical problem of pluralism for Dewey is "working toward a type of society in which we can at once respect and even celebrate differences and plurality but always strive to understand and *seek* a common ground with what is other and different" (Bernstein, 1987b, p. 521). This is the vision Bernstein thinks we can recover and must seek new ways to embody, amid the current fluctuation between concern about radical differences and celebration of them.

The literature on Dewey and pluralism focuses on conflicts between experimentalism and traditional authority rather than the questions of oppression that have dominated recent discussions (although cultural oppression represents a common theme). Since the goal of radical democracy is a state of affairs in which no individual uses "another so as to get desired results, without reference to the emotional and intellectual disposition and consent of those used," Dewey's theories encourage struggles to end oppression (Dewey, 1916, p. 5).

Some theorists have explored the implications of Dewey's political and educational views for feminism. In a 1975 article based partly on Dewey's correspondence with Scudder Klyce in which Dewey defended Charlotte Perkins Gilman's feminist views, Boydston declared Dewey "the philosopher of the common woman" as well as the common man (Boydston, 1975, p. 441). Recent work has been more equivocal. Sherman (1984) argues that Dewey's aesthetic theory is "genderist" (assumes the superiority of gender-specific traits), while Miranda (1980) finds in Dewey's work both "radical implications" for feminist theory and a "conservative influence" (p. 197). Miranda credits Dewey with attempting to de-genderize the virtues of the domestic feminist's service ideal by making them an aspect of his concept of democracy. For Dewey, freedom was the power to take part in a democratic community in which one could participate with others in determining the conditions of life. Thus, he linked power and sharing, features that are often regarded as separate aspects of experience. Instead of dividing society into those who are in control and those who serve, Dewey viewed power and service as shared in democratic participation.

However, Miranda (1980) finds problematic aspects to this synthesis. The effect of Dewey's definition of power was "to make coercion and confrontation illegitimate strategies of change" (p. 200). Thus, women teachers, influenced by the service ideal, reacted to 20th-century bureaucratic intrusions on their autonomy by trying to bring about democratic change through building community and character (i.e., through moral reform). As a result, women are blamed for the lack of full professional status for teachers because of their "passivity." The domestic "virtues" Dewey sought to de-genderize have come to be regarded as vices. "Dewey's questions of how qualities valued in the school house become 'forces' in the world still face the profession" (Miranda, 1980, p. 201). The concepts of service and power that Dewey sought to unite continue to be disjoint.

Noting that Dewey's major writings do not discuss his views on feminism, Laird examines his sparse writings on coeducation with a feminist eye. (These works consist of two unpublished statements to William Rainey Harper, president of the University of Chicago, and three nonphilosophical articles.) Laird concludes that the record of the Dewey school compiled by Mayhew and Edwards gives much more serious attention to women's education than any of Dewey's writings on coeducation. She argues that the "experimental practice [of the Dewey School] was radically and consciously *coeducational*, indeed far more dramatically so than almost any school practices today" (1988, p. 115). A consciously planned curriculum avoided typical sex segregation (cooking for girls and carpentry for boys, for example) and stereotyping. She faults Dewey for not having developed a theory that expressed the best practice of the School. Indeed, Laird argues, his published views on coeducation support gender divisions of labor effectively denied by the practice of the School. She does, however, believe that we should "read earnestly egalitarian intentions in the gender-neutrality of his [Dewey's] better-known thinking and writing on education" and that Deweyan thought can be helpful in thinking about contemporary coeducation "once women and gender become visible within the philosophic tradition" (1988, pp. 128–129).

Dewey and Marx

Around the time of Dewey's death in 1952, he was simultaneously denounced by Friedrich A. Hayek as "the leading philosopher of American left-wingism" and by Harry K. Wells as "the leading philosopher of U.S. imperialism" (Campbell, 1988, p. 119). Such contradictory charges were possible because Dewey opposed both capitalism and Marxism as he experienced them. Dewey and Marx had similar analyses of the effects of capitalism on the possibilities of a democratic politics. Dewey agreed with Marx about the importance of economic factors for political operations (although he did not think that *only* economic factors were important), and that capitalism was adverse to democratic freedom. Dewey rejected the two major defenses of capitalism: that capitalism was "natural" and that it worked. Dewey did not believe that capitalism was somehow grounded in human nature. He held that it simply reflected choices about how to organize labor and capital, not nature itself. The transformation of social relations was not precluded by unalterable human dispositions; rather, those dispositions were themselves the result of present social relations and could be different under different conditions. Thus, Dewey agreed with Marx about there being considerable plasticity to "human nature." And Dewey argued that capitalism was actually *in*efficient, because it sought to increase profits rather than advance the common good and because it prevented coordinated social planning. Furthermore, capitalism was responsible for the insecurity of unemployment. And such employment as there was, for many, brought no meaning or intrinsic reward, only service to machines in the interest of a leisure of bodily excitation. Most significant, capitalism was undemocratic. Political relations could not be equal so long as disparities in wealth meant that the majority had no control over the conditions of their own existence (Campbell, 1988).

It is widely agreed that Dewey was no scholar of Marx's work (Campbell, 1988; Manicas, 1988b; McBride, 1988; Tilman, 1984; West, 1989). In explaining why, Campbell mentions the low regard Dewey's contemporaries had for Marx as a thinker, the fact that some crucial works by Marx did not become available in the United States until after Dewey's death, the negative example of Marxism provided by the Soviet Union, and the American Communist Party's support of Leninism. Nevertheless, despite his failure to read Marx, Dewey followed closely the fortunes of Marxist movements and wrote extensively about them. From the point of view of Dewey's instrumentalism, it seems fitting that he would judge ideas by their consequences when set loose in the world. From this perspective, what Marx said is less important than what Marxists did (McDermott, 1984; Tilman, 1984). Thus, when Dewey wrote about Marxism, he had in mind, philosophically, "Second International Marxism," influenced by Georgi Plekhanov's monist philosophy of history, and, politically, Leninism and the doctrine of the "Vanguard Party" (Manicas, 1988b, p. 158). He did not distinguish Soviet policy from theoretical Marxism. Thus, a common complaint about Dewey's criticisms of Marxism is that they are correct with respect to various formulations of Marxism but miss the mark when applied to Marx himself (Brodsky, 1988; McBride, 1988).

Dewey rejected several features of Marxism, as he understood it. (For Dewey's critique of Marxism, see Dewey, 1939a. For discussions of Dewey's views, see Bien, 1980; Gavin, 1988; Giarelli, 1982, 1986; Lichtenstein, 1985; Manicas, 1988a; Moreno & Frey, 1985).

1. *Historical inevitability.* Dewey saw a contradiction in the efforts of Marxists to take action to determine the course of history while regarding its outcomes as inevitable. He also found it ironic that Marxism itself was an excellent example of the power of ideas, yet denied that they had any efficacy.

2. *Economic determinism.* As we have seen, Dewey recognized the importance of economic factors in social life, but he did not believe that economics was "*the* cause of *all* social change" (Dewey, 1939a, p. 76). He rejected the doctrine of economic determinism in favor of a pluralistic view that social change could be the product of mores, art, science, and so forth, as well as economics. But Dewey did hold that in the American society of his day, the economic factors were the most significant ones. Even so, however, he thought that it was our culture that *made* the economic factor primary, not that our culture was determined by the economic factor. And he did not think that state action always coincided with the interests of capitalists:

The Marxist theory holds that government in so-called democratic states is only the organ of a capitalist class, using legislatures, courts, army and police to do its will and maintain its class supremacy. But the effect of constant criticism of governmental action; of more than one political party in formulating rival policies; of frequent elections; of the discussion and public education that attend majority rule, and above all the fact that political action is but one factor in the interplay of a number of cultural factors, have a value that critics of partial democracy have not realized. (Dewey, 1939a, p. 94)

3. *Lack of experimentalism.* Dewey regarded the theses of historical inevitability and economic determinism as false. But the primary difficulty he had with them was that Marxists held them to be a priori truths rather than hypotheses open to experimental verification. "In the name of science, a thoroughly anti-scientific procedure was formulated, in accord with which a generalization is made having the nature of ultimate 'truth,' and hence holding good at all times and places" (Dewey, 1939a, p. 87). In this respect, communism was like a religion in being a body of fixed dogmas. This a priorism had the unfortunate consequence of making its proponents blind to experience; they failed to examine the actual course of events to uncover specific causal connections. Facts that did not conform to the party line were ignored.

Scientific method in operating with working hypotheses instead of with fixed and final Truth is not forced to have an Inner Council to declare just what is the Truth nor to develop a system of exegesis which rivals the ancient theological way of explaining away apparent inconsistencies. It welcomes a clash of "incompatible opinions" as long as they can produce observed facts in their support. (Dewey, 1939a, pp. 97–98)

4. *Class struggle, violence, and the dictatorship of the proletariat.* Dewey rejected class struggle and violence as the necessary mode of radical social change and the dictatorship of the proletariat as an intermediate goal. Dewey thought Marxists' division of society into two classes, the owners of the means of production and the laborers, the producers of surplus value, was simplistic. Sharp separation of social classes is an abstraction that fails to fit with the overlapping interests of actual individuals. Dewey rejected violence because he believed that democratic ends could only be brought about by democratic means. "Democratic methods are . . . indispensable to effecting economic change in the interest of freedom" (Dewey, 1939a, p. 93; see also Dewey, 1935). On similar grounds, Dewey rejected the intermediate goal of the dictatorship of the proletariat as totalitarian and nondemocratic, as involving the denial of civil liberties.

As we have seen, Dewey believed that there could be peaceful evolution toward the common good if action were directed by intelligence. Dewey acknowledged that the benefits of economic life were being appropriated by a few and that the economic system had as its end the maximization of personal profit rather than the benefit of all. But the cause of this state of affairs, he thought, was the unexamined persistence of traditions and institutions that supported capitalism. Given wide dissemination of relevant information and the application of collective intelligence, radical change could be brought about through peaceful means.

Furthermore, Dewey did not believe that ending capitalism would end social conflict. Adjudicating conflicts among the interests of various groups would require strong democratic institutions and dispositions. Since the "public" acts always through the hands of individuals, how can it be assumed that public ownership will be in the public interest? How we eliminate capitalism is as important as its elimination.

Institutions and habits of democratic interaction must be in place first; and, consequently, democracy should be our primary social value. . . . For the Marxists, the creation of a just, egalitarian society through the destruction of capitalism was the necessary prelude to effective and lasting democracy. For Dewey, justice could be established and defended only if the political system was antecedently democratic. (Campbell, 1988, p. 138; see also Manicas, 1981)

Marxists agree with Dewey's goal of a radical democratic culture; they acknowledge that Dewey's heart was in the right place. But they criticize both his analysis of what has prevented us from reaching that goal and the means he proposed for getting there. "Dewey's idea that 'coercion and oppression on a large scale exist' because 'of the perpetuation of old institutions and patterns not touched by scientific method' is patently fallacious," asserts Manicas (1988b, p. 166). Marxists hold that oppression is not the result of traditional moral outlooks, for example, that justify the actions of capitalists, but rather is the outcome of deliberate exploitation. Those who benefit from the current system will not negotiate away their interests. "Marxists," says Manicas, "were not so foolish as to suppose that the lions, the finance capitalists, would sit down with the lambs and 'adjudicate' away their privileged power" (1988b, p. 167). Dewey's preferred means of education and dialogue presumed that there are no conflicts of interests, and no divisions of structure and power, that cannot be overcome by dialogue (West, 1989). But from a Marxist perspective, "attempts to reason with these individuals [capitalists] through 'cooperative inquiry' will only allow them to continue in their privileged status" (Campbell, 1988, p. 140). Social struggle, not cooperative inquiry, is the way to end oppression.

Some Marxists agree with Niebuhr's position that this struggle requires firm conviction reinforced by emotionally powerful slogans and symbols, not the cool experimentalism of Dewey:

Contending factions in a social struggle require morale; and morale is created by the right dogmas, symbols and emotionally potent oversimplifications. . . . No class of industrial

workers will ever win freedom from the dominant classes if they give themselves completely
to the "experimental techniques" of the modern educators. They will have to believe rather
more firmly in the justice and in the probable triumph of their cause, than any impartial
science would give them the right to believe, if they are to have enough energy to contest
the power of the strong. (Niebuhr, cited in Campbell, 1988, p. 140)

Obviously, from Dewey's point of view, Niebuhr's position amounts to
manipulation of people, not their democratic liberation. But, says Manicas, if Marxists had the problem of developing revolutionary consciousness among the workers, Dewey had the problem of how to develop it
among the citizens: Dewey's solution to economic and political oppression required the existence of a public that can apply the method of
intelligence in social reconstruction. But Dewey himself showed why
such publics are lost and "eclipsed" and can no longer recognize themselves. Hence, "creative intelligence . . . lacks institutions" (Manicas,
1988b, p. 167).

Yet Manicas concludes that

what is now needed . . . is to renew the possibilities of democratic politics which acknowledges the insights of Marx but yet strips Marxism of the idea that history is on the side of
emancipation. We need, that is, to combine the best in Marx and Dewey. (Manicas, 1988b,
p. 170)

What can we do? Manicas suggests taking every opportunity we can to
bring publics into existence, to create communities by revealing common
goods that can be achieved through joint participation. But how, on his
own analysis, are we to do this? A "Marxist understanding of what is
happening to us and why" may make it possible to "take advantage of
opportunities . . . to build some incipient but progressively growing democratic publics" (Manicas, 1988b, p. 170). But does not the Marxist analysis require us to use nondemocratic means to reach democratic ends,
and is not that just what Dewey tells us will not work in the creation of
a democratic culture?

Assessments of Dewey's work by Marxist scholars have undergone a
transition from wholesale rejection to a perception that there may be
something to be gained from combining the best in Marx and Dewey. In
general, this change in opinion has accompanied a transition in emphasis
in Marxism from deterministic theories to ones that place greater emphasis on the power of human agency. A hybrid theory would take from
Marx his understanding and analysis of the dynamic of power in capitalist
America, as well as his emphasis on the need for radical means to effect
real change. Dewey's contribution lies in his understanding better than
Marx the cultural problem of creating a social space within which various
publics can find common ground by engaging in dialogue (i.e., the problem

of developing a radically democratic culture). (See West, 1989, and Gavin, 1988. For dissenting opinions, see Paringer, 1990, and Gonzalez, 1982. Paringer suggests that the radicals' use of Dewey may be simply a political tactic designed to enlist the support of the less radical and to counter conservative influence by working from within the larger sphere of liberalism. While this claim may be correct in the case of some individuals, the textual evidence seems to me to suggest serious interest in Dewey's work on the part of those surveyed here.)

In trying to assess Dewey's views about the role of intelligence in bringing about radical economic change, we might think about whether "intelligence" necessarily omits "confrontational politics and agitational social struggle," as West asserts (1989, p. 102). Dewey's own commitments to intelligence were embodied in concrete proposals and actions such as the forming of the American Civil Liberties Union (ACLU) and his attempts to form third-party movements. There is, Dewey stated,

nothing in the nature of liberalism that makes it a milk-water doctrine, committed to compromise and minor "reforms." . . . The question of method in formation and execution of policies is the central thing in liberalism. The method indicated is that of maximum reliance upon intelligence. . . . A genuine liberal will emphasize as crucial the complete correlation between the means used and the consequences that follow. . . . The emphasis of liberalism upon the method of intelligence does not commit it to unqualified pacifism, but to the unremitting use of every method of intelligence that conditions permit, and to search for all that are possible. (Dewey, 1946/1958, pp. 138–139)

Confrontational politics can be educational, in that they make plain to all the negative consequences of targeted policies and attitudes for a "public" struggling to organize itself. Dewey thought that in deciding on a course of action, its effects on others and their reactions were part of the consequences one had to consider. For example, in expressing his sympathy with the Pullman strike in a letter to his wife Alice, Dewey wrote:

The exhibition of what the unions might accomplish, if organized and working together, has . . . given the public mind an object lesson that it won't soon forget. I think the few freight cars burned up a pretty cheap price to pay—it was the stimulus necessary to direct attention, and it might easily have taken more to get the social organism thinking. (Dewey, cited in Westbrook, 1991, p. 88)

An alternative to fitting such actions within the framework of Dewey's conception of inquiry would be to emphasize the "hermeneutic" dimensions of Dewey's outlook (Campbell, 1984a; Greene, 1989; Smiley, 1990). Confrontational politics at its best, as in the civil rights movement, for example, attempts to interpret and reconstruct social meanings and values through stimulating a reflective public conversation. Attention to Dewey's ideas about meaning and value, and the part communication plays in their transmission and reconstruction, might provide help in acknowledging the political dimensions of democratic community without forsaking Dewey's insight that freedom must be achieved through democratic means. And when confrontational strategies are developed as part of a program of education, they may have the potential for combining Marx's notions of the need for struggle with Dewey's rejection of violence and manipulation.

CONCLUSION

Implications for Research

Some possible directions for future research have emerged in the course of this review. I have suggested that our evaluation of Dewey's radically democratic vision could profit from further study of the hermeneutic themes in his work, his theory of ethics, his proposals for the resolution of social conflicts, and his conception of social science. Work in each of these areas has promise for enriching the standard interpretation of Dewey as endorsing a scientific, problem-solving approach to social dilemmas. While, in my view, Dewey's focus on developing social intelligence should not be discounted in reconstructions of his social and political philosophy, a fuller, and probably more adequate, theory emerges when attention is paid to his emphasis on the importance of meaning in human life and his account of how meaning is created, transmitted, and reconstructed through communication in communities of loyalty and attachment. We need to understand better how Dewey attempted to mediate "between the modern and the counter-modern," between "the breadth and certainty of the scientific outlook" and "the vividness and personal meanings of the village order" (Featherstone, 1979, p. 12). Similarly, new work on Dewey's ethics casts doubt on standard interpretations of Dewey as holding that evaluative propositions express matters of fact, thus encouraging a more complex interpretation of how Dewey thought valuation propositions could receive factual warrant through social inquiry. A better understanding of Dewey's theory of the reinterpretation of meaning through communication and of the role of inquiry in the formation of social values might, then, put us in a better position to judge his hopes for education and dialogue as the primary means of conflict resolution. And Dewey's conception of social science deserves further study in our attempt to understand how democratic politics could be enlightened, but not supplanted, by scientific research. Finally, as suggested in the section on multiculturalism, in assessing the viability of Dewey's social and political philosophy, we must come to terms with claims that Dewey's commitments to experimentalism and radical democracy illegitimately privilege a Eurocentric modernist outlook.

Democratic Education

In *Experience and Education*, Dewey noted the greater need of reformers for a philosophy of education. Traditionalists could rely on established institutional habits to do the real work and thus needed only "a few fine-sounding words to justify existing practices" (Dewey, 1938a, p. 29). Reformers, by contrast, required new ideas that would lead to new practices. Can Dewey's philosophy of education provide needed guidance for our current reform efforts? Is Dewey's educational vision still viable?

The argument of this paper has been that how we answer that question depends, in part, on whether we share Dewey's commitments to developing the powers of inquiry of all citizens and their capacities for democratic participation. While both "critical thinking" and "citizenship" are often cited as important educational aims, it may be doubted whether their usual interpretation matches Dewey's. Maxine Greene (1988) reads Dewey as concerned with teaching students a

kind of critical interpretation that gives content to the idea of freedom, that reveals lacks and deficiencies, and that may open the way to surpassing and repair. . . . [He] encourages free and informed choosing within a social context where ideas could be developed "in the open air of public discussion and communication." (p. 6)

And she notes:

The language of contemporary schooling and, indeed, of proposed reforms emphasizes something quite different from such interpretive thinking. Rather than being challenged to attend to the actualities of their lived lives, students are urged to attend to what is "given" in the outside world—whether in the form of "high technology" or the information presumably required for what is called "cultural literacy." There is, in consequence, an implicit encouragement of the tendency to accede to the given, to view what exists around us as an objective "reality," impervious to individual interpretations. (Greene, 1988, pp. 6–7)

Giroux and McLaren also find an absence in current reform proposals of discussions of the role of schooling in "advancing democratic practice" and "critical citizenship." The focus, instead, has been on adapting to the needs of the labor market, they claim (Giroux & McLaren, 1986, p. 215; see also Aronowitz & Giroux, 1985). Even practices that sound Deweyan, such as cooperative learning, often seem more focused on enhancing learning effectiveness than on developing what Dewey (1916) called "a social direction of disposition" (p. 39; for an alternative view, see Popp, 1987).

Of course, what Greene and Giroux and McLaren are pointing out is that Dewey's commitments do not appear to be what is guiding either current practice or mainstream efforts at reform. But this does not mean that his vision ought not to be adopted; their position, emphatically, is that it should be. And there are many good reasons for believing that, as Westbrook (1991) concludes, "Dewey's philosophy . . . merits another, closer look" (p. 552). Concerns about the vitality of American political life, citizen apathy, growing disparities among citizens in wealth and power, and conflict among racial and ethnic groups make Dewey's conception of public life attractive. Communities in which all share in the creation and enjoyment of common goods, in which each person's flourishing is thought necessary for the full flourishing of the others and the individuality of each is respected, and in which conflicts are brought out in the open and resolved through public discourse surely are a worthy goal. If we were to commit ourselves to education for radical democracy, what would be a plausible agenda for schooling? (For discussions of democratic schooling, see Gutman, 1987; Pratte, 1986; Preskill, 1990). How would current policy proposals fare when judged by the criterion of potential for enhancing democratic participation and critical citizenship? While no one could claim with certainty to be speaking in the spirit of Dewey in imaging how he might respond to our current situation, the following observations seem to me to be consistent with his outlook.

First, a caveat. While Dewey's aspirations for democratic schooling may be in eclipse at present, another aspect of Dewey's legacy has a firm foothold. Lazerson argues that the Deweyans' emphasis on breaking down the barriers between school and society and on the need of the school to adjust to social change made relevance to social needs and outcomes the primary criterion for judging educational success. They destroyed the autonomy of the school as a separate institution and made it responsive to social demand. The result was a tendency to see the school as a social panacea, the central instrument of social policy (Lazerson, 1984). Perhaps it is now time to heed Dewey's views about the relationship between school and society that emerged in his later work. Without underestimating the importance of good schools, Dewey would have us acknowledge that efforts to change schools to make them centers of democratic learning will require changes in work, the economy, and political culture. Educators would need to look outward, not only for cues about what to teach, but to become advocates for the social changes required if schooling is to succeed in its primary purposes (see Arnstine, 1982; Wirth, 1982).

That said, it is hard to imagine Dewey wavering in his commitment to public education and to public schools that include within the same environment something of the diversity represented in the community at large. Difficult though communication across ethnic, racial, religious, gender, and social class divisions may now be, the need for "a broader horizon than is visible to the members of any group while it is isolated" remains at least as great as it was in Dewey's day (Dewey, 1916, p. 21).

Even if children learn math or science as readily when embedded in their own communities as when part of a more inclusive group, it seems less likely that they could acquire the social attitudes Dewey favored: the willingness to rethink one's own beliefs and traditions when encountering others who think and act differently, a commitment to considering the bearing on the community as a whole of one's own actions, a belief that one's own good cannot be separated from the good of others, the willingness to make every effort to resolve conflict through public discourse. Furthermore, as Tyack and Hansot point out, public schools are one of the few institutions in American society open to citizen participation. Thus, "discourse and action concerning public schools provide an opportunity for citizens to become concerned not simply about what is good for themselves or their own children but also what is necessary to bring about a more just and effective society" (Tyack & Hansot, 1982, p. 261). The groups that form around shared concerns for children's education within inclusive public schools might, then, become local communities of the sort Dewey sought. Whether or not they do will depend on whether ways can be found to encourage individuals to transcend narrow self-interest and consider what would be required to improve the education of all (i.e., on whether or not such groups can become sites for the cultivation of civic virtue).

Of course, democratic attitudes and dispositions are not created by the mere fact of inclusive public schools. It matters what the school itself is like. Dewey thought that in order to build democratic character, the school should itself be an embryonic democratic community. The school should realize the ideal of democracy as a form of community life by engaging students in joint enterprises sustained by common commitment, mutual communication, and shared social inquiry. Teachers, too, were to be fully participating members of the school community, helping to formulate the educational policies of the school and sharing in the development of the instructional concepts and methods that directed their own teaching (see Dewey, 1946/1958).

Thus, among current instructional and organizational reforms, "cooperative learning" and "teacher empowerment" have some promise of embodying Dewey's vision. In each of these cases, however, there is the possibility that the similarities to Dewey's ideas will be merely superficial. For example, Slavin describes cooperative learning methods as getting students to help each other in learning "academic material" as a supplement to teacher instruction. Reversing Dewey's conception of the relationship between doing and learning, Slavin (1991) writes that "in Student Team Learning [a particular cooperative learning method] the students' tasks are not to *do* something as a team but to *learn* something as a team" (p. 73). Evaluations of cooperative learning techniques have explored their impact on academic achievement, self-esteem, and friendships be-tween students of different ethnic groups (Slavin, 1991). While all of these are important outcomes, they do not necessarily entail achievement of Dewey's aim of democratic community and the growth of social intelli-gence. And the packaging of cooperative learning techniques for teachers, like other curricular and methods packages, runs the risk Dewey noted of inducing a teacher "to dispense with exercise of his [or her] own judg-ment" rather than making his or her "personal reaction more intelligent" (Dewey, 1916, p. 172). But it is enhancing the teacher's ability and op-portunity to exercise intelligence that "empowerment" presumably aims at. Nevertheless, current interest in cooperative methods and teacher participation in decision making provides an opportunity for realizing Dewey's belief that we must seek the good with others and not for them, that a good society requires the freely chosen participation of all con-cerned (Westbrook, 1991).

Similar analyses could be made of other proposals, analyses that would show both a resemblance between these proposals and Dewey's ideas and substantial differences in meaning and direction. For example, in 1899 Dewey noted "the necessity for free play between the school and the needs and forces of industry" (Dewey, 1899/1980, pp. 45–46). But he added that he did not mean

that the school is to prepare the child for any particular business, but that there should be a natural connection of the everyday life of the child with the business environment about him, and that it is the affair of the school to clarify and liberalize this connection. (Dewey, 1899/1980, pp. 45–46)

Partnerships between business and schools might take the direction of training students or of enlightening them about the role of business in everyday life. My point in including such examples is to call attention both to the potential for our gaining guidance from Dewey's vision and what appears to me to be a substantial difference between his outlook and the tenor of many reform proposals. My hypothesis has been that what explains this discrepancy is that the development of social intelli-gence and participatory democracy do not rank as high on the national agenda as they did for Dewey. And, of course, it should be said that accomplishing Dewey's goals would not be easy even with firm commit-ment. It would require substantial transformation in the culture of schools as well as the culture of the broader society.

Westbrook argues that Dewey should be viewed "as a minority, not a majority, spokesman within the liberal community, a social philosopher whose democratic vision failed to find a secure place in liberal ideology— in short, a more radical voice than has been generally assumed" (Westbrook, 1991, p. xiv). Within education, Dewey in one sense surely has a secure place; his ideas are a continuing source of inspiration for many. But in order for his vision to become the guiding force for educational policy, the American people will have to come to terms with his radically democratic voice. But even without being mainstream, Dewey's vision continues to stand as a critical conscience that draws attention to the far-from-completed agenda of making all students masters of their economic and social fate.

REFERENCES

Appleton, N. (1983). Democracy and cultural pluralism: Ideals in conflict. In D. H. Kerr (Ed.), *Proceedings of the Philosophy of Education Society 1982* (pp. 151–158). Normal, IL: Philosophy of Education Society.

Arnstine, D. (1982). How practical are Dewey's proposals for educational and industrial reform? In D. R. DeNicola (Ed.), *Proceedings of the Philosophy of Education Society 1981* (pp. 286–290). Normal, IL: Philosophy of Education Society.

Aronowitz, S., & Giroux, H. A. (1985). *Education under siege: The conservative, liberal and radical debate over schooling.* South Hadley, MA: Bergin & Garvey.

Beardsley, M. (1966). *Aesthetics from classical Greece to the present: A short history.* New York: Macmillan.

Bernstein, R. J. (1980). Philosophy in the conversation of mankind. *Review of Metaphysics, 33,* 745–775.

Bernstein, R. J. (1985). Dewey, democracy: The task ahead of us. In J. Rajchman & C. West (Eds.), *Post-analytic philosophy* (pp. 48–59). New York: Columbia University Press.

Bernstein, R. J. (1987a). One step forward, two steps backward, Richard Rorty on liberal democracy and philosophy. *Political Theory, 15,* 538–563.

Bernstein, R. J. (1987b). The varieties of pluralism. *American Journal of Education, 95,* 509–525.

Bien, J. (1980). Dewey and Marx: Two notions of community. *Philosophy Today, 24,* 318–324.

Bloom, A. (1987). *The closing of the American mind.* New York: Simon and Schuster.

Borrowman, M. (1981). The school and society: Vermont in 1860, Chicago in 1890, Idaho in 1950, California in 1980. *Educational Studies, 11,* 377–392.

Bowers, C. A. (1987). *Elements of a post-liberal theory of education.* New York: Teachers College Press.

Boydston, J. A. (1975). John Dewey and the new feminism. *Teachers College Record, 76,* 441–448.

Boydston, J. A., & Andresen, R. L. (1969). *John Dewey: A checklist of translations, 1900–1967.* Carbondale: Southern Illinois University.

Brodsky, G. M. (1988). Politics, culture and society in Marx and Dewey. In W. J. Gavin (Ed.), *Context over foundation, Dewey and Marx* (pp. 77–118). Dordrecht, The Netherlands: D. Reidel.

Burnett, J. R. (1988). Dewey's educational thought and his mature philosophy. *Educational Theory, 38,* 203–211.

Campbell, J. (1984a). Politics and conceptual reconstruction. *Philosophy and Rhetoric, 17,* 156–170.

Campbell, J. (1984b). Rorty's use of Dewey. *Southern Journal of Philosophy, 22,* 175–187.

Campbell, J. (1987). Optimism, meliorism, faith. *History of Philosophy Quarterly, 4,* 93–113.

Campbell, J. (1988). Dewey's understanding of Marx and Marxism. In W. J. Gavin (Ed.), *Context over foundation, Dewey and Marx* (pp. 119–145). Dordrecht, The Netherlands: D. Reidel.

Caspary, W. R. (1990). Judgments of value in John Dewey's theory of ethics. *Educational Theory, 40,* 155–169.

Chaloupka, W. (1987). John Dewey's social aesthetics as a precedent for environmental thought. *Environmental Ethics, 9,* 243–260.

Colwell, T. (1983). The significance of ecology for philosophy of education. In D. H. Kerr (Ed.), *Proceedings of the Philosophy of Education Society 1982* (pp. 177–188). Normal, IL: Philosophy of Education Society.

Colwell, T. (1985). The ecological perspective in John Dewey's philosophy of education. *Educational Theory, 35,* 255–266.

Crunden, R. M. (1982). *Ministers of reform, The progressives' achievement in American civilization 1889–1920.* New York: Basic Books.

Dewey, J. (1916). *Democracy and education.* New York: The Free Press.

Dewey, J. (1929). *Experience and nature.* La Salle, IL: Open Court.

Dewey, J. (1930). *Individualism old and new.* New York: Minton, Balch & Company.

Dewey, J. (1933). *How we think.* Boston: Heath.

Dewey, J. (1934a). *A common faith.* New Haven, CT: Yale University Press.

Dewey, J. (1934b). *Art as experience.* New York: Capricorn Books.

Dewey, J. (1935). *Liberalism and social action.* New York: G. P. Putnam's Sons.

Dewey, J. (1938a). *Experience and education.* New York: Collier Books.

Dewey, J. (1938b). *Logic, The theory of inquiry.* New York: Henry Holt and Company.

Dewey, J. (1939a). *Freedom and culture.* New York: G. P. Putnam's Sons.

Dewey, J. (1939b). *Theory of valuation.* Chicago: University of Chicago Press.

Dewey, J. (1954). *The public and its problems.* Chicago: Swallow Press. (Original work published 1927)

Dewey, J. (1957). *Reconstruction in philosophy.* Boston: Beacon Press. (Original work published 1920; enlarged edition published 1948)

Dewey, J. (1958). *Philosophy of education (problems of men).* Totowa, NJ: Littlefield, Adams & Co. (Original work published 1946)

Dewey, J. (1980). *The school and society* (J. A. Boydston, ed.). Carbondale: Southern Illinois University Press. (Original work published 1899)

Dewey, J. (1988). *The quest for certainty.* Carbondale: Southern Illinois University Press. (Original work published 1929)

Eisele, J. C. (1983). Dewey's concept of cultural pluralism. *Educational Theory, 33,* 149–155.

Featherstone, J. (1972). Reconsideration: John Dewey. *New Republic, 167,* 27–32.

Featherstone, J. (1979). John Dewey and David Riesman: From the lost individual to the lonely crowd. In H. J. Gans, N. Glazer, J. R. Gusfield, & C. Jencks (Eds.), *On the making of Americans: Essays in honor of David Riesman* (pp. 3–39). Philadelphia: University of Pennsylvania Press.

Fisher, J. (1989). Some remarks on what happened to John Dewey. *Journal of Aesthetic Education, 23,* 54–59.

Frankel, C. (1977). John Dewey's social philosophy. In S. M. Cahn (Ed.), *New studies in the philosophy of John Dewey* (pp. 3–44). Hanover, NH: University Press of New England.

French, R. S. (1980). The humanities and the challenge of the new ecological consciousness. *American Studies International, 19,* 17–38.

Garrison, J. W. (1988). Democracy, scientific knowledge, and teacher empowerment. *Teachers College Record, 89,* 487–504.

Gavin, W. J. (Ed.). (1988). *Context over foundation: Dewey and Marx.* Dordrecht, The Netherlands: D. Reidel.

Gelb, L. H. (1991, July 3). Why the political mess? *New York Times,* p. A19.

Giarelli, J. M. (1982). Review of *Progressive education: A Marxist interpretation* by Gilbert G. Gonzalez. *Educational Studies, 13,* 464–471.

Giarelli, J. M. (1986). Review of *Education under siege: The conservative, liberal and radical debate over schooling* by Stanley Aronowitz and Henry A. Giroux. *Harvard Educational Review, 56,* 318–323.

Giroux, H. A., & McLaren, P. (1986). Teacher education and the politics of engagement: The case for democratic schooling. *Harvard Educational Review, 56,* 213–238.

Gonzalez, G. G. (1982). *Progressive education: A Marxist interpretation.* Minneapolis, MN: Marxist Educational Press.

Gordon, D. (1984). *The myths of school self-renewal.* New York: Teachers College Press.

Gouinlock, J. (1989). Dewey. In R. J. Cavalier, J. Gouinlock, & J. P. Sterba (Eds.), *Ethics in the history of Western philosophy* (pp. 306–334). New York: St. Martin's Press.

Greene, M. (1986). In search of a critical pedagogy. *Harvard Educational Review, 56,* 427–441.

Greene, M. (1988). *The dialectic of freedom.* New York: Teachers College Press.

Greene, M. (1989). The teacher in John Dewey's works. In P. W. Jackson & S. Haroutunian-Gordon (Eds.), *From Socrates to software: The teacher as text and the text as teacher* (Eighty-ninth Yearbook of the National Society for the Study of Education, Part I, pp. 24–35). Chicago: University of Chicago Press.

Gutman, A. (1987). *Democratic education.* Princeton, NJ: Princeton University Press.

Holton, S. M. (1981). Should public education yield to Christian fundamentalism? *The High School Journal, 64,* 229–231.

Ichimura, T. (1984). The Protestant assumption in progressive educational thought. *Teachers College Record, 85,* 445–457.

Itzkoff, S. (1969). *Cultural pluralism and American education.* Scranton, PA: International Textbook Company.

Kaufman-Osborn, T. V. (1984). John Dewey and the liberal science of community. *Journal of Politics, 46,* 1142–1165.

Kaufman-Osborn, T. V. (1985). Pragmatism, policy science, and the state. *American Journal of Political Science, 29,* 827–849.

Kliebard, H. M. (1986). *The struggle for the American curriculum 1893–1958.* Boston: Routledge & Kegan Paul.

Kliebard, H. M. (1987). The question of Dewey's impact on curriculum practice. *Teachers College Record, 89,* 139–141.

Kolenda, K. (1986). Rorty's Dewey. *Journal of Value Inquiry, 20,* 57–62.

Lagemann, E. C. (1989). The plural worlds of educational research. *History of Education Quarterly, 29*, 185–214.

Laird, S. (1988). Women and gender in John Dewey's philosophy of education. *Educational Theory, 38*, 111–129.

Lawler, P. A. (1980). Pragmatism, existentialism, and the crisis in American political thought. *International Philosophical Quarterly, 20*, 327–338.

Lazerson, M. (1984). If all the world were Chicago: American education in the twentieth century. *History of Education Quarterly, 24*, 165–179.

Lichtenstein, P. M. (1985). Radical liberalism and radical education: A synthesis and critical evaluation of Illich, Freire, and Dewey. *American Journal of Economics and Sociology, 44*, 39–54.

Manicas, P. T. (1981). John Dewey and the problem of justice. *Journal of Value Inquiry, 15*, 279–291.

Manicas, P. T. (1988a). Dewey and the class struggle. In M. G. Murphey & I. Berg (Eds.), *Values and value theory in twentieth century America* (pp. 67–81). Philadelphia: Temple University Press.

Manicas, P. T. (1988b). Philosophy and politics: A historical approach to Marx and Dewey. In W. J. Gavin (Ed.), *Context over foundation: Dewey and Marx* (pp. 147–175). Dordrecht, The Netherlands: D. Reidel.

Maxcy, S. J. (1984). Ethnic pluralism, cultural pluralism, and John Dewey's program of cultural reform: A response to Eisele. *Educational Theory, 34*, 301–305.

McBride, W. L. (1988). Science, psychology, and human values in the context of Dewey's critique of Marx. In W. J. Gavin (Ed.), *Context over foundation: Dewey and Marx* (pp. 37–47). Dordrecht, The Netherlands: D. Reidel.

McDermott, J. J. (1984). Classical American philosophy: A reflective bequest to the twenty-first century. *Journal of Philosophy, 18*, 663–675.

Miranda, W. R. (1980). Implications in Dewey for feminist theory in education. *Educational Horizons, 58*, 197–202.

Miranda, W. R. (1984). Limitations in the ideological analysis of Dewey's logic. *Journal of Educational Thought, 18*, 69–75.

Moreno, J. D., & Frey, R. S. (1985). Dewey's critique of Marxism. *Sociological Quarterly, 26*, 21–34.

Noble, C. (1978). A common misunderstanding of Dewey on the nature of value judgments. *Journal of Value Inquiry, 12*, 53–63.

Paringer, W. A. (1990). *John Dewey and the paradox of liberal reform.* Albany: State University of New York Press.

Passow, A. H. (1982). John Dewey's influence on education around the world. *Teachers College Record, 83*, 401–418.

Peters, R. S. (1977). John Dewey's philosophy of education. In R. S. Peters (Ed.), *John Dewey reconsidered* (pp. 102–121). London: Routledge & Kegan Paul.

Popp, J. A. (1987). If you see John Dewey, tell him we did it. *Educational Theory, 37*, 145–152.

Pratte, R. (1986). Civism and its problems. In D. Nyberg (Ed.), *Proceedings of the Philosophy of Education Society 1985* (pp. 113–122). Normal, IL: Philosophy of Education Society.

Preskill, S. (1990). Strong democracy, the ethic of caring and civic education. In R. Page (Ed.), *Proceedings of the Philosophy of Education Society 1989* (pp. 217–225). Normal, IL: Philosophy of Education Society.

Proefriedt, W. (1980). Socialist criticism of education in the United States: Problems and possibilities. *Harvard Educational Review, 50*, 467–480.

Pruitt, R. (1984). The biases of pluralism. *Journal of Philosophy, 18,* 676–677.

Radest, H. B. (1980). Schooling and the search for a usable politics. In M. Wohlgelernter (Ed.), *History, religion, and spiritual democracy: Essays in honor of Joseph L. Blau* (pp. 317–340). New York: Columbia University Press.

Rorty, R. (1981). Method, social science, and social hope. *Canadian Journal of Philosophy, 4,* 569–588.

Rorty, R. (1982). *Consequences of pragmatism (essays: 1972–1980).* Minneapolis: University of Minnesota Press.

Rorty, R. (1988). That old-time philosophy. *The New Republic, 198,* 28–33.

Rorty, R. (1989). *Contingency, irony, and solidarity.* Cambridge, England: Cambridge University Press.

Rorty, R. (1991, July 25). Just one more species doing its best. *London Review of Books,* pp. 3–7.

Sarlos, B. (1983). Cultural pluralism in a new key. In D. H. Kerr (Ed.), *Proceedings of the Philosophy of Education Society 1982* (pp. 159–162). Normal, IL: Philosophy of Education Society.

Schwartz, R. (1988). Whatever happened to pragmatism? In M. G. Murphey & I. Berg (Eds.), *Values and value theory in twentieth-century America: Essays in honor of Elizabeth Flower* (pp. 37–45). Philadelphia: Temple University Press.

Sherman, A. L. (1984). Genderism and the reconstruction of philosophy of education. *Educational Theory, 34,* 321–325.

Shusterman, R. (1989). Why Dewey now? *Journal of Aesthetic Education, 23,* 60–67.

Slavin, R. E. (1991). Synthesis of research on cooperative learning. *Educational Leadership, 48,* 71–82.

Sleeper, R. W. (1986). *The necessity of pragmatism, John Dewey's conception of philosophy.* New Haven, CT: Yale University Press.

Sleeter, C. E. (Ed.). (1991). *Empowerment through multicultural education.* Albany: State University of New York Press.

Smiley, M. (1990). Pragmatic inquiry and social conflict: A critical reconstruction of Dewey's model of democracy. *Praxis International, 9,* 365–380.

Tanner, L. (1987). Events that happen—and "unhappen." *Teachers College Record, 89,* 133–138.

Tiles, J. E. (1988). *Dewey.* London: Routledge.

Tilman, R. (1984). Dewey's liberalism versus Veblen's radicalism: A reappraisal of the unity of progressive social thought. *Journal of Economic Issues, 18,* 745–769.

Tyack, D., & Hansot, E. (1982). *Managers of virtue.* New York: Basic Books.

West, C. (1989). *The American evasion of philosophy.* Madison: University of Wisconsin Press.

Westbrook, R. B. (1991). *John Dewey and American democracy.* Ithaca, NY: Cornell University Press.

Wirth, A. G. (1981). An alternative image of "school and society" and the Deweyan tradition (1950–1980): A reply to Merle Borrowman. *Educational Studies, 11,* 393–400.

Wirth, A. (1982). John Dewey on the relations between industrial democracy and education. In D. R. DeNicola (Ed.), *Proceedings of the Philosophy of Education Society 1981* (pp. 273–285). Normal, IL: Philosophy of Education Society.

III.
TEACHER EDUCATION

Chapter 9

Craft Knowledge and the Education of Teachers

PETER P. GRIMMETT and ALLAN M. MACKINNON
Simon Fraser University
Burnaby, British Columbia, Canada

I knew a teacher once
With words as soft
As moths on summer screens:
Brittle—bright—and
Cruel was not his style.
As others barked,
His whispers touched the dark
Inside your soul
And seemed to echo there.
The way was sure.
He always took the time:
Refused the rush
Of world reports for poems—
And pushed aside
The weight of dusty tomes
To scratch his nose
And passed around the mints.
He seemed alive.
You couldn't put him on.
He'd take a book
And make it yours and his
In magic ways
That made your breath come quick.
His wink was slight.
The eyes were bright and clear,
A hush of blues.
You'd watch the pause of smile,
A patient blink
That let the question hang.

We would like to acknowledge the helpful insights and critique given by our colleagues
Gaalen Erickson (University of British Columbia) and Pat Holborn (Simon Fraser University). We also acknowledge the supportively rigorous critiques offered by Mary Kennedy
(Michigan State University) and Ken Zeichner (University of Wisconsin) in their capacity
as chapter reviewers.

His tease would make
You more than eyes and ears:
It often made
Your insides twist and think.
I guess he liked
His work enough to make
It play for us.

Other Voices, Other Worlds
—William Strong[1]

How do some teachers produce this response in students? What is it they do that engages learners in such a provocative and captivating manner? How do teachers orchestrate the experiences of learning so that students find the engagement attractive and stimulating? And what is it about the poem *Other Voices, Other Worlds* that seems to capture a sense of the "craft of teaching"—"the pause of smile, a patient blink that let the question hang"—that somehow transcends the intellectualist tradition of "wait-time," for example? When we think about these sort of questions, we begin to turn our attention away from the science or, say, the discipline of teaching, and we look toward a different notion of teaching as an art form, or teaching as craft. In this chapter, we will review a body of literature that informs this "alternative conception of knowledge" for the practical purpose of educating teachers.

Discussion about teacher education has focused, in recent years, around the idea that there are certain bases of knowledge required for the practice of teaching and, therefore, seen as essential components of a teacher preparation curriculum (see Houston, 1990; Reynolds, 1989; Wittrock, 1986). Wilson, Shulman, and Richert (1987) characterize "knowledge base" as a term usually associated with applied science:

It refers to the set of rules, definitions, and strategies needed by a computer to perform as an expert would in a given task environment. That set of rules is usually rather specific to a particular domain or task. . . . In teaching, the knowledge base is the body of understanding, knowledge, skills, and dispositions that a teacher needs to perform effectively in a given teaching situation. (pp. 106–107)

On the basis of this definition, Shulman (1987b) has characterized the knowledge base for teaching as consisting of the following categories: content knowledge, general pedagogical knowledge, curriculum knowledge, pedagogical content knowledge, knowledge of learners, knowledge of educational contexts, and knowledge of educational ends, purposes, and values, as well as their philosophical and historical grounds. His point in this characterization was to make a case for the amalgam of content knowledge and general pedagogical knowledge to form *pedagogical con-*

tent knowledge. For example, in order to teach algebra a teacher needs not only to understand the subject matter but also how students typically grapple with such abstract content. We would argue that pedagogical content knowledge, as set out by Shulman, is epistemologically different from the other six categories posited.[2] These other categories are sometimes discussed as though they contain a set of principles derived from research that *could* dictate practice (see Reynolds, 1989). Moreover, the applied science approach has been known to influence the design of teacher education programs in which principles of teaching are learned in the lecture hall and later applied in practice during student teaching (Fullan & Connelly, 1987; Schön, 1987). By contrast, pedagogical content knowledge is derived from a considered response to experience in the practice setting, and, though related to knowledge that can be taught in the lecture hall, it is formed over time in the minds of teachers through reflection. In our view, Shulman's rendition of pedagogical content knowledge is more analogous to a craft conception of teaching than to one of teaching as an applied science.

From our review of literature that portrays teaching as craft, we see the need in this discussion for a further amalgam between general pedagogical knowledge and knowledge of learners, which we would call *pedagogical learner knowledge*. Whereas pedagogical content knowledge concerns itself with teachers' representations of subject matter content in terms of how it might be effectively taught, pedagogical learner knowledge revolves around procedural ways in which teachers deal rigorously and supportively with learners. Although the "maxims" of craft knowledge are useful in guiding practice, they cannot replace the role of experience in the development of craft. Thus, pedagogical learner knowledge can be defined as pedagogical procedural information useful in enhancing learner-focused teaching in the dailiness of classroom action.[3]

We wish to make three important points at the outset. First, craft knowledge consists of pedagogical content and pedagogical learner knowledge derived from considered experience in the practice setting. Second, pedagogical content and pedagogical learner knowledge are radically different forms of knowledge from the other six categories posited by Shulman. Third, and most important, craft knowledge represents teachers' judgment in apprehending the events of practice from their own perspectives as students of teaching and learning, much as a "glue" that brings all of the knowledge bases to bear on the act of teaching. As such, this chapter focuses on the "wealth of teaching information that very skilled practitioners have about their own practice . . . [a] deep, sensitive . . . contextualized knowledge" (Leinhardt, 1990, pp. 18–19) derived from the "wisdom of practice" (Shulman, 1987a).

INTRODUCING THE CONCEPT OF CRAFT

It is useful to begin with Kurshan's (1987) story of Eleazer Hull, a New England sea captain who had no training in navigation but was hired by merchants throughout the Northeast:

> Once when Hull was asked how he steered his ships with such a sure hand through the hazards and unknowns of the seas, he replied, "I go up on the deck, listen to the wind in the rigging, get a drift of the sea, gaze at a star and set my course." One year the State Commissioner for Navigation discovered that Hull was unlicensed and untrained in the science of navigation and told him he would have to get an education and upgrade his skills. Hull consented, graduated at the top of his class, received a certificate of achievement, and was told he was now licensed to return to sea. He did so, and when he returned from his first two-year-long voyage, his friends asked him how it felt to sail by the new scientific principles he had learned from books. Hull replied, "Whenever I needed to chart my course I pulled out my maps, followed the equations, and calculated my location with mathematical precision. Then I went up on the deck, listened to the wind in the rigging, got the drift of the sea, gazed at a star, and corrected my computations." (p. 124)

Craft knowledge, or occupational "savvy," is what made Hull into the outstanding sea captain that he was. This "know-how" enabled him not only to understand with proficiency the scientific principles he was deemed to be lacking but also to determine when his mathematically precise computations were not adequate to the arduous task of steering the ship through difficult waters. We do not use this illustration to argue that Hull's craft knowledge was superior to knowledge based on scientific principles. Nor do we use it to suggest that all craft knowledge is inherently good. On the contrary, we are of the view (as we elaborate below) that some versions of craft knowledge do not improve practice from the learner's perspective. What intrigues us about the story of Hull is that merchants in the Northeast recognized in his practice a proficiency and skill that, according to the story, was not based on scientific principles. Moreover, the story tells of the importance of experience in making sense of the way principles can or cannot be helpful in practice. We believe that these insights have profound implications for the education of practitioners.

That craft knowledge exists as a powerful determinant of teachers' practice is neither new nor controversial. What *is* new is the possibility that such knowledge could become an integral part of teacher education; what *is* controversial is the debate over whether this would be a productive direction to take. We argue here that, in addition to codified knowledge bases framed around university-based research, teacher education could benefit from the contribution that craft knowledge can make to the formation of skillful, reflective, and empowered teachers.[4] In this chapter, we define craft knowledge, grapple with its validity and morality as a form of knowing, and, through a review of research and writings, attempt to develop a framework for understanding the essential contribution it could make to the development of inquiring minds in teachers. Finally, we attempt to grapple with the thorny question of how it might be of value in the education of teachers.

TRADITIONS OF THINKING ABOUT CRAFT KNOWLEDGE

Liston and Zeichner (1991) note that "our understanding of the role of the teacher and the activity of teaching is conceptually dependent on particular communities" (p. 43). Their purpose is to grapple with the ways in which the role of teacher and the activity of teaching is differentially understood according to three traditions of educational thought, namely, "the conservative, progressive, and radical traditions" (p. 46).[5] We regard an understanding of craft knowledge as being analogous to the case of the role of teacher and the activity of teaching. Craft knowledge is understood differentially by those who subscribe to each of these respective traditions.

The Conservative Tradition

The conservative tradition essentially views the educational process as a means of cultural transmission (Liston & Zeichner, 1991, p. 47). The teacher's role is either to initiate students into predetermined and distinct forms of public knowledge (Hirst, 1965) or to reestablish a common cultural heritage that is shared by all Americans (Hirsch, 1988). When the term *craft* is associated with classroom practice emanating from this tradition of educational thought, it is frequently used in a pejorative sense. Broudy (1956), Gage (1985), and Scheffler (1960) view craft knowledge in this tradition as having an inherently antiscientific bent and rue its effects on the professional status (or lack thereof) of teaching. Within the British context—one in which teacher education has recently come under concerted attack—Turner (1990) characterizes craft as representing "an apprenticeship in which good practice from the past could be perpetuated" (p. 5). He equates an emphasis on the craft of teaching with the application of certain fashionable political and social ideas, which, together with harsh government intervention, has undermined the credibility of teacher education.[6] Thus, Stones (1990) describes such an emphasis as leading toward "the deprecation of pedagogy [by holding forth] practical experience . . . as the means by which teaching quality is to be enhanced" (p. 1).

Teaching as craft in the conservative tradition, then, is equated by some British writers to the mindless imitation of practice (Hartnett & Naish, 1980; Stones, 1990; Turner, 1990). Such an assumption appears to be central to the work of those in North America who attempt to discredit a craft conception of teaching (see Broudy, 1956; Schaefer, 1970). Indeed, Howsam, Corrigan, Denemark, and Nash (1976) strongly contend that "professionals cannot exist without an undergirding science. . . . [To do so] is to restrict the occupation to the level of a craft" (p. 11). There is no doubt that some teaching is far from theoretically based and represents a form of mindless, habitual practice (see Hargreaves & Dawe, 1989). The quality of practice observed in the majority of experienced teachers within this tradition has been found by researchers such as Goodlad (1984) and Sizer (1984) to leave much to be desired. Cuban's (1984) historical analysis of teachers' pedagogical practices suggests that the continued dominance of teacher-centered didacticism in schools is the consequence of "situationally-constrained choice" (p. 251). He maintains that school and classroom structures generate an invisible, encompassing environment that few teachers recognize as potentially shaping what they can and cannot do on a daily basis in classrooms:

Chairs in rows, recitations, whole group instruction, worksheets, and textbook assignments need to be viewed as a series of successful solutions invented by teachers to solve daily problems of managing a score or more of students while they also acquired information and values. Coping with these structures, teachers constructed workable pedagogical solutions that have proved useful in personally maintaining control while carrying out instruction. (Cuban, 1984, p. 250)

Cuban further suggests that those teachers (particularly elementary ones), who were able to change this traditional pattern of pedagogy and teach in a more learner-focused way, were helped by two important factors: First, they did not hold to conservative educational beliefs, and, second, the organizational culture of the elementary school (instructional time slots were larger, with a greater emphasis on skill development than on content coverage) was not as constraining as that of the secondary school. Feiman-Nemser and Buchmann (1985) suggest that such a change occurs when teachers transcend the "unquestioned familiarity" of experience (p. 56) that limits their frameworks to the uncritical "folkways of teaching" (Buchmann, 1987, p. 156).[7] Wood (1988) further contends that action research plays an important role in helping teachers to question their pedagogical thinking and classroom action (p. 147).

Consequently, the growing evidence that some teachers learn considerably by reflecting on their practice (see Clift, Houston, & Pugach, 1990; Grimmett & Erickson, 1988) in a manner that is belief challenging, instructive, and theoretically sensitive is a manifestation of these different educational beliefs at work. When this occurs, teaching as craft essentially represents transformed experience. It is this latter view of craft, based as it is in the progressive and radical traditions of educational thought, that provides the focus for this review.

The Progressive and Radical Traditions

The central aim of the progressive tradition is "for students to become competent inquirers, capable of reflecting on and critically examining their everyday world and involved in a continual reconstruction of their experience" (Liston & Zeichner, 1991, p. 48). Unlike the conservative tradition, with its emphasis on teacher-centered didacticism, the progressive tradition assumes that a process of inquiry revolving around students' interests gives rise to mastery of subject matter.

The radical tradition assumes that men and women "are capable of being free and equal members of a just, democratic, and caring society . . . [it starts] from a position of critique arguing that our public schools do not support or develop these capacities equally for all children" (Liston & Zeichner, 1991, p. 49). Consequently, the central aim of the radical tradition is to change the current nature of schools as a means of bringing about equity and social justice.

Craft as a metaphor for teaching (in the progressive and radical, non-conservative sense) has recently received more serious and nonpejorative attention. Tom (1980, 1984) and Tom and Valli (1990) characterize teaching as a moral craft. Blumberg (1989) attempts to change the language used to describe administration from one of science to one of progressive craft. His intent is to focus attention on the implicit, tacit, and intuitive aspects of administration—the ways in which school principals develop a certain know-how. Cohen (1977) argues that understanding educational phenomena in a manner consistent with a progressive and radical craft orientation will increase the likelihood of reforming practice in schools, and Perrone (1989, 1991)—a progressive educator—states clearly that teachers need to learn from practice in order to refine their craft. Major works by Kohl (1986, 1988a), Greene (1984), and Lieberman (1984) have the craft metaphor as a central feature. Furthermore, the way teachers learn their craft would seem to be consistent with Resnick's (1991) and Liston and Zeichner's (1991) theory of situated practice,[8] because it leads to "action-based situated knowledge of teaching" (Leinhardt, 1990, p. 23). How, then, can craft knowledge be understood?

Toward a Definition

Blumberg (1989) builds his definition of craft around the work of Collingwood (1938) and Howard (1982). He draws on Collingwood (1938) to suggest that "craft always involves a distinction between means and end, each clearly conceived as something distinct from the other but related to it. . . . The end is always thought out first, and afterwards the means are thought out" (pp. 15–16; cited in Blumberg, 1989, p. 24). Craft, according to Blumberg, is inherent in a knowledge and mastery of the means that are appropriate for realizing a given end. He asks, "[But] how does a man get a 'nose' for something? And how can this nose be used?" (p. 58; Howard, 1982; Wittgenstein, 1958). This essentially raises the idea of know-how that Blumberg sees as central to an understanding of craft. Kohl (1988a) is much more definitive about what constitutes such know-how as it relates to the craft of teaching:

The craft of teaching has a number of aspects. It relates to the organization of content and the structuring of space and time so that student learning will be fostered. It requires an understanding of students' levels of sophistication and the modes of learning they are accustomed to using. But most centrally, the craft of teaching requires what can be called teaching sensibility. This sensibility develops over a career of teaching and has to do with knowing how to help students focus their energy on learning and growth. (p. 57)

According to Kohl (1986), craft involves the proficient use of some basic skills, including "balancing teacher-initiated ideas with student-initiated ones, . . . doing research as a teacher, and observing and responding to what actually succeeded with the students. This is quite different from following a set curriculum or using standard texts, and involves *planning, research, observation, and response*" (p. 54).

Tom (1984) sets forth a definition by Popkewitz and Wehlage (1973, p. 52) that, though lacking in a moral dimension, encompasses the essential features of craft: "Teaching should be viewed as a craft that includes a reflective approach toward problems, a cultivation of imagination, and a playfulness toward words, relationships, and experiences" (p. 113). He goes on to show how the addition of a moral dimension essentially helps a craft conception of teaching avoid the trap of representing an unthinking approach to trial and error. Zeichner, Tabachnick, and Densmore (1987) use craft knowledge broadly as

a rubric for describing a number of different approaches to studying the psychological context of teaching, from the more conventional studies of teachers' attitudes and beliefs to the more recent attempts to describe the "implicit theories" of teachers, from the teachers' point of view and in the teachers' own language. (p. 21)

Greene (1984) appeals to the basic definition of craft offered by Kohl to grapple with how educators think about their craft:

To speak of craft is to presume a knowledge of and a certain range of skills and proficiencies. It is to imagine an educated capacity to attain a desired end-in-view or to bring about a desired result. . . . We may find ourselves reconstructing familiar techniques, honing a set of unused skills, and—significantly—using our imaginations in what turns out to be an effort to improve our craft . . . [which] is in many ways an example of personal knowledge at work; it is also a reminder that mastering a craft goes far beyond repetitions and routines. (pp. 55, 61)

Teaching as craft, then, assumes certain skills, proficiencies, and dispositions among accomplished teachers—in brief, it suggests an emphasis on a special kind of pedagogical content and learner know-how, a "teaching sensibility," rather than a knowledge of propositions.

Earlier, we defined pedagogical learner knowledge as pedagogical procedural information useful in enhancing learner-focused teaching in the dailiness of classroom action. In other words, we are arguing that when teachers reflect on pedagogical matters from the perspective of the learner, they generate knowledge[9] that represents the nonconservative conception of craft as transformed experience.

At the same time, we are arguing that when teachers reflect on pedagogical matters from the perspective of connecting subject matter content to the minds of learners, they generate another form of the nonconservative conception of craft as transformed experience (i.e., pedagogical content knowledge; see Grossman, 1991; Leinhardt, 1990; Shulman, 1986, 1987a, 1987b). This is a form of craft knowledge that

embodies the aspects of content most germane to its teachability. Within the category of pedagogical content knowledge I include, for the most regularly taught topics in one's subject area, the most useful forms of representation of those ideas, the most powerful analogies, illustrations, examples, explanations, and demonstrations—in a word, the ways of representing and formulating a subject that make it comprehensible to others. . . . [It] also includes an understanding of what makes the learning of specific topics easy or difficult: the conceptions and preconceptions that students of different ages and backgrounds bring with them to the learning. (Shulman, 1986, p. 9)

Craft knowledge, therefore, concerns itself both with teachers' representations of the declarative knowledge contained in subject matter content and with teachers' tacit instantiations of procedural ways of dealing rigorously and supportively with learners. As a form of professional expertise, craft knowledge is neither technical skill, the application of theory or general principles to practice, nor critical analysis; rather, it represents the construction of situated, learner-focused, procedural and content-related pedagogical knowledge through "deliberate action" (Kennedy, 1987).

Validity and Morality of Craft Know-How

Schön (1983, 1987, 1988, 1991) argues for the importance of the sort of competence practitioners display in "divergent" settings (i.e., those indeterminate situations that do not conform to the standard "textbook-like" problems of practice). He focuses instead on the "intuitive perfor-mance" of the actions of everyday life and puts forth the construct of "knowledge-in-action." Similarly, in his classic book, *The Concept of Mind*, Ryle (1949) distinguishes between "knowing how" and "knowing that." Schön and Ryle are not arguing for a kind of mindless knowing in action sometimes referred to as "kinesthetic." Rather, they argue that there is nothing in common sense to suggest that knowing how to do something well consists of knowing rules and plans that we entertain in the mind prior to actions.

What distinguishes sensible from silly operations is not their parentage but their procedure, and this holds no less for intellectual than for practical performance. "Intelligent" cannot be defined in terms of "intellectual" or "know how" in terms of "knowing that"; "thinking about what I am doing" does not connote "both thinking what to do and doing it." When I do something intelligently, I am doing one thing, not two. My performance has a special procedure or manner, not special antecedents. (Ryle, 1949, p. 32)

The assertion that intelligent performance does not necessarily depend on the consideration of a prior set of procedures also is crucial to Schön's analysis. However, this is not to say that thinking in advance of a situation has no bearing on how one subsequently executes action. As Ryle elab-orated:

Certainly, we often do not only reflect before we act but reflect in order to act properly. The chess-player may require some time in which to plan his moves before he makes them. Yet the general assertion that all intelligent performance requires to be prefaced by the consideration of appropriate propositions rings unplausibly, even when it is apologetically conceded that the required consideration is often very swift and may go quite unmarked by the agent. I shall argue that the intellectualist legend is false and that when we describe a performance as intelligent, this does not entail the double operation of considering and executing. (1949, pp. 29–30)

Ryle's example of the "wit" provides a good illustration:

The wit, when challenged to cite the maxims, or canons by which he constructs and ap-preciates jokes, is unable to answer. He knows how to make good jokes and how to detect bad ones, but he cannot tell us or himself any recipes for them. So the practice of humor is not a client of its theory. (1949, p. 30)

For Ryle, and for Schön, the existence and importance of "knowing how"—what we call craft knowledge—is central to the matter of learning a practice. This is not to say that "learning by doing" without analysis is a substitute for "learning by doing" with a theory. Rather, it is to say that "learning by doing" mitigates some of the difficulties that arise when a practitioner is unable to analyze competent performance and articulate its features. Certainly, intuitive actions are subject to thoughtful consideration and further articulation and understanding. The point is, however, that the reverse is not the case; intelligent performance does not follow automatically from knowing a set of procedures and maxims. As Ryle puts it:

Efficient practice precedes the theory of it; methodologies presuppose the application of the methods, of the critical investigation of which they are the products. It was because Aristotle found himself and others reasoning now intelligently and now stupidly and it was because Izaak Walton found himself angling sometimes effectively and sometimes ineffectively that both were able to give their pupils the maxims and prescriptions of the arts. It is therefore plausible for people intelligently to perform some sorts of operations when they are not yet able to consider propositions enjoining how they should be performed. Some intelligent performances are not controlled by any anterior acknowledgements of the principles applied in them. (1949, p. 30)

So far we have argued that craft knowledge represents intelligent and sensible know-how in the action setting. But we would also argue that it is more than this. Craft knowledge contains certain criteria that comment on the "rightness" of such intelligent and sensible action. We find Gutman's (1987) elaboration of what constitutes democratic education to be persuasive and useful in establishing guidelines for morally appropriate intelligent and sensible action.[10] Gutman maintains that a democratic education must honor three basic principles: First, it must develop a deliberative, democratic character in students; second, it cannot repress rational deliberation; and, third, it cannot discriminate against any group of children. Her reasoning is that, within a democracy, we are collectively committed to re-creating the society we share:

As citizens, we aspire to a set of educational practices and authorities [a core commitment to conscious social reproduction] of which the following can be said: these are the practices and authority of which we, acting collectively as a society, have consciously agreed. It follows that a society that supports conscious social reproduction must educate all educable children to be capable of participating in collectively shaping their society. (Gutman, 1987, p. 39)

A society thus committed to conscious social reproduction through the broad educational aim of critical deliberation, however, requires some limits to preserve the rights of all its democratic members. Gutman (1987) suggests two such limits: (a) nonrepression, which "prevents the state, and any group within it, from using education to restrict rational deliberation of competing conceptions of the good life and the good society" (p. 44), and (b) nondiscrimination, which prevents the state and families from "excluding entire groups of children from schooling or denying them an education conducive to deliberation among conceptions of the good life and the good society" (p. 45). Following Gutman, then, we would claim that intelligent and sensible action is morally appropriate when it is nonrepressive and nondiscriminatory in its attempts at developing critical deliberation in students.

Thus, craft knowledge of teaching is not substantive, subject matter knowledge, nor is it syntactical knowledge (that knowledge that derives from the disciplines and enables people to know how to acquire further disciplinary knowledge); rather, it is a particular form of morally appropriate intelligent and sensible know-how that is constructed by teachers, holding progressive and radical educational beliefs, in the context of their lived experiences and work around issues of content-related and learner-focused pedagogy. In the final analysis, the essential validity and morality of craft knowledge resides in readers "living" the life of particular teachers through stories, narrative, case studies, and other forms of vicarious experience.[11]

THE SUBSTANCE OF CRAFT KNOWLEDGE

The substance of craft knowledge is made up of first- and second-order abstractions of pedagogical content and learner knowledge. By first-order abstraction we mean research and writing that is undertaken by practicing teachers in the action setting. By second-order abstraction we mean studies that are undertaken by university researchers who are themselves highly sensitive to and rigorously curious about teachers' pedagogical content and learner knowledge. We do not, however, propose to present this review along clearly differentiated first- and second-order lines of investigation. Our purpose is to bring first-order abstractions into the teacher education curriculum equation, not to downplay university research; thus, our focus is on the substance of craft knowledge and its potential use in the education of teachers, not on its derivation. Furthermore, we see the distinction between pedagogical content and learner knowledge as important for the purpose of delimiting the definition of craft knowledge but not useful as an analytical device for presenting the actual research.[12] The body of this review, then, will consist of two broad sections. The first presents five interrelated (and rather loosely categorized) genres of research representing pedagogical content and learner knowledge that have furthered our understanding of craft knowledge in teaching.[13] These are put forth as appropriate avenues for investigating and portraying the craft knowledge of teachers. The second section examines the writings of three major practitioners whose work has made important contributions to such knowledge.[14]

Research Contributions to Craft Knowledge

This section of the review will examine studies conducted in the genres of pedagogical content knowledge, reflective practice, narrative/story telling (and the cinema), teachers' lives, and teacher research.[15]

Pedagogical Content Knowledge

There is considerable research in the science education community[16] that investigates the ways in which students make sense of science concepts and science classroom phenomena. Some of the well-known projects include the Project for Enhancing Effective Learning (Baird & Mitchell, 1987) in Australia, the Learning in Science Project (Osborne & Freyberg, 1985) in New Zealand, the Children's Learning in Science Project (Driver, 1988; Driver, Guesne, & Tiberghien, 1985) in England, the Students' Intuitions and Science Instruction project (G. Erickson, 1987; MacKinnon & Erickson, 1988) in Canada, and Tobin's science and mathematics teacher development studies (Tobin, 1990; Tobin & Espinet, 1987).

Each of these research centers is investigating and documenting "student conceptions" of the subject matter of science, often together with the ways in which teachers develop a capacity to recognize these student conceptions and take them into consideration in the design of instructional practices. For example, MacKinnon and Erickson (1988) write about a student teacher, Barry, and his practicum supervising teacher, Mr. Kelly, who inquire into their students' learning about ways of graphically representing motion in physics. They show, in excerpts of supervision dialogue, how Mr. Kelly recognized that some students had misinterpreted the slope of a distance versus time graph (which represented uniform velocity with a straight line rising to the right) as meaning acceleration. The account is a rather good example of a teacher's pedagogical content knowledge—Mr. Kelly understands from his prior experience in presenting this topic how some students interpret the rising slope of the graph as increasing rather than uniform speed. This knowledge is revealed when Mr. Kelly says to Barry, "Well, I think one of the things that might have been done is to say, 'Yes, this is a straight line . . . yes, something is increasing, but distance is increasing and we call that speed. Later on, we're going to look at how you represent acceleration' " (p. 125).

The extensive work that has investigated the understandings children have of subject matter in science is informative in a pedagogical sense. It has given rise to knowledge of how children sometimes (in some cases, often) conceive of concepts such as heat, temperature, gravity, mechanics, sound, light, animal, and so forth—all in the effort to seek out better instructional design. In short, this knowledge helps teachers prepare subject matter knowledge to be comprehensible and fruitful to the learner.

Another genre of educational research that contributes to our understanding of craft knowledge but puts more emphasis on processes, as distinct from products of inquiry, is research in the area of reflective practice.

Reflective Practice

The genre of work in reflective practice is directed to conceptualizing and illustrating those times when teachers contemplate about their practice and begin to see their work in a new way. For example, teachers may begin to see a student's inattentiveness as a lack of interest in the lessons rather than evidence for the particular learning style once attributed to him or her. They may begin to appreciate a student's wisecracking in terms of the child's character as a comedian rather than as a saboteur. Or they may begin to search diligently for an understanding of why pupils say that the opposite of *work* is *home*, rather than *play*, as teachers might be given to think. At times such as these, contemplation leads to insight, new possibilities for understanding complex events, and new ideas for future practice.

Many of the writers in the reflective practice genre seem to hold assumptions about the nature of competence in teaching that recognize the value of craft knowledge. They speak about practitioners' "intuitive feel for situations," "learning at the elbows of masters," "knowledge-in-action," "reflection-in-action," and so on.

Although much of the work in reflective practice can be traced to the writings of John Dewey (1929, 1933, 1944), Grimmett, MacKinnon, Erickson, and Riecken (1990) note, in their review of literature on the nature of reflective teaching, that there is a diversity of meanings that are attached to reflection and "little agreement on what conditions may be required to foster reflective teaching" (p. 20). They give examples of work that reveal researchers' conceptions of reflection as instrumental mediation of action, deliberating among competing views of teaching, and reconstructing experiences—including the reconstruction of action situations, "self-as-teacher" in terms of the cultural milieu of teaching, and taken-for-granted assumptions about teaching (derived from a critical-theoretical stance).

One category of inquiry in teacher reflection includes works that investigate how teachers represent and explain practice situations—how they see certain events, as well as the significance they attach to these events. Schön's (1983) work on reflection-in-action and professional education for reflective practice (Schön, 1987, 1988, 1991) has influenced several researchers in this area. Some have investigated the character and quality of reflection as it can be seen to occur in teachers' dialogue about recorded lessons and particular classroom events (e.g., G. Erickson & MacKinnon, 1991; Grimmett & Crehan, 1990; Grimmett & Erickson, 1988; Kilbourn, 1986, 1990; MacKinnon, 1987, 1989). In some cases, the view of reflection developed is associated with the tensions that often arise when a teacher is concerned about the perceived integrity of the subject matter on the one hand, and recognizing pupils' current understandings, purposes, and interests on the other. Other examples investigate the use of metaphors by teachers in making sense of their work and cultural milieu (e.g., Russell, 1984; Russell & Munby, 1991; Russell, Munby, Spafford, & Johnston, 1988). Bullough, Knowles, and Crow (1991) document a quest for compelling and fitting metaphors that represent who beginning teachers imagine themselves to be as teachers.

Another arena of work in reflective teacher education draws upon critical theory (Habermas, 1973) in formulating a view of reflection as an emancipatory activity. Critical reflection allows a practitioner to articulate and, ultimately, to eliminate the social, political, and cultural conditions that frustrate and constrain self-understanding (e.g., Carr & Kemmis, 1983; Smyth, 1986; van Manen, 1977). Such critical reflection, however, is not merely based on the thinking of Habermas. Liston and Zeichner (1991, pp. 26–34) discuss several strands of critical reflection in the social reconstructionist tradition. These include the work of Adler and Goodman (1986), whose social studies methods course enabled prospective teachers to critique school and social contexts of education; Smyth and Gitlin's (1989) emancipatory approach to supervision; Zeichner and Liston's (1987) "inquiry orientation" to student teaching; Ginsburg's (1988) proposal for more progressive political activity by teacher educators; and Maher and Rathbone's (1986) feminist-inspired proposals to correct gender inequities in schools and society.

Ultimately, the aim of research in teacher reflection is not only to understand the contexts and processes about which reflection seems to occur, but to create the conditions that evoke a capacity or disposition for reflection among teachers. Richardson (1990) reviews the evolution of reflective teaching and teacher education. She portrays the failure of competency-based teacher education—the view that prospective teachers should be able to display the behaviors that are seen to be necessary for teaching to be effective (determined by researchers on teaching effectiveness who related these behaviors to student achievement on standardized tests). For Richardson, this failure has occurred because "the nature of teaching seemed more than the sum of a set of behaviors learned in teacher education programs, even though those behaviors were shown to be performed by effective teachers" (1990, p. 15). She continues:

Schön's concepts of knowledge and reflection-in-action have both validating and legitimizing qualities. They describe a way of thinking in action that makes sense to practitioners and that applies to practitioners in many different professions, not just teaching. . . . We must avoid turning the notion of the reflective teacher into a technical approach to teacher education, as well as resist pressure for research and evaluation approaches that examine the reflective teacher education process in a positivistic manner. We in teacher education should, instead, concern ourselves with the content of teachers' reflections—with what teachers will view as problematic. (Richardson, 1990, p. 16)

A recent analysis of "working knowledge" by Yinger, Hedrick-Lee, and Johnston (1991) is of particular interest here because it is very similar to the notion of reflective practice. In moving away from the technical (positivistic, Cartesian) conception of knowledge that has dominated social science and educational theory in this century, Yinger et al. propose the notion of "ecological intelligence," that is, knowledge inherent in cultural, material and physical, social, historical, and personal systems. They suggest the knowledge inherent in each of these systems can best be understood when it is in action:

When a person engages in activity, the knowledge inherent in different systems helps to determine structure and meaning in the activity . . . we offer three propositions about "ecological intelligence": (1) Knowledge is inherent and widely dispersed across complex systems of information and action—cultural, physical, social, historical, and personal; (2) Knowledge within these systems becomes available as working knowledge in particular activities and events; (3) Working knowledge is constructed jointly through responsive interaction (conversation) among all the participants (systems) in an activity. (pp. 10–11)

Yinger (1987, 1990) and Lave (1988) use the notion of "improvisation" to explain the continuity people experience when they are engaged in working knowledge. Thus, working knowledge, as it is manifest in action, consists in a dialectical relationship among the participants and materials of a situation—in this sense, knowledge resides in the interaction of systems, not solely in the mind of the individual. Yinger et al. (1991) extend this idea even further to suggest that in the context of the classroom there are ongoing "conversations-of-the-moment" within the cultural, physical, social, historical, and personal systems that are present, and among all of the systems:

The special role or responsibility of the teacher in the classroom is to understand the conversations which are occurring within and among all the systems and to recognize which are appropriate for the classroom activity. The teacher acts as guide to and translator of the structure, action, and information contained in each system. The degree to which each system is contributing to the activity determines the authenticity of the conversation. The systems may not be contributing equally in every situation. Not all of the systems necessarily have contributions to make to the activity ("the conversation-of-the-moment"); however, over time all systems have vital contributions to make. . . . The effective teacher finds the appropriate conversation for the moment, one that continues prior conversations in each of the systems and connects these to ongoing conversations and sets the stage for future conversations. It is this notion of connectedness between past, present, and future as well as among the systems that defines appropriateness of the conversation. (pp. 34–35)

Although in a somewhat different language, these very issues emerge in a variety of professional contexts (many of which are related to educational issues) in Schön's (1991) recent publication, *The Reflective Turn*. This is a book of 14 case studies from diverse practice settings—from psychotherapy and social psychology to counseling hearing-impaired children, from town planning to organizational development in an industrial firm, from managers and professional staff in a not-for-profit organization to the educational context of schools. G. Erickson and MacKinnon's (1991) case draws from a 5-year collaboration with a group of three science teachers working together to make school science more meaningful to children. Their chapter focuses on the work of one of the teachers, Colin, with a student teacher, Rosie, and their efforts to understand the problems a group of students had in learning about electricity. Bamberger's (1991) case deals with her work with a group of teachers from Graham and Parks Elementary School in Cambridge, Massachusetts, where they developed the Laboratory for Making Things. The account is about students and teachers moving between material objects and the computer—between "hand knowledge" and "symbolic knowledge." Russell and Munby's (1991) cases deal mainly with teacher education—focusing on the puzzles of practice that confront two teachers and give rise to their "reframing" of experiences with their own intervention of pupils, involving parents in a Grade 1 reading program, attempts at inquiry-oriented science teaching versus traditional teaching, and their understanding of students' conceptions in science. Clandinin and Connelly (1991) report on an elementary school principal, Phil Bingham, whose "personal narrative" came to bear in their interpretation of his attempts to establish an atmosphere of trust among the staff and community at Bay Street School in Toronto. Newberg's (1991) chapter deals with his work with parents, teachers, and administrators in coming to a new understanding of a massive dropout problem in Philadelphia. All of these case studies, and other chapters in *The Reflective Turn*, attempt to "give practitioners reason" by the authors' attempts to understand teachers' complex worlds and ways of making sense. As Schön (1991) puts it:

> What the authors of these cases do *not* do is to regard practice mainly as a field of applications. They do not try to subsume practitioners' decisions under a research-based theory or model—decision theory or microeconomics, for example. When they bring an explicit theoretical framework to their studies, they use it to guide observation, description, or analysis of what practitioners already know or how they already learn in the context of their own practice. . . . The reflective turn . . . offers, as a first order answer to the question, What do practitioners need to know?, reflection on the understandings already built into the skillful actions of everyday practice. (p. 5)

Thus, reflective practice is one rubric of educational research that looks carefully at the ways teachers make sense of their everyday activities and learn from their practices. Another area of research that informs a view of teaching as craft is narrative inquiry.

Narrative/Story Telling and the Cinema

Teachers' stories are often highly emotive and complex—not readily amenable to the usual form of reporting in educational research journals. One avenue that seems appropriate for communicating teachers' narratives and stories is the popular cinema.

In an invited address to the annual meeting of the American Educational Research Association, Lee Shulman (1989) lectured, in part, on the topic of pedagogical content knowledge. He drew upon the popular film *Stand and Deliver* to develop an illustration of this special form of teachers' knowledge—a powerful image of a frustrated mathematics teacher trying to get his class of rough-and-tumble high school students to learn calculus and thus better their chances to liberate themselves from the depressed conditions of life in south-central Los Angeles. Shulman described an episode from the film in which the teacher attempted to get a particularly "hard-to-reach" student to see how a negative number can be added to a positive number of the same absolute value (thereby coming to a new understanding of the concept *zero*). The teacher used the analogy of digging a hole in the sand—the amount of sand taken from the hole represented $+2$ and the amount of sand in the hole was represented by -2. With this analogy, the student was able to think about filling the hole and say that -2 plus $+2$ equals zero. This particular episode from *Stand and Deliver* is a compelling example of Shulman's idea of pedagogical content knowledge, as is the film in its whole a compelling portrayal of this teacher's craft. Interestingly, excerpts from *Stand and Deliver* are now being used for teacher recruitment commercials on national television networks in the United States.

A similar focus is central to the film *Dead Poets' Society*—an account of a teacher of English literature who dares to defy the tired and tiresome curriculum that has been used much as a matter of tradition in a prestigious private school for boys. Indeed, this teacher directs pupils to rip the first chapter out of their literature textbook on the grounds that it is unwarranted as a scheme for making judgments about the quality of literary works. In an unrelenting effort to step out of the stoic tradition of the school and make literature meaningful to these students, the teacher lives out his motto, *Carpe Diem* ("Seize the day"), and, through "unorthodox" teaching approaches, encourages his young students to "come alive" and stand up for their beliefs and aspirations (part of which involves their formation of the "Dead Poets' Society," a group of boys who sneak out of their dormitory and off to a secret place in the woods for poetry readings). One of the boys develops a passion for theater and begins to deceive his father, who sees no place in his son's education for such "frivolous" deviations from school work as taking on a leading role in a local Shakespearean production. The father is infuriated when he discovers that his son has deceived him and, after the boy's brilliant performance, tells him he will be leaving the school for a military academy. Later that night, in a state of extreme remorse and defiance the boy takes his own life—an event that leads to the dismissal of his teacher and the tragic portrayal of a lost battle—meaningful learning having been plowed under by the weighty and powerful conservativism and dogmatism that, in this case, is associated with a traditional private school. As an audience of educators, some of us are given to refurbish our own answer to the calling of teaching, and we revitalize our commitment to make schooling meaningful to our young people, despite the unwielding inertia of institutions.

Another popular film that tells a great deal about craft knowledge is *Madame Sousatzka*, a story based on a novel of the same title written by Sam Howard. The film portrays powerful feelings about teachers, not so much in terms of their passionate views about their purposes and their commitment to make the process of education meaningful to their clients, but of their love and care for students. Madame Sousatzka is a gifted piano teacher who prepares concert pianists. She struggles to harbor and develop her students' talents and strength in her own studio until they are "ready" for the concert stage. Sousatzka, herself, was paralyzed in her own career as a pianist, largely because she aspired so strongly to play in the concert hall that she did so before she had the "cybernetic skill" to deal properly with the anxiety that can bind any artist in hesitation and self-doubt. Frozen in a passage of music that she could not finish, she fled from the stage, never to return. Now, after teaching for many years, part of her special talent is in knowing when her students have the strength to face an audience. Much of the film tells the story of Madame Sousatzka's tutelage of a young boy named Manic. Predictably, Manic reaches the point when he is approached by agents who would eagerly schedule concert tours for him—he is a brilliant pianist, and he desperately aspires to the concert stage. The boy's mother, too, is enormously impatient for the proceeds that are inevitable from her son's impending fame as a prodigy. Yet, Sousatzka knows that Manic is not ready, and much of the film portrays her feeling for the boy and how she struggles, despite the pressures from agents and his mother, to nurture the strength he requires to enter the professional concert circuit successfully.

There is a wonderfully sensitive moment in the film when a former student of Sousatzka—Edward, now an accomplished concert pianist himself—comes to visit her:

Edward: How old is he [Manic]?
Sousatzka: The same age as you were when you left me.

Edward: I never left you . . . you know that. Even now . . . today, when I'm playing in a concert . . . somehow you're with me. When things get tough—when the piano behaves like a monster . . . and the audience is like another monster—then I hear you . . . yelling at me . . . telling me to get on with it. And I listen. And I know. I know that all those hours with you have formed a little core of strength in me . . . and that gets me through it.

Sousatzka: You know what they say about me? They say I made terrible scenes when you left me for Leo [another piano teacher—connected with the concert circuit]. Like a jealous mistress. That I was in love with you. Oh boy!! *Well, of course I was in love with you. Isn't every mother in love with the son she creates? So why shouldn't I love in the same way?*

The film is a gripping account of a teacher's love for her students, a love that seems to be driven by Sousatzka's desperate attempt to prevent her students' premature entry into the concert scene and the subsequent heartbreak—that she knew so well—of failing to live out a dream. Moreover, one gets a special feel for Madame Sousatzka's craft as a pianist and teacher. Edward speaks of this element so well when we hear how Sousatzka comes to him when the piano and the audience are behaving like monsters. The image of her craft has formed a core of strength in him that he recognizes in times of difficulty—he sees his response to these difficult situations in terms of the image of Sousatzka as a pianist and teacher: a strikingly rich portrayal of her craft knowledge and how it has become a part of the repertoire of her student as he deals with the materials of situations.

We have included examples of film and poetry in our review as important media for communicating about craft knowledge in teaching.[17] The images that can be expressed by these media seem to be rich enough to capture craft.

Similarly, narrative inquiry (Connelly & Clandinin, 1990) is most informative of the ways in which teachers make sense of their lives and roles in educational contexts. Essentially, the study of narrative is the study of the ways humans experience the world. Accordingly, education is seen to be "the construction and reconstruction of personal and social stories; teachers and learners are storytellers and characters in their own and other's stories" (Connelly & Clandinin, 1990, p. 2).

Teachers' stories are seen by Connelly and Clandinin as much more than mere accounts of their lives—they are actually seen to be a critical medium by which teachers make sense of their work:

We are continually trying to give an account of the multiple levels (which are temporally continuous and socially interactive) at which the inquiry proceeds. The central task is evident when it is grasped that people are both living their stories in an ongoing experiential text and telling their stories in words as they reflect upon life and explain themselves to others. For the researcher, this is a portion of the complexity of narrative, because a life is also a matter of growth toward an imagined future and, therefore, involves retelling stories and attempts at reliving stories. A person is, at once, engaged in living, telling, retelling and reliving stories. . . . Seeing and describing story in the everyday actions of teachers, students, administrators, and others requires a subtle twist of mind on behalf of the inquirer. It is in the telling and retellings that entanglements become acute, for it is here that temporal and social, cultural horizons are set and reset. (Connelly & Clandinin, 1990, p. 4)

Narrative inquiry encompasses the close examination of "personal practical knowledge" of teachers—"a term designed to capture the idea of experience in a way that allows us to talk about teachers as knowledgeable and knowing persons" (Connelly & Clandinin, 1988, p. 25). Knowledge, in this sense, is seen by Connelly and Clandinin as residing in both mind and body, manifest in our practices. Personal practical knowledge is "a particular way of reconstructing the past and the intentions for the future to deal with the exigencies of a present situation" (p. 25). The stories that teachers tell and retell (and live by) are, therefore, seen to provide a critical purchase on their personal practical knowledge (generally speaking, their craft knowledge). The essential value of telling and listening to teachers' stories follows directly from this idea:

We need to listen closely to teachers and other learners and to the stories of their lives in and out of classrooms. We also need to tell our own stories as we live our own collaborative researcher/teacher lives. Our own work then becomes one of learning to tell and live a new mutually constructed account of inquiry in teaching and learning. What emerges from this mutual relationship are new stories of teachers and learners as curriculum makers, stories that hold new possibilities for both researchers and teachers and for those who read their stories. (Connelly & Clandinin, 1988, p. 12)

Teachers' stories are inextricably bound with teachers' lives. The study of teachers' lives enables us to understand the biographical influences that affect the development of craft knowledge.

Teachers' Lives

The study of teachers' lives is undertaken to understand the socializing influence of the full range of life experiences on classroom practice, the context in which teachers formulate their craft knowledge. Thus, it represents an important genre of research for understanding why some teachers generate craft knowledge in a manner consistent with the progressive and radical traditions and why others construct an essentially conservative appreciation of their craft. Goodson (1991) argues that, if practice is the primary arena in which teachers develop craft knowledge, then the most promising point of entry for researchers is not in an examination of practice per se (which, for Goodson, represents the area of maximum vulnerability for teachers) but in the study of teachers' work in the context of their lives. By beginning here, researchers find the voice of the teacher not simply as a practitioner but as a striving, purposeful person as well. Issues of class, gender, life-style, and life cycle are all formative influences upon the teacher and his or her teaching. They are influences that should be acknowledged and deserve a hearing. In understanding something as intensely personal as teachers' craft knowledge, Goodson says, it is critical that we know something about the person the teacher is. By tracing a teacher's life over time, it becomes possible to view the changes and underlying forces that influence that person at work—to estimate the part that teaching plays within the overall life of the teacher. Woods (1979) frames it thus: "The social scientist has to begin to develop a perspective that enables him to develop the connection between macro-sociological and historical processes on the one hand and individual biographies on the other" (p. 3).

A useful exemplar of this aspect of teachers' craft is the research program looking at teachers' life cycles reported by Huberman (1989, 1991).[18] His interview study conducted between 1982 and 1986 with 160 secondary Swiss teachers revealed two career routes hitherto not reported in the literature: harmony recovered and positive and/or negative renewal.

Harmony recovered essentially maps a trajectory from painful beginnings through a period of stabilization to experimentation. In other words, the initial stage of survival and discovery involved much more survival than it did discovery. Nevertheless, close to half the Swiss teachers interviewed by Huberman transcended the pain of survival to enter into a period of experimentation rather than reassessment after the stabilization stage. Hence, it was not inevitable that teachers who had painful beginnings to their teaching career would be given to reassessment, conservatism, and bitter disengagement.

Renewal with a positive or negative issue represents an expansion of what was previously understood by disengagement. Renewal was a stage found to lead into three broad types of disengagement: positive focusing, defensive focusing, and disenchantment. Positive focusing represents teachers who disengage with serenity. Renewal has been expansive for them and they now wish to contract by focusing on a preferred grade level or subject matter or by disinvesting in schoolwork and increasing outside interests. Defensive focusing has a similar reduction of commitments, but the tone is different. Teachers in this category begrudgingly underwent renewal and now, having paid their dues to educational change, as it were, wish to be left alone. Disenchantment carries with it a sense of distress and gloom. It corresponds with bitter disengagement. Most of the teachers in this trajectory approved of the educational changes in the renewal stage but are not sanguine about the outcomes. They are critical of "bureaupathological" (Hodgkinson, 1978) administrators, mindless paperwork, and ill-advised policies. They talk in terms of "lassitude and fatigue" and are considerably vexed by "turncoat" power brokers who they thought would bring the reforms to fruition.

The intriguing finding from Huberman's research is that he was able to distinguish which career factors associated with each of the three stages following renewal. Positive focusers tended to steer clear of schoolwide innovation. They defined and stuck with their areas of professional (and outside) interests, while at the same time investing heavily in classroom-level "tinkering" to make small changes and improvements in their own way and in their own time. By contrast, negative focusers and those who were disenchanted were heavily involved in schoolwide innovation. Huberman (1991) sums it up thus: "Tending one's private garden, pedagogically speaking, seems to have more payoff in the long haul than land reform, although the latter is perceived as stimulating and enriching while it is happening" (p. 183). He concludes from this study that the best scenario for satisfactory career development is through a "craft" model— one that encourages and supports teachers in their tinkering around in their classrooms as a way of expanding and improving their instructional repertoires. Teachers who followed this model were more inclined toward a progressive or radical formulation of their pedagogical content and learner knowledge than those who had become disenchanted by widespread school renewal.

Nias (1984) attempted to understand the personal dispositions that teachers bring into the classroom context. She discovered that teachers have a "deeply held substantial view of self" (p. 268) that they felt had to be defended and affirmed in contemporary classroom contexts. She speculates that this need to protect their "substantial self" can frequently lead teachers to avoid situations that might bring disagreement or differences, thereby reinforcing norms of isolationism and privacy in the workplace. We would argue that such a tendency is associated more readily with teachers who hold to the conservative, as distinct from the progressive and radical, traditions of thinking about teaching as craft.

Butt and Raymond (1989) and Raymond, Butt, and Townsend (1991) explore the influence of teachers' life stories on craft knowledge through collaborative autobiography. They illustrate how approaches to teaching are profoundly influenced by such factors as ethnic background, social class origins, experience of living in different cultures, gender influences, and range and type of previous teaching experiences. Early personal experiences were found to be particularly formative, with teachers repeatedly invoking previously occurring significant others, events, or relationships to infuse meaning and purpose into current classroom realities. They suggest that teachers' personal biographies not only influence their responses to context and opportunities, but also help to frame the search for specific professional development. This is particularly the case when teachers possess the capacity and confidence to reinterpret external mandates for curriculum change in terms that are compatible with their own grounded dispositions and their students' needs as they see them. Such grounded dispositions and interpretations of students' learning needs are frequently a condition of and target for teacher research.

Teacher (Action) Research

Teacher research—the "systematic, intentional inquiry by teachers" (Lytle & Cochran-Smith, 1990, p. 85) into their craft—has received considerable attention of late[19] in the works of Bissex and Bullock (1987); Cole (1989); Elbaz (1991); Elliott (1990); Ellwood (1991); Gomez (1990); Gore and Zeichner (1990); Goswami and Stillman (1987); Hustler, Cassidy, and Cuff (1986); Lampert (1985); McNiff (1988); Miller (1990); Mohr and MacLean (1987); Oja and Smulyan (1989); Rorschach and Whitney (1986); Ross (1987); Ruddick (1985); Ruddick and Hopkins (1985); Strieb (1985); Tabachnick and Zeichner (1991); Tikunoff, Ward, and Griffin (1979); and Wood (1988).[20] Publications such as *Hands On* (the Foxfire Fund, Inc., Rabun, Georgia), *Outlook* (the Mountain View Publishing Company, Boulder, Colorado), *Democratic Education* (the Institute for Democratic Education at Ohio University), and *Rethinking Schools, Working Teacher*, and *Our Schools Ourselves*, published respectively by groups of teachers in Milwaukee, Vancouver (Canada), and Toronto, all serve as outlets for teachers' accounts of their practice. In addition, there have also been three important reviews of teacher research in Cochran-Smith and Lytle (1990) and Lytle and Cochran-Smith (1990, 1991).[21] Accordingly, this subsection will examine new examples of teachers' writings and studies that have not previously been included in a review.

A new journal has recently started up under the auspices of Brown University and Ted Sizer's Coalition for Essential Schools. *The Teacher's Journal*, edited by Joseph Check and Joseph McDonald, attempts to provide a forum for developing the voice of teachers in the educational process. In a similar vein, the British Columbia Teachers' Federation in Canada has instituted the Program for Quality Teaching, designed to encourage teacher inquiry in the practice setting, and *Voices of Teaching*, a series of monographs edited by Ted Aoki that provides teachers with opportunities to share their "lived meanings of teaching" (Aoki, 1990, p. 1) with their colleagues far and near. The Langston Hughes School-Based Research Group (1988), a group of teacher researchers at the Langston Hughes Intermediate School in the Fairfax County Public Schools, has also engaged in inquiry in the practice setting. Their inquiries, published in a 1988 monograph, examined different strategies for helping minority and underachieving students learn in their school. This subsection will draw primarily on these sources as exemplars of teacher research.

The ambivalence and uncertainty associated with teaching causes teachers to be constantly searching for ways to improve their craft. They continually raise questions about issues or dilemmas that they face in the dailiness of teaching. For Saul (1990), a mathematics teacher of 20 years standing, the central question is not "How do we teach? It is instead, What do we teach in the face of all that we might teach?" (p. 2). This question is important to him as he struggles to decide what not to teach when deliberately setting out "to cover less material in more time" (p. 5). Lewis (1990) takes the question of curriculum content to a deeper level. For him, the searching question is existential and has to do with how teachers can transcend the burden of prescribed curriculum content to connect with the lives and minds of students:

How can I be both teacher of a curriculum and distance myself from it personally [as is his preference]; how can I be both with the students and not with them in the same instant? Where is the "transformative intellectual" that Giroux so passionately argues for? (Lewis, 1990, p. 49)

Lewis concludes that teaching is a paradox and that teachers face dilemmas that require choices. There are no answers, only tensions:

To live in paradox is to live in ambiguity, in openness, in difference, in conflict, in difficulty, in between. Between opposites, between heaven and earth, between birth and death, between freedom and authority, between subject and object . . . to live in paradox is to concede a tension. . . . Being human we need to recognize our inevitable situation of life as being paradoxical but within which we can and must exercise our choice. Being teachers we need to recognize our choice and our responsibility to follow the curriculum or to make our own course. (1990, p. 50)

Case (1990) contends that the choice open to teachers is neither one of following the curriculum nor one of charting their own course, arguing that "the idea that a teacher is given a curriculum to teach is too simple" (p. 21). This is because the curriculum, whether prescribed or teacher developed, "is subject not only to action upon it but to interaction with it, and thereby to constant and unplanned modification" (p. 21). In other words, even when the content of the curriculum remains unchanged from that which is prescribed or locally developed, its meaning does change as a result of the interaction that takes place when teachers encourage a diversity of student responses. "The meaning of curriculum changes not because new substance is added but because new interactions take place between the book [a literary novel] and the reader, between the curriculum and the students" (p. 22).

Case relates how the meanings ascribed by students and by himself to the novels in his high school English classes changed during the course of dialogue and interaction. He notes that in order to encourage student input and interpretation, he, as teacher, had to withhold his own ideas about what was happening in the various novels and what the different events could mean. He defines education as being fundamentally interactive and argues that

successful teaching engages the students with the subject, and their engagement affects the teacher and the subject, as well as the students themselves. Discourse is often discussion, and learning is just as likely to be the discovery of relationships as the acquisition of information. The number of questions with right and wrong answers is minimized; questions instead tend to focus on responses, reaction, opinions, criticisms, and judgments. The most valuable responses are those that are different enough to provoke thought and more discussion. (Case, 1990, p. 28)

Mohr (1988) also probes the curriculum in the classroom. In her dual capacity as colleague teacher and consultant to the Langston Hughes School-Based Research Group, she observed nine teacher researchers at work and drew insightful conclusions about the expectations and configurations of the classroom curriculum. First, she notes that there are four curricula present in every classroom and that teachers work with all four versions simultaneously to bring about an intersection:

Official authorized curriculum and unofficial, more personal curriculum exist in every classroom. Teachers and students contribute to both. They each have perceptions of the official curriculum, they each have their own unofficial curriculum, and they each are making use of both at any one time in the classroom—four curricula in all. . . . Teachers work with all four curricula at once, learning the students' official and unofficial curriculum and managing their own, while looking for and seizing opportunities to cause as many students as possible to absorb the official authorized curriculum. . . . When all four curricula intersect at once, the classroom curriculum becomes functional. At the intersection, student learning of the teacher's official authorized curriculum can take place. (Mohr, 1988, pp. 64, 66)

This intersection brings the official curriculum alive by connecting it with students' interests and intentions. The path to such a functional classroom curriculum runs through careful use of the unofficial curricula:

The unofficial curricula provide the path to the official curriculum and both students and teachers take advantage of this pathway. . . . Students who do not achieve as well as is expected of them are not making enough connections enough times to become proficient at the official curriculum. (Mohr, 1988, pp. 67, 68)

For students to reach this point of connections, teachers need to learn how to integrate the official and unofficial curricula in their classrooms. Mohr suggests that in-service programs could take on such a focus in order to help teachers tap into the intentions and interests[22] that students bring with them into the classroom:

Inservice and staff development programs need to integrate curriculum, management, and child development (all the unofficial and official curricula) to present material in recognition of the actual curriculum that exists in classrooms. . . . Effective teachers of official curriculum acknowledge all the curriculum alive in the classroom, even that of students who bring alien unofficial curriculum to the classroom and see the official curriculum as far removed from their needs. All students, but especially those who are experienced at failing, benefit from teacher strategies that make curricula intersecting possible. (1988, pp. 69, 70)

She ends with the recognition that a teacher's effectiveness in working in this way with individual underachieving students is dependent on a structural constraint: "The size of a class affects the teacher's ability to assist individual students in becoming successful at the official curriculum" (p. 71).

Meeting the individual needs of minority and underachieving students was an important focus of the Langston Hughes School-Based Research Group. Christian (1988) explored the extent to which student achievement in science was enhanced through individualized attention. She found that student attitude toward learning is critical to success and that teachers can exert considerable influence on student attitudes through individualized attention. She concludes that "when we, as teachers, buy into meeting individual needs, students will in turn buy into learning" (p. 8). Doyle's (1988) project with learning-disabled students in Basic Skills Review (BSR) produced a similar finding. Student attitude was again found to be crucial to academic achievement; however, Doyle also discovered that by providing learning-disabled students with priority writing time on the computer, she gave them the motivational force to begin to compete on an academic level with other students. A similar motivational finding was reported by Johnson (1988). She explored what happened when students in seventh-grade science wrote consistently in learning logs. She found that students were eager to share what they had written, that they were surprised at how much they found to write about, that they prepared differently for a writing test because they viewed writing with pleasure, and that minority and less successful students enjoyed success in the writing and sharing process. All of this, she reports, gave her more immediate feedback from the students, and she, in turn, observed an improved attitude toward science and toward her as teacher. Lloyd's (1988) experiment with communicating high expectations with similar kinds of students resulted in an important motivational insight for her as teacher. She found that the students needed to know, on a daily basis, what she expected of them.

Meeting the needs of minority and underachieving students can also be undertaken collectively. Jeffrey (1988) chose to explore what happens to such students in social studies when they learn in groups. She found that the use of group work in this context produced improved student understandings of both the curriculum content and the group processes. She suggests that teachers need to use more group work in the classroom but cautions that the key to its success lies in students' writing about the group process and in their evaluation of the groups themselves. She concludes that group work is a process that can be applied equally well to minority and nonminority members of the class:

It gives minority students the opportunity to develop leadership skills and to grow in confidence and self-awareness. Without singling out minority students, they are allowed to become comfortable and active members of a group. I offer that feeling comfortable within one's academic environment is one key to successful learning. (Jeffrey, 1988, p. 36)

A study that provided opportunities for teachers to begin to feel comfortable and active in their educational environment was the work of the Boston Women's Teachers' Group, *The Effect of Teaching on Teachers* (Freedman, Jackson, & Boles, 1986). This group of teachers met as a support group for 3 years. They were committed to helping one another cope with the stresses of their daily work. At the same time, they decided to study how their work conditions affected them as teachers. They were particularly interested in how their sense of isolation from one another and from other organizational conditions influenced their sense of efficacy, their self-esteem, and their attitude toward work. One of their important findings was that teachers work "in an institution which supposedly prepares its clients for adulthood, but which views those entrusted with this task as incapable of mature judgment" (Freedman, Jackson, & Boles, 1983, p. 263). One of the consequences of this research was that its findings were used to combat the individualistic bias in the school reform of the 1980s that served to direct teachers' sense of frustration with their work conditions away from a critical analysis of the structure of schools to a preoccupation with their own individual failures (Liston & Zeichner, 1991, p. 150).

Smith's (1988) project exploring what happened to the work conditions of underachieving seventh-grade students when they were taught mathematics using a project/gaming approach revealed a different aspect of comfort within one's academic environment. He found that the most successful projects and games were those that the students could integrate into their own lives. The students were essentially provided with a process that they could relate to and own. They were part of the process. Smith was not the teacher handing out homework or disciplining the students; he became another player, but one who communicated and reinforced high expectations for the project or game that provided the process for learning. The nature of the learning process, then, was highly dynamic and interactive.

Reeves (1990) focuses her case study on a dimension of interaction, namely, teacher listening. She relates how, in her first year of teaching, the students had numerous stories to tell, both within and without formal class instruction. As a beginning teacher, Reeves attempted to maintain a strict task focus to her lessons. But the students were naturally curious about her, asking many and varied questions. In the process of fending off such inquisitive attitudes to return the lesson to its curriculum focus, Reeves discovered that the students needed to talk to her, to tell her their problems, their successes, their dreams. The important point she makes is that "in the listening is the teaching" (p. 20). That is, the more she listened, the more the students opened up and the more she subsequently did too. The consequence of listening, then, is a truly interactive approach to classroom instruction. It also brought about in Reeves certain insights about the craft of teaching:

As I listened to my students, I found myself opening up to them more and more. At the Christmas break, I was to be married, and they eagerly questioned me about my wedding and my future husband. Instead of chastising them, I found that my willingness to reveal my life, my other "self," caused me to appear more human to them. As we laughed and talked at the end of the day, I became less and less, yet more and more, of what a teacher is. Instead of a teacher and pupils, we became a group committed to respecting one another, learning and listening, and picking up the mathematics along the way. During that year I learned a bit about the place of dignity, not the dignity of the teacher, but the dignity of even the weakest student. . . . What all of them [the students] lacked was the recognition of their basic humanness, their worth as human beings. (Reeves, 1990, pp. 22–23)

Law (1990) extends the notion of listening in teaching. Listening is important to teaching and learning because its true meaning is to be found in silence—"the silence necessary to truly hear the self and others" (p. 24). Silence, then, is the essence of listening. Law contends that it is silence that enables one to make sense of dialogue. The silence enables one to learn of oneself and of others because, when one is silent, one is listening truthfully:

The self does not respond, but allows us the time and space to make our own responses. Yet, it is not a passive silence; it is a very active voice. As we question who we are, what we think and what we do, this silence is our guide. It is our line of self-truth drawn through struggle and time. It will accept no lie. But to neglect this silence is to do ourselves and others a great disservice. For if we cannot hear ourselves, can we truly hear others? (Law, 1990, p. 24)

Law (1990) recounts a time when he had to relearn the value of silence.

He had given his English 12 class a public speaking assignment that proved too much for one of his students:

Lorna was a top academic student, a basketball star, and an excellent horsewoman with honors in several competitions. She was always talkative and animated and it was with genuine surprise that I received her request to be let out of the public speaking assignment. Quietly, nervously, she came after school to make her request. She said simply, "I can't." I laughed and said I'd always had a problem with public speaking, too, and the fact that I did it every day made it no easier. It was her turn to be surprised. So we talked about what made public speaking difficult and how we might accommodate some of her concerns. I gave her all the reasons for seeing public speaking as a valuable experience. . . . After considerable discussion . . . we decided that Lorna would give her speech privately after class, with three friends she would choose. I was puffed with professional pride. I felt I had given where needed and had gained a great deal for Lorna. . . . Next morning Lorna arrived early to class. . . . She said she had decided to do the speech in class after all. I hailed myself a hero for my good work. I magnanimously allowed her to do the speech the next day . . . [when] she came to class early, but this time to say again, "I can't." I was not ready to lose my hero status and refused to listen silently. I chose to make light of her anxiety. She persisted, talking as quickly as she could and giving all her reasons in a rush of words, and I could see in her eyes that nothing I could say would change her mind. The mantle of hero fell. Without thought or tact, I flippantly called her "wimp." . . . The silence that fell was a silence of no sound, of no breath. It was the silence of time stopped, a silence of pain. Tears sprang instantly to Lorna's eyes. She had asked me to listen and I had insisted on talking. I had not allowed her to be heard. As Lorna rushed from the room, silence remained. . . . My failure to be silent (to listen truthfully) was jarringly apparent. . . . I was listening to my story, not to Lorna's. By insisting on talking I had left no silence with which to hear Lorna. She had not asked for answers, only to be allowed the time and space to find her own understanding. (pp. 24–25)

Tsujimoto's (1990) experiences as a student were quite different from Lorna's. He recounts the story of his life as a student in an attempt to grapple with a critical question that troubles him now as a teacher: "What makes a teacher memorable or great?" (p. 8). He tells his story as a way of addressing the issue:

It seems that all his life he had been searching for a theory that would explain the mystery of his behavior in the classroom, since, upon reflection, much of what he did did not seem rational. That is, much of his life had been consumed in the search for a language that, in addition to improving his teaching, would clarify what it was that he did right; that explained why his students found him an effective teacher. For it was said that he was strange, distant, and tyrannical. . . . In his early life . . . he had the fortune of witnessing . . . passion in certain of his [own] teachers—an angry passion, a quiet passion, a querulous passion— apart from the particular subjects they taught, apart from the singular theories according to which they operated. He recognized that, with this passion, they became, in a sense, what they taught, transcending personality, transporting the students to a world that was finer, purer, and nobler than any they had previously known. So he too aspired to the creating of worlds. (Tsujimoto, 1990, p. 6)

After 7 years of obsessive striving as a classroom teacher, the insight suddenly strikes. The story of his life is telling Tsujimoto that what is ultimately taught is the teacher, the one who embodies the idea that students have the power to create their own worlds. . . . The great teacher—or, as is more often the case, the instances of great teaching by good teachers—transcends personality and philosophy, which are merely vehicles for the illumination of wisdom—or knowledge made beautiful. In such instances, the teacher is so immersed in his reading, or in his speaking, or in his characterization that he sheds all consciousness of self. And the students see embodied before them the very spirit of learning: the very subject to be learned. It may then be said, "What is ultimately learned is the teacher." (1990, pp. 7–8)

Tsujimoto (1990) ends his story by asking himself a further question. But this one is not so much a matter of serious inquiry as it is a challenge to other educators: "How can the mind be reached except through the heart?" (p. 13).

Hogue (1990) wants to reach the minds of students and is convinced that it will not happen if we naively assume that the world created for children in school serves as a sufficient replacement for the real world. Her narrative concerns the ways in which she attempted to take her students out into the real world and bring the real world from outside into the classroom. This, she contends, is a teacher's pedagogical responsibility:

There is a real world which alone gives life to school and beckons us forward to knowledge. It is a rich, mysterious and challenging world, untidily packaged and unexpectedly revealing. It is this world we must welcome into our schools as our pedagogic responsibility. . . . It is not enough for us to be conscious of the world beyond the classroom. Mindfulness of it must live in the classroom, in our pedagogic action and in the lives of children. . . . Pedagogic action calls us to guide children, but the guiding also calls us to pause with them. Lingering with children in the novelty of experience clarifies new understanding and reaffirms our being in the world. (Hogue, 1990, pp. 45–46)

In New York City, Debbie Meier and the faculty of Central Park East Secondary School (CPESS) have been involved in pausing with students as a way of transforming a traditional secondary school curriculum into one that values and respects the ideas of adolescent learners and attempts, thereby, to make the process of learning meaningful and relevant. They set out to engage the students by trying to understand the world in which they lived and by attempting to make connections between what is studied in school and the students' world. Being alert to students' interests and using them as starting points for the content being examined led the teachers at CPESS to cover curriculum content in the form of classroom themes. To ensure this greater depth in their teaching, they organized a multigrade integrated curriculum around two courses: (a) humanities and (b) math, science, and technology. During the school's first year, teachers chose to focus on "explorers and exploration," a fitting theme offering innumerable possibilities in both areas of study: the development of new ideas, literary forms, scientific discoveries, technological breakthroughs, music, and architecture; and the setting of new styles, new ways of thinking, and new ways of knowing. The teachers' sense of efficacy (and the students') so increased that the entire faculty committed itself to a statement of purpose and mission. It is called *The Promise*:

At CPESS we make an important promise to every student—one we know we can keep. We promise our students that when they graduate from CPESS, they will have learned to use their minds—and to use their minds well. In every class, in every subject, students will learn to ask and answer these questions: *(1) From whose viewpoint are we seeing or reading or hearing? From what angle or perspective?; (2) How do we know what we know? What's the evidence, and how reliable is it?; (3) How are things, events or people connected to each other? What is the cause and what the effect? How do they "fit" together?; (4) So what? Why does it matter? What does it all mean? Who cares?* [italics added] We are committed to the idea that a diploma is a meaningful piece of paper, not one that says only that a student has "stuck it out" through high school. A CPESS diploma tells the student—and the world—that the student has not only mastered specific fields of study but is curious and thoughtful, above all, has learned "how to learn" and to use his/her learning to deal with new issues and problems. (Central Park East Secondary School, 1988)

The faculty at CPESS have chosen to enshrine for all to see their passionate care for students and their learning. The drive to teach students to be critical by first seeing learning from their perspective and then connecting the content of the curriculum with the outside world is a feature of teacher research. What is not always seen is the "inside" perspective of what it means to teachers to engage in the process. The Langston Hughes School-Based Research Group monograph includes three reports that analyze what the process of teacher research does for the teachers themselves.

Harrison (1988) examined what happened when nine teachers in the Langston Hughes Intermediate School engaged in classroom research in a collaborative manner. He found that the process required considerable personal definition on the part of the teacher researchers, and this proved to be difficult. However, until the teacher researchers defined the process for themselves and became comfortable with it, a focused schoolwide plan of action was inappropriate and unlikely to occur. Moreover, Harrison found that most of the teacher researchers at Langston Hughes were initially familiar only with quantitative research designs. Consequently, the use of qualitative data required most of the teacher researchers to modify their thinking about research and to learn new skills before conclusions about their efforts could be reached. These experiences and the process of collaborative sharing brought with them an underlying sense of anxiety and struggle as teacher researchers grappled with the self-examination inherent in the process:

Teachers who wish to conduct research in their classrooms need to know that it is normal and okay to struggle, to feel anxiety, to be frustrated as a result of their efforts. (Harrison, 1988, p. 99)

The search for answers to [their] questions was both exhausting and invigorating . . . teacher research involves not only reflecting on teaching practices and theory but facing oneself honestly. Self-exploration and self-exposure are as much a part of the teacher research process as selecting and answering a research question. (Nocerino, 1988, p. 90)

Cricchi (1988) discovered that the monthly sharing seminars placed most teachers on the horns of a dilemma. One teacher wrote:

The monthly seminars were both good and bad for me. I loved the intellectual stimulation and sharing of ideas. I was always inspired by or learned something from (or both) each one. However, being out of my classroom was a problem for me. (p. 75)

Despite this dilemma, Cricchi found that the Langston Hughes teacher researchers developed a new awareness of the complexity of individual needs and a warehouse of strategies to help all students as they engaged in the process of collaborative sharing. Teacher research is therefore like a double-edged sword, helping teachers to understand and develop their learner-focused craft and, at the same time, revealing much about themselves as persons and learners. The role of collaboration and support from a group of colleague teachers was an important aspect of a process that was both stimulating and anxiety ridden.

Many of the teachers whose writings have been cited are, in some way or other, associated with attempts to change pedagogical practices in schools. The Coalition for Essential Schools, based at Brown University, is one such attempt formulated around nine common principles (see Appendix A) derived from Sizer's (1984) study of American high schools. The Coalition encourages teacher research and attempts to provide the necessary organizational support for such inquiry through a network of schools.[23] The senior researcher for the Coalition is Patricia Wasley, who recently (1991b) documented an insightful case study of three teachers' attempts at changing their teaching practices. The changes in practice fell into three broad categories: reconceptualizing curriculum, assessment changes, and role changes. In reconceptualizing curriculum, the teachers began to plan from questions and problems, not from textbooks. Wasley describes the sessions in which the teachers determined the essential questions (such as classroom themes) that would guide their teaching in an interdisciplinary fashion. She notes how these sessions serve as "excellent examples of teachers engaged in sophisticated, intellectual discussions about their craft" (p. 42). Such intense dialogue made them realize that "by using the text as one resource among many they were teaching students a very important skill—the ability to find information when they needed it'' (p. 42). This, in turn, led to the realization that students needed to practice important skills ''in more realistic contexts if they were to develop the ability to use those skills in their own lives'' (p. 42). Consequently, the teachers' reconceptualization of the curriculum involved them in making interdisciplinary connections and shifting consciously from an emphasis on coverage to one on learning in depth. This entire process brought them a ''clearer sense of the interrelatedness of curriculum, pedagogy, and assessment'' and the belief that ''they had to work on all three at once'' (p. 42).

The assessment changes reported by Wasley evidenced the teachers rethinking their earlier practices. Assessment became an opportunity for students to demonstrate their learnings. Teachers graded less and diagnosed more. Because they expected students to demonstrate what they had learned through performance and analysis, the teachers realized they had to teach the students the requisite skills. So assessment became inextricably linked with instruction. The role changes undertaken by the teachers were significant. Not only did they act in less isolated fashion with colleagues, but they also changed their approach within the classroom. They moved ''from quarterback to coach'':

This change represented a significant shift in practice. The teachers developed a new way of seeing the classroom and their roles in it. They began to see their students in a different light—no longer as passive receivers but as learners in need of powerful and engaging experiences. They began to understand that students are capable of carrying the ball, participating in the action, and that they need to do so if they are to learn. . . . Now they worked hard to get everything done before class—organizing the lesson, preparing descriptions of the projects, getting the props ready, anticipating problems and questions, upgrading their background knowledge—so that their time in class was free to watch and listen to the students. They had to learn to be quiet, not answer all of the questions, not to give information every time but to turn the learning back to students. This was difficult, as they all admitted that they liked the performance aspect of teaching. Still, they each described a very important moment of insight when they realized that their students needed the same opportunity to prepare and organize the material that they as teachers did. They discovered that they needed to do their own preparatory learning before class so that the students could fill that role in class. (Wasley, 1991b, p. 44)

As a consequence, teachers and students became inquirers, and teaching was viewed as a complex partnership wherein students became more responsible for their learning and were the recipients of increased respect. Building this kind of classroom collaboration takes time, but Wasley reports that all three teachers ''believed that their students were working at a more sophisticated level and that their students were generating more rigorous work'' (p. 44). Wasley (1991a) suggests that the logical consequence of teacher empowerment is less teacher-centered instruction and an improved intellectual environment in the school. Such results should lead teachers, she argues, to examine the legitimate role of students in the classroom in a manner similar to the one in which they have examined their own role in the educational enterprise. This would entail giving students a voice. She describes an experience that foreshadows what this might be like:

> I was working with a third grade class as they wrote and raised money for the printing of a book for children at a local hospital. We were struggling with the title; I didn't particularly like their suggestions, and so kept pushing them. Finally, an eight year old looked up at me and said, "Pat, figuring out the title of our book is our responsibility, not yours!" I immediately breathed in to bestow upon them my most authoritative, teacher-like voice when I stopped. They were right. They were learning the power of their own voices. Teachers deserve no less. (Wasley, 1991a, p. 30)

Major Practitioners Contributing to Craft Knowledge

This section of the review will examine closely the works of practitioners who have made important contributions to an understanding of craft knowledge.[24] Three authors—Herbert Kohl, Vivian Gussin Paley, and Eliot Wigginton[25]—have been selected on the basis of each being a recognized successful and experienced scholarly oriented practitioner. They also represent differences in gender and race[26] and exemplify the progressive and radical ways of thinking about craft. Their writings represent exemplary cases in which the wisdom of practice is documented. Wigginton represents a progressive scholarly practitioner whose writings embody wisdom gained over approximately the last 25 years. Paley is a female progressive educator whose work deals extensively with radical-like issues of social justice and cultural differences. Kohl is a radical scholarly practitioner who moves with apparent ease between the academic world of universities and the challenging rigors of classroom practice. It is with his writings on craft knowledge in teaching that we begin.

Herbert Kohl

Kohl has written three major works about the craft of teaching: *36 Children* (1988b), *On Teaching* (1986), and *Growing Minds: On Becoming a Teacher* (1988a). His definition of the craft of teaching undergirds the thinking and writing of both Greene (1984) and Tom (1980, 1984, 1991). Kohl is passionate about teaching. He believes that "good teaching consists to a large degree with being obsessed with helping others grow" (1988a, p. 6). Furthermore, he is committed to teachers taking control of their work through the refining of their craft:

> It is a wonderfully difficult thing to teach well, and the work is hardly rewarded other than by the flowering of our students. . . . We must take control of our work, hone the skills of our craft, and strive for the fullest development of all children. (1986, p. vii)

But it is not just control of their work that teachers must have. It is how they control their work that is important. They must also possess some important characteristics (see Appendix B). For Kohl, teachers who are curious about the lives and worlds of their students; who like spending time with them and helping them grow; who are just, fair, and who hold high expectations for themselves and for students; and who consistently enjoy teaching and can articulate their own educational philosophy are the ones who challenge and motivate students because their craft epitomizes teaching sensibility.

Developing teaching sensibility is the essential focus of Kohl's books *On Teaching* (1986) and *Growing Minds: On Becoming a Teacher* (1988a). The craft of teaching he characterizes as revolving around recurring problems of practice. The ones that Kohl (1986) discusses are gathering resources; tools for teaching; developing a curriculum theme; integrating the curriculum; observing and listening to students; setting limits for student behavior; developing the learning environment; discovering hidden resources in oneself, the students, and the community; making transitions; convincing parents of the value of one's work; dealing with one's own and the students' feelings; losing control; being out of control; fatigue; and private spaces.

An examination of how Kohl (1986) addresses "developing a curriculum theme" (pp. 37–54) will illustrate the vitality of his thinking and writing. He relates how he had decided to begin the school year with the theme of "circus-time." He had chosen this particular theme because the circus conveyed an aura of magic to almost everyone, combining danger and discipline with farce, high seriousness, and wildness. Before introducing the theme, Kohl undertook two things. First, he visited the circus to remind himself of what it was all about and, second, he did some analysis of what he had observed and pondered the things that could be taught in conjunction with such a focus. He started his planning by mapping out eight different aspects of the circus: animals; clowns; freaks; parades; posters, art, and music; pitching tents moving from place to place; tightrope walkers; and jugglers. These eight aspects then became subthemes that he began to develop in much the same way. (The curriculum mapping activities are illustrated and described in detail in the text.) He shows how he developed the subthemes of animals, balancing, and clowning in such a way as to teach other aspects of life beyond the circus (e.g., from balancing on the tightrope to balancing the checkbook, to weights and balances, to making balancing toys, etc.). For each subtheme (e.g., making balancing toys), he developed sets of specific resources and activities that can be used in teaching the concept of balance. He readily admits that this may seem like an extensive amount of preparation but contends that teachers learn a good deal about themselves when they engage in this type of planning. They also gather more material than they can possibly use in any one year of teaching. However, having undertaken all this preparation, Kohl demonstrates how he was prepared to put it to one side when introducing the theme to students because the ideas for the curriculum activities around the theme of circus have to be derived together with the students themselves. As he documents (1986, p. 43), when he developed the theme of circus with a group of first graders it came out very differently from what he had planned. The subthemes involved elephants, trick animals, midgets, men on stilts, clowns, trapeze artists, and beautiful costumes and parades. He concludes this exposition of the circus theme by noting that it is always satisfying to the students to end the unit with a performance.

Clearly, some of Kohl's originally planned ideas found their way into the mix but, equally clearly, some did not. Why, then, would teachers go to such lengths in preparation knowing that only some of their planning would ever see fruition in the classroom? An answer to this conundrum was given when the senior author discussed with Kohl the dilemma of attempting to review craft knowledge without the resulting synthesis losing the essential features of teaching as craft. Kohl (personal communication, March 17 and April 8, 1991) reported that the purpose of such a synthesis (he could just as easily have related it to preparation) was to enable teachers to develop and increase their "repertoire of responses, understandings, and magical tricks," based on knowledge of good practices that have worked in similar situations previously.

The persuasive point that Kohl (personal communication, March 17 and April 8, 1991) makes about these repertoires is that teachers who possess them have a hard time articulating their discipline procedures because they do not use them. They are so caught up in thinking about and using their craft that the kind of conflict that causes teachers to cope with student discipline problems rarely arises. Rather, such negative conflict is prevented by teachers assiduously using their craft to set limits— protecting some students, redirecting others, and so on. In Kohl's terms, teachers who set limits to establish classroom order in this way become, in the nonpejorative sense of the word, "crafty" about teaching and learning.

Growing Minds (Kohl, 1988a) is devoted to unpacking what teaching sensibility means in the action setting of learning experiences. For Kohl, teaching sensibility finds its expression in the idea of loving students as learners. Loving students as learners is finding strengths in students where others see nothing but weakness and failure. Kohl provides a living example of this tough love from his own teaching experience. It is the story of a 14-year-old who had consistently defied all his teachers and could explode with an uncontrolled and undirected rage that frightened people, including Kohl:

[For the first month] he had been remote . . . politely bored with my attempts to help him read. On this day, however, he was clearly angry. I could sense an impending explosion from the way he held his thumbs tightly in his fists and looked straight down at the floor instead of at me or at the book I was trying to get him to read. After a few minutes he did explode and knocked the manuscript of a book I was working on off my desk. I exploded quicker than he did at that. Next to the people I love, my manuscripts are the most important things in my life and I told him so. I ranted on about how important writing and books were to me. He tried to pick up the manuscript, but I let him know that at that moment I couldn't trust him to do it. As I calmed down I noticed for the first time that he was afraid, almost in tears, actually trembling. I put the manuscript together and explained that books and writing were not small school things to me but central to life and understanding, that it was no joke not to be able to read, that it was a form of poverty, and that he didn't have a right to not read. I doubt that he had ever experienced an adult express so much care for learning and books—not for some relationship to a reading test or grade, but for books themselves. Anyway, his whole attitude toward reading began to change. I felt that my love of learning and my pride in teaching him gave him a very different perception of himself as a learner. (1988a, p. 65)

For the next few weeks, Kohl and the boy talked about what was contained in different books. Kohl, as teacher, read sections from books that dealt with subjects of interest to the 14-year-old. Slowly, the boy took to reading simple books again and continued the process of learning to read. Kohl felt a pride in being part of the process. However, when the boy was on track and reading well, they parted company no more friends than when they had begun their relationship. For both Kohl and the boy, the parting was matter-of-fact, not one of sweet sorrow. But what pleased Kohl was the job-related affection he had been able to demonstrate in loving the boy as a learner; this love of the student as learner had been the key that unlocked the door to the boy focusing his previously undisciplined energy on the task of learning to read. Loving the student as a learner led Kohl (1988a)

to study him carefully and build on the strengths and personal interests I could tease out of him. It required that my personal feelings about him be subordinated to my feelings about him as a learner. Teachers have preferences and can't be expected to like every student equally. . . . Nevertheless, a teacher has an obligation to care about every student as a learner. (p. 66)

Care and compassion for all students characterizes Kohl's (1988a) rendering of teaching sensibility. Kohl, himself, displays teaching sensibility both in his life as a teacher and in his writings about teaching. His original piece about his second teaching experience in a Harlem school during 1962–1963 documents the ways in which he struggled to find ways of helping his 36 children focus their energy on learning and growth. He had been transferred involuntarily to this school because he had had the temerity to criticize the reading program in his previous school on the west side of Manhattan. The Harlem school was known for its difficult and unmanageable students, and the idea was to drive the transferred teachers out of the profession. In Kohl's case, he felt like Brer Rabbit thrown into the briar patch. But he is not a conformist and does not encourage it in his students:

Wildness and the brilliance that often accompanies it offer a challenge that I am still enchanted with in my teaching. . . . Children who question, who do not accept uncritically the things they are asked to do by their teachers, are not necessarily bad students or troublemakers. For the most part they are young people whose minds have not yet been seduced into conformity and whose wills are not yet broken. If you can reach these creative and intelligent students, you have done what my grandfather called "a job of work." (Kohl, 1988b, pp. vi, vii)

And this is exactly what Kohl set out to do during that year in Harlem; *36 Children* recounts the high and low points of his beginning struggle to develop teaching sensibility and his Sisyphus-like commitment to make himself available to his students "without hope and without cynicism" (p. 224). It also documents the difficult conditions of social inequity and cultural differences under which Kohl learned his craft, for example, the time when he took some of his students to see the opulent end of Park Avenue:

We set out from 120th Street and Park Avenue, passing the covered markets at 116th, the smelly streets down to 110th, and the dismal row upon row of slum clearance projects all the way to 99th Street. . . . We ascended from 99th to 96th, reaching the summit of that glorious hill where the tracks sink into the bowels of the city and Park Avenue is metamorphized into a rich man's fairyland. . . . The children couldn't, wouldn't believe it . . . where are the ash cans? This can't be Park Avenue . . . something's wrong. . . . It was Pamela, not angry but sad and confused. . . . The city was transformed for me through the eyes of the children. I saw a cruel contradictory New York and wanted to offer something less harsh to the children. (pp. 102–103)

Kohl's radicalism was born in experiences like this. The exploration of cultural differences and how they affect both teacher and students is the theme of Vivian Paley's work.

Vivian Paley

Paley, a teacher in the University of Chicago Laboratory School, has written two books about the craft of teaching: *Wally's Stories* (1981) and *White Teacher* (1989). The former is written from a progressive educator's perspective, while the latter deals with cross-cultural communication and understanding. Because of the increasingly serious problem in North America of a mostly White teaching force being faced with teaching an increasingly non-White student population, we choose to focus this subsection on Paley's *White Teacher* (1989). We also choose this book because, after being originally published in 1979, it was reissued in 1989 to coincide with Paley's receiving of a prestigious MacArthur Foundation fellowship, the first classroom teacher to be thus honored.

Growing up a White Jewish woman in a gentile world, Paley was aware of what it means to be different. This sensitivity comes out in her vivid and refreshingly honest observations and reactions to teaching urban Black kindergarten children. The book details Paley's struggle with understanding the African-American culture. She became concerned about Black children having difficulty approaching and relating with her. She notes in the preface to the book that school frequently amounted to Black children attempting to cover up their differences:

> Suddenly a stranger called ''teacher'' is trying to find out not who he is but what he knows. The further away the teacher is from the child's cultural or temperamental background, the more likely it is that the wrong questions will be asked. The child instinctively knows the questions are inappropriate but soon figures out that he must be the one who is inappropriate. Thus he begins the energy consuming task of trying to cover up his differences. (Paley, 1989, pp. xiii–xiv)

But students were not the only ones engaged in covering up their differences. Teachers (including Paley initially) also equated questions of equity and social justice to the denial of cultural differences. When teachers came into her room under various pretexts, they singled out Fred, a Black boy who had joined an aggressive group of six White kids, as a potential problem. At the first faculty meeting, Paley raised the issue that, even though all the children in Fred's group behaved as they did, the teachers had singled out Fred because of his color. After vigorous discussion, the faculty reached what was for Paley a confusing consensus:

> More than ever we must take care to ignore color. We must *only* look at behavior, and since a black child will be more prominent in a white classroom, we must bend over backward to see no color, hear no color, speak no color. . . . We showed respect by completely ignoring black people as black people. Color blindness was the essence of the creed. (1989, p. 9)

Such a stance obscured the need for an understanding of how Black children's cultural backgrounds might be very different from those of White children. As Paley acknowledges, when a child shares something that is compatible with her own Jewish background, she receives instant messages about his or her intelligence that a non-Jewish teacher might have missed. However, she probably misses much of the intelligence of the children who are from another culture. Consequently, when something goes awry with a child, she does not assume that the cause necessarily rests with the child:

> With most white children the smallest clue reveals a totality of characteristics I recognize. I am not aware, for example, of all the implications of a black child's saying "I don't eat pig. Only white people eat pig." . . . I think I am missing part of the picture presented by many black children by not being familiar with the context within which certain simple statements are made. (Paley, 1989, p. 77)

This lack of familiarity with the context prevents Paley from understanding the children in her care, thereby frustrating her instructional purposes. *White Teacher* chronicles the development of her learning from and with the students over a period of 5 years. She writes about her surprise and dismay when Steven, a Black student, deliberately labels her "White":

> Was I reacting to this affront to my authority as a teacher or to my authority as a white? . . . If Steven had said, "I won't listen to you," this would have been an ordinary situation. I was of course, reacting to the hostile use of "white" by this black child. If he had yelled "Jew" instead of "white" I would have been more upset. This thought gave me courage. I do feel more Jewish than white. . . . But Steven Sherman sees me only as a white lady. I can't crawl into my Jewish role. I must react as a white, so Steven will know I'm not worried about our differences. But I *was* reacting as a white and I *was* worried about our difference. (1989, pp. 14–15)

We also read of Kenny, who did not wish to take off his jacket because the teacher might not like what his T-shirt said: "SUPERNIGGER, guardian of the oppressed." We read of Claire, from a West Indian family, who was in such unfamiliar cultural territory that she was almost diagnosed as retarded. We meet another boy named Kenny, whose acrobatic movements and confidence in the schoolyard belied his timid speechlessness if required to perform some small task in front of the class. Paley's point in all this is that cultural differences between teacher and child not only affect the teacher and her educational actions, but that they also affect the child's reactions to school.

Over the course of the book, Paley develops from initially attempting to "spare" students' feelings by avoiding a discussion of such cultural differences to a position where she can talk openly about these differences in a nonjudgmental manner. She comes to the realization that to refuse to see color in the classroom represents a refusal to see children. Her students were then able to understand that people could be proud of their color and their differences and that the teacher could facilitate this process of self-discovery. Throughout her cultural discovery, Paley sought the insights of her Black students' parents and learned a good deal from Janet, an African-American student teacher. She was able to use such culturally diverse colleagues as learning resources because she respected them as persons and valued their opinions. In so doing, she was alerting herself to deeply embedded patterns of prejudice and discrimination and attempting to see the school experience through the eyes of a "cultural other."

A scholarly practitioner whose craft espouses methods of teaching and assumptions about student learning similar to those made by Kohl and Paley is Eliot Wigginton, the initiator of the Foxfire project.

Eliot Wigginton

For 25 years, the students at Rabun Gap High School in Georgia have published *Foxfire* books and magazines. Conceived in Eliot Wigginton's English classes, the project's publication of oral history has grown into a form of cultural journalism. The Foxfire project has received national recognition and its financial success has led to the establishment of the Foxfire Fund, Inc., which publishes *Hands On*—a journal for teachers— and supports teacher projects, networks, and in-service courses. Wigginton, the initiator of the original Foxfire project, serves as the President of the Foxfire Fund, Inc., and has, in recent years, analyzed what has happened in the Foxfire experience. His analyses are documented in *Sometimes a Shining Moment* (1985), "Foxfire Grows Up" (1989), articles published in *Hands On*, and a new book, *Foxfire: 25 Years* (1991a).

When Wigginton first started teaching at the Rabun County High School in 1966, he was frustrated at his inability to make the teaching of English (and Shakespeare in particular) relevant and meaningful to his students. He therefore set out to discover what would be meaningful to them and still fulfill the expectations of the Georgia State Education Department. He began by asking the students to generate ideas about good teachers. Statements such as the following emerged: Good teachers never embarrass their students in class; they treat student work seriously; they know their subject well and are excited about it; and they are fair with students, giving them a second chance. He then asked them to think about memorable experiences they had had. Students developed the following list: There was an audience for what we did; it was our choice—our creation; nobody got left out, we worked in small groups and teams; we got to teach each other; we made something real; we all did things we had never done before; we were trusted; and what we made was really useful to someone. Wigginton (1985) himself undertook the same writing assignment when some of his students had difficulty recalling memorable school experiences. He realized that his own memorable experiences were few and far between:

times when there were visitors to our class from the world outside the classroom . . . times when as students we left the classroom on assignments or field trips . . . times when things we did had an audience beyond the teacher . . . times where we, as students, were given responsibility of an adult nature, and were trusted to fulfill it . . . times when we as students took on major independent research projects that went far beyond simply copying something out of an encyclopedia, or involved ourselves in periods of intense personal creativity and action. (p. 308)

The result of Wigginton's reflections and those of his students brought about the beginnings of the Foxfire project. Wigginton and his students mutually negotiated that they would undertake projects that had connections with the real world. Living in an agricultural community located in the Appalachian mountain range in northeast Georgia, the students opted to study their roots and the local history of the region by interviewing the older members of the community. The *Foxfire* books, first published in 1972, contain the oral history reports based on interviews with the town's older citizens, many of whom were related to the students doing the project. The important point about student projects for Wigginton is that they are chosen by students and carefully matched to curriculum objectives by the teacher. Without this matching of the project to the statewide curriculum objectives set out for (in his case) the subject of English, Wigginton considers the process to be fundamentally dishonest. The essential character of Foxfire, then, is that the student-driven, real-world-connected project work is a means whereby students acquire all the skills and competencies outlined in the state's curriculum expectations.[27]

Wigginton (1985, 1989) grapples with the basic questions that drive the Foxfire project: How can teachers get students to come together for a common cause? How can they integrate innovative projects into the normal curriculum? How do schools relate to the outside community? How can teachers help adolescents understand the problems of the world outside the school? How can teachers help students move beyond themselves and their understandings into a caring and active relationship with others? How can teachers find compelling activities that serve all the goals of education? These questions enable Wigginton to explore students' concerns about teaching, teachers' assumptions about student learning, and the school curriculum. They also set the scene for the careful articulation of the solid educational philosophy that had evolved around the Foxfire experiences (Wigginton, 1985, 1989, 1991b).

The goal of schooling—and of this approach to instruction—is a more effective and humane democratic society. Individual development through schooling is a means to that goal. Often given rhetoric approval while being ignored in practice, that goal should infuse every teaching strategy and classroom activity. . . . As students become more thoughtful participants in their own education, our goal must be to help them become increasingly able and willing to guide their own learning, fearlessly, for the rest of their lives. Through constant evaluation of the experience, and examination and application of the curriculum, they approach a state of independence, of responsible behavior, and even, in the best of all worlds, of something called wisdom. (1991b, p. 1)

These perspectives on learning frame the strategies and classroom activities that constitute the Foxfire approach. The core practices of Foxfire (see Appendix C), first described in Wigginton (1985) as "overarching truths" and refined in Wigginton (1989) as "core practices," are contained in their most recent form in Wigginton (1991b). These core practices essentially represent a framework of broad principles that chart a progressive student-centered direction for teaching rather than a set of prescriptions for action. All work must have academic integrity and flow from student desire and student concerns. The teacher's role is that of collaborator and team leader, and the work is characterized by student action. Such work must be connected with the real world outside the classroom and have an audience beyond the teacher. Students should always be challenged to do something they have not previously mastered, and ongoing evaluation and reflection must accompany the work.

According to Wigginton (1989), when these core practices are followed "the result is often the most memorable, formative, intense, educative experience of [students'] lives" (p. 30).

And there is the point: if we can only become more thoughtful and analytical about our pedagogy, and if we can somehow divorce ourselves from the conviction that memorization of sacred lists of sequential bits and pieces of trivia equates with good education, we will find that we can accomplish many worthy goals *simultaneously* without adding substantially to our work. (1989, p. 31)

Wigginton (1989) further describes a course for teachers that enables them to learn how to incorporate the Foxfire core practices into their pedagogical repertoire. He demonstrates how teacher networks can support this kind of approach. Without thoughtful practice and support from colleagues, teachers' attempts to emulate the Foxfire project will reproduce the procedures but fail to understand the essential philosophy behind the core practices.

UNDERSTANDING CRAFT KNOWLEDGE

Craft knowledge is essentially the accumulated wisdom derived from teachers' and practice-oriented researchers' understandings of the meanings ascribed to the many dilemmas inherent in teaching. As such, craft knowledge emphasizes judgment—often in aesthetic terms—rather than following the maxims of research-generated knowledge. It relies heavily on intuition, care, and empathy for pupils. It is steeped in morality and ever critical in its search for meaningful schooling and benefit for pupils.[28] Understandings derived from craft knowledge appear to revolve around the purposes of teaching, the context of work within which learning takes place, teachers' sentiments about their role as facilitators of learning, and their need to be heard during a tumultuous time of restructuring.

A primary purpose of the craft of teaching is to understand and engage the minds of all learners. Teaching is not telling, talking, cajoling, or coercing; nor is it a case of teachers merely teaching and students merely learning. Crafty (in the dexterous, ingenious sense) teachers seek to know their students, to listen and reach out to them with care and understanding. Such understanding provides the basis for teacher-student engagement. This engagement, in turn, enables teachers to create opportunities and capacities for students to reach beyond what they currently know toward what is yet to be known.

Crafty teachers set out to foster an insatiable desire for learning, a zestful curiosity about events, encounters, and experiences. This purpose is based on the assumption that all students have legitimate ways of making sense of phenomena and experiences and that curriculum content can be organized in ways that "validate" students' ideas. Such reorganization is often thematic and interdisciplinary. Together with students, teachers frame the essential questions of inquiry and focus directly on what it takes to make learning memorable to students.

Learning becomes memorable when all students become active explorers and teachers themselves. To promote this purpose, however, it is not always possible to cover every aspect of the curriculum content. As Whitehead (1957) points out:

Let the main ideas which are introduced into a child's education be few and important, and let them be thrown into every combination possible. The child should make them his own, and should understand their application here and now in the circumstances of his own life. (p. 14)

Understanding, not speed, is the essential aim of crafty teachers, because students who learn to understand their world also acquire the skills necessary for searching out new information when it is needed. Thus, a main purpose associated with craft knowledge is that of becoming an advocate for all students and their learning. This typically means refusing to believe that any student cannot learn and striving with colleagues to ensure that no student is disadvantaged by unfair diagnosis, labeling, or institutional pressure. Crafty teachers set out to transform classrooms— and ultimately schools—into places in which all students become celebrated learners controlling their own inquiries.

An important aspect of craft knowledge in teaching is that the work context of inquiry and student learning is collaborative in the sense that teachers and students negotiate meaning and work activities together. Furthermore, the work context in the classroom is intimately related to the "real world" outside, as crafty teachers see connecting points everywhere. To function in the work context in a manner consistent with the accumulated wisdom of craft knowledge, teachers have to prepare assiduously by anticipating some of the situations that could arise in classroom teaching. They gather materials, note down ideas and activities, hunt out different kinds of teaching resources, and generally prepare much more than they could ever use in any given lesson, unit, or year. This preparation gives them a repertoire upon which they can draw as they attempt to engage the minds of all students and negotiate with them potential work activities. It also permits teachers to act flexibly within the work context as they engage in "rolling" planning.

Anticipatory preparation happens before classroom teaching begins. It equates to Jackson's (1968) notion of preactive teaching. Rolling planning occurs when the teacher is actively involved in the engagement process with students in the classroom—Jackson's (1968) notion of interactive teaching. In other words, anticipatory preparation before teaching is the foundation for rolling planning during classroom interaction. Such preparation is not a prescription for action (such as a lesson plan) that determines what the students can and cannot do; rather, its function is to prepare the teacher (not student work) for what could happen in the classroom. Once instruction is under way, the anticipatory preparation becomes a safety net of ideas and activities that the teacher can fall back on as he or she "rolls with the punches," as it were, in negotiating work activities with the students. Rolling planning therefore involves teachers in observing, analyzing, hypothesizing, and responding as they attempt to create curriculum on the spot with their students. It consists of balancing teacher-initiated ideas with student-initiated ones so that the work context of inquiry becomes one that truly facilitates student learning.

Exceptional teachers assume that all students have within them an inherent desire to learn. Of utmost importance to them is finding ways of channeling that desire within the classroom. Consequently, it becomes sacrosanct for crafty teachers to ensure that all classroom work is infused by student choice, student volition, and student action. This perspective attempts to treat students not as potentially wayward children but as able members of society.

Teaching is not just a job for crafty teachers; it is a vocation for which they have developed a passion. They are obsessed by the urge to help others learn and grow. Caring for students and nurturing them as persons is a theme that runs through much of the writings on teaching as a craft. Kohl crystallizes this in his phrase "loving students as learners." Unlike an undifferentiated sense of caring and nurturing, such a sentiment is neither unfocused nor languid; rather, it is the clear communication of teachers' passion for learning.

The purposes of teaching that crafty teachers pursue, the collaborative context within which they do their work, and the sentiments they hold inevitably lead to an emerging moral voice. These teachers speak on behalf of students. Their language is not confined to the expression of educational aims and expectations; its syntax is existential with words such as *caring, loving, nurturing, listening, empathic understanding*, and *connecting*. They do not write about technical aspects of teaching; rather, their focus is on ways of relating with all students in a manner that promotes learning. They make promises to students they intend to keep; they reach out to difficult or withdrawn children to include them in the group; they involve students in curriculum making; they insist that the work be that of the students and not an imposition of their own agendas; they are prepared to take risks and face ridicule to contend for student-oriented opportunities to learn. In short, they bring a finely tuned conscience to bear on all of their actions in the classroom.

This moral voice is not to be confused with professional ethics. The latter is more teacher centered, whereas the former is most definitely student centered. Thus, it is possible for a crafty teacher to become morally outraged at the talk and actions of a colleague who denigrates students. Such a moral standpoint is closely linked to criticism of taken-for-granted assumptions in the status quo. Traugh et al. (1986) refer to this as "having a voice that is critical and political" (p. 4).

Teachers gain a critical voice when they begin to think about and discuss with others the complex and embedded social, political, and moral frameworks of schooling. They begin to challenge many of the instrumental assumptions embedded in schooling and curriculum guides—they are quick to point out fallacious assumptions about students as learners. They critique the "persistency of individualism, presentism, and privacy" (Little, 1990) among teachers in school and call for a genuine collegiality that permits individuality but not isolation. They rue the conditions of the workplace that militate against teacher reflection, particularly where such lack of reflection condemns teachers to repeat their past inadequacies. They understand the power that resides in the beliefs and values that constitute the normative basis for action in the school culture, and they argue vehemently against those kinds of beliefs and values that do not respect the esteem and learning needs of students. They essentially act in this way because they are committed to changing the way society functions from a meretricious, unforgiving collectivity to an authentic, caring, and responsible community.

The empowering of teachers and the development of their moral and critical voice is not, however, unproblematic. Zeichner (1990) has pointed to the contradictions and tensions inherent in the professionalization of teaching and the democratization of schools. Giving teachers more opportunities to shape their work conditions can easily lead to an intensification of the workplace.

Teacher empowerment does not necessarily have to lead to a situation in which the achievement of the school's academic mission is undermined or teachers are overstressed, but it can, unless efforts are made to incorporate their participation in schoolwide decision making *into* their work instead of adding it *to* their work. (Zeichner, 1990, p. 367)

At the same time, an uncritical acceptance of teacher empowerment can "serve under some circumstances to undercut important connections between schools and their communities" (p. 367), leading to teachers using their increased professionalism to further distance parents and communities from attaining a meaningful voice in school affairs. Zeichner (1990) thus maintains that most of the second-wave educational reforms "fail to acknowledge the need for . . . the kinds of economic, social, and political changes outside of schools that will be needed to complement the democratic educational projects advocated for within schools—the social preconditions for educational reform" (p. 374). At the same time, he argues that

there is little hope of achieving this ideal [of democratic school governance] without linking this project to efforts in other spheres of society that are directed toward the elimination of inequalities based on gender, race, social class, sexual preference, physical condition, and so forth, no matter how noble our intentions. We cannot create democratic school communities in an undemocratic society. We cannot build "tomorrow's schools" in today's unequal society. (Zeichner, 1990, p. 374)

Such a direction inevitably involves the cultivation of a political voice for teachers who are committed to a less-teacher-centered approach to instruction and to the democratization of schools.

In a time when there appears to be no shortage of advice on how teachers should do their jobs, their political voice stands up to external imposition of others' initiatives, such as standardized testing, a national core curriculum, and basic skills in kindergarten. Perrone (1991) succinctly sums up the reasons why teachers need to develop their political voice:

Children and young people . . . need our best efforts. But . . . teachers need to construct for themselves a more powerful voice. . . . I remain convinced that we would not have the same external pressures—for accountability rooted in standardized tests, a regulatory orientation to schools governed by persons or groups who stand far away from particular schools and their students—if teachers themselves were clearer and more articulate about their purposes, speaking and writing about their hopes for children, young people and communities. (p. 133)

USING CRAFT KNOWLEDGE IN TEACHER EDUCATION

Kennedy (1990) posits that there are two broad goals in professional education. One is the goal of providing as much codified knowledge as possible for teachers so that they can be armed with the conceptual, methodological, and curricular-instructional knowledge that is foundational to good teaching. The other is the goal of enhancing independent thought and analysis, based on the assumption that best strategies have not yet been discovered or that they are too situation specific to be prescribed and, thus, practitioners must learn to create solutions for themselves.

According to Kennedy, teacher education is marked by an ambivalence between these two goals. She claims that most professions choose to emphasize one goal over the other, and that when one goal is chosen, the other is left to become the province of a small counterculture within the profession. Furthermore she claims that professions that do not have a single overarching goal (e.g., social work, nursing, journalism, teaching) are less well regarded than those that do (e.g., medicine, law, engineering, architecture). She contends that if teaching is to make progress as a profession, it will have to choose one goal over the other as its primary emphasis. Having a single primary goal, she says, "enables both educators and practitioners to share a coherent set of beliefs and values and, thereby . . . develop unified professional standards, guidelines, and curricula" (Kennedy, 1990, p. 823). Only in this way does an occupation such as teaching begin to take on the intellectual identity associated with a profession. Such an identity generates

a public perception that practitioners really do possess a special kind of expertise that justifies a special status. Professions without such strongly held views do not have clear intellectual identities, and their work is often not perceived as having a particularly demanding intellectual quality. The choice between providing volumes of knowledge and fostering independent thought and analysis is a hard one to make, for either option necessarily means that students do not receive all the intellectual tools they need. But failure to choose denies students an intellectual identity and, consequently, all of the social and personal benefits that accompany that identity. (Kennedy, 1990, p. 823)

Given the current emphasis in teacher education on "codified knowledge, based on university research" by such groups as the Holmes Group and the Carnegie Commission, it could be argued that focusing on craft knowledge will, at best, exacerbate the ambivalence that daunts teacher education and, at worst, merely fly the flag of the countercultural forces in education.

However, we would contend that there are disciplined, practical arts of inquiry present in all "professional activities," including those of medicine and law, that warrant serious attention in professional schools. Our point is that a craft conception of teaching could be included in any formulation of the so-called "knowledge bases" that inform the act of teaching. The following ways represent a sketch of how craft knowledge could be used in teacher education.

One obvious way is to include more writing and research (actual studies, not reviews) of teachers in the reading material of teacher education courses. This is something that Ken Zeichner and colleagues at the University of Wisconsin have developed very well. An extension of this is to engage student teachers in a form of active inquiry into their own classroom teaching. Teacher educators can model this process by conducting action research into their own teaching of student teachers, with a particular focus on the role they play in fostering student teacher classroom research. The initial focus could be on exploring dilemmas of practice, with a subsequent focus on the classroom, school, and societal context of teaching and learning. In this way student teachers and teacher educators begin to connect their classroom research to issues of equity and social justice and grapple with how such issues constrain student learning.

A different way of incorporating craft knowledge into teacher education would be to follow Clark's (1991) principles of design for self-directed professional development. Briefly, this would involve student teachers in writing their own credo for teaching. They would start with their strengths and proceed to make a long-term plan. They would "look in your own backyard," that is, look to other teachers rather than outside experts for useful instructional strategies and information. They would be encouraged to ask for support, "go first class" in everything they do, and ultimately "blow their own horn." These principles would be used to encourage student teachers to begin with themselves (Hunt, 1987) as sources of vital and viable information about the craft of teaching.

A further way of incorporating craft knowledge in teacher education would be to get student teachers to develop their metaphors of teaching. Lakoff and Johnson (1980) state that "the essence of a metaphor is understanding and experiencing one kind of thing in terms of another" (p. 5). It would therefore be possible to frame activities around research strategies found in Russell et al. (1988) and Munby and Russell (1989) as a way of having student teachers develop metaphors that reframe their classroom experiences. Developing metaphors might, however, appeal only to the more linguistically inclined student teachers. For the more visually oriented student teacher, it would be possible to have them draw pictures of such ideas as "school," "classroom," "students," "good teaching," and "memorable learning experiences." This is not dissimilar to what Wigginton proposes with students in the Foxfire project courses, except that it is done visually and not orally. A further example of visual expression for incorporating craft knowledge into teacher education can be found in Weade and Ernst (1989). They ask student teachers to take photographs of their field experiences as a kind of visual ethnography. These photographs then become an instrument for expressing personal meaning. This approach is advantageous for student teachers who are visually inclined but disadvantageous in that this "technical" approach can, if not carefully monitored, lead to the objectification of experience.

We think that three other approaches merit serious consideration. One is using imagination, visualization, and guided fantasy. Allender (1982) uses what she calls the "4th Grade Fantasy Activity" to help student teachers become aware of experiences that inhibit or promote learning.

I tell everyone to find a comfortable spot, to close their eyes and I then turn off the lights. (This always produces a lot of stirrings and giggling. Remember that most of these students are used to a traditional lecture series). I then proceed to have them quietly breathe deeply for 2–3 minutes, listening to the air flowing in and out of their bodies, and then 2–3 minutes tensing and untensing each part of their bodies beginning from the feet up to the head. When I feel they are relaxed I begin the activity. (Allender, 1982, pp. 37–38)

In their imagination, Allender takes the students back to their elementary school. They are told they are fourth graders and that it is time for classes to begin. The teachers spend 10 minutes in their imaginary fourth-grade classrooms; Allender suggests they look closely at the room arrangement, the wall coverings, the placement of students, the general climate, and so forth. After 10 minutes, they return to the present and draw a picture or write a story about the classroom they saw in their mind's recollection. In small groups, they exchange their pictures and stories. In the large group, they then share what they discussed in small groups. This activity can then be repeated with a focus on the "ideal" classroom.

The second approach revolves around showing movies, such as the ones reviewed earlier. A structured task would accompany the showing. Student teachers would be asked to characterize the view of learning embedded within the movie. They would also be asked to respond educationally to some of the dilemmas presented in the film and asked to compare these with their own experiences in classrooms. We have used this approach ourselves in a recent course. The movie *Dead Poets' Society* was shown and the following represents one student's response:

I remember watching the film and was, I am sure like many, initially captivated by the engaging performance of Robin Williams in his portrayal of the charismatic, seemingly unconventional English teacher who touches the souls of his students. For some reason, however, the movie gave us [a group of students] feelings of discomfort. . . . What kind of teacher is portrayed in this film? In what sense is he unconventional? The word which leaps to mind in trying to characterize the man is charismatic. The teacher appeals to us because he is humorous, personally engaging, and does strange, unorthodox things like stand on desks and speak in odd accents. I think, however, if we remove the gloss that what we are left with is a teacher who, like many of his more "conservative" counterparts, is the centre of instruction, the fount of knowledge and inspiration for his students, one who uses the classroom to play out his own psychodramas. It is interesting that in the film the metaphor commonly employed by the young men is that of the "captain," the leader, the officer of highest rank, not the craftsman. My interpretation, albeit somewhat harsh, helps me to explain why the young man, at the end of the movie, unable to cope with his problems, commits suicide. How can we expect him to cope? He has been presented with an alternative vision but has never been given the tools with which to construct it; in the end his education has failed him. There is a scene when the star-crossed student approaches his teacher; the streets are snow-covered; the student is obviously distraught; the teacher is in somewhat of a hurry. When he is asked for advice, the teacher responds by telling the student to talk it over with his father; the teacher drives away with his girlfriend while the young man goes home and shoots himself.

A third approach to teacher education that would express a view of teaching as craft concerns the design of classes in teaching methodologies. Such "methods classes," as they are referred to in many institutions, are often taught in terms of the disciplines of the school curriculum (i.e., science methods, math methods, language arts methods, etc.). Traditionally, these classes in teaching methodology have taken place on campus, in the absence of children. At best, they have engaged "microteaching" of peers as a way of introducing the analysis of actual teaching. A craft conception of teaching would justify undertaking some of this work in school settings, where professors of education would join education students in the actual teaching of children. MacKinnon and Grunau (1991) document a school-based teacher education program in which a group of education students in their student teaching year took most of their course work in a school setting. They write about the development of a community among these student teachers, and about the learnings that occurred when student teachers taught alongside their professors and their peers and when videotapes of that teaching were critiqued. The central argument of this work is that all the participants—beginning and experienced teachers, as well as professors—learned a great deal about teaching at one another's elbows.

Any or all of these approaches could be used to incorporate craft knowledge into the teacher education program. We have a predilection toward using teacher research reports, active classroom inquiry, the showing of movies, and school-based "methods" courses, but we are convinced that the other methods provide useful avenues for exploring craft knowledge in teacher education. In the final analysis, finding ways of incorporating craft knowledge about teaching into teacher education is itself a craft that can only be acquired through rigorous and reflective practice on the part of teacher educators.

Reviewing the Review

This chapter began with the poem *Other Voices, Other Worlds*, a rather powerful account of the work of a teacher who obviously had a tremendous impact on at least one student: the writer of the poem, William Strong. We confess that we chose to begin with this poem not only because of its portrayal of the craft of a teacher in a way that is difficult, if not impossible, to put in the form of "research says . . . ," but because it gives us a feeling for this teacher. In a very important sense, the poem almost allows us to experience his classroom.

In all of the writings we have reviewed we are given the same message— that we learn about the craft of teaching through experience. Certainly, analysis and considered response to practical experiences are also necessary. But craft knowledge is vastly different from the packaged and glossy maxims that govern the "science of education"—at the very least, the expectation that rules and "findings" can drive practice. Craft knowledge has a different sort of rigor, one that places more confidence in the judgments of teachers, their feel for their work, their love for students and learning, and so on, almost on aesthetic grounds. Ryle and Schön reminded us that there are good reasons for distinguishing between "knowing that" and "knowing how," suggesting that craft is something that is acquired "at the elbows" rather than in books. In that sense, our review of craft knowledge has been a bit of an oxymoron.

Yet we have attempted to include in our review sources that reveal craft knowledge, both in terms of the genres of educational research that seem appropriate for getting a handle on the craft of teachers and the works of major writers on teaching as a craft. We went so far as to include a review of popular films that portray something of the craft of teachers.

As a result of this synthesis, we have generated a framework in which we have attempted to document how craft knowledge forms around the specific purposes that some teachers pursue in their classroom practice, around the particular themes they associate with a positive work context of inquiry and the sentiments derived therein, and around teachers' moral, critical, and political voices.

As we stated at the outset of this chapter, we believe that the insights of those who write about teaching as a craft have much to offer researchers and teacher educators alike. To that end, we are hopeful that others will find our synthesis and resulting framework useful. Indeed, this chapter has visited "other voices, other worlds." Again, it is hard to imagine how craft could be captured in the form of "what research says. . . ." But on this very point, we would ask readers and ourselves, Is there room for this alternative conception of knowledge in our teacher education programs? And, if there is, what would that mean? Would the portrayal of craft be left to individual teachers? Or are there guiding principles and ways of thinking about teaching as craft that would enhance the learning of students of education?

We think that craft knowledge has an important role to play in the formation of skillful, reflective, and empowered teachers. We have attempted to articulate different ways in which craft knowledge of teaching can be incorporated into teacher education. At the same time, we believe in the power of demonstration of good practice by outstanding teachers that is carefully debriefed and reflectively analyzed. We also would resist the temptation to formulate craft knowledge in a propositional manner. How, then, can craft knowledge be used productively in teacher education?

We believe that craft knowledge can act as a sensitizing framework to teachers' collaborative explorations with students of how learning occurs. Such a framework, we argue, would constitute a broadly conceived set of principles framed around such themes as the purposes of teaching, the work context of inquiry in teaching, the sentiments derived by teachers in such a positive work context, and the moral, critical, and political voices that teachers develop in pursuing the purposes of teaching in a work context of inquiry. These principles would provoke discussion and intellectual ferment; they would stimulate teachers to reflect on why they enact certain classroom practices and resist others. The principles also require a situationally based, contextually relevant interpretation. Craft knowledge, therefore, has a powerful contribution to make to teacher education programs that are practice based and that have the goal of instilling in teachers a love of learning and a disposition toward student-focused reflective inquiry. Metaphorically speaking, this would be an instance when teacher educators would provide the occasion for future and existing teachers to "go up on deck, listen to the wind in the rigging, get a drift of the sea, gaze at a star, and set [their] course" (Kurshan, 1987, p. 124).

NOTES

[1] This poem was received by one of the authors from a student in one of his classes in the Faculty of Education, University of Toronto. The full bibliographic reference for the poem is not known.

[2] We are not arguing against the usefulness of these categories of knowledge in the education of teachers. Indeed, we believe that teachers can develop a craft understanding of teaching that is related to each of these knowledge bases, but that this typically occurs in an individualistic fashion that reinforces conservative practices. What we are attempting to do in this review is to draw together a coherent body of literature about teaching as craft from the progressive and radical

traditions that might help us move beyond the applied science approach to the development of practice.

[3] We see pedagogical learner knowledge in teaching as a special case of what Sternberg and Caruso (1985) call practical knowledge. Their definition imposed two criterial limitations on what they were prepared to call practical knowledge. First, it is procedural, as distinct from declarative, and second, it is relevant to one's everyday life. Practical knowledge is required to be procedural because it is "knowledge of and for use" (Sternberg & Caruso, 1985, p. 134). Practical knowledge is required to be relevant because "knowledge becomes practical only by virtue of its relation to the knower and the knower's environment" (Sternberg & Caruso, 1985, p. 136). We see knowledge derived from the considered experience of teaching as being not only practical in the Sternberg and Caruso sense but also pedagogical (in terms of skills and proficiencies) and learner focused (in terms of dispositions). Consequently, we have added two further criteria to pedagogical learner knowledge—that it relates to *pedagogy* that emphasizes *learner-focused classroom action*. We require such knowledge to be pedagogical because the procedural information of and for use in the dailiness of classroom action concerns itself with different ways of teaching simple and complex concepts. Pedagogical, procedural knowledge is required to be learner focused to prevent it from becoming technicist, formulaic, and unreflective.

[4] Schneider (1987) notes that, historically, the study of pedagogy has been considered an integral part of philosophy. When philosophy and psychology divided into two distinct disciplines, educational psychology took up "didactic thinking" and philosophy assumed the moral and ethical aspects of pedagogy. Neither one grappled with questions of "how to teach," leaving teacher education to pick up the technical aspects of pedagogy. This is the reason, Schneider maintains, why teacher education has been preoccupied with defining "a legitimate substantive knowledge base focused on training" (p. 214). Our argument is that craft knowledge can contribute to teacher education's search for what Schneider has termed "a legitimate substantive knowledge base focused on training." Moreover, we agree with the position of Cochran-Smith and Lytle (1990) and Liston and Zeichner (1991) that it would be a mistake to limit the knowledge base for teaching and teacher education to those topics and foci that university-based researchers have chosen to study and write about. We acknowledge that many of the university-based researchers in the progressive and radical traditions make every effort to be responsive to the questions, issues, and dilemmas that practitioners face but would argue that there needs to be a greater appreciation within the established research community of the important contributions that inquiries conducted by practitioners can make to the process of teacher education.

[5] Liston and Zeichner (1991) make a delightful characterization of how representatives of these respective traditions argue the correctness of their own position and complain about students being misled and miseducated by the other traditions. "Conservatives rail against the antiintellectual orientation of the progressive tradition, and they deride the utopianism and foggy thinking of the radical tradition. Progressives accuse conservatives of seeing only the intellectual powers of children and thereby misconstruing students' engagement in learning, and they criticize radicals for their indoctrinatory practices. Radicals condemn conservatives for their ideological obfuscation and elitist politics, and they chide progressives for their romantic view of the child and liberal view of society" (p. 53).

[6] It should be mentioned that the British alternative route to teacher certification is a concrete manifestation of an antiprofessional studies attitude that main-

tains that there is nothing to teach about pedagogy beyond that which can be learned from an apprenticeship experience in a school. Those who hold this position (one with quite a history in Britain) believe that the accumulation of content knowledge is all that is needed for good teaching. It is perfectly understandable, therefore, that British pedagogues interpret craft as unreflective practice, whereas others (e.g., Blumberg, Tom, Zeichner), removed from the austere, governmental imposition of such a damning view of pedagogy, see craft more positively.

[7] Buchmann (1987) describes the folkways of teaching as eliminating "troublesome inquiries by offering ready-made directions for use, [replacing] truth hard to attain by comfortable truisms, and [substituting] the self-explanatory for the questionable" (p. 156), thereby causing teachers in this tradition to learn their craft through mindless reliance on imitation, custom, habit, and tradition.

[8] Liston and Zeichner (1991) define teaching as situated practice as "a view of teachers as social actors engaged in practices within a particular context" (p. 122).

[9] Eisner (1985, p. 32) makes the useful point that, in our culture, knowledge is considered something that one discovers, not something that one makes. He points out, in a delightful footnote (p. 36), how even some constructionist views of knowledge reinforce this misconception through the titles of major works (e.g., Glaser & Strauss, 1967, *The Discovery of Grounded Theory*, and Popper, 1968, *The Logic of Scientific Discovery*).

[10] We are most appreciative of Ken Zeichner's supportive critique that first brought the work of Amy Gutman to our attention.

[11] We wish to state an important caveat here. It goes without saying that a review cannot provide this vicarious experience in the same depth as the original sources can and do. We therefore urge readers to consult the original sources because, whereas the review provides an organizing framework, the original sources provide the essential vicarious experience.

[12] We view the task of reviewing the craft knowledge of teachers as being fraught with a dilemma—namely, how do we frame it in such a way that it retains the essential features of craft and does not become another prescriptive knowledge base? The tacit character of craft knowledge makes its attempted explication an oxymoron. Tom and Valli (1990) crystallize this very dilemma thus: "In what ways can tacit knowledge from the craft tradition be codified? Which forms of codification make this knowledge accessible to other practitioners? Or is the codification of craft knowledge, knowledge sensitive to various contexts and to contrasting conceptions of good teaching, a contradiction in terms? Is the very idea of codification, that is a knowledge base, appropriate only in the case of positivism, an orientation that presumes that practice can be derived from knowledge? Can craft knowledge ever be viewed as a systematic way of knowing, with its characteristic methods of inquiry, rules of evidence, and forms of knowledge, so that we can talk about an epistemology of craft knowledge? What is the warrant for craft knowledge?" (p. 390). Further questions come to mind. If craft knowledge/teacher research produces the questions that spawn "searchings" (F. Erickson, 1991) and not "findings" (the usual grist for the "knowledge base" mill), how does one come to a communal understanding of accounts of practice that deliberately make the familiar strange, thereby increasing the alienation and continual uncertainty inherent in teaching? Can "the sweet poison of searching," as F. Erickson (1991) characterizes it, be captured in anything other than a story or a case study? Can one synthesize craft knowledge without colonizing it? These thorny questions pinpoint the difficulty of our task but do not propose any po-

tential resolution to the dilemma. To attempt to codify the "searchings" of craft knowledge in a manner similar to a positivistic knowledge base would "strip it of its meaning and vitality" (Yinger, 1987, p. 309) because, in the final analysis, teacher education is more a process of moral development than a process of building a knowledge base, skills, and expertise (Sirotnik, 1990). Consequently, we do not present craft knowledge as a knowledge base as such, but as a framework for helping prospective and experienced teachers develop their "repertoire of responses, understandings, and magical tricks" (H. Kohl, personal communication, March 17 and April 8, 1991). This course of action precludes the possibility of differentiating this section according to categories, such as pedagogical content and pedagogical learner knowledge, that we would see as undermining the essential features of craft in the knowledge that we review.

[13] Although similar in focus, the work by Fenstermacher (1986, 1987, 1988) on teachers' practical arguments and the work by Clark and Peterson (1986) and Clark and Yinger (1987) on teacher thinking and teacher planning are not included in this review because we see the purpose as different from that which we associate with craft knowledge. While the focus of these works is most definitely on teachers' knowledge born of practical experience and not on knowledge created by scientific research, the purpose is to examine links between such research-validated knowledge and teachers' thinking and action. Furthermore, none of these represents the study of teaching as situated practice, paying attention to issues of gender, race, and class, and giving "voice" to practicing teachers (Liston & Zeichner, 1991, pp. 136–137).

[14] These pieces may be regarded in some circles as rhetorical and not empirical contributions to craft knowledge of teaching. Lytle and Cochran-Smith (1990) address this point by outlining four different categories of teacher research: teachers' journals, essays by teachers, oral inquiry processes, and classroom studies. They make the point clearly that all four types constitute legitimate research contributions.

[15] This section of the review does not include studies of the psychological context of teaching that Zeichner, Tabachnick, and Densmore (1987) have previously reviewed. The reader is directed toward that work for a review of craft knowledge constituted by studies of teachers' attitudes and beliefs and teachers' "implicit theories."

[16] Although studies of pedagogical content knowledge bear on teachers' craft knowledge in an important way, we believe this genre merits a research review in its own right. The work that has been conducted in the area of science education is exemplary of programmed research and is presented here for illustrative purposes. A great deal of work has been carried out on pedagogical content knowledge both as a general area of study (e.g., Grossman & Richert, 1988; Shulman, 1986, 1987b) and with a specific focus on the teaching of mathematics (e.g., Carpenter et al., 1988; Cobb, 1987; Cobb & Steefe, 1983; Leinhardt & Smith, 1985; Steinberg et al., 1985; Wheatley, 1989), English, and social studies (e.g., Grossman, 1991; Gudmundsdottir, 1989; Wilson & Wineburg, 1988).

[17] We could easily have included Tracy Kidder's best-seller *Among School-children* (1989) in this section on cinema and media, for we view the book as a popular account of 1 year in the life of a very ordinary teacher named Chris Zajak. Like Ayers (1990), we find it "well-written, highly praised [yet] depressingly familiar . . . all style, no substance" (p. 3). Because it does not evidence the wisdom of a superior practitioner, the book is not included. Indeed, to include it would take away from the experimental nature of craft knowledge as we have

come to understand it. Similarly, there are many popular films about teaching in addition to those that we have included in this review (such as *Lean on Me, The Karate Kid*, and *Educating Rita*), but space did not permit their inclusion.

[18] The following stages of life cycle are documented by Huberman (1991). *Survival and discovery* occur at career entry. Survival has to do with reality shock in confronting for the first time the complexity and simultaneity of classroom teaching. Discovery has to do with the initial enthusiasm of finally having one's own class. *Stabilization* corresponds to a personal choice (commitment to teaching as a career) and to an administrative act (an official appointment or the granting of tenure). Teachers have developed their instructional repertoire over the first 4–5 years of teaching to the point where they are confident they can handle most classroom situations, even seize the "teachable moment." *Experimentation/activism* represents attempts on the part of "stable" teachers to increase the impact of their repertoire on student learning. Experiments take place with different materials, different groupings, different sequencing, and so forth. Such experiments emerge from a consciousness that they potentially face the prospect of becoming stale. They also make teachers aware of institutional barriers at the level of school, district, and society that are constraining their attempts to increase their impact on student learning. *Taking stock: self-doubts* occurs when some form of "midcareer crisis" follows on the stage of stabilization and/or experimentation/activism. This ranges from a sense of ennui with daily routine to a radical reevaluation of whether or not to stay in teaching. The more extreme reaction appears to stem from disenchantment arising from frustrated attempts to restructure practice at the school or district level. In the main, however, this stage represents discernible moments of taking stock during a relatively calm career transition more so than a midlife crisis. *Serenity* represents the stage in which teachers switch from near-manic activism to a period of more relaxed and self-accepting activity in class. Gradual loss of energy and enthusiasm is replaced by a greater sense of confidence and self-acceptance. *Conservatism* is a stage in which teachers tend to complain a good deal. They effect unfavorable comparisons of new and old generations of students, they regret the negative public image of educators and the pragmatic opportunism of administrators, and they rue the lack of serious commitment to teaching of their younger colleagues. There is a greater concern with holding onto what one has than with getting what one wants. *Disengagement* happens toward the end of a teacher's career and represents a gradual easing of one's energies away from work and toward other pursuits. This may well be a response to pressures in the work environment to "hand over the torch" to younger colleagues and their fresh ideas and can be undertaken with serenity or bitterness. Huberman (1989, 1991) suggests that there are two broad-brush models of the teaching career: One is the experimentation-serenity-serene disengagement route, and the other is the reassessment-conservatism-bitter disengagement trajectory.

[19] A corollary to this is the fact that two of the most recent major compilations of research on teaching (Reynolds, 1989; Wittrock, 1986) include few, if any, references to teacher research. To their credit, these sources do at least focus on topics, such as teacher thinking, teacher cultures, and so forth, that recognize the intentionality and purposefulness in teachers' work, something that Cochran-Smith and Lytle (1990) see as a step in the right direction from the emphasis of Travers (1973) on the examination of teacher behavior. However, these compilations still perpetuate the view that the knowledge that makes teaching a profession comes from authorities (i.e., university researchers, not teachers) outside

the profession itself. Such a view is exclusionary and disenfranchising. "It stipulates that knowing the knowledge base for teaching—what university researchers have discovered—is *the* privileged way to know about teaching. Knowing the knowledge base is, as the preface to the [AACTE (Reynolds, 1989, p. ix)] volume suggests, what 'distinguishes more productive teachers from less productive ones' " (Lytle & Cochran-Smith, 1991, p. 3).

[20] Although similar to studies in the tradition of "narrative inquiry," work in this genre is less biographical in character, and it is not regarded as a "tool" by which teachers make sense of their storied lives and practice but as accounts of the practical dilemmas they encounter in their work contexts.

[21] Lytle and Cochran-Smith (1991) argue strongly for a different epistemology, one that regards inquiry by teachers themselves as a distinctive and important way of knowing about teaching. They claim that this would "not simply add new knowers to the same knowledge base, but would *redefine the notion of knowledge for teaching and alter the locus of the knowledge base and the practitioner's stance in relationship to knowledge generation* in the field" (p. 28). This point notwithstanding, all three reviews present teacher research as a relatively recent phenomenon. This is correct inasmuch as the recency of teacher research refers to a resurgence of interest in teachers' writings about and investigations into their school contexts and classroom practice. However, as Perrone (1991, p. 90) so aptly points out, during the first decades of this century it was common for teachers in the progressive schools to write about their experiences. In reminding us of the historical context of teacher research, Perrone alerts us to the works of progressives from the beginning of the 20th century to the present.

[22] Zola (1991) refers to this as the "real curriculum," the learning that takes place in students' minds. While using different terms from Mohr (he talks about the "actual" and "hidden" curricula), Zola distinguishes the "real curriculum" from what Mohr is calling the official and unofficial versions.

[23] Coulter (1991) documents a similar mechanism of support for teacher research in schools but at the level of a school district.

[24] It is not possible in one review to examine the entire works of all the authors mentioned. Accordingly, this section presents portions of their work, selected on the basis of their contribution to an understanding of craft knowledge in teaching.

[25] The temptation was to include other authors, such as Maxine Greene, Alan Tom, Ann Lieberman, Lee Shulman, Vito Perrone, Eleanor Duckworth, David Hawkins, William Ayers, and Estelle Fuchs, who also write about craft knowledge. They were excluded because, despite their strong practice orientation, not one is a recognized school teacher. Furthermore, the three practitioners chosen were deemed to be representative of the work in craft knowledge.

[26] We are again grateful to Ken Zeichner for bringing our attention to the need to include the work of Vivian Gussin Paley to illustrate these criteria.

[27] While there appears to be considerable similarity between Wigginton and Kohl in their approach to teaching, their purposes are fundamentally different. Whereas Wigginton uses his novel methods instrumentally to help students learn what the state prescribes, Kohl clearly uses his radical approach in teaching to question the system and, ultimately, sets out to change it.

[28] Like Shulman (1987a) and Leinhardt (1990), we recognize that craft knowledge can sometimes be faulty and misleading in contexts in which teachers' subject matter knowledge is less than adequate. However, we see this as further evidence of the tentativeness and uncertainty that, for us, characterizes craft knowledge. We do not hold it up as an example of perfection, but rather as a

framework of teachers' searchings for and implicit theorizing about the nurturing of students and their learning.

REFERENCES

Adler, S., & Goodman, J. (1986). Critical theory as a foundation for methods courses. *Journal of Teacher Education, 37*, 2–8.

Allender, J. A. (1982). Fourth grade fantasy. *Journal of Humanistic Education, 6*, 36–41.

Aoki, T. T. (Ed.). (1990). *Voices of teaching*. Vancouver: British Columbia Teachers' Federation.

Ayers, W. (1990). Classroom spaces, teacher choices. *Hungry Mind Review, 15*, 3–21.

Baird, J., & Mitchell, I. (Eds.). (1987). *Improving the quality of teaching and learning: An Australian case study: The PEEL project*. Melbourne, Victoria: Monash University Press.

Bamberger, J. (1991). The laboratory for making things: Developing multiple representations of knowledge. In D. A. Schön (Ed.), *The reflective turn: Case studies in and on educational practice* (pp. 37–62). New York: Teachers College Press.

Bissex, G., & Bullock, R. (1987). *Seeing for ourselves: Case study research by teachers of writing*. Portsmouth, NH: Heinemann.

Blumberg, A. (1989). *School administration as a craft: Foundations of practice*. Boston: Allyn & Bacon.

Broudy, H. S. (1956). Teaching—Craft or profession? *Educational Forum, 20*, 175–184.

Buchmann, M. (1987). Teaching knowledge: The lights that teachers live by. *Oxford Review of Education, 13*(2), 151–164.

Bullough, R. V., Knowles, J. G., & Crow, N. A. (1991). *Emerging as a teacher*. Unpublished manuscript, University of Utah, Salt Lake City.

Butt, R. L., & Raymond, D. (1989). Studying the nature and development of teachers' knowledge using collaborative autobiography. *International Journal of Educational Research, 13*, 403–419.

Carpenter, T., Fennema, E., Peterson, P., & Carey, D. (1988). Teachers' pedagogical content knowledge of students' problem solving in elementary arithmetic. *Journal for Research in Mathematics Education, 19*, 385–401.

Carr, C., & Kemmis, S. (1983). *Becoming critical: Knowing through action research*. Victoria: Deakin University Press.

Case, J. H. (1990). Unexpected responses: Interaction in the classroom. *The Teacher's Journal, 3*, 18–28.

Central Park East Secondary School. (1988). *The promise*. New York: Author.

Christian, B. F. (1988). Enhanced achievement through individualized attention. In Langston Hughes School-Based Research Group, *Teacher research on student learning* (pp. 3–8). Fairfax, VA: Fairfax County Public Schools.

Clandinin, D. J., & Connelly, F. M. (1991). Narrative and story in practice and research. In D. A. Schön (Ed.), *The reflective turn: Case studies in and on educational practice* (pp. 258–281). New York: Teachers College Press.

Clark, C. M. (1991). Teachers as designers in self-directed professional development. In A. Hargreaves & M. Fullan (Eds.), *Understanding teacher development* (pp. 105–120). London: Cassells.

Clark, C. M., & Peterson, P. L. (1986). Teachers' thought processes. In M. Wittrock (Ed.), *Handbook of research on teaching* (3rd ed., pp. 255–296). New York: Macmillan.

Clark, C. M., & Yinger, R. (1987). Teacher planning. In D. C. Berliner & B. V. Rosenshine (Eds.), *Talks to teachers* (pp. 342–365). New York: Random House.

Clift, R. T., Houston, W. R., & Pugach, M. C. (Eds.). (1990). *Encouraging reflective practice in education: An analysis of issues and programs.* New York: Teachers College Press.

Cobb, P. (1987). Information processing psychology and mathematics education: A constructivist perspective. *Journal of Mathematical Behaviour, 6,* 3–40.

Cobb, P., & Steefe, L. P. (1983). The constructivist researcher as teacher and model builder. *Journal for Research in Mathematics Education, 14,* 83–94.

Cochran-Smith, M., & Lytle, S. L. (1990). Research on teaching and teacher research: The issues that divide. *Educational Researcher, 19*(2), 2–11.

Cohen, D. (1977). *Ideas and action: Social science and craft in educational practice.* Chicago: Center for New Schools.

Cole, A. (1989). Researcher and teacher: Partners in theory building. *Journal of Education for Teaching, 15,* 225–237.

Collingwood, R. G. (1938). *The principles of art.* Indianapolis, IN: Hackett.

Connelly, F. M., & Clandinin, D. J. (1988). *Teachers as curriculum planners: Narratives of experience.* New York: Teachers College Press.

Connelly, F. M., & Clandinin, D. J. (1990). Stories of experience and narrative inquiry. *Educational Researcher, 19*(5), 2–24.

Coulter, D. (1991, February). *Organizational support for teacher researchers.* Paper presented at the Second International Conference on Teacher Development, Vancouver, BC.

Cricchi, A. (1988). A collaborative school-based research project. In Langston Hughes School-Based Research Group, *Teacher research on student learning* (pp. 73–82). Fairfax, VA: Fairfax County Public Schools.

Cuban, L. (1984). *How teachers taught: Constancy and change in American classrooms, 1890–1980.* New York: Longman.

Dewey, J. (1929). *The sources of a science of education.* New York: Horace Liveright.

Dewey, J. (1933). *How we think.* New York: Heath and Company.

Dewey, J. (1944). *Democracy and education.* New York: Free Press.

Doyle, L. A. (1988). Working to improve the self-concept of learning disabled students in an intermediate school. In Langston Hughes School-Based Research Group, *Teacher research on student learning* (pp. 9–16). Fairfax, VA: Fairfax County Public Schools.

Driver, R. (1988). Theory into practice II: A constructivist approach to curriculum development. In P. Fensham (Ed.), *Developments and dilemmas in science education* (pp. 133–149). London: Falmer Press.

Driver, R., Guesne, E., & Tiberghien, A. (Eds.). (1985). *Children's learning in science.* Philadelphia: Milton Keynes/Open University Press.

Eisner, E. (1985). Aesthetic modes of knowing. In E. Eisner (Ed.), *Learning and teaching the ways of knowing: Eighty-fourth yearbook of the National Society for the Study of Education* (pp. 23–36). Chicago: University of Chicago Press.

Elbaz, F. (1991). Research on teachers' knowledge: The evolution of a discourse. *Journal of Curriculum Studies, 23*(1), 1–19.

Elliott, J. (1990). Teachers as researchers: Implications for supervision and for teacher education. *Teaching and Teacher Education, 6*(1), 1–26.

Ellwood, C. (1991, April). *Can we really look through our students' eyes? An urban teacher's perspective.* Paper presented at the annual meeting of the American Educational Research Association, Chicago.

Erickson, F. (1991, April). *Teacher research and research on teaching: Perspectives and paradoxes.* Symposium held at the annual meeting of the American Educational Research Association, Chicago.

Erickson, G. (1987, April). *Constructivist epistemology and the professional development of teachers.* Paper presented at the annual meeting of the American Educational Research Association, Washington, DC.

Erickson, G. L., & MacKinnon, A. M. (1991). Seeing classrooms in new ways: On becoming a science teacher. In D. A. Schön (Ed.), *The reflective turn: Case studies in and on educational practice* (pp. 15–36). New York: Teachers College Press.

Feiman-Nemser, S., & Buchmann, M. (1985). Pitfalls of experience in teacher preparation. *Teachers College Record, 87,* 53–65.

Fenstermacher, G. D. (1986). Philosophy of research on teaching: Three aspects. In M. Wittrock (Ed.), *Handbook of research on teaching* (3rd ed., pp. 37–49). New York: Macmillan.

Fenstermacher, G. D. (1987). A reply to my critics. *Educational Theory, 37,* 413–421.

Fenstermacher, G. D. (1988). The place of science and epistemology in Schön's conception of reflective practice. In P. P. Grimmett & G. L. Erickson (Eds.), *Reflection in teacher education* (pp. 39–46). New York: Teachers College Press.

Freedman, S., Jackson, J., & Boles, K. (1983). Teaching: An imperiled profession. In L. Shulman & G. Sykes (Eds.), *Handbook of teaching and policy* (pp. 261–299). New York: Longman.

Freedman, S., Jackson, J., & Boles, K. (1986). *The effect of teaching on teachers.* Grand Forks: University of North Dakota Press.

Fullan, M. G., & Connelly, F. M. (1987). *Teacher education in Ontario.* Toronto: Queen's Printer.

Gage, N. L. (1985). *Hard gains in the soft sciences: The case of pedagogy.* Bloomington, IN: Phi Delta Kappa.

Ginsburg, M. (1988). *Contradictions in teacher education and society: A critical analysis.* Lewes, England: Falmer Press.

Gomez, M. L. (1990). Reflections on research for teaching: Collaborative inquiry with a novice teacher. *Journal of Education for Teaching, 16,* 45–56.

Goodlad, J. (1984). *A place called school.* Chicago: University of Chicago Press.

Goodson, I. F. (1991). Sponsoring the teacher's voice: Teachers' lives and teacher development. In A. Hargreaves & M. Fullan (Eds.), *Understanding teacher development* (pp. 154–170). London: Cassells.

Gore, J., & Zeichner, K. (1990, April). *Action research and reflective teaching in preservice teacher education.* Paper presented at the annual meeting of the American Educational Research Association, Boston.

Goswami, D., & Stillman, P. (1987). *Reclaiming the classroom: Teachers research as an agency for change.* Upper Montclair, NJ: Boynton/Cook.

Greene, M. (1984). How do we think about our craft? In A. Lieberman (Ed.), *Rethinking school improvement: Research, craft, and concept* (pp. 13–25). New York: Teachers College Press.

Grimmett, P. P., & Crehan, E. P. (1990). Barry: A case study of teacher reflection in clinical supervision. *Journal of Curriculum and Supervision, 5,* 214–235.

Grimmett, P. P., & Erickson, G. L. (Eds.). (1988). *Reflection in teacher education.* New York: Teachers College Press.

Grimmett, P. P., MacKinnon, A. M., Erickson, G. L., & Riecken, T. J. (1990). Reflective practice in teacher education. In R. Clift, R. Houston, & M. Pugach (Eds.), *Encouraging reflective practice in education: An analysis of issues and programs* (pp. 20–38). New York: Teachers College Press.

Grossman, P. (1991). *The making of a teacher*. New York: Teachers College Press.

Grossman, P., & Richert, R. (1988). Unacknowledged knowledge growth: A reexamination of the effects of teacher education. *Teaching and Teacher Education, 4*, 53–62.

Gudmundsdottir, S. (1989). *Knowledge use among experienced teachers: Four case studies of high school teaching.* Unpublished doctoral dissertation, Stanford University, Stanford, CA.

Gutman, A. (1987). *Democratic education.* Princeton, NJ: Princeton University Press.

Habermas, J. (1973). *Knowledge and human interest.* London: Heinemann.

Hargreaves, A., & Dawe, R. (1989, March). *Coaching as unreflective practice: Contrived collegiality or collaborative culture?* Paper presented at the annual meeting of the American Educational Research Association, San Francisco.

Harrison, M. (1988). Teacher-research project in an intermediate school: Impact beyond participants. In Langston Hughes School-Based Research Group, *Teacher research on student learning* (pp. 91–100). Fairfax, VA: Fairfax County Public Schools.

Hartnett, A., & Naish, M. (1980). Technicians or social bandits? Some moral and political issues in the education of teachers. In P. Woods (Ed.), *Teacher strategies: Explorations in the sociology of the school* (pp. 254–274). London: Croom Helm.

Hirsch, E. D. (1988). *Cultural literacy.* New York: Vintage Books.

Hirst, P. (1965). Liberal education and the nature of knowledge. In R. D. Archambault (Ed.), *Philosophical analysis of education* (pp. 113–138). New York: Humanities Press.

Hodgkinson, C. (1978). *Toward a philosophy of administration.* Oxford, England: Blackwell.

Hogue, N. (1990). Pause that invites us to living pedagogy. In T. Aoki (Ed.), *Voices of teaching* (pp. 43–46). Vancouver: British Columbia Teachers' Federation.

Houston, W. R. (Ed.). (1990). *Handbook of research in teacher education.* New York and Washington, DC: Macmillan and Association of Teacher Educators.

Howard, V. (1982). *Artistry: The work of artists.* Indianapolis, IN: Hackett.

Howsam, R. B., Corrigan, D. C., Denemark, G. W., & Nash, R. J. (1976). *Educating a profession.* Washington, DC: American Association of Colleges for Teacher Education.

Huberman, M. (1989). The professional life cycle of teachers. *Teachers College Record, 91*, 31–57.

Huberman, M. (1991). Teacher development and instructional mastery. In A. Hargreaves & M. Fullan (Eds.), *Understanding teacher development* (pp. 171–195). London: Cassells.

Hunt, D. E. (1987). *Beginning with ourselves: Practice, theory, and human affairs.* Toronto: OISE Press.

Hustler, D., Cassidy, T., & Cuff, T. (Eds.). (1986). *Action research in classrooms and schools.* London: Allen & Unwin.

Jackson, P. W. (1968). *Life in classrooms.* New York: Teachers College Press.

Jeffrey, S. G. (1988). When students learn in groups. In Langston Hughes School-Based Research Group, *Teacher research on student learning* (pp. 23–36). Fairfax, VA: Fairfax County Public Schools.

Johnson, R. W. (1988). Improving minority achievement through writing in science. In Langston Hughes School-Based Research Group, *Teacher research on student learning* (pp. 37–44). Fairfax, VA: Fairfax County Public Schools.

Kennedy, M. M. (1987). Inexact sciences: Professional education and the development of expertise. In E. Z. Rothkopf (Ed.), *Review of research in education* (Vol. 14, pp. 133–167). Washington, DC: American Educational Research Association.

Kennedy, M. M. (1990). Choosing a goal for professional education. In W. R. Houston (Ed.), *Handbook of research in teacher education* (pp. 813–825). New York and Washington, DC: Macmillan and Association of Teacher Educators.

Kidder, T. (1989). *Among schoolchildren*. Boston: Houghton Mifflin.

Kilbourn, B. (1986). Situational analysis of teaching in clinical supervision. In W. J. Smyth (Ed.), *Learning about teaching through clinical supervision* (pp. 111–136). London: Croom Helm.

Kilbourn, B. (1990). *Constructive feedback: Learning the art*. Toronto: OISE Press.

Kohl, H. R. (1986). *On teaching*. New York: Schocken Books.

Kohl, H. R. (1988a). *Growing minds: On becoming a teacher*. New York: Harper & Row.

Kohl, H. R. (1988b). *36 children*. New York: New American Library.

Kurshan, N. (1987). *Raising your child to be a Mensch*. New York: Ivy Books.

Lakoff, G., & Johnson, M. (1980). *Metaphors we live by*. Chicago: University of Chicago Press.

Lampert, M. (1985). How do teachers manage to teach? Perspectives on problems in practice. *Harvard Educational Review, 55*, 178–194.

Langston Hughes School-Based Research Group. (1988). *Teacher research on student learning*. Fairfax, VA: Fairfax County Public Schools.

Lave, J. (1988). *Cognition in practice: Mind, mathematics and culture in everyday life*. New York: Cambridge University Press.

Law, J. (1990). Pedagogical silence as a mode of being with students. In T. Aoki (Ed.), *Voices of teaching* (pp. 24–26). Vancouver: British Columbia Teachers' Federation.

Leinhardt, G. (1990). Capturing craft knowledge in teaching. *Educational Researcher, 19*(2), 18–25.

Leinhardt, G., & Smith, D. (1985). Expertise in mathematics instruction: Subject matter knowledge. *Journal of Educational Psychology, 77*, 247–271.

Lewis, H. (1990). Pedagogical reaching in the midst of paradoxes. In T. Aoki (Ed.), *Voices of teaching* (pp. 47–51). Vancouver: British Columbia Teachers' Federation.

Lieberman, A. (Ed.). (1984). *Rethinking school improvement: Research, craft, and concept*. New York: Teachers College Press.

Liston, D. P., & Zeichner, K. M. (1991). *Teacher education and the social conditions of schooling*. New York: Routledge.

Little, J. W. (1990). The persistence of privacy: Autonomy and initiative in teachers' professional relations. *Teachers College Record, 91*, 509–536.

Lloyd, J. (1988). What happens to students when a teacher communicates high expectations? In Langston Hughes School-Based Research Group, *Teacher research on student learning* (pp. 45–48). Fairfax, VA: Fairfax County Public Schools.

Lytle, S. L., & Cochran-Smith, M. (1990). Learning from teacher research: A working typology. *Teachers College Record, 92,* 83–103.

Lytle, S. L., & Cochran-Smith, M. (1991). *Teacher research as a way of knowing.* Unpublished manuscript, University of Pennsylvania, Philadelphia.

MacKinnon, A. M. (1987). Detecting reflection-in-action among preservice elementary science teachers. *Teaching and Teacher Education, 3,* 135–145.

MacKinnon, A. M. (1989). Conceptualizing a "hall of mirrors" in a science teaching practicum. *Journal of Curriculum and Supervision, 5*(1), 41–59.

MacKinnon, A. M., & Erickson, G. L. (1988). Taking Schön's ideas to a science teaching practicum. In P. P. Grimmett & G. L. Erickson (Eds.), *Reflection in teacher education* (pp. 113–135). New York: Teachers College Press.

MacKinnon, A. M., & Grunau, H. (1991, April). *Teacher development through reflection, community, and discourse.* Paper presented at the annual meeting of the American Educational Research Association, Chicago.

Maher, F., & Rathbone, C. (1986). Teacher education and feminist theory: Some implications for practice. *American Journal of Education, 94,* 214–235.

McNiff, J. (1988). *Action research: Principles and practice.* London: Macmillan Education.

Miller, J. (1990). *Creating spaces and finding voices: Teachers collaborating for empowerment.* Albany: State University of New York Press.

Mohr, M. (1988). Classroom curriculum: Expectations and configurations. In Langston Hughes School-Based Research Group, *Teacher research on student learning* (pp. 63–72). Fairfax, VA: Fairfax County Public Schools.

Mohr, M., & MacLean, M. (1987). *Working together: A guide for teacher-researchers.* Urbana, IL: National Council of Teachers of English.

Munby, H., & Russell, T. (1989, March). *Metaphor in the study of teachers' professional knowledge.* Paper presented at the annual meeting of the American Educational Research Association, San Francisco.

Newberg, N. A. (1991). Bridging the gap: An organizational inquiry into an urban school system. In D. A. Schön (Ed.), *The reflective turn: Case studies in and on educational practice* (pp. 65–83). New York: Teachers College Press.

Nias, J. (1984). The definition and maintenance of self in primary schools. *British Journal of Sociology of Education, 5,* 266–277.

Nocerino, M. A. (1988). Teacher research: A look at the process. In Langston Hughes School-Based Research Group, *Teacher research on student learning* (pp. 83–90). Fairfax, VA: Fairfax County Public Schools.

Oja, S. N., & Smulyan, L. (1989). *Collaborative action research: A developmental approach.* Lewes, England: Falmer Press.

Osborne, R., & Freyberg, P. (1985). *Learning in science: The implications of children's science.* London: Heinemann.

Paley, V. G. (1981). *Wally's stories.* Cambridge, MA: Harvard University Press.

Paley, V. G. (1989). *White teacher.* Cambridge, MA: Harvard University Press.

Perrone, V. (1989). *Working papers: Reflections on teachers, schools, and communities.* New York: Teachers College Press.

Perrone, V. (1991). *A letter to teachers: Reflections on schooling and the art of teaching.* San Francisco: Jossey-Bass.

Popkewitz, T. S., & Wehlage, G. G. (1973). Accountability: Critique and alternative perspective. *Interchange, 4*(2), 48–62.

Raymond, D., Butt, R., & Townsend, D. (1991). Contexts for teacher development: Insights from teachers' stories. In A. Hargreaves & M. Fullan (Eds.), *Understanding teacher development* (pp. 196–221). London: Cassells.

Reeves, M. (1990). I learned to smile before Christmas: A beginning teacher's early experience. In T. Aoki (Ed.), *Voices of teaching* (pp. 19–23). Vancouver: British Columbia Teachers' Federation.

Resnick, L. (1991, April). *Situations for learning and thinking.* Recipient's address for the award for distinguished contributions to educational research 1990 presented at the annual meeting of the American Educational Research Association, Chicago.

Reynolds, M. (Ed.). (1989). *Knowledge base for the beginning teacher.* New York: Pergamon Press.

Richardson, V. (1990). The evolution of reflective teaching and teacher education. In R. Clift, R. Houston, & M. Pugach (Eds.), *Encouraging reflective practice in education: An analysis of issues and programs* (pp. 3–19). New York: Teachers College Press.

Rorschach, E., & Whitney, R. (1986). Relearning to teach: Peer observation as a means of professional development for teachers. *English Education, 18*, 159–172.

Ross, D. (1987). Action research for preservice teachers: A description of why and how. *Peabody Journal of Education, 64*, 131–150.

Ruddick, J. (1985). Teacher research and research-based teacher education. *Journal of Education for Teaching, 11*, 281–289.

Ruddick, J., & Hopkins, D. (1985). *Research as a basis for teaching: Readings from the work of Lawrence Stenhouse.* London: Heinemann.

Russell, T. (1984). The importance and the challenge of reflection-in-action by teachers. In P. P. Grimmett (Ed.), *Research in teacher education: Current problems and future prospects in Canada* (pp. 21–31). Vancouver: Centre for the Study of Teacher Education and Centre for the Study of Curriculum and Instruction, University of British Columbia.

Russell, T., & Munby, H. (1991). Reframing: The role of experience in developing teachers' professional knowledge. In D. A. Schön (Ed.), *The reflective turn: Case studies in and on educational practice* (pp. 164–187). New York: Teachers College Press.

Russell, T., Munby, H., Spafford, C., & Johnston, P. (1988). Learning the professional knowledge of teaching: Metaphors, puzzles, and the theory-practice relationship. In P. P. Grimmett & G. L. Erickson (Eds.), *Reflection in teacher education* (pp. 67–90). New York: Teachers College Press.

Ryle, G. (1949). *The concept of mind.* London: Hutchinson & Co.

Saul, M. E. (1990). What we are teaching: Soul searching and mathematics. *The Teacher's Journal, 3*, 1–5.

Schaefer, R. J. (1970). Teacher education in the United States. In A. Yates (Ed.), *Current problems of teacher education* (pp. 156–185). Hamburg: UNESCO Institute for Education.

Scheffler, I. (1960). *The language of education.* Springfield, IL: Charles C Thomas.

Schneider, B. (1987). Tracing the provenance of teacher education. In T. Popkewitz (Ed.), *Critical studies in teacher education* (pp. 211–241). Lewes, England: Falmer Press.

Schön, D. A. (1983). *The reflective practitioner: How professionals think in action.* New York: Basic Books.

Schön, D. A. (1987). *Educating the reflective practitioner: Toward a new design for teaching and learning in the professions.* San Francisco: Jossey-Bass.

Schön, D. A. (1988). Coaching reflective practice. In P. Grimmett & G. Erickson

(Eds.), *Reflection in teacher education* (pp. 19–29). New York: Teachers College Press.

Schön, D. A. (Ed.). (1991). *The reflective turn: Case studies in and on educational practice.* New York: Teachers College Press.

Shulman, L. S. (1986). Those who understand: Knowledge growth in teaching. *Educational Researcher, 15*(2), 4–14.

Shulman, L. S. (1987a). The wisdom of practice: Managing complexity in medicine and teaching. In D. C. Berliner & B. V. Rosenshine (Eds.), *Talks to teachers* (pp. 369–386). New York: Random House.

Shulman, L. S. (1987b). Knowledge and teaching: Foundations of the new reform. *Harvard Educational Review, 57*, 114–135.

Shulman, L. S. (1989, March). *An end to substance abuse: Reclaiming the content for teacher education and supervision.* Invited address to the annual meeting of the American Educational Research Association, San Francisco.

Sirotnik, K. (1990). Society, schooling, teaching and preparing to teach. In J. Goddlad, R. Soder, & K. Sirotnik (Eds.), *The moral dimensions of teaching* (pp. 296–328). San Francisco: Jossey-Bass.

Sizer, T. S. (1984). *Horace's compromise: The dilemma of the American high school.* Boston: Houghton Mifflin.

Smith, B. (1988). Projects and games in the intermediate classroom. In Langston Hughes School-Based Research Group, *Teacher research on student learning* (pp. 53–56). Fairfax, VA: Fairfax County Public Schools.

Smyth, J., & Gitlin, A. (1989). *Teacher evaluation: Educative alternatives.* Lewes, England: Falmer Press.

Smyth, W. J. (1986). *Reflection in action.* Victoria: Deakin University Press.

Steinberg, R., Haymore, J., & Marks, R. (1985, March). *Teachers' knowledge and structuring content in mathematics.* Paper presented at the annual meeting of the American Educational Research Association, Chicago.

Sternberg, R. J., & Caruso, D. R. (1985). Practical modes of knowing. In E. Eisner (Ed.), *Learning and teaching the ways of knowing: Eighty-fourth yearbook of the National Society for the Study of Education* (pp. 133–158). Chicago: University of Chicago Press.

Stones, E. (Ed.). (1990). *A new agenda for teacher education.* Birmingham, England: Carfax.

Strieb, L. (1985). *A (Philadelphia) teacher's journal.* Grand Forks: Center for Teaching and Learning, University of North Dakota.

Tabachnick, B. R., & Zeichner, K. M. (1991). *Issues and practices in inquiry-oriented teacher education.* Lewes, England: Falmer Press.

Tikunoff, W. J., Ward, B. A., & Griffin, G. A. (1979). *Interactive research and development on teaching study: Final report.* San Francisco: Far West Regional Laboratory for Educational Research and Development.

Tobin, K. (1990, April). *Constructivist perspectives on teacher change.* 1989 Cattell Early Career Award invited address presented at the annual meeting of the American Educational Research Association, Boston.

Tobin, K. G., & Espinet, M. (1987). *Teachers helping teachers to improve high school mathematics teaching.* Occasional paper, Florida State University, Tallahassee.

Tom, A. R. (1980). Teaching as a moral craft: A metaphor for teaching and teacher education. *Curriculum Inquiry, 10*, 317–323.

Tom, A. R. (1984). *Teaching as a moral craft.* New York: Longman.

Tom, A. R. (1991, February). *Stirring the embers: Reinventing the structure and curriculum of teacher education.* Paper presented at the Second International Conference on Teacher Development, Vancouver, Canada.

Tom, A. R., & Valli, L. (1990). Professional knowledge for teachers. In W. R. Houston (Ed.), *Handbook of research in teacher education* (pp. 373–392). New York and Washington, DC: Macmillan and Association of Teacher Educators.

Traugh, C., Kanesky, R., Martin, A., Seletsky, A., Woolf, K., & Strieb, L. (1986). *Speaking out: Teachers and teaching.* Grand Forks: North Dakota Study Group on Evaluation.

Travers, R. (1973). *The second handbook of research on teaching.* Chicago: Rand McNally.

Tsujimoto, J. I. (1990). The affective teacher. *The Teacher's Journal, 3,* 6–14.

Turner, J. (1990). Teacher education under threat. In E. Stones (Ed.), *A new agenda for teacher education* (pp. 5–10). Birmingham, England: Carfax.

Van Manen, M. (1977). Linking ways of knowing with ways of being practical. *Curriculum Inquiry, 6,* 205–228.

Wasley, P. A. (1990). *Stirring the chalkdust: Three teachers in the midst of change.* Providence, RI: Coalition for Essential Schools, Brown University.

Wasley, P. A. (1991a). The practical work of teacher leaders: Assumptions, attitudes, and acrophobia. In A. Lieberman (Ed.), *Staff development for education in the 90s: New demands, new realities, new perspectives* (pp. 158–183). New York: Teachers College Press.

Wasley, P. A. (1991b). Stirring the chalkdust: Changing practices in essential schools. *Teachers College Record, 93*(1), 28–58.

Weade, R., & Ernst, G. (1989, March). *Through the camera's lens: Pictures of classroom life and the search for metaphors to frame them.* Paper presented at the annual meeting of the American Educational Research Association, San Francisco.

Wheatley, G. (1989, November). *Constructivist perspectives on science and mathematics learning.* Paper presented at the History and Philosophy of Science in Science Teaching First International Conference, Florida State University, Tallahassee.

Whitehead, A. N. (1957). *The aims of education: And other essays.* New York: Macmillan.

Wigginton, E. (1985). *Sometimes a shining moment: The Foxfire experience.* Garden City, NY: Doubleday.

Wigginton, E. (1989). Foxfire grows up. *Harvard Educational Review, 59,* 24–49.

Wigginton, E. (1991a). *Foxfire: 25 years.* Garden City, NY: Doubleday.

Wigginton, E. (1991b). *The Foxfire approach: Perspectives and core practices.* Rabun Gap, GA: The Foxfire Fund, Inc.

Wilson, S., & Wineburg, S. (1988). Peering at history through different lenses: The role of disciplinary perspectives in teaching history. *Teachers College Record, 89,* 525–539.

Wilson, S. M., Shulman, L. S., & Richert, A. E. (1987). "150 different ways" of knowing: Representations of knowledge in teaching. In J. Calderhead (Ed.), *Exploring teachers' thinking* (pp. 104–124). London: Cassells.

Wittgenstein, L. (1958). *Philosophical investigations* (2nd ed.). Oxford, England: Basil Blackwell.

Wittrock, M. (Ed.). (1986). *Handbook of research on teaching* (3rd ed.). New York: Macmillan.

Wood, P. (1988). Action research: A field perspective. *Journal of Education for Teaching, 14,* 135–150.

Woods, P. (1979). *The divided school.* London: Routledge & Kegan Paul.

Yinger, R. (1987). Learning the language of practice. *Curriculum Inquiry, 17,* 293–318.

Yinger, R. (1990). The conversation of practice. In R. Clift, R. Houston, & M. Pugach (Eds.), *Encouraging reflective practice in education: An analysis of issues and programs* (pp. 73–94). New York: Teachers College Press.

Yinger, R., Hedricks-Lee, M., & Johnston, S. (1991, April). *The character of working knowledge.* Paper presented at the annual meeting of the American Educational Research Association, Chicago.

Zeichner, K. M. (1990). Contradictions and tensions in the professionalization of teaching and the democratization of schools. *Teachers College Record, 92,* 363–379.

Zeichner, K. M., & Liston, D. P. (1987). Teaching student teachers to reflect. *Harvard Educational Review, 57,* 1–22.

Zeichner, K. M., Tabachnick, B. R., & Densmore, K. (1987). Individual, institutional, and cultural influences on the development of teachers' craft knowledge. In J. Calderhead (Ed.), *Exploring teachers' thinking* (pp. 21–59). London: Cassells.

Zola, M. (1991, July). *What I cannot tell my mother is not fit for me to know: The making of a teacher.* Public lecture presented at the Summer Institute in Teacher Education, Simon Fraser University, Burnaby, BC.

APPENDIX A
The Nine Common Principles of The Coalition of Essential Schools

1. The school should focus on helping adolescents learn to use their minds well. Schools should not attempt to be "comprehensive" if such a claim is made at the expense of the school's central intellectual purpose.

2. The school's goals should be simple: that each student master a limited number of essential skills and areas of knowledge. While these skills and areas will, to varying degrees, reflect the traditional academic disciplines, the program's design should be shaped by the intellectual and imaginative powers and competencies that students need, rather than necessarily by "subjects" as conventionally defined. The aphorism "Less Is More" should dominate: curricular decisions should be guided by the aim of thorough student mastery and achievement rather than by an effort merely to cover content.

3. The school's goals should apply to all students, while the means to these goals will vary as those students themselves vary. School practice should be tailor-made to meet the needs of every group or class of adolescents.

4. Teaching and learning should be personalized to the maximum feasible extent. Efforts should be directed toward a goal that no teacher have direct responsibility for more than 80 students. To capitalize on this personalization, decisions about the details of the course of study, the use of students' and teachers' time and the choice of teaching materials and specific pedagogies must be unreservedly placed in the hands of the principal and staff.

5. The governing practical metaphor of the school should be student-as-worker, rather than the more familiar metaphor of teacher-as-deliverer-of-instructional-services. Accordingly, a prominent pedagogy will be coaching, to provoke students to learn how to learn and thus to teach themselves.

6. Students entering secondary school studies are those who can show competence

---

Note. From *Stirring the Chalkdust: Three Teachers in the Midst of Change* (p. 42) by P. A. Wasley, 1990, Providence, RI: Coalition for Essential Schools, Brown University. Copyright 1990 by the Coalition for Essential Schools. Reprinted by permission.

in language and elementary mathematics. Students of traditional high school age but not yet at appropriate levels of competence to enter secondary school studies will be provided intensive remedial work to assist them to meet these standards. The diploma should be awarded upon a successful final demonstration of mastery for graduation—an "Exhibition." This Exhibition by the student of his or her grasp of the central skills and knowledge of the school's program may be jointly administered by the faculty and higher authorities. As the diploma is awarded when earned, the school's program proceeds with no strict age grading and with no system of "credits earned" by "time spent" in class. The emphasis is on the students' demonstration that they can do important things.

7. The tone of the school should explicitly and self-consciously stress values of unanxious expectation ("I won't threaten you but I expect much of you"), of trust (until abused) and of decency (the values of fairness, generosity, and tolerance). Incentives appropriate to the school's particular students and teachers should be emphasized, and parents should be treated as essential collaborators.

8. The principal and teachers should perceive themselves as generalists first (teachers and scholars in general education) and specialists second (experts in but one particular discipline). Staff should expect multiple obligations (teacher-counselor-manager) and a sense of commitment to the whole school.

9. Ultimate administrative and budget targets should include, in addition to total student loads per teacher of 80 or fewer pupils, substantial time for collective planning by teachers, competitive salaries for staff and an ultimate per pupil cost not to exceed that at traditional schools by more than 10 percent. To accomplish this, administrative plans may have to show the phased reduction or elimination of some services now provided students in many traditional comprehensive secondary schools.

APPENDIX B
Kohl's Important Characteristics of Teaching

Teaching well is the fundamental basis for long-term teacher power and educational leadership. . . . Some of the most important characteristics of a good teacher . . . are:

—curiosity about the lives and cultures of your students and a willingness to explore their worlds. This exploration leads to an understanding of the way your particular students learn and their usual modes of communication. It also helps develop insight into the rhythms of their lives and the things they enjoy doing outside of school.

—affection for most of your students (unfortunately, unless you're a saint, you can't love them all) and pleasure in spending time with them.

—a sense of fairness and justice that operates in all of your work. This implies refusing to play favorites or use privileges to control behavior.

—high expectations for all students regardless of their past school experiences. If a teacher has high expectations for students it is impossible to accept the explanation that some children are unable to learn because of so-called hyperactivity or educational or social handicaps. Instead the teacher will modify the classroom environment and the curriculum in the search for more effective learning.

—consistent and compassionate behavior standards. There is a delicate line between imposing order on young people and maintaining an environment where creative and thoughtful learning can occur.

—knowledge of subject matter as well as the ability to shape it for particular students. Teachers have to have some mastery of the subjects they teach, so they can invent new explanations of concepts and processes if the ones they are accustomed to don't work. Of course, it is impossible for any elementary school teacher to master every subject she or he is expected to teach. However being an active learner in

Note. From *On Teaching* (pp. viii–ix) by H. R. Kohl, 1986, New York: Schocken Books. Copyright 1986 by the author. Reprinted by permission.

many of them will add to the richness of learning in the classroom and compensate for the unevenness of one's knowledge.

—liking teaching and getting rewards from the work. There is a lot of talk about teacher burnout and not enough about the sustaining rewards of teaching. The pleasure of contributing to the growth of others is the greatest reward of teaching well.

—having educational ideas of your own, developing an articulated philosophy of teaching and learning. Thinking about teaching and learning cannot be left to the professors and researchers. It should be an integral part of one's work and the basis for communication with parents, students, and community people.

APPENDIX C
Wigginton's Foxfire Core Practices

1. *All the work teachers and students do together must flow from student desire, student concerns.* It must be infused from the beginning with student choice, design, revision, execution, reflection and evaluation. Teachers . . . are still responsible for assessing and ministering to their students' development needs. . . . Most problems that arise during classroom activities must be solved in collaboration with students. When one asks, "Here's a situation that just came up. I don't know what to do about it. What should I do?" the teacher turns that question back to the class to wrestle with and solve, rather than simply answering it. Students are trusted continually, and all are led to the point where they embrace responsibility.

2. Therefore, *the role of the teacher must be that of collaborator and team leader and guide* rather than boss. The teacher monitors the academic and social growth of every student, leading each into new areas of understanding and competence. And the teacher's attitude toward students, toward the work of the class, and toward the content area being taught must model the attitudes expected of students—attitudes and values required to function thoughtfully and responsibly in a democratic society.

3. *The academic integrity of the work must be absolutely clear.* Each teacher should embrace state- or local-mandated skill content lists as "givens" to be engaged by the class, accomplish them to the level of mastery in the course of executing the class's plan, but go far beyond their normally narrow confines to discover the value and potential inherent in the content area being taught and its connections to other disciplines.

4. *The work is characterized by student action,* rather than [by] passive receipt of processed information. Rather than students doing what they already know how to do, all must be led continually into new work and unfamiliar territory. Once skills are "won," they must be reapplied to new problems in new ways. Because in such classrooms students are always operating at the very edge of their competence, it must also be made clear to them that the consequences of mistakes is not failure, but positive, constructive scrutiny of those mistakes by the rest of the class in an atmosphere where students will never be embarrassed.

5. A constant feature of the process is its *emphasis on peer teaching, small group work and teamwork.* Every student in the room is not only included, but needed, and in the end, each student can identify his or her specific stamp upon the effort. In a classroom thus structured, discipline tends to take care of itself and ceases to be an issue.

6. *Connections between the classroom work and surrounding communities and the real world outside the classroom are clear.* The content of all courses is connected to the world in which the students live. For many students, the process will engage them for the first time in identifying and characterizing the communities in which they reside.

Note. From *The Foxfire Approach: Perspectives and Core Practices* (pp. 1–2) by E. Wigginton, 1991, Rabun Gap, GA: The Foxfire Fund, Inc. Copyright 1991 by The Foxfire Fund, Inc. Reprinted by permission.

Whenever students research larger issues like climate patterns, or acid rain, or prejudice, or AIDS, they must "bring them home," identifying attitudes about and illustrations and implications of those issues in their own environments.

7. *There must be an audience beyond the teacher for student work.* It may be another individual, or a small group, or the community, but it must be an audience the students want to serve, or engage, or impress. The audience, in turn, must affirm that the work is important and is needed and is worth doing—and it should, indeed, *be* all of those.

8. As the year progresses, *new activities should spiral gracefully out of the old,* incorporating lessons learned from past experiences, building on skills and understandings that can now be amplified. Rather than a finished product being regarded as the conclusion of a series of activities, it should be regarded as the starting point for a new series. The questions that should characterize each moment of closure or completion should be, "Now what? What do we know now, and know how to do now, that we didn't know when we started out together? How can we use those skills and that information in some new, more complex and interesting ways? What's next?"

9. As teachers, *we must acknowledge the worth of aesthetic experience*, model that attitude in our interactions with students, and resist the momentum of policies and practices that deprive students of the chance to use their imaginations. We should help students produce work that is aesthetically satisfying, and help them derive the principles we employ to create beautiful work . . .

10. *Reflection*—some conscious, thoughtful time to stand apart from the work itself—is an essential activity that must take place at key points throughout the work. It is the activity that evokes insights and nurtures revisions in our plans. It is also the activity we are least accustomed to doing, and therefore the activity we will have to be most rigorous in including, and for which we will have to help students develop skills.

11. *The work must include unstintingly honest, ongoing evaluation for skills and content, and changes to student attitude.* A variety of strategies should be employed, in combination with pre- and post-testing, ranging from simple tests of recall of simple facts through much more complex instruments involving student participation in the creation of demonstrations that answer the teacher challenge, "In what ways will you prove to me at the end of this program that you have mastered the objectives it has been designed to serve?"

Students should be trained to monitor their own progress and devise their own remediation plans, and they should be brought to the point where they can understand that the progress of each student is the concern of every student in the room.

Chapter 10

Teacher Education and the Case Idea

GARY SYKES and TOM BIRD
Michigan State University

Some teacher educators are proposing and practicing "case teaching."
Those proposals and practices center on cases: narratives and descriptions of teaching that were constructed specifically for use in a teacher's
education. Typically, teacher educators or teachers write the cases, but
sometimes they choose and adapt material from other literature such as
research case studies. Occasionally, they ask student teachers to write
cases that organize and reflect on their own experience, apply educational
theory to those experiences, or both.

Doyle (1990a) contrasts case methods both with propositions about
teaching and learning and with opportunities to teach in laboratory or
field settings, thus suggesting that case teaching is focused more on particular situations than on general principles, findings, and rules, but also
that the student's encounter with those situations typically is vicarious
rather than direct. Although the encounter with cases is an indirect contact
with teaching, it is not passive. Work with a case typically is intended to
draw the student into the situations, problems, and roles that are represented in the case. An engaging discussion of a case can become like role
playing or simulations, which sometimes are used to explore cases.

In this chapter, we aim to survey the variety of cases, descriptions of
case teaching, and arguments being made about case teaching in teacher
education. At present, there is little research or theory specific to case
teaching in teacher education; therefore, we will join the proponents and
practitioners in looking at other literature that raises issues, questions,
and possibilities. We will begin with a sampling of cases and arguments
about them, then discuss cases as part of the teacher education curric-

We are grateful to Mary Kennedy, Judith Kleinfeld, and Lee Shulman for their detailed
commentary and many helpful suggestions on an earlier draft of this essay. We also profited
from discussions with Deborah Ball, Christopher Clark, Walter Doyle, Susan Florio-Ruane,
Magdalene Lampert, Katherine Merseth, Rita Silverman, Rand Spiro, and Steven Tozer.
Finally, we gratefully acknowledge the assistance of Teresa Scurto in reviewing literature,
sharing ideas, and asking good questions.

ulum, discuss learning to teach from cases, and conclude with proposals for research and development.

A SAMPLE OF CASES AND CASE LITERATURE

Case teaching has histories in several fields of professional education. Harvard University figures prominently in several of those histories; there, case teaching emerged in law in the 1870s, in medicine around the turn of the century, and in business in the 1930s (Doyle, 1990a). Proponents of case teaching for teacher education frequently argue by analogy to case teaching in other fields of professional education. Masoner (1988) provides an "audit" of the few studies of case teaching in several fields. Christensen and Hansen's (1987) book of cases and articles on case teaching, which grew out of experience at Harvard's business school, is a valuable and frequently cited resource. Katherine Merseth (1991, in press) often relies on analogies to case teaching in business education to construct and teach cases in preparatory teacher education. Kathy Carter has explored analogies to case teaching in legal education (Carter & Unklesbay, 1989). However, teacher education is not completely without its own history of case teaching. Doyle (1990a) notes a proposal for case teaching from 1864 and a report on the use of cases at New Jersey State Teachers College from the 1930s. McAninch (1991) mentions several predecessors, the earliest from 1925, of the casebooks published recently.

A systematic historical comparison of case methods in professional education would be both interesting and useful for teacher educators, but we have not undertaken that task. Rather, we have followed the maxims that the subject matters and the context counts. Although occupational comparisons are useful, each occupation also faces a distinctive set of problems, employs ideas related to those problems, and constitutes a materially distinctive environment for case teaching. To some large extent, case teaching must be constructed or reconstructed for teacher education and its setting.

For the most part, the current literature for teacher education comprises cases, descriptions of case teaching, and arguments about the virtues and uses of cases. Beyond the common features described earlier—narratives of teaching are constructed specifically for use in teacher education and are used at a distance from actual teaching situations—case teaching encompasses considerable variation in the situations, actors, acts, thoughts, and feelings reported by the cases; the ways in which the case materials are organized; the media in which they are presented; the activities in which they are used; the ways in which teacher educators participate in those activities; the rationales given for case teaching; and the aims pursued.

Textbook Cases

Cases figure prominently in a series of teacher education textbooks from Teachers College Press (Feinberg & Soltis, 1985; Fenstermacher & Soltis, 1986; Phillips & Soltis, 1985; Strike & Soltis, 1985; Walker & Soltis, 1986). In the first chapter of their textbook on the ethics of teaching, Kenneth Strike and Jonas Soltis (1985) offer a case in which a teacher finds that truth telling and caring for children are not always easy to reconcile. In less than two pages of narrative with dialogue, they place an elementary school teacher in a recognizable dilemma. She stopped a fight between some boys. For one of these boys, she thought, the fight was more than an incident; it was another sign of some kind of trouble that ought to be discussed with the boy's parents. In an interview with the boy's father, she began to fear that the father was violent and would do violence to the boy for his part in the fight. She heard herself describing the fight to the father in terms that might protect the boy but that did not fully square with the circumstances as she knew them. Thinking about the interview afterward, the teacher was distraught. Was the boy actually in danger? Even if he was, could she justify her fabrication of the fight?

Strike and Soltis then go on to report that one tradition in ethics strives to evaluate conduct in light of its consequences, whereas another strives to evaluate conduct in relation to universal principles or rules. The authors reflect on the case they have offered, suggesting how each moral strategy might be employed. This discussion exposes the different ways of reasoning and their respective difficulties, such as knowing or anticipating the consequences of a given act. Seemingly, Strike and Soltis anticipate that their readers may react to the difficulties of settling the case and deciding between the two ethical positions by retreating into mere personal preference. They propose to explore, in several chapters on ethical issues in teaching, whether "objective" arguments can be made on moral questions. Each of those chapters uses cases much like that described above. Two chapters at the end of the book provide a number of "cases" and "disputes" for thinking and discussion.

Casebooks

Several casebooks for teacher education recently have been or soon will be published (Greenwood & Parkay, 1989; Kowalski, Weaver, & Hensen, 1990; J. Shulman & Colbert, 1987, 1988; Shuman, 1989; Silverman, Welty, & Lyon, 1991). In their book of "case studies for teacher decision making," Gordon Greenwood and Forrest Parkay (1989) provide 30 cases that range from 4 to 10 pages in length and are grouped under the headings of curriculum, instruction, group motivation and discipline, pupil adjustment, and conditions of work. The authors say that the book presents "high-frequency teaching situations that have been constructed from survey data supplied by real teachers in six states regarding their 'most troublesome' and 'most enduring' teaching situations" (p. ix).

One of the cases on group motivation and discipline is "Two Different Worlds," 10 pages of narrative, dialogue, and data followed by a page of discussion questions. A new art teacher is assigned to teach mornings at one junior high and afternoons at another. The first school she finds to be welcoming, well-equipped, well-maintained, and well-stocked with art supplies, and the students there to be compliant, cooperative, and motivated. The second school she finds to be forbidding, chaotic, run down, and poor, and the students to be uncompliant, uncooperative, and abusive to her and each other. After 3 months, the teacher describes herself as being locked into a battle for control of her classes in the latter school; she is resorting to threats. A teacher from that school asks her how she "really feels" about her students. The discussion questions included with "Two Different Worlds" speak of school climate, handling confrontations with individual students, intrinsic and extrinsic motivation, social and economic environments, teaching objectives in art and their propriety for different students, classroom management models, parental involvement, and self-fulfilling prophecies.

In their introduction, Greenwood and Parkay cast teachers as decision makers and propose a strategy for decision making: Examine the situation and decide to deal with it, gather data about the situation, interpret the data, generate alternatives and choose among them, examine the decision for consistency and feasibility, and then execute and evaluate the decision. They provide sample analyses—a "psychological analysis," a "behavioral analysis," and "an analysis based on classroom-management theories"—of a case titled "Joe Defies Authority." They want students to strengthen their theoretical knowledge by applying it to situations and to become more aware of their decision making.

Other authors also provide assistance in using cases. Silverman et al. (1991) offer an instructor's manual with their case collection. Judith Shulman (in press) has assembled a volume in which several teacher educators present cases and discuss their use; in the same volume, Lee Shulman (in press) provides a discussion of pedagogy with cases.

Conversations, Preconceptions, and Videotape

Community and conversation are prominent themes in an introductory teacher education course that Tom Bird (1991) organizes around three videotapes of teaching. The tapes depict, respectively, a direct instruction lesson on sentence writing for eighth graders, an open classroom for second and third graders, and a teacher-led mathematical discussion involving fifth graders. A transcript of an interview with one of the teachers and a journal article written by another provide contexts for the videotapes as well as access to the teachers' thinking. After the first viewing of each tape, the education students report and discuss their reactions in order to notice and begin evaluating the ideas about teaching that they brought with them to the course. The tapes provide much to talk about; some of the teaching shown on the tapes is likely to be familiar and appealing to student teachers, but other teaching is likely to be unfamiliar and vexing to the point that some student teachers declare that it is not teaching, or is not good teaching. Bird matches each of the videotapes with an article or essay that raises an important problem of teaching, provides language for analyzing the videotape, and provides rationales for the teaching shown there. The student teachers study and discuss these articles intensively, often in small groups, with the aim of mastering the arguments well enough to use them to interpret the tapes, which they attempt to do after a second viewing. To promote interaction between their prior conceptions of teaching and the arguments they encounter in their reading, students write a "conversation" about each videotape in which they employ three voices: "Myself as Experienced Student," "Myself as Inexperienced Teacher," and "Myself as [the author of the article being studied]." Within this framework of activities about videotapes, Bird pursues objectives such as challenging students' tendency to think of teaching only as telling, introducing students to arguments about classroom organization, and socializing prospective teachers to pointed discussions of practice.

In this instance, the cases used are videotapes, found and adapted for teacher education, augmented by other material such as the interview with the teacher in one videotape, and brought into relation with foundational texts by means of a writing assignment designed for that purpose. Such strategies for combining materials and activities in work with cases can become quite elaborate, as we will describe later.

Subject-Specific Cases: Mathematics

Cases can take all sorts of teaching problems as their subjects, and the recent rediscovery of the subject matter is reflected in the development of cases. Carne Barnett (1991) describes a project to develop cases for mathematics teacher education; its premise is that "good teaching requires highly developed subject-specific thinking and reasoning skills that allow teachers to quickly generate alternative strategies and evaluate those strategies, based on continually shifting conjectures about students' thinking, motivations, and beliefs" (p. 1). Barnett illustrates her argument by reporting on some conversations about the following case written by Kim Tolley, a classroom teacher:

There's No One-Half Here!

I had been using cube blocks with my fifth graders to demonstrate the concept of multiplying fractions such as ⅔ × 6. First, I'd have students take six cubes and lay them on the table. Next, I'd ask them to divide their cubes into thirds. Finally, I'd ask them to pick up two-thirds of the blocks and be ready to tell me how many they were holding. After lots of practice, students could easily demonstrate the following types of problems:

⅜ × 32

⅘ of 40

Several days later, I decided to use the cubes again to demonstrate problems like ¾ × ½. I asked the students to take eight cubes and set them on the table. I asked them to pick up one-half of the cubes, which they quickly did. Next, I asked them to show me three-fourths of the one-half they were holding. Immediately, I sensed confusion everywhere. "Ms. Tolley, what do you mean, three-fourths of the one-half we're holding?" asked one girl. "We're holding four cubes, not one-half."

This lesson seemed to parallel the other lessons so perfectly that I was surprised to encounter such confusion, especially with my above-average students. What was the source of their confusion? Weren't my directions clear? Should I have selected another manipulative for this lesson?

Barnett presents findings from discussions of this case with four groups of practicing teachers having a wide range of experience. As is common in case teaching, "the role of the facilitators was to pose strategic questions, press for analysis of alternative strategies and their consequences, and provoke challenges of opinions, ideas, and beliefs expressed by members of the group" (Barnett, 1991, p. 8). Barnett reports that four issues were prominent in all discussions of this case. Should the teacher have used a continuous model or a discrete model for fractions? What representations, other than concrete materials, could the teacher have used? Could the students have been confused by the teacher's language? Is it developmentally appropriate to teach multiplication of fractions to fifth graders? The facilitator's guide for this case, Barnett says, will be organized around those issues. Barnett reports that, even with inexperienced teachers who had limited exposure to professional development programs, case discussions were deep and rich, with minimal input from the discussion leader; she notes data suggesting that similar results are possible with preservice teachers.

Context-Specific Cases

Although most published cases clearly are intended for widespread use, their relevance in a given teacher education program may, as we have seen in "Two Different Worlds," be defined by features of the situations they describe. Moreover, some cases are intended for particular contexts. Judith Kleinfeld is editing a series of cases written by teachers about particular problems of teaching in isolated Alaskan communities (Allen, 1990; Carey, 1989; Finley, 1988, 1990). These cases are centerpieces in a teacher education program, Teachers for Alaska, that Kleinfeld (in press) describes as a "problem-centered, case-based approach to teacher education" that has "done away with the traditional sequence of foundations courses followed by methods courses." Instead, students study and discuss problems and cases in small groups where "research, theory, and methods are introduced as such material becomes useful in understanding and dealing with the classic problems of professional practice." The program is organized in "thematic blocks" focused on "the dilemmas of teaching particular subjects to culturally diverse student groups."

The literature and literacy block is organized around "Malaise of the Spirit" (Finley, 1988), a case written by a teacher. This case runs to 50 single-spaced pages that draw the reader into the personal, professional, and instructional problems of an English teacher working in an isolated Alaskan community. One of Kleinfeld's aims in using this case is to absorb her students in a teaching situation for which, in many instances, their previous experience cannot have prepared them.

"Malaise of the Spirit" is a complex narrative, cast as the teacher's retrospection about his successful experience working in isolated Alaskan communities and his recent move to a new post where things have not gone nearly as well as before—in the classroom, in relations with colleagues, or in the morale of the school. The first and longest part of the case describes a combination of problems that bring the teacher to a crisis in which he is, briefly, too distraught to work. The second part describes the thinking and the actions by which the teacher begins to improve the situation, both in his English class and in the school. The third part provides background information about the community and the school where the teacher works.

In using such cases, Kleinfeld aims to give novices

vicarious experience with the kinds of problematic situations characteristic of teaching, a model of how an expert teacher goes about framing and constructing educational problems, a model of how a sophisticated teacher inquires about and reflects on such problems, a stock of educational strategies for use in analogous problem situations, and a sense that teaching is an inherently ambiguous activity requiring continuous reflection. (Kleinfeld, in press)

The use of cases in the Teachers for Alaska program continues to evolve. Regarding length of cases, it is worth noting that Kleinfeld also uses cases as short as a paragraph to foster discussion. The case topics vary as well. The literature and literacy block also employs a case of censorship and selectivity, a case about teaching Native myth, and a six-part case that takes student teachers through a series of decisions that a teacher made in teaching *Hamlet* to diverse students. The latter cases are intended to help student teachers to build their ability to teach particular subjects to culturally diverse students. Kleinfeld argues that, as with literature, close attention to the particulars of good cases may provide students not only with exposure to particular settings, but also with glimpses of universal problems in teaching (J. Kleinfeld, personal communication, September 20, 1991).

Transition

As we have begun to describe (and will continue to do), case workers around the country are beginning to generate, test, and use cases of various sorts in teacher education. Strike and Soltis used cases to make distinctions in an ethics textbook and to support discussions in classes that use it. In their casebook, Greenwood and Parkay surveyed teachers to find and describe common problems that enable students to exercise their knowledge and to practice decision making. Kleinfeld works with teachers to construct a set of extended, first-person narratives that, among other things, immerse prospective teachers in difficult and unfamiliar problems of teaching in Alaska's rural villages. Bird tried to combine videotape, related material, and educational literature in order to set up an interaction between students' prior ideas about teaching and ideas they will encounter in their teacher education programs. Barnett works with teachers to portray problems in mathematical pedagogy and to work out the questions and procedures for fruitful discussions of those problems.

These projects raise a variety of questions. For example, around what conception of a case is development organized? Our opening description of teaching cases surfaced a number of dimensions, including the medium, the genre, the length, whether the case is actual or contrived, and others. The choices here relate to the case developer's purposes and uses for cases as well as to theoretical hunches about learning from cases. Instructors may use extended, detailed narratives as centerpieces in courses, revisiting them repeatedly in conjunction with other material and experiences. Or they may use many, brief vignettes in a course, spending little time on any one but building a pattern of understanding across multiple instances. The first use treats the case as a richly contextualized, multilayered account containing a variety of issues for analysis. The second treats cases as instances across which to build nuanced understanding of concepts, issues, or practices.

Likewise, some developers rely on videotapes as an essential case medium through which to portray vivid instances of instruction. The old maxim that a picture is worth a thousand words applies here: It is better to view a case of cooperative learning (and its vicissitudes) in action than to read an account of it. Written cases, however, allow portrayal of problematic situations that cannot easily be captured on film while providing simultaneous access through narrative to thoughts, feelings, and actions. One may suspect that emerging practice will combine materials: transcripts of videotaped instruction that intersperse the teacher's thoughts with the written record of the verbal interactions on tape.

The case idea harbors diverse materials, arguments, and possibilities. Proponents and practitioners of cases and case teaching pursue a variety of aims, construct cases on many topics and problems of teaching in different media, and use those cases in different ways. They hold different ideas about what prospective teachers should know or know how to do and about the kinds of conversations and reasoning into which they should be inducted. They would organize the curriculum of teacher education differently, and would teach it differently. We will next consider the curriculum, and then turn to the teaching.

CASES AND THE CURRICULUM OF TEACHER EDUCATION

Compared with the theory-into-practice idea that has been so prominent in teacher education, the case idea implies a shift in emphasis from the theories to the practices and a shift in genre from exposition to narrative. In these and other respects, the case idea might be counted as a part of the broad "refiguration" of social thought (Geertz, 1981) in which teacher education has been participating. Educational work reflects a set of connected (and relative) trends, from a law-seeking to an interpretive aspiration in inquiry; from a concern for universal principle to a concern for particular relationships; from the positivistic stance of an observer on the scene to the pragmatic stance of the actor in the situation; from authoritative transmission to mutual exploration of knowledge; from conditioned behavior to meaningful action as a model for teaching and learning; from a cooler appraisal of teaching as knowledge and technique to a more passionate consideration of teaching as moral agency; from vision as a metaphor for knowledge to speech as the literal means for constructing meaning; and from lecture to conversation as the mode of interaction between professors and teachers.

There is ferment about what should count as knowing, about what should count as useful writing or persuasive conversation, and about what kinds of thinking should be respected. Teacher education participates in that ferment, and the teacher education curriculum is at issue. One question is, What topics, issues, aspects, or problems of teaching should prospective teachers study in the teacher education curriculum? The case idea does not bear directly on this question; as we have seen, cases could and do address many topics and problems in teaching. Another question is, To what modes of thinking and conversation about teaching should prospective teachers be introduced and inducted? The case idea has extensive implications here, which we will pursue.

In this section, we will consider four kinds of conversation and reasoning that might be desirable in teaching and teacher education; to use Soltis's term, these are four kinds of "community of practice" that might be organized about teaching. The first kind honors theory and treats teaching as a matter of applying theory to practice. In this community, cases appear as instances of theory. The second kind also is concerned with the relation between theory and practice but reverses the emphasis; this community assigns priority to the situated problems of practice and employs abstractions in the course of action on those problems. In this community, cases are problems for deliberate and reflective action. The third kind of community relies on stories and other narrative modes of knowing and communicating; community members enact, tell, recall, and ponder stories about practice. Here, cases are literature, as well as a kind of knowledge that theory cannot supply. The fourth kind of conversation resembles the tradition of moral casuistry; members of the community reason from case to case by analogy—without resort to theory. Here, cases are a body of knowledge in themselves.

These differences among conversations, communities, and their use of cases raise questions about the character of cases in teacher education. What is a case of teaching? How do teachers work on a case? What do case materials provide, if anything, that other curriculum materials do not? Where and how would cases appear in a teacher education curriculum? Why would teacher educators use them? We will take the four kinds of communities in turn, discuss their respective use of cases, and, at what we hope are interesting junctures, describe current work on cases and case teaching in teacher education.

Foundations for a Profession of Teaching

The incumbent kind of conversation and reasoning in teacher education might be called foundational, as in "the foundations of teacher education." Where the normal school curriculum had emphasized subject matter knowledge and uniform methods of teaching, college and university teacher education programs since the 1940s increasingly have emphasized the importance of foundational disciplines. They have rejected what they regarded as technical training in favor of a more academic program that would prepare "practitioners who were as theoretically informed as practitioners in other professions" (Tozer, Anderson, & Armbruster, 1990, p. 294). The community for teaching and teacher education would be a professional enclave modeled on more prestigious occupations, including medicine and law. Its core knowledge would be theoretical and scientific and would govern practice. That aspiration has extended to include the body of research on teaching produced in the last two decades.

We note that the term *foundation* has a number of related meanings. Within the branches of philosophy that treat problems of knowing and valuing, the term refers to some set of first principles, assumptions, or methods upon which may be erected a system of knowledge and values. Within teacher education, the familiar division into foundations and methods courses reflects a distinction between discipline-based perspectives on the history, aims, and social and cultural contexts of schooling and general and particular methods of instruction. These two usages share a conception of the relation of theory to practice that sees the latter resting on or derived from the former. A foundational approach to teacher education organizes the enterprise in terms of "applying" theory to practice, where "theory" might include various conceptions of the aims of education or methods of instruction derived from conceptual and empirical work in cognitive psychology. The notion of a foundational approach to teacher education, then, includes both the curricular division into the two types of courses and an epistemological stance toward the theory-practice relationship that is represented in both foundations and methods courses.

Generally, the foundational approach claims priority for knowledge that is general, propositional, and organized along the lines of the foundational subdisciplines—the psychology, history, philosophy, and sociology of education. In the typical program sequence for teacher education—foundations, methods, and practice work—prospective teachers are expected to become familiar with the concerns and vocabularies of theorists and researchers before they turn to the practical problems of the occupation that they want to enter; they are expected to learn the theoretical parts of teaching before they address its problems as practical wholes.

Moreover, teaching methods tend to be treated as prescriptions for, rather than descriptions of, practice; these prescriptions are developed with the aim of replacing existing practices (Doyle, 1990a). Confidence in a theoretical or scientific approach to teaching has combined with concern about or contempt for common current practice to generate a missionary outlook among teacher educators; they are likely to perceive that their task is to lift the dead hand of the past, overcome the folklore of teaching, and break the continuity of traditional practice. One consequence has been "neglect, if not devaluing, of both context and the common forms of schooling in building a body of knowledge about teaching" (Doyle, 1990a, p. 348).

Some teacher educators now propose to modify the foundational idea so as to deal more adequately with situated practice. In their introduction to a recent symposium on foundational studies in teacher education, Tozer et al. (1990) report:

Many of the articles explicitly criticize contemporary practices in foundations of education for their persistently inadequate integration of foundational theory and knowledge into teacher education programs. . . . Nearly every article in this volume responds to that perceived inadequacy by arguing for attention to the practical contexts of teaching as a guide to integration of research knowledge and theory into the teacher education curriculum. (p. 296)

Although those criticisms and arguments might be interpreted as a limited call for better curriculum design, instruction, or use of practice work, in the context of the symposium they are more basic.

The authors included in the symposium seek alternatives to the metaphor of foundations for teaching; they want a different conception that "assumes not a primary but instead a complementary role for the study of mind and culture" (Tozer et al., 1990, p. 296). In their rethinking of foundational studies, they appear to be moving from a conception of the psychology, history, sociology, and philosophy of education as the bases and sources for practice to a conception of those subdisciplines as resources for dealing with practice or as useful ways to talk about practice. They present the initial and continuing education of teachers as a "dialogue" (Peterson, Clark, & Dickson, 1990) or conversation in a "community of practice" (Soltis, 1990) or in a "community of learners" (Leck, 1990). That community might eventually harbor a "consensus of the learned" that forms around "paradigm cases" of practices (Broudy, 1990). Short of that consensus, the members of this community of practice would share "common language, concepts, and interests," which they use to discuss cases of practice (Soltis, 1990).

Some of the participants in the symposium suggest or assert that schoolteachers bring more to a conversation with professors than a need to know what principles to follow or a need to have their work interpreted for them. Rather, they present teaching practice and teaching problems as an independent domain of situated action in which teachers can construct "classroom knowledge" (Doyle, 1990a) and accumulate "wisdom of practice" (L. Shulman, 1987, 1990) that deserves respect in its own right. Doyle is explicit that such knowledge is "event-structured" knowledge or case knowledge; he asserts that such knowledge should be accumulated, studied, and taught as "the core foundation" of teaching. He states that "all other disciplines have utility for teaching as resources in the continuing development of classroom knowledge" (Doyle, 1990a, p. 358).

Although the symposium contributors retain their conviction that the psychology, history, sociology, and philosophy of education are defensible, enlightening, persuasive, and useful ways to talk about practice, they also are assigning greater epistemic, psychological, and social weight to schoolteachers' experience, thought processes, and know-how. They want to integrate the foundational subdisciplines both with subject matter knowledge and with classroom knowledge.

Cases as Instances of Theory

Although we have highlighted some current discontent with the foundational approach to teacher education, it remains firmly institutionalized in the division of courses into foundations and methods and in the dominant mode of conceiving and enacting the theory-practice connection. Within this mode, some cases fit comfortably because they constitute the particular representations of "practice" to which theory is applied; for example, recall the "Two Different Worlds" case from Greenwood and Parkay's casebook, described earlier. One approach to cases in the curriculum, then, consists of the interweaving of theoretical material, in the form of course readings, lectures, and other presentations, with cases. The instructor leads a discussion the goal of which is to apply the theoretical ideas to the case particulars, to illuminate the case by the light of theory. Case discussion serves to extend, test, and consolidate students' knowledge of theory, which is accorded pride of place because of its broad explanatory or interpretive power.

Within this mode of curriculum organization, a single case might instantiate two or more theoretical perspectives that might conflict or compete, be compatible, or simply draw attention to different features of the case. Within the foundational approach, a key design decision is the selection and sequencing of theoretical material with cases, to promote what L. Shulman (1986b) refers to as "strategic understanding," the wise application of knowledge to situations where principles conflict and no simple solution is possible. Through skillful curricular arrangement, instructors may cultivate strategic knowledge in students while exploring the problematic nature of applying theory to practice, in cases where theories present conflicts, trade-offs, or dilemmas, and where theories are silent, multivocal, or ambiguous. The foundational approach to cases does not presuppose any tidy correspondence between theory and cases, but aims to cultivate analytic skills in the application of ideas and to convey theoretical knowledge in a form useful to the interpretation of situations, the making of decisions, the choice of actions, and the formation of plans and designs.

"Theory into practice" is a versatile curricular principle that may be employed in both methods and social foundations courses. For example, in a reading course, students might review case profiles of students, such as those rendered by Bussis, Chittenden, Amarel, and Klausner (1985), and then discuss the pros and cons of using particular theories of literacy with the presenting cases. Or, in a school and society course, the instructor might first acquaint students with contending theories of social justice, comparing, for example, the work of John Rawls (1971) with that of Robert Nozick (1974) or others, and then supply written cases that present justice-oriented issues in education such as the policy of affirmative action, the financing of schools, or a proposal to expand parental choice of schools. Case discussion examines such issues from the perspective of the contending theories, but the aims of instruction may also include helping students to clarify and justify their own views about social justice, in addition to learning to apply theory to practice.

Pragmatic Action and Deliberation

To this point we have used the familiar phrase "theory into practice" to indicate a common view of the relationship. But this formulation is just what the antifoundationalists have criticized; for some, the case idea provides one useful corrective. They resort to an indigenous tradition in education that makes use of theory but grants pride of place to case particulars. The pragmatic tradition, with its distinctively American roots in the writing of Pierce, Mead, James, and Dewey (see Scheffler, 1974) and its contemporary extensions in the work of Richard Rorty (1979, 1982), poses a challenge to the basic assumptions of unreconstructed foundationalism.

For our purposes, the pragmatic tradition might be traced to *The Quest for Certainty*, where John Dewey (1929) posed a problem that arises when abstractive studies become disconnected from concrete situations and alternatives for action in those situations. Dewey wrote that "objection comes in . . . when the results of the abstractive operations are given a standing which belongs only to the total situation from which they have been selected. All specialization breeds a familiarity which tends to create an illusion" (p. 173). Like Dewey, Joseph Schwab regarded the social sciences as providing our "fullest and most reliable knowledge" about various aspects of education (Schwab, 1978), but he also pointed to a tendency for abstractive work to tear practical situations apart:

In the course of . . . readying a subject of scientific, or theoretic inquiry, the principles which distinguish it from the whole tend to confer on the partial subject an appearance of wholeness and unity. The connecting and entwining threads which originally made it one aspect of a larger whole are smoothed down and covered over. . . . The bodies of knowledge are themselves separated, each couched in its own set of terms. Only a few terms of each set have connections with terms of another set. Hence, the bodies of knowledge that we inherit from the behavioral sciences are, taken separately, only imperfectly applicable to practical problems, problems which arise in the whole web of the original complexity. (Schwab, 1978, p. 330)

Important connections disappear in abstractive work, Schwab argued; in his proposals for "eclectic arts" and "practical deliberations" on particular problematic situations (cases?), he was trying to construct a program to restore those connections.

Pragmatists also have sought to correct what they perceived as an amoral tendency in social science, the drive to purge disciplinary vocabularies of terms carrying moral significance. According to Dewey, "the problem of restoring integration and cooperation between man's beliefs about the world in which he lives and his beliefs about the values and purposes that should direct his conduct is the deepest problem of modern life" (Dewey, 1929, p. 204). In the effort to become scientific, Rorty argued, the social sciences have pursued a futile and misleading objective. In contrast, he proposed two requirements for the vocabulary of the social sciences. First, it should contain "descriptions of situations which facilitate their prediction and control." Second, it should contain "descriptions which help one decide what to do" (Rorty, 1982, p. 197). This second criterion necessarily introduces moral considerations and judgments into social science accounts.

Cases as Problematic Situations

Alike, Dewey, Schwab, and Rorty sought to deal with problems of illusion, partiality, distortion, and amorality in abstractive work by restoring situated problems of practice to a central position, by taking up the stance of the actor in the situation, and by treating research findings, theories, and principles as part of the actor's equipment. The confrontation with problems of practice—with cases—would force a reconnection between moral and other arguments. Likewise, that confrontation would tend to expose the partiality and distortion of the various theoretical arguments that might be made about the problem, and so provide a corrective for academic illusion.

The pragmatic stance alters the interplay of theory and cases within curricula in subtle but important ways, for now the case serves as the basic unit of deliberation and action, while theory supplies useful but limited and fallible tools for work on cases. The aim of instruction shifts from applying theory to practice to acting, reflecting, and deliberating on problematic situations with the aid of various theories. In work with cases, students do this vicariously. Through case deliberations students develop a proper appreciation for the value of theory, learning its uses in wise, self-reliant ways. The flow of practical deliberation moves back and forth, revealing aspects of the case from the perspective of theory and exposing the limits, lacunae, biases, and interconnections of theory in its encounter with cases. Within the pragmatic mode the case becomes the primary focus of curriculum, and the artful construction and arrangement of cases becomes the central act of curriculum development. This contrasts with a foundationalist approach that begins with theory, then selects cases as an afterthought (if at all), as merely the material upon which to practice the application of theory.

Some researchers have begun to explore how to select and sequence cases for purposes of generating or testing generalizations. For example, Collins and Stevens (1982) have analyzed transcripts of inquiry teaching to understand the interplay of cases in instruction, where the cases are examples or instances rather than extended written or video texts. Their work surfaces the tactics teachers use in selecting and sequencing particular types of example (e.g., positive and negative exemplars, counterexamples, hypotheticals), but as yet investigators have paid little attention to larger patterns of instruction with cases.

The selection and sequencing of cases, together with tactics for interweaving presentation of principles and theory with case activities and other elements of teacher education, present a complex curricular issue. Some current uses are quite conventional in presenting written cases for discussion in university-based courses. Other approaches, however, use cases in conjunction with field experiences to focus reflection on issues that emerge in novice teaching (Florio-Ruane & Clark, 1990). If cases are themselves devices for situating knowledge, so are they variously situated in relation to other instructional activities, but the principles for such coordination have yet to be worked out. For example, advocates argue that cases can present subject matter knowledge in the context of teaching and so help acquaint students with pedagogical perspectives on such knowledge (e.g., see Barnett, 1991; Wilson, in press). But students may engage with such cases in university classrooms, in encounters with supervising teachers, in interaction with their initial forays into classrooms, and elsewhere.

The thoroughgoing integration of case with theoretical material may help both to relieve the argued shortcomings of the foundational approach and to abet the pragmatic program, because cases may be vehicles for representing classroom knowledge and problems directly in the curriculum of teacher education. Cases, though, might serve purposes other than to make the theory-practice relation. To explore some alternatives, we turn to two other modes of conversation about teaching that might be cultivated in a teacher education curriculum.

Narrative Knowing as a Natural Kind

The recent writing of Jerome Bruner (1985, 1986, 1990) describes a bold project to rebalance if not reorient the study of mind; if accepted, his proposals would have substantial implications for discourse in a community of teaching practice. Bruner proposes that there are two fundamental modes of knowing that he labels the "paradigmatic" and the "narrative," associated, respectively, with science and with humanism. The paradigmatic mode refers to the logico-scientific enterprise in the Western tradition and encompasses the disciplines of logic, mathematics, and the sciences. "The imaginative application of the paradigmatic mode," he writes (1985, p. 98), "leads to good theory, tight analysis, logical proof, and empirical discovery grounded by reasoned hypothesis." The foundational approach to teacher education is largely based on paradigmatic knowing as Bruner describes it, and leads naturally to the formulation of applying theory to practice. By contrast, the narrative mode deals in good stories, stirring drama, and richly portrayed historical accounts.

Bruner argues that each mode constitutes a natural kind of knowing that cannot be reduced to or subsumed by the other. Each has its own operating principles, criteria, and procedures for establishing truth, but their aims are fundamentally different. In the paradigmatic mode, one seeks to know the truth by establishing causal relations. In the narrative mode, one pursues questions of human intention or the meaning of experience. Science making and narrative making each constitute ways of constructing the world. Bruner comments:

Science creates a world that has an "existence" linked to the invariance of things and events across transformations in life conditions of those who seek to understand. . . . The humanities seek to understand the world as it reflects the requirements of living in it. In the jargon of linguistics, a work of literature or of literary criticism achieves universality through context sensitivity, a work of science through context independence. (1986, p. 50)

Bruner is but the most prominent of a wide range of social scientists studying the role of narrative in inquiry and in human affairs (for a brief review, see Connelly & Clandinin, 1990). He argues that until recently psychology, beguiled by the cultural authority of science, has concentrated on paradigmatic ways of knowing while ignoring the central role of narrative in human affairs. His perspective leads to study of the interaction between individual readers, with their different histories, and texts rendered in a variety of communal genre. He urges the social sciences, psychology in particular, to take up study of narrative knowing, concentrating on the dual landscapes of action and of consciousness—what those involved in action know, think, or feel. He would explore, for example, how elements of genre influence interpretation. Drawing on contemporary literary theorists, Bruner describes how texts recruit the imagination of the reader in constructing a "virtual" text in response to the actual. "Literary texts initiate 'performances of meaning' rather than actually formulating meaning themselves" (Bruner, 1986, p. 25).

These ideas have several implications for the curriculum of teacher education. They clearly provide broad theoretical support for learning through engagement with narratives. They call into question the lopsided attention in teacher education to paradigmatic modes that seek, in the name of "effectiveness," to convey scientifically validated rules of practice to aspiring teachers. Rather (or also), these ideas suggest, teaching is a hermeneutic act that requires skills of interpretation and analysis, skills that may be developed through encounters with narrative. Any situation in teaching is open to interpretation from multiple perspectives, using multiple frameworks composed of concepts, ideas, and values; teacher educators aim to help novices to extend and enrich their ways of seeing life in classrooms. Well-wrought cases supply texts for such analyses that include the interior perspectives of actors in the situation together with accounts of their actions.

Cases as Literature

Bruner's work also raises questions about the nature and use of cases, for he draws attention to the literary properties of texts and their effects. If cases are to provide opportunities for interpreting human intentions, then elements of genre may come into play, and poems, novels, or short stories may have unique value. So it is not simply any narrative that fascinates Bruner, but rather literary texts whose indeterminate "meaning" invites, even requires, the reader to "write" the text—that is, to render an interpretation.

Case advocates (e.g., Kleinfeld, 1990, in press; Merseth, 1991, in press) have noted the virtue of texts that can be revisited repeatedly to deepen understanding and to develop alternative views. The instructional value of such cases rests on their verisimilitude or lifelikeness. Many proponents of case teaching will use only cases drawn from actual situations, and they strive for compelling lifelikeness in the writing. However, Bruner and his colleagues are careful to note that

literature is not quite life, and the procedures for the interpretation of intentionality in texts and in human interaction may differ in some important ways. In other words, where we say that an action means such and such and that a text means such and such, we invoke two different notions of meaning—and it is not yet clear how different they are. (Feldman, Bruner, Renderer, & Spitzer, 1990, p. 3)

Narratives provide powerful advantages in simulating and representing complex, multidimensional realities; reading and discussing rich narratives of teaching in relative tranquility may help prospective teachers to gain the understanding they need in teaching. At the same time, if teacher education is to prepare for action as well as thought, then novices must gain direct experience. Writing cases, as we will discuss later, might help them to make sense of action situations, as simulations may help them to learn how to act in those situations.

The work of Bruner and his colleagues indicates that literary properties of texts influence their interpretation by the reader—or learner. Bruner emphasizes how literature uniquely conveys the interplay of the inner world of the protagonists—conscience and consciousness—with the outer world of action and events. Some initial experiments (see Bruner, 1986; Feldman et al., 1990) demonstrate that properties of text do shape interpretation. When students read different versions of the same short story— the original and an altered version that lacks figures of speech and mental state terms—subtle but unmistakable differences emerge in the interpretation of intentionality. The storyteller's art, this exploratory research suggests, stimulates the imagination of the reader in ways surprisingly measurable.

The significance of narratives and learning from narratives may depend, then, on artful, literary uses of language. In particular, as one experienced teacher educator speculated, literary texts may serve best in developing an appreciation for the moral dimensions of teaching, because literature situates moral dilemmas in the larger contexts of character and culture while provoking strong emotions in response to matters of value (S. Florio-Ruane, personal communication, April 9, 1991). How, for example, might a teacher educator provoke not only an intellectual understanding of educational inequity, but a sense of outrage and the resolve to respond? We read literature for many reasons, not least for arousing our passions, for discovering our values, and for understanding the human condition outside the narrow confines of our own experience. Such purposes, and materials that serve them, have a place in the education of teachers.

The influence of the genre-related aspects of cases has several implications. Case writing might systematically incorporate the dual landscapes of consciousness and action, and teacher educators might be encouraged to produce literary narratives. Alternatively, teacher educators might rely more often on literature to supply broad perspectives on educating; there are certainly many texts to choose from if the aim is to illuminate the lives of children, to explore issues of character and morality, or to understand the social and historical milieus in which teaching takes place.

Bruner suggests that narratives are not simply occasions for deploying paradigmatic knowledge, as foundationalists might construe it, but constitute in themselves a kind and form of knowledge worthy in its own right and on its own terms. Is it stretching his point to claim a place for such knowledge in teacher education? Do we really mean that teacher candidates might read poems as part of their preparation to teach? Literature currently has little place in professional education, but psychiatrist Robert Coles is famous for his literature-based courses in Harvard's Schools of Law, Medicine, Business, Divinity, and Education (for a description, see Coles, 1989). Literary narratives supply unique access to some of the most profound matters confronting human beings; they have a place in teacher education, and the case idea might embrace literary as well as other texts.

Casuistry: Case Reasoning Independent From Theory

The narrative mode is not the only alternative to a paradigmatic procedure of theory into practice. Advocates of the case idea have not overlooked the fact that moral philosophy harbors an ancient tradition of reasoning from and with cases—without resort to theoretical principles. Here we will again encounter "paradigm," but in a quite different sense. Casuistry is a method of reasoning about problems of moral conduct by considering precedents: cases that tend to command attention whenever a particular kind of moral problem arises. Such precedents are called "paradigmatic cases" or "paradigm cases."

Faced with a new situation or problem, the casuist undertakes a careful exploration of the features of the case in relation to paradigm cases, particularly noting exceptional circumstances such that precedents either apply ambiguously or conflict with one another, making a resolution difficult. Such reasoning requires and is supported by knowledge of the tradition or history of the issue as represented in the precedent cases. This approach to practical reasoning drew support from no less an eminence than Aristotle, who regarded practical wisdom as flowing from the particulars of a present case to a provisional resolution grounded in precedents but subject to rebuttals based on exceptional circumstances. Aristotle regarded ethics not as a science but as a practical matter calling for informed prudence.

In Jonsen and Toulmin's (1988) history of casuistry, this venerable mode of moral reasoning was disgraced in medieval times by corrupt French clerics who resorted to selective use of precedents and to tortured analogies among cases in order to justify ecclesiastical favors for wealthy patrons. Jonsen and Toulmin argue that it was not the method itself, but its misuse by a corrupt community, that gave a bad name to casuistry. That history invokes a thoroughly contemporary problem. If, as Dewey (1929) and Rorty (1979) have argued, epistemology's long "quest for certainty" has not paid off in universal rules for discerning truth, then much depends on the communities of inquiry:

There are no constraints on inquiry save conversational ones—no wholesale constraints derived from the nature of the objects, or of the mind, or of language, but only those retail constraints provided by the remarks of our fellow-inquirers. (Rorty, 1982, p. 165)

Conversely, when there is no prospect of effective criticism—bracing remarks—then survey data, short stories, and case reasoning all can be

distorted. Much depends on having a community of practice that culti-
vates a "critical consciousness" (Geertz, 1981) about the modes of inquiry—including case discussion—that it employs.

In the Enlightenment period, the corrupted practice of casuistry provoked a stinging attack by Blaise Pascal. Subsequently, moral reasoning
followed his lead in seeking foundational principles that might form axioms in a moral geometry. Moral reasoning would resemble rigorous deductive logic, following the sequence from major and minor premises to
a necessary conclusion about a present instance. But that project did not
pay off either, argue Jonsen and Toulmin; rather, the effort to establish
a rule-based science of ethics crucially misdirected the field of moral
philosophy for several centuries. It is high time, they reckon, to resurrect
a worthy method from the dustbin of history.

Cases as a Body of Knowledge

Modern casuists have explored such classic problems as what constitutes a "just"war (Walzer, 1979) and the circumstances under which lying
or secrecy are morally permissible (Bok, 1978, 1983). And policy analysts
(e.g., Neustadt & May, 1986) have proposed similar methods of reasoning
by analogy from historical cases to present instances of foreign and domestic policy, seeking to lightly systematize the uses of the past for current policy deliberations. Their practice might suggest how teacher educators could approach cases. Jonsen and Toulmin (1988) note that "one
indispensable instrument for helping to resolve moral problems in practice
. . . is a detailed map of morally *significant likenesses and differences*
[italics added]: what may be called a moral taxonomy" (p. 14). In her
book on lying, for example, Sissela Bok provides a taxonomy that distinguishes the following types of case: "white" lies, excuses, lies in a
crisis, lying to liars, lying to enemies, lies protecting peers and clients,
lies for the public good, lies to the sick and dying, paternalistic lies, and
deceptive social science research. This list suggests the range of cases
within which the temptation to lie is strong and for which justification
may be thought to exist. Her analysis of lying proceeds case by case;
rather than arguing from biblical injunctions, Kantian categoricals, or
other general principles, she explores the significance of special circumstances that differentiate the case types.

In its developed form, then, casuistry comprises a tradition of discourse
about a high stakes moral issue that is embodied in and bounded by a
differentiated set of paradigm cases. These paradigm cases command attention whenever the issue arises, but their application to current issues
may be problematic, because each case contains peculiar aspects. In this
mode of moral and practical reasoning, the individual does not resort to
general principles but to a developed taxonomy of the morally relevant

differences and similarities among the cases. Crucially, the casuist does not wrestle with cases alone. The voice of conscience harkens to its root meaning: *conscientia* or "knowing together." Knowledge of the tradition of cases and participation in the community that practices reasoning by cases is a vital cultural resource.

Is the model of casuistry applicable to reasoning in teaching? What would constitute enduring, delimited problem domains equivalent to the moral issues of lying, war, usury, or others? What might emerge as paradigm cases that command attention as each uniquely situated problem of teaching arises? Broudy (1990) and his colleagues are working to develop cases that qualify as paradigms, at least to the extent that they are clearly recognizable and recommended by practicing teachers as central and recurring problems of their work. By virtue of their own surveys of teachers, Greenwood and Parkay (1989) claim similar status for the entries in their casebook. Neither project, however, takes up casuistry's machinery of taxonomies and reasoning by analogy.

Transition

We have sketched four kinds of conversation and thinking that might go on in a community of practice for teaching and therefore in a teacher education program. In the foundational mode, the members of that community emphasize social science knowledge and seek to apply it to practice; cases appear as instances of theory, if at all, and are peripheral equipment. In the pragmatic mode, the members of the community emphasize situated problems of the practice; they explore these problems by way of action, employing abstract knowledge as a useful but limited resource in their reflections and deliberations. Cases are central to this discussion. In the narrative mode, the members of that community are storytellers who share tales of practice and speak of the insight they offer about intention, action, and its meaning to the actors. The members draw on the stories they have heard to act out stories to be told. In the casuistical mode, the members of the community share a tradition of influential cases and a taxonomy of important similarities and differences among those cases that together embody their ongoing concerns. The members of this community speak of resemblances between the traditional cases and the case at hand, seeking a prudent resolution that may in its turn be added to, or even alter the significance of, the developing tradition of cases.

We might treat this array of conversations and reasoning as an occasion for choice: Which is best, soundest, most useful? Before choosing, however, we should consider that teacher education did, in effect, make such a choice in this century. The consequent complaint is that teacher education has overinvested in the foundational mode, in the stance of the observer over the stance of the actor, and in a dominant substantive account of teaching—the psychology of learning. Although teacher educators might wish to speak more uniformly, so as to speak more authoritatively, a choice among modes might not be prudent. Exploration seems to be in order.

Each of the four modes of conversation, reasoning, and action discussed above puts case development in a different light and so suggests possibilities for exploration and development work. Case development depends on the context of use and on the part cases will play in the knowledge of the field. Cases might be employed within the foundational approach, expanding the array of examples already used in the teacher education curriculum as it is currently organized and taught. But in the pragmatic, narrative, and casuistic approaches, the cases would be regarded as knowledge in themselves and would provide not only the backbone but also much of the body of the teacher education curriculum. The differences among the approaches provide a great deal for case developers to consider. Therefore, we will now take up topics for research and development on cases and the curriculum, before turning to the equally complex issues of learning to teach from cases.

DEVELOPING CASES FOR TEACHER EDUCATION

Three important matters arise in constructing a teacher education curriculum that employs cases. First, what is a case, or what does a case represent? In some accounts, cases describe situations that a knowledgeable person can work on, or that a person who is gaining knowledge can work on for practice. Knowledge exists outside cases, and is brought to bear on them. In other accounts, cases represent knowledge or constitute knowledge in themselves; a person can gain or construct knowledge from working with them. Second, why is a case significant in a community of practice for teaching? How are cases specified theoretically and related to other cases or to other knowledge? What devices either organize cases as a body of knowledge or tie them into a body of knowledge for teaching? In what account or conception of teaching do the cases reside? In a casuistic approach, cases constitute a body of knowledge that is organized by analogies among and taxonomies of cases. In a foundationalist approach, cases have meaning within, and are organized by, distinct bodies of theory. Third, what understandings of and conventions for using cases enable a community of practice to employ them fruitfully? How do the members of that community know what to do with cases?

If knowledge and meaning are constructed, co-constructed, and negotiated, then those three matters are inseparable. What a case is, or is a case of, depends on the ideas used in constructing it and the purposes for doing so. The case resides in something akin to a culture, of which it is a tool. A case would have shared meaning and value in the context of its joint use by members of a community and would be bound up in their interactions, relationships, and understandings of what they are doing. In case teaching, as in school teaching, the entangling of these issues shifts attention from discrete variables of activity to larger patterns of activity—routines, practices, and designs—that incorporate the substance of the case, the larger pattern of meaning in which it is significant, and the communal activity in which it is used. Walter Doyle (1990a) has proposed three "frameworks for using cases in teacher education" that associate the first two issues—the substance of cases and their place in a body of knowledge—and have implications for the third. He labels those frameworks "precept and practice," "problem solving and decision making," and "knowledge and understanding." We will use Doyle's labels to organize this section of the chapter, examine his account of each of the three frameworks, and call attention to the third issue of conventions in a community of practice.

Precept and Practice

Doyle argues that most teacher education employs a framework that he calls precept and practice, in which teachers are given information that is presumed to have practical applications and then are assisted to practice using it. This knowledge comprises both "propositions from basic disciplines" and "maxims, aphorisms, and tips distilled from practical experience." Teaching is regarded as a process of applying rules or skills; to learn to teach, one learns the rules and skills and then practices applying them.

In this framework, cases are "rhetorical devices" used to clarify precepts and make them interesting. The function of the case, which may be as limited as an example in a text or as extended as a 45-minute videotape, is to exemplify the desired principle, theory, or instructional technique. Cases—instances of practice—are models to be emulated (or avoided). "Within this theory of practice," Doyle argues, "cases are hardly an essential component of teacher education" (p. 10). Cases come into their own, he would argue, when learning to teach is regarded as a matter of learning to recognize and solve problems that arise in the classroom or as a matter of acquiring "classroom knowledge" needed to enact particular components of the school curriculum.

Other teacher educators might wish to argue with Doyle here, on the grounds that his account of case teaching is entangled not only with his argument for the value of teachers' classroom knowledge but also with a polemic against renditions of theory into practice that are simplistic, narrow, or oblivious to context. A sophisticated foundationalist might wish to use cases in a precept and practice mode because it is not easy to exemplify any substantial proposition, whether theoretical or aphoristic, particularly if the aim is to have students apply that precept with due regard for its limitations, for the relevance of other precepts, and for the unavoidable complications of context in every application. Such an interest in cases would further increase if the aim is to convey a precept that itself is a complex network of propositions that has been applied in a complex system of instruction like group work. In this instance, cases might help both to convey that the context of application matters and to suggest how the network of underlying propositions may be resolved in any particular application of this class of instructional designs.

In Doyle's account, precept and practice is a matter of transmission from those who know better to those who do not. Although he does not actually mention lecture halls and passive students, they do come to mind. Lee Shulman (in press) explicitly invokes that image in suggesting that case proponents are trying to cast out the "twin demons of lectures and textbooks." However, he also provides persuasive examples to show that there is no necessary relation between cases as material for teaching and the discussion method of teaching. He reports that both James B. Conant and Joseph Schwab aimed to change the representation of science to undergraduates, in Schwab's terms, from a "rhetoric of conclusions" to a "narrative of inquiry"; however, Conant used cases in a lecture series and Schwab used them in Socratic dialogue. There is room to explore how cases combine with various approaches to instruction.

Teacher educators using the precept and practice approach well might prize particular cases because they vividly portray teaching (or educational research) as a "narrative of inquiry," validly exemplify given conceptual distinctions or educational designs, reliably raise particular theoretical issues or applications, consistently place students in particular moral or technical dilemmas, or typically promote lively discussion among student teachers. Conceptual and empirical work will be needed to establish these properties of particular cases; studies like Bruner's work on the literary properties of narrative may provide more general guidance for case construction.

Problem Solving and Decision Making

A second framework for using cases Doyle labels problem solving and decision making. For Doyle, this framework is shaped by the "emphasis on cognition in teaching," which has "called attention to the complexity of the process of learning to teach and to the intricate processes involved in connecting knowledge to situations" (p. 10). In this approach, cases present some range of problematic situations within a general class such as management and discipline.

Citing L. Shulman (1986a), Doyle calls these cases "precedents," in that a case "exemplifies not only how a lesson was conducted but also what the problematics of the performance were. As a precedent a case is a more complex and less well-formed representation of an instance of teaching than an example or demonstration model" (p. 10). Such cases are likely to represent "an actual instance of practice presented in much of its complexity, rather than an episode constructed to illustrate a point" (p. 10). With such cases, students can practice such professional skills as interpreting situations, framing problems, generating various solutions to the problems posed, and choosing among them.

Problem Solving With Casebooks

Commercially produced casebooks (e.g., Greenwood & Parkay, 1989; Kowalski et al., 1990) take a problem-solving and decision-making approach, at least to the extent that they present problems for students to work on. The cases are argued to be "real" problems by virtue of the interviewing and surveying techniques used to develop them. In light of Doyle's discussion, we might ask whether cases bring out the "problematics" of the performance. McAninch's (1991) reaction to the casebook by Kowalski et al. (1990) is that "most of the cases reduce potentially significant concerns about teaching or schooling to narrow instructional issues or 'sticky situations' in dealing with colleagues, parents, and administrators" (p. 346). She argues that the "text supports an intellectually narrow view of the teacher as a receiver of knowledge who must then apply that knowledge to technical problems in order to reach decisions" (p. 347). She is disappointed that this casebook's list of suggested readings—which presumably contain the knowledge that prospective teachers were to receive—included mostly popular journals rather than scholarly and theoretical outlets. Regarding Greenwood and Parkay's (1989) casebook, McAninch notes that the authors conceive of teachers as decision makers who use a variety of sources of knowledge, including theoretical and personal knowledge, but suggests that "it is never made quite clear how those different types of knowledge play out in actual decisionmaking" (p. 348). She notes their proposal for a six-step decision-making strategy but wonders what, other than the experience of the authors, recommends that strategy.

Another avenue for development of cases and case teaching would be to examine cases as used. What images of teaching do students derive from various case materials, and from the tasks in which those materials are used? What forms and processes of reasoning do they employ in dealing with cases? What range of issues do the cases raise for students, given their prior experience with teaching? What properties of cases or of the activities surrounding cases help students recognize "significant concerns"? What counts as good solutions to problems or good decisions about situations? Barnett's (1991) study of teachers' discussions of cases of mathematical pedagogy, described in the introductory section, examined such issues.

A Search for Compelling Cases

Another question that arises from McAninch's criticisms of the two casebooks is, How should teacher educators find or nominate substantial cases that deserve and bear attention? In other professional fields, the identification of a case seems somewhat more clear-cut than in teaching, since some domains of practice in those fields present naturally occurring units of deliberation and action that both command attention and "generate case records as a by-product of the occupation" (McAninch, 1991, p. 350). The appellate court decision, the patient with presenting symptoms, the manufacturing firm tendering its annual report, the building together with its blueprints, and the chess or bridge game are instances. In teaching, the current approaches to defining and developing cases vary considerably. The indefinite (or diverse) character of cases in teaching may help to explain why the case idea has not taken hold more powerfully, despite efforts over the years to develop a case literature.

At the University of Illinois, Champaign-Urbana, Steven Tozer, Harry Broudy, and colleagues (see Broudy, 1990) are approaching case development in a manner that appears similar to that used to develop the two casebooks mentioned above but is more detailed and thorough. They began with pilot work in several sites, interviewing teachers to generate a set of roughly 100 typical problems of teaching. These were refined into a survey that was mailed to a large sample of teachers, who were asked to rate each problem in terms of its incidence or frequency and in terms of its seriousness. The researchers categorized the problems identified as both typical and serious into such broad areas as discipline, parent-teacher interactions, peer interactions, and others. With their survey of problems in hand, the authors began preparing cases that represent these empirically derived crucial problems. One case, for example, consists of a videotaped interview of a minority student, together with his teachers, parents, and counselor, that prompts questions about how to interpret and respond to the student's relation to schooling in terms of motivation, cultural and family background, prior educational experiences, and other perspectives. The prototype cases so far do not represent exemplary practice, but instead depict central problems upon which professional knowledge may have substantial bearing.

In connection with their cases, the researchers also developed a set of readings that provided theoretical illumination, then tested these materials with teacher candidates, working back and forth between the video cases and the readings. The project suggests steps that might be taken to support problem solving and decision making informed by educational research.

Domain Specificity

Doyle argues that the problem-solving and decision-making framework has a significant limitation:

> At best, the framework underscores the importance of the general decision processes teachers engage in and even provides categories within which such decisions fall, e.g., students, activities, routines, subject matter, resources, and management. But work in this area has done little to explicate the domain-specific knowledge that teachers use to define problem-spaces or forge decisions. To say that teachers make decisions about students or about activities tells little about the knowledge teachers use in making those decisions. (Doyle, 1990a, p. 11)

In the above passage Doyle is acknowledging and adopting a stance toward the debate over whether it is more fruitful to conceive and teach reasoning, problem solving, and decision making as generalized skills that apply across many domains or as domain-bounded operations that rely on knowledge in the domain (see Perkins & Solomon, 1989, for discussion of this issue). Doyle adopts the latter position, as we shall see. However, while we agree that some approaches to problem solving and decision making, such as the casebooks discussed above, do assume generalized reasoning processes, we also doubt that that is a necessary element of the framework. If prospective teachers work on particular cases of teaching and use particular ideas from educational literature to identify and solve problems in those cases, they are engaged in domain-specific activity. Such activity might or might not draw the prospective teachers' attention to the "general decision processes teachers engage in." That would seem to depend on whether the case portrayed a teacher engaged in such decision-making processes and whether the teacher educator, in a metacognitive move, called the prospective teachers' attention to the general characteristics either of their own thinking or to that of the teacher portrayed in the case. In any event, recognition of general decision processes might not be the aim. An alternative objective is that the prospective teachers gain particular interpretive vocabularies and learn to use them in framing and solving problems in lifelike cases that contain situational complications.

Developmental research here might compare case work with field observation as alternatives for fostering problem-framing, problem-solving, and decision-making powers in prospective teachers. Several proponents of cases suggest that cases may be preferable because they afford the teacher educator considerable control of the phenomena that education students encounter, enable the teacher educator to prepare specifically for the discussion or other activity that is employed with the case, and may be easier to arrange. The general question, perhaps, is whether or in what circumstances case work poses authentic (if vicarious) tasks of teaching.

Knowledge and Understanding

In Doyle's third framework, knowledge and understanding, a case represents knowledge that practicing teachers gain in the classroom and that prospective teachers can gain, in part, by work with those cases. Doyle's characterization of this framework is informed by the study of teachers' knowledge, which, he says,

rests on the premise that what teachers do reflects their understandings of action-situation relationships in classroom environments. . . . The teacher's knowledge is organized around *tasks* related to solving the problems of order and learning in the classroom environment and around the *events* in which those tasks are accomplished. . . . Since classroom tasks are seldom well-formed, teachers' knowledge is likely to be organized in complex, conditional networks. (1990a, p. 11)

In this framework a case might reveal, for example, teachers' knowledge in action to achieve order in the classroom. Cases might include both positive and negative exemplars, the study of which "develop[s] the knowledge structures that enable teachers to recognize novel events, understand them, and devise sensible and educative ways of acting" (p. 13). Within this approach, teaching knowledge would be represented in the case, not brought to bear from outside it as in the other two frameworks. The case would be organized around a task, as distinct from the dilemmas that appear so often in cases, so that cases might depict routine and typical aspects of practice as well as the unusual and problematic ones.

In this framework, Doyle argues, cases would play a vital part in teacher education, as prototypes that "instantiate theoretical knowledge about teaching." According to Doyle, "the essential task of teacher education is one of content representation" (p. 12), and cases "become a way of knowing" (p. 13) or of portraying knowledge in teaching. To accomplish these purposes, Doyle argues, the case must provide "sufficient detail to enable someone to experience the complexity of the original situation" (p. 13).

Case and Theory

Detail or specificity alone is not enough, however. It also is necessary to locate the case in some broader framework of ideas, some conception of teaching.

In addition, the case must be theoretically specified, i.e., it must be a case of something within a theoretical framework. Rather than simply providing opportunities to practice analysis and decision-making, theoretically specified cases supply a foundation for developing the knowledge structures necessary for comprehending classroom realities. . . . Given that much of teachers' knowledge is conditional and context-specific, multiple representations will be needed to help teachers develop the professional knowledge needed for practical reasoning about classroom tasks. (Doyle, 1990a, p. 13)

What would count as "theoretical specification" for cases? From his and others' work on classroom organization, Doyle (1990b, p. 355) proposes that teachers "organize their thinking during planning and enactment around specific chunks of content." For each chunk of content, the teacher must integrate some worthy academic task with a workable structure for student participation in order to create a "program of action" that provides both humane order and learning. These ideas suggest cases of curriculum enactment. The case "There's No One-Half Here!" described earlier (Barnett, 1991), might be regarded as such a case to the extent that it does portray a mathematical task and a participation structure. One might ask whether it satisfies the criterion of providing sufficient detail to allow a person to experience the complexity of the original situation.

Of course, curriculum enactment is not the only rubric under which cases might be developed and theoretically specified. To date, case development has been guided by the particular theoretical and substantive interests of individual investigators rather than any achieved consensus within a broad professional community. Theoretical specification is thereby achieved on a local basis, often in the context of particular projects with tightly focused interests such as teaching mathematics for conceptual understanding, responding to the characteristics of urban learners, coping with rural Alaskan communities, or creating a productive and orderly classroom environment.

Context as a Knowledge Issue

Taken together, the dual demands for details about complexity in a case and for theoretical specification of a case raise the question of how to treat context as a knowledge issue. If knowledge in teaching is highly contextualized, as many investigators now believe, what aspects or dimensions of context should be represented in cases and related materials? A videotape of a single lesson inevitably provokes questions about what went before and what will come after the lesson portrayed, as a context for interpreting the teacher's actions on the tape. Likewise, teachers react to particular students based on substantial background knowledge that constitutes a context for their decisions and actions. When such contextual cues are stripped down or absent, brief cases contain the potential to mislead and miseducate about the nature of teacher decision making. Case specification, then, requires not only an overarching substantive account of teaching but a theory of how context operates in teachers' case knowledge; to date this crucial issue has received little attention.

Negotiating the Knowledge in Cases

The claim for cases as knowledge asserts there is a "wisdom of practice" or a body of craft knowledge possessed by expert teachers that has been ignored or underrepresented in traditional social science approaches to teaching. According to Leinhardt (1990a):

There exists a natural tension between general, subject-based, principled knowledge in a discipline and the specific, eclectic, particular knowledge acquired in the practice of a related craft. Teachers as both professionals and practitioners are caught up in this tension. Their professional training in institutions of higher learning emphasizes theory as an efficient, universal, cohesive truth filter for disorganized, practical experience. . . . However, teachers also appear to learn in their profession and to communicate with their colleagues and others in the language of craft and practice—in fact, in the language of the particular. (p. 18)

Many of the initiatives described in this chapter engage teachers in identifying problems and in discussing their responses to them. Judith Shulman and Joel Colbert (1987, 1988) have produced two casebooks on problems that mentor and intern teachers face. Each contains a series of brief vignettes. In their use, a vignette is

a story about a particular event, experience, or relationship. [It] describes what led up to the event and the consequences that followed the event. To the extent possible, it also describes how the participants in the event were thinking and feeling. (p. 81)

The vignettes were written by mentor or intern teachers in collaboration with the casebook editors; Judith Shulman (in press) describes persuasively and specifically how much there is to learn if such collaborations are to be successful.

In these two casebooks, the vignettes are treated as cases within a class of characteristic issues such as teaching with minimal content knowledge or interacting with students. In organizing their casebook, Shulman and Colbert treated the vignettes as data, in the manner of critical incidents, subsequently categorizing and grouping vignettes into sections of the casebooks based on their familiarity with the literature on coaching, advising, mentoring, and related topics. Thus, as McAninch (1991) notes in her review, "the contribution of theory to this casebook was primarily in the organization and selection of vignettes, rather than in the explication and interpretation of the events initially described by teachers" (p. 351). However, accompanying the vignettes are reactions—commentaries—by other teachers or researchers who supply various perspectives on the incidents described.

Collaboration between teachers and researchers in case writing and incorporation of teachers' and researchers' commentaries with teacher-written cases both suggest a negotiation of the knowledge in or about cases. The case development advocated by Doyle (1990a, 1990b) and described by Leinhardt (1990a) in the context of teacher assessment clearly moves beyond description and analysis to representation of knowledge that aspires to normative and prescriptive status. This is a controversial ambition in the face of suspicions about best or exemplary practices. However, such work might aim at the more modest goal of representing defensible practice that admits defensible alternatives, so the objective is not to rule out alternatives via some experimental, hypothesis-testing procedure but to identify alternative, effective practices under specified conditions and assumptions.

Leinhardt (1990a) reviews several efforts to extract craft knowledge, discussing such methodological issues as how to identify experts, how to delimit the boundaries of their expertise, how to codify their knowledge, and how to treat the validity of their knowledge. Whereas some investigators rely on methods such as surveys, which cast teachers as subjects from whom to elicit information, the methods she reviews involve teachers in shared, public discussions about practice, draw on theory to "find patterned consistency in what is happening in a successful episode of teaching" (p. 19), and rely on performance verification across a range of situations to test whether teachers actually do what they say.

In these procedures, for which few precedents or principles exist, case development resembles a complex hybrid of applied research and curriculum development, linking questions about how to test and represent knowledge in cases to questions about how to construct curriculum to promote learning. Such development work further requires new forms of supportive organization that join teachers, teacher educators, and teaching researchers in productive relations. How such case development might fruitfully be organized and how the status of cases might be negotiated are issues for study.

The limited experience to date does suggest that case development is a promising activity around which to structure professional community. The question here concerns who participates in case development through what methods and how knowledge represented in cases is generated and tested. Case advocates argue that case development is a natural collaborative activity through which to join the knowledge of university-based researchers and teacher educators and the craft wisdom of experienced teachers. The promise of such collaboration rests on a sociopolitical claim as well as an assertion about case and craft knowledge. Through engagement in the generation and testing of knowledge in teaching, case development may accord practitioners greater status and authority within their professional community. If teachers merely implement high-status knowledge that emanates from the university and is represented in mandates, training and instructional materials, and policy instruments, their status is defined as that of bureaucratic functionaries working under the direction of others. But if teachers aspire to professional status, then their involvement in the production of knowledge useful to teaching is critical. Case development is one vehicle through which teachers might participate in such knowledge production.

A Curriculum of Cases?

One option for organizing a professional education curriculum, evident in other fields, is to create a curriculum that is entirely case based: To some high degree, the problems of and thinking about the field would be embodied in a set of cases, and work in the field would revolve around those cases, as would interaction among persons who participate in it. Kleinfeld (in press) reports that Teachers for Alaska, which we described briefly in our introductory section, is such a program.

Law school is the most obvious example of a case-based curriculum, but some business schools have developed strong case traditions as well. Christensen and Hansen's (1987) volume on case teaching at Harvard suggests that when all the faculty of a professional school rely on cases and case teaching, strong conventions develop that are shared by faculty and students. The professional community supports the use of cases and contributes to their power and utility as tools for learning. Constituent elements of that community include a set of cases held in common as well as a set of understandings and conventions about what is a case, how cases are constructed and evaluated, and how cases are taught. A discourse community develops into which new members are inducted through repeated engagement with cases across courses, so that students receive powerful reinforcement in the skills of analysis and deliberation. A case-based curriculum gains its power as much through the shared conventions and traditions within which cases are employed as through particular characteristics of the cases themselves. Experience with case analysis is cumulative over the course of professional education.

Although a fully case-based approach seems unlikely in the near future for teacher education, the approach draws attention to the importance of the cultural and social elements of a community of practice—traditions, conventions, discourse norms, shared values, and induction processes. It is too early to tell whether a set of case conventions will take root in teacher education; a range of practices has emerged, and teacher educators are beginning to offer accounts of their work. These early developments reveal teacher educators seeking to integrate cases with all aspects of the teacher education curriculum. Teacher educators, as we have noted, report using cases to teach subject matter (Wilson, in press; J. Kleinfeld, personal communication, May 17, 1991), to convey understanding about cultural characteristics of students (Kleinfeld, 1990), to impart instructional methods including classroom management (Carter, in press), to encourage reflection on field experiences (Florio-Ruane & Clark, 1990), to focus mentor-novice conversations (Carter, 1988), and to analyze experiences of both mentors (J. Shulman & Colbert, 1987) and interns (J. Shulman & Colbert, 1988). These emerging practices do not reveal any shared set of conventions but rather an eclectic mix of approaches; there are more questions than answers about how to use cases effectively in conjunction with other instructional methods and experiences.

Transition

Advocates of the case idea are anxious to enlist more teacher educators in their cause, but the methods for producing cases for use in instruction are not settled, nor is it clear that they will be settled soon. The advocacy literature on the case idea projects bold, revisionist hopes, but what faces case proponents is a major curriculum development effort. That effort might be guided by a view of the kind of community of practice to which prospective teachers should be inducted and the part that cases might play in it. It should also be guided by views of the learner and of learning to teach, to which we will turn next.

LEARNING TO TEACH FROM CASES

As cases might be employed in several kinds of conversation and reasoning about teaching, they also might be employed in several strategies for helping prospective teachers learn to teach. In this section, we will explore the case idea in relation to a number of prominent theories of learning that are influencing how teacher educators approach their work. We aim to set the description of the case idea into an account of intellectual currents in teacher education, against the backdrop of the problems facing the enterprise.

Contemporary research on teacher education portrays teacher educators struggling to surface, explore, and ultimately change novices' beliefs and knowledge on a wide range of issues, as the starting point for desirable practice (e.g., see Carter, 1990; Feiman-Nemser & Buchmann, 1989; Hollingsworth, 1989; Kennedy, 1991). Much evidence amply documents that novice teachers have faulty, incomplete, biased, or limited knowledge about the subjects they will teach, the pedagogies involved, the nature of knowledge and of learning, the nature of teaching as role and activity, and the nature of the learners they will encounter. On the basis of their long involvement in schooling, novice teachers have unusually rich images and powerful ideas that are resistant to change. Teacher educators explicitly frame their aims in terms of seeking to induce cognitive change. The themes of breaking with experience (Buchmann & Schwille, 1983; Floden, Buchmann, & Schwille, 1987), reflective practice, and cognitive or conceptual change tilt teacher education toward reform of practice and toward independent thinking as the goal of professional education (for a cautious reappraisal, see Buchmann, 1989, 1991).

Because many teacher educators are critical of current practice and increasingly view teaching as complex work, they appreciate more pointedly than ever how difficult is their mission. Cases might serve teacher educators' orientation toward reform and analysis and aid them in some of their difficulties. For example, videotaped and written cases may allow the portrayal of exemplary practices, could provide vivid counterimages to conventional practice, and might encourage multiple interpretations and problem-solving discussions.

To explore such prospects, we will attempt here to overlay two sets of sketches, that is, brief synopses of relevant arguments. One set of sketches concerns teacher learning; it includes arguments that suggest what prospective teachers have learned by the time of entry to teacher education programs, what they need to learn to become teachers, and how that learning occurs. We will begin with arguments that depict learning to teach as a matter of learning to think about teaching activity and proceed to arguments that depict learning to teach as a matter of learning to act as a teacher. We will include arguments under the headings of prior knowledge and conceptual change, cognitive flexibility in the use of ill-structured knowledge, situated knowledge and action, reflection in and on practice, and expertise in teaching.

The second set of sketches concerns the kinds of encounters that prospective teachers might have with case materials, case methods, and case teachers. As suggested earlier, learning from cases may depend on the nature of the cases and on how cases are integrated into curricular patterns and activities; variation also is emerging in instruction with cases. While strong conventions of case-based instruction exist in other fields (see Christensen & Hansen, 1987), education is exploring. We will consider approaches through which teacher educators teach with cases. One kind of encounter is the case discussion, in which a group of students explores interpretations of a written or video case under the tutelage of an expert leader. A second kind of encounter relies on combinations of computers and media that allow students singly or in groups to repeatedly study and reconstruct a multimedia case. With such "hypermedia," students can literally reconstruct a videodisc text. A third kind of encounter is the student production of cases. Increasingly in teacher education, instructors have students create their own written cases based on their experiences in classrooms. And a fourth kind of encounter, participation in a simulation, is a method related to but distinct from case teaching as commonly defined. We stretch our topic somewhat here in noting that some teacher educators use simulated experiences as a form of encounter with what arguably might be termed cases. Each approach provides a distinctive form of encounter with cases, within which the student constructs meaning and builds knowledge, but the constructivist principles vary. Each approach attempts to take the prospective teacher closer to practice, but under conditions that avoid the pitfalls of unmediated experience (Feiman-Nemser & Buchmann, 1985).

By shuffling the synopses of arguments about learning to teach with the synopses of encounters with cases, we intend to explore the possibilities in the case idea. To help keep order in this discussion, we will mark off the arguments about learning to teach with the headings listed above, and we will designate the various approaches to cases as "close encounters of the first, second, third, and fourth kinds."

Prior Knowledge and Conceptual Change

One influential line of contemporary research asks how knowledge grows and changes in the mind. Such research begins with the observation that individuals' prior knowledge of various phenomena often impedes more accurate and valid understanding, so that the aim of instruction must be to modify prior beliefs rather than simply fill up the empty vessel of the mind. There has been an explosion of research of this sort around children's conceptions of mathematics, of scientific concepts—of school knowledge and skills in general. This work strongly demonstrates that prior conceptions exert a powerful hold and are difficult to alter. Contemporary instructional aims include inducing conceptual change as a central preoccupation.

Not surprisingly, research on learning to teach has adopted this conceptual change orientation, exploring the prior knowledge that novices bring to teaching and tracking the effects of teacher education in modifying these prior beliefs. Investigators have studied novices' knowledge of and beliefs about subject matter, pedagogy, students, the teacher's role, and the practice of teaching (see the volume by Brophy, in press; Carter, 1990; Hollingsworth, 1989; see also the research sponsored under the auspices of the National Center for Research on Teacher Education [NCRTE], particularly Ball, 1990a; Florio-Ruane & Lensmire, 1990; Kennedy, 1990; McDiarmid, 1990; and Paine, 1989). This work demonstrates that individuals enter teaching with ideas that are well-formed, powerful, resistant to change, and often, from the perspective of teacher educators, wrong-headed.

A brief sampling of findings from the program of research carried out within NCRTE is suggestive (Kennedy, 1991). The student teachers in the NCRTE study sample were primarily young, White women from small, homogeneous, lower-middle-class communities who had little experience with people different from themselves. They had little understanding of learners who might respond to school subjects differently than they did. They tended to view subjects, in this case mathematics and writing, as sets of fixed rules and procedures with few connections among them or to the world outside the classroom. With respect to teaching and learning, the researchers found that most undergraduates hold a limited view of their role as teacher, regard teaching as telling, and view learning as absorbing and reciting back what the teacher has told. With respect to student diversity, most students had two simple values—to treat all students alike and to meet individual differences—but little sense of the apparent contradiction or of how to act on either view.

The evidence also suggests that teacher education does not significantly modify these prior beliefs. Clearly, the apprenticeship of observation so widely noted plays a lead role:

Because teachers have logged over 3,000 days as classroom participant observers, they have not only developed strongly entrenched beliefs about teaching and learning but have also developed a strongly entrenched belief that they already know what teaching is all about and that they have little to learn. (Kennedy, 1991, p. 9)

Conceptual change theory (e.g., Posner, Strike, Hewson, & Gertzog, 1982) posits that to modify strongly held beliefs requires the introduction of discrepant images and information to provoke dissonance. But the alternatives proposed must be plausible and vivid in order to make an impression and must be accompanied by questions or experiences that force the teacher candidates to recognize the discrepancy between their present beliefs or knowledge and the alternatives. Supplying a vivid case by itself, the evidence suggests, will not provoke change, for a common dissonance-reduction technique is to distort or deny the conflicting information.

The research on teachers' prior knowledge and beliefs and the resiliency of such knowledge constitutes a major challenge to teacher education. The conceptual change perspective suggests a central role for cases as primary sources of stimulation for change. Video cases that portray unconventional approaches to teaching may serve as starting points for the process of change. Narrative accounts such as Vivian Paley's also challenge common beliefs. In her book *White Teacher* (1979), for example,

Paley describes the evolution of her thinking about managing race relations in a kindergarten classroom. Her initial impulse to be color-blind gave way gradually to an open acknowledgment of race and a celebration of racial and cultural differences among the children. When teacher candidates read this case, they often recognize that they would respond as Paley did initially, but come to recognize how this stance backfires and why Paley's discoveries and subsequent actions are wise. The case serves both to convey knowledge about how to deal with race in the classroom and to portray how a sensitive teacher studied and modified her own beliefs.

Close Encounters of the First Kind: Case Discussion

Of the four kinds of encounters that prospective teachers might have with cases, case discussion has been most thoroughly practiced, using both written and video cases. To date, the most common aims of case discussion either have been to apply theory to cases or to practice making decisions about problematic situations that arise in teaching, where the basis and justification for decisions often are weakly specified (for discussion of this problem, see McAninch, 1991; Sykes, 1989). But cases clearly hold potential as devices for helping to induce change in prospective teachers. Most obviously, if teacher educators hold transformative aims and seek to promote new instructional practices and social ideas that are not widely available for observation in schools, then cases might constitute one bridge between hortatory pronouncements and new practices and attitudes. Video cases, in particular, may have value in presenting vivid, concrete images of desirable instructional practices that may help change the minds of prospective teachers.

However, there are competing images of what such change-oriented discussion might resemble. Some researchers, such as Leinhardt (1990b) and Berliner (1988), advocate video cases that demonstrate exemplary teaching. Such cases help to bridge between theories of instruction and their implementation under real conditions. For example, students can observe process approaches to writing, successful instances of mainstreaming special-needs children into regular classrooms, science teaching for conceptual understanding, or cooperative learning techniques. The case, in this image, represents an instance of desirable or effective practice that the novice might be expected to admire and to emulate. What may provoke change, then, is not only the existence of a workable alternative to conventional instruction but the portrayal of a master teacher exhibiting great skill in modeling effective practice. Case discussion would analyze various complexities in the teaching portrayed; the aim would not be to demand slavish imitation but to induce change in the direction of the exemplar.

Other investigators, however, frame a somewhat different set of purposes. Rather than demonstrating exemplary practice, video cases

> would be treated in a manner analogous to a piece of literature or an historical event to be understood from a variety of perspectives. . . . The teacher educators and the prospective teacher would be engaged . . . in active reflection and research. (Lampert & Ball, 1990, p. 6)

Rather than applying abstract theory or mastering techniques, these researchers propose that students engage in inquiry as the fundamental act of learning to teach: "Teacher education students will have the capacity to do research on their own questions about how teaching and learning proceed in classrooms" (Lampert & Ball, 1990, p. 7).

In this second instance, the modeling occurs as much within the discussion as through presentation of text or video images. The teacher educator models ways of reasoning about teaching and learning, ways of observing and of bringing relevant considerations to bear. Here, too, the teacher educator may take great care in selecting exemplars (i.e., examples of interesting phenomena rather than normative models or techniques). In a mathematics methods course, for example, the instructor might provide several cases of teachers responding to students' incorrect answers, raising for discussion the issue of how to treat "wrong" answers in mathematics. Novices might be expected to quickly dismiss such answers and to issue swift correction directly or by calling on other students of known ability; through discussion, however, the instructor might surface the assumptions about mathematics that such pedagogy entails and explore various uses of incorrect answers to foster understanding and to build a genuine community of mathematical discourse. And, as we see next, such use of interactive video also aims to promote the flexible use of knowledge in the fluid context of classroom teaching.

Cognitive Flexibility and Ill-Structured Knowledge

Rand Spiro and his co-workers have developed a line of research that bears directly on learning from cases (see Spiro, Coulson, Feltovich, & Anderson, 1988; Spiro, Feltovich, Coulson, & Anderson, 1989; Spiro & Jehng, 1990; Spiro, Vispoel, Schmitz, Samarapungavan, & Boerger, 1987). They propose a basic distinction between well- and ill-structured knowledge, concentrating on how to promote concept use in the latter area. By "ill-structured," Spiro et al. (1987) refer to a set of overlapping characteristics: (a) There are no rules or principles of sufficient generality to cover most of the cases, nor are there defining characteristics for determining the actions appropriate for a given case. (b) Hierarchical relations of dominance and subsumption are inverted from case to case.

(c) Prototypes tend to often be misleading. (d) The same features assume different patterns of significance when placed in different contexts. (e) An explosion of higher order interactions among many relevant features introduces aspects of case novelty (p. 184). Under these conditions, approaches to teaching that assume routine application of general principles or concepts to cases probably will be ineffective, because students will be unable to apply abstractions to the range of irregular cases that arise. Furthermore, teaching itself is likely to contribute to misunderstanding by supplying single representations or analogies that validly convey some aspects of complex ideas and situations but fail to convey, or misconvey, other important aspects of those same ideas and situations.

Spiro et al. illustrate their theory with examples drawn from medical education, 20th-century literary criticism, and military strategy, among others, but language learning provides the most basic example. Many words have multiple meanings and many shades of nuance. We build an understanding of how to use language through repeated practice in speech, reading, and writing. In company with the situated cognition theorists (see the discussion below), Spiro believes that all knowledge is like language, and he seeks a method for reproducing the principles of language acquisition and use in the procedures and materials of formal instruction.

Spiro et al. further distinguish knowledge reproduction from knowledge use, associating the latter process with another major topic in cognitive psychology, the transfer of knowledge to new situations. Their theory of cognitive flexibility focuses on the application (or transfer) of knowledge in ill-structured domains, and they indicate that most school learning deals with reproducing, not using, knowledge in well-structured domains. Their work has led to explicit principles and practices—in fact, a technology—of case instruction. Their aim is to promote flexible use of knowledge in complex and ill-structured domains such as medical education.

Spiro and colleagues describe experiments with two instructional tactics. One is to provide multiple representations and analogies to counteract the misconceptions that arise when teachers rely on single representations (see Spiro et al., 1989). Here, Spiro and his colleagues are concerned with ways to promote deep understanding of complex concepts. They argue that too often teachers rely on single, simple analogies that create problems through incomplete or biased imagery. To illustrate, they describe various efforts to portray the functioning of muscle fibers by means of a range of representations (e.g., muscle contractions as a rowing crew, a turnbuckle, or a Chinese finger cuff). These different analogies convey differential images and understanding of muscle activity. Each representation by itself is inadequate; taken together, they correct each other's partiality and distortion and convey more usable understandings of muscle activity. Single analogies are like single cases as instantiations of concepts or principles; multiple cases correct errors of over-generalization from any case, increase the range of valid connections between abstractions and cases, and so build the student's cognitive flexibility.

A second tactic, perhaps only a variation on the first, is to decompose complex, multidimensional cases into many "minicases" through which students can build an understanding of the larger case (see Spiro & Jehng, 1990). Spiro and colleagues illustrate this tactic with a literary example, the analysis of themes in the movie *Citizen Kane*. The entire film constitutes the case, but individual scenes and episodes serve as minicases for analysis. To construct an interpretive matrix for the film, the minicases are cross-referenced with dramatic themes typically used to interpret the film as a whole, so that a student can explore a given theme's differential applications to several minicases or compare the application of different themes to a given minicase.

Close Encounters of the Second Kind: Exploring Cases in Hypermedia

This tactic of assembling nuanced knowledge of the whole from repeated visits to its parts is crucially abetted by new technologies: Scenes from *Citizen Kane* are stored on a videodisc linked to a computer that allows students to select the scenes and themes for study. The student can choose a dramatic theme and read commentaries about how that theme applies to scenes throughout the film, or choose a scene and study commentaries that discuss how several dramatic themes are played out in it. The students are working in a matrix that is defined by scenes and themes and is filled by commentaries that link them. Following Wittgenstein (1953), Spiro and colleagues refer to work in this matrix as "crisscrossing a conceptual terrain," and this becomes their central metaphor for learning.

This last example illustrates several theoretical hunches. One is that cases or examples, like the scenes of the film, must be studied in context (the film as a whole), not as stripped-down textbook vignettes that oversimplify the application of principles to cases. Another is that whole cases are frequently too complex to serve as the proper unit of instruction. Minicases are more manageable and more numerous in promoting the construction of flexible knowledge, for in economical fashion students gain experience with a variety of cases. Finally, Spiro and colleagues argue for extensive and specific interaction between powerful generalizations that help to interpret a domain and the rich particulars of case knowledge. "Our programs neither neglect cases to teach concepts," Spiro and Jehng (1990) comment, "nor concepts to teach cases—both are taught in the context of the other. Learning is situated but abstract knowledge is not ignored. Our approach teaches concepts and cases simultaneously, not separately: concepts-in-practice" (p. 199).

To apply such reasoning and techniques to cases in teaching, teacher educators would have to decide how to mark off a case of teaching, to distinguish the minicases within it, and to relate useful principles and generalizations to those minicases. For example, it appears that a detailed story of teaching a unit on the negotiation of the U.S. Constitution might, like *Citizen Kane*, serve as a case. Each day's lesson or each activity within a lesson might be nominated, like a scene from the film, as an interesting minicase. A variety of arguments regarding the purposes of teaching political history, the lines of historical interpretation pertinent to the U.S. Constitution, students' prior conceptions of politics and government, conceptual learning, classroom organization, and other topics might usefully be brought into connection with those minicases to complete an accessible object of study that is rich both in descriptive detail and in interpretive equipment.

The second kind of encounter with cases, then, would involve students in the use of videodisc technology and hypermedia programs that allow students to explore and reconstruct cases that could include both moving and still pictures with multiple audio tracks and text. This experimental approach is one notch more active than construction through discussion, because students can actually manipulate the case to juxtapose scenes and episodes repeatedly in order to explore different themes and ideas, patterns and sequences of action. Some exploratory work is under way that tests the power of this technology in conjunction with video cases. In addition to the work already cited by Spiro and Jehng (1990), the Cognition and Technology Group at Vanderbilt (1990) is experimenting with hypermedia applications in an elementary school; they report enhanced learning in relation to such outcomes as writing stories, making use of historical information, and solving mathematics problems. They also report positive effects on motivation and on transfer of learning.

To our knowledge, investigators have not yet tested case-based instruction with this technology in teacher education, but work is under way in the Mathematics and Teaching Through Hypermedia (MATH) Project at Michigan State University (see Lampert & Ball, 1990). Magdalene Lampert and Deborah Ball are engaged in a long-term development project that involves extensive documentation of teaching and learning in their own elementary mathematics classrooms, where each teaches on a regular basis. Over the course of a year, they have assembled a rich storehouse of case material that includes videotapes of their instruction, journals of reflection on their teaching, lesson plans and instructional materials, videotaped interviews with their students that explore their mathematical understanding, and a range of student-produced materials, including student journals. The videotape library eventually will be catalogued, stored, and accessed via hypermedia technology for use in teacher education.

This project might be considered as two yearlong cases of mathematics teaching portrayed in unusually extensive detail from the multiple perspectives of teacher thinking and planning, the instructional record, and student learning. But the documentation subsequently may be assembled in a variety of ways to constitute a curriculum for teacher education where videotaped instructional episodes, linked to other sources of data, provide minicases for exploration of a range of topics that could be organized by central problems (e.g., "what to do in this situation"), by task (e.g., "how to open a lesson on the equivalence of fractions"), or by aspects of instruction derived from theory (e.g., "norms of discourse in mathematical communities" and "use of multiple representations in conveying mathematical ideas"). Finally, these investigators frame their use of this case material in terms of inquiry, rather than as instances of model teaching to be emulated by novices. Developmental work of this sort is slow and costly but holds promise for creating a new pedagogy in teacher education that possesses powerful advantages. Even if hypermedia applications of cognitive flexibility theory do not achieve widespread application, these ideas about knowledge acquisition, transfer, and use provide provocative leads for the design of teaching cases.

Authentic Learning and Situated Actions

Another influential theory argues that all knowledge is situated in—and grows from—the contexts of its use (see Brown, Collins, & Duguid, 1989; Collins, Brown, & Newman, 1989). "Situations co-produce knowledge through activity," claim the authors (Brown et al., 1989, p. 32). The activities and contexts within which individuals learn provide vital, necessary support for learning and for action; when the context of knowledge use is stripped down, as too often occurs in school, the learning is impoverished.

The view that learning is fundamentally situated derives from a number of examples; again, the most prominent is language acquisition. In everyday speech, many words point to a part of the situation in which speech is occurring. Such words as *here, now, next*, and *this* serve to index or refer to the situation of the speech act. Brown et al. make the central claim that all knowledge is like language in indexing or referring to situations. Thus, to promote conceptual understanding requires attention to situations of use. Furthermore, abstract concepts contain many nuances of meaning so that conceptual learning proceeds with multiple encounters with concepts across situations to facilitate the progressive construction of understanding. Ordinary speech naturally provides occasions for acquiring such nuanced understanding, but in specialized domains of knowledge that include technical concepts, the occasions for use that naturally index concepts may not be routinely available to build understanding.

Coupled with the indexical character of conceptual knowledge is an emphasis on the activity within which concepts are used. Brown et al. underline this insight by comparing concepts with tools used by craftspersons in guilds where they learn tool use through participating in the community of tool users working on authentic tasks. Entry to such communities of practice often occurs through apprenticeship, another idea explicitly invoked in their theory. Authentic activity comprises the ordinary practices of the culture. In the case of language, it is general culture. In the case of specialized conceptual knowledge, it is the technical subculture formed by disciplinary or professional fields.

The problem of school learning, and perhaps of teacher education, then, is twofold: The situations of knowledge use supply neither rich, indexical references nor genuine activities and contexts that produce learning:

Teaching methods often try to impart abstracted concepts as fixed, well-defined, independent entities that can be explored in prototypical examples and textbook exercises. But such exemplification cannot provide the important insights into either the culture or the authentic activities that learners need. (Brown et al., 1989, p. 33)

While the situated cognition theorists acknowledge the partial and speculative character of their ideas, they also propose several specific instructional strategies, including "cognitive apprenticeships" (Collins et al., 1989) and collaborative activities within which knowledge is socially constructed.

These ideas shed disturbing light on learning to teach. First, if future teachers' own learning is thin, abstracted, and inauthentic, then they will have difficulty teaching authentically, because their own understanding of subject matter is likely to be limited, and they will have few models to work from. Second, many of the concepts conveyed in teacher education course work, such as "metacognition," "schemata," "cooperative learning," and "reciprocal teaching," are indexed to situations of use in classrooms (see Kennedy, 1991). In terms of situated cognition theory, this means that such concepts are best learned in the context of authentic activities in schools and classrooms. It follows that for such concepts to become part of the working vocabularies and repertoires of teachers, theoretical knowledge should be connected to situations of use, and this insight leads naturally to an interest in apprentice and intern arrangements, simulations, and other forms of guided clinical experience. Cases of various kinds might then constitute learning tools within field experiences.

Finally, the situated cognition perspective draws on the image of apprenticeship in a guild or professional community as a powerful form of learning. But this image requires a stable, satisfactory practice that the novice can join. If the aim of teacher education is a reformed practice that is not readily available, and if there is no reinforcing culture to support such practice, then the basic imagery of apprenticeship seems to break down. Teachers' knowledge is situated, but this truism creates a puzzle for reform. Through what activities and situations do teachers learn new practices that may not be routinely reinforced in the work setting?

Close Encounters of the Third Kind: Cases, Simulations, and Field Experiences

If the most common form of encounter with cases is through discussion, another form is through direct participation in a case constructed to simulate core tasks or problems of practice. And if one form of case is a narrative, another form might be an exercise that requires problem framing and solving by students under conditions that approximate the real thing. Simulations might be designed around such events as an IEP (Individual Educational Plan) meeting, a parent-teacher conference, the development and presentation of a school improvement plan, or the response to a desegregation order that will alter the student composition of a school. The simulation materials might include background information on the school and community, test scores and other data on students, district or state policy that bears on the problem or task, and resource readings that supply theoretical perspectives.

In a simulation, students gain experience once removed in such skills as analyzing problems, interpreting data, working in groups, making presentations, dealing with interpersonal and group conflicts, and bringing theoretical knowledge to bear. Simulations can be carefully designed and controlled to focus on issues that instructors deem important. This contrasts with clinical experiences in the field, where the presenting problems are naturally occurring rather than under the control of a teacher educator. However, there are also some options for standardizing problem-based clinical practice, including, for example, the use of portfolios that set tasks for students to accomplish and document during field assignments.

In drawing brief attention to simulations and the possibilities for structuring clinical experience, we want to set the case idea into a larger family of practices in teacher education that have a long history (see Cruickshank, 1988) but that may enjoy renewed interest in light of the new ideas in cognitive psychology. In some fields, such as medicine and the military, simulations are computer based. There may be analogs in education, but a more likely prospect are exercises in which students work in groups on problems and tasks replete with rich case particulars.

Simulations hold promise for introducing situations and contexts for use of the disciplinary knowledge that makes up education course work. Some cognitive psychologists now argue that when learners acquire information and skills in the context of problems of practice rather than within the frame of the disciplines, they retain and use such knowledge more effectively (see Bransford, Franks, Vye, & Sherwood, 1989). This theoretical perspective on the problem-based organization of knowledge supports the case idea in its extensions to simulations and to guided clinical experience.

Reflection in and on Action

While all of the arguments considered so far have been concerned with the interaction between thought and action, the emphasis has been shifting from thought to action. In the next argument to be considered, the balance tips toward action. Donald Schon (1983, 1987, 1991) has explored the processes through which professionals think productively about their work, setting and solving problems through the interplay of thought and action. In his cross-professional analyses, Schon follows closely the footsteps of John Dewey in identifying problematic situations as fertile ground for thought. He begins with a critique of academic or school-based approaches to practice, which he characterizes as "technical rationality"— the deployment of rigorous methods on well-formed problems to yield determinate solutions. He goes on to argue that professionals confront many problems that are not of this kind. Rather, "situations of uncertainty, instability, uniqueness, and value conflict" (1983, p. 49) are common—what operations researchers used to call "messes" as opposed to problems, or what Spiro refers to as "ill-structuredness." In response, Schon proposes to abstract from close study of professionals at work on such messy problems some descriptions, guidelines, and heuristics that may prove generally helpful. Again following Dewey, he argues that professionals have no recourse but to engage in a process of inquiry made up of informal experiments, trials, revisions, and "conversations with the situation." His studies of several instances of professional practice (e.g., architecture, psychotherapy, town planning) yield his basic claim of discovering "a fundamental structure of professional inquiry which underlies the many varieties of design or therapy advocated by contending schools of practice" (1983, p. 130).

Within Schon's account of reflection in action, cases play a dual role. They serve first as the natural units of deliberation and action through which the practitioner builds expertise. The case is a unit of practice; cases are organized into sets that share family resemblances: "cases of the measles," "cases of libel," or "cases of schizophrenia." Practitioners gradually accumulate experience with the variations of a relatively small set of case types and so develop routines and repertoires. Schon writes:

As long as his practice is stable, in the sense that it brings him the same types of cases, he becomes less and less subject to surprise. His knowing-in-practice tends to become increasingly tacit, spontaneous, and automatic, thereby conferring on him and his clients the benefits of specialization. (1983, p. 60)

Schon goes on to argue, however, that repeated engagement with the "typical" cases of practice can produce overlearning, rigidity, blindness, and boredom. The antidote is reflection, through which practitioners can surface and critique tacit understandings and begin to notice situations of uniqueness and uncertainty that they have denied or avoided. In this process, anomalous cases serve as strategic sites for development, where the practitioner constructs "a theory of the unique case." Crucial to such reflective inquiry is the process of "seeing as," in which the practitioner draws on the standard repertoire of past cases, identifying similarities and differences with the unique case. The case is a fundamental unit of learning for routine practice, and potentially for reflective practice, but the latter, Schon implies, must be cultivated as actively as the former.

Schon's account resembles casuistry, in that the practitioner reasons from case to case rather than from theory to case. However, there is an important difference: The process of casuistry locates interaction in thought, between precedents and current cases, whereas in Schon's argument the interaction occurs between thought and action as the practitioner experiments informally.

One final feature of Schon's work is worth comment. Under the rubric of "repertoire-building research" (1983, p. 315), he advocates development of cases that reveal not only "the starting situation, the actions taken, and the results achieved," but also the process of inquiry, as the practitioner turns over in his or her mind alternatives, competing considerations, and various moves to make. Such cases can serve as double exemplars, revealing both the precedents with their solutions and the ways of thinking about them.

The emphasis on reflection and inquiry into one's own practice as a disposition and complex set of skills may suggest a bridge between the transformative aims of teacher educators and current practice in the schools. Teacher educators might create "virtual worlds" in laboratory settings within which to convey exemplary practices, as scholars have urged (e.g., Berliner, 1985). And they might seek to encourage an inquiry orientation to practice through which teachers learn to study and continuously modify their teaching. If inquiry and continuous improvement are established as norms and dispositions with associated knowledge and skills, this supplies indirect impetus to the long-term improvement of teaching and the transformative goals of teacher education. Cases, then, might portray master teachers reflecting in action and modeling the dispositions, norms, and skills involved in this stance toward teaching.

A case literature of this kind has begun to emerge in teaching (e.g., see Ball, 1990b; Lampert, 1985, 1990; Paley, 1979, 1984) in which teachers both describe episodes, lessons, or aspects of their teaching and report their unfolding thoughts about the teaching portrayed. What is not yet clear is how to use such cases to help novices become reflective. If reflection is regarded not only as an ability, but also as a habit that typically must be practiced in the face of a multitude of competing demands, one might argue that (vicarious) reflection on cases, as preparation for reflection on practice work, ought to be a staple in teacher education.

Close Encounters of the Fourth Kind: Students Writing Cases

Each of the encounters described to this point—case discussion, use of hypermedia, and simulations—can promote the aims, dispositions, and methods of reflection, but a fourth form of encounter with cases seems specially pertinent. Some teacher educators involve students in writing their own cases as a quintessential method for reflecting on experience. Student-produced cases can also be used with future students, so this instructional activity has multiple benefits. Teacher educators also provide instruction in case writing, sometimes linking case development to field-based data collection (see Florio-Ruane & Clark, 1990; Laboskey, in press) and sometimes to writing stories of personal experience in student teaching (see Kleinfeld, 1991). Case writing is a demanding, intensive, complex activity that is most often used strategically as a culminating or consolidating assignment.

Here too a variety of practices has emerged, and the case idea is but one lead to the use of student narratives in teacher education. Schon's volumes stimulated substantial interest in reflection, but subsequent work has both challenged (e.g., Grimmett & Erickson, 1988) and extended his insights (e.g., Clift, Houston, & Pugach, 1991). Recent work has begun to distinguish among the aims and frames for reflection. Gore and Zeichner (1991), for example, note that generic approaches to reflective teaching are largely silent about

what it is that teachers ought to be reflecting about, the kinds of criteria that should come into play during the process of reflection (e.g., what distinguishes good from unacceptable educational practice), and the degree to which teachers' deliberations should incorporate a critique of the institutional contexts in which they work. (p. 120)

There is great variety in the substance and methods of reflective inquiry in teacher education. Gore and Zeichner (1991) go on to identify four broad topics for reflection that have deep roots in teacher education and that continue to receive attention today: the representation of subject matter to promote student understanding; the thoughtful application of particular teaching strategies and principles derived from research on teaching; the interests, thinking, and development of students as a basis for sensitive, responsive teaching; and the critical scrutiny of the social and political context of schooling and the assessment of school and classroom processes from the perspectives of equity, social justice, and humanity (for extended treatment of these traditions, see Liston & Zeichner, 1991).

Likewise, teacher educators pursue a variety of methodological approaches. Some teacher educators (e.g., Richert, in press) describe student writing in the language of the case idea. Others draw on the tradition of action research to frame student assignments and engage students in inquiry during field assignments (see Gore, in press; Noffke & Brennan, in press). Still others (e.g., Mattingly, 1991) are exploring story telling as a mode of action research through which practitioners frame and share cases with colleagues.

Evidence about the effects on student learning of case write-ups, action research projects, and other forms of reflective inquiry is mixed. Zeichner and Liston (1987) and Gore and Zeichner (1991) report disappointing results from various studies of one inquiry-oriented teacher education program. On the other hand, Kleinfeld (1991) analyzed her students' narratives of beginning teaching in terms of initial and concluding concept maps. She reports:

> The structure of students' thinking changed as they reflected upon their experience and wrote a case about it. Students typically began with a map of the world that was rigid, simplistic, and implicit. They ended the case with a map of the world that was much more complex, conditional, contextual, and explicit. Students often ended the case with new problems, questions they were now asking of experience. (p. 2)

Expertise in Teaching

In the final argument we consider, the emphasis shifts yet more strongly from thought to action, locating knowledge and skill more squarely in the acts—and particularly the routines—of teaching. The theories described to this point all divide their terrain into two broad regions: routine and nonroutine; well-structured and ill-structured; certain and automatic versus uncertain and problematic (geographical metaphors abound across the literature describing knowledge and practice). They choose to focus on the nonroutine, ill-structured, and problematic region, commenting along the way how knowledge and skill from the well-structured, routine, and paradigmatic region serves as a resource, to be used with judgment on ill-structured problems. Their work also focuses on thought, although Schon portrays the interplay of thought and action.

But we might now ask, How important are the routines in teaching?

Are they easy to acquire, establish, and use? How do teachers learn to act as well as to think? And are cases helpful in acquiring action routines? These questions occupy another line of research that is worth juxtaposing to the arguments so far covered; a balanced view of learning to teach requires an account of action as well as of inquiry and reflection.

Investigators interested in such matters have drawn their ideas from the branch of cognitive psychology that explores artificial intelligence (AI) and the creation of "machines who think" (McCorduck, 1979). The analogy of mind to machine is an old one that has received stimulus in the modern era with the invention of the computer, and within the AI field, a central distinction suggests how action can become a central topic in the study of mind. In the early days of AI work, argument broke out over whether to program "expert systems" via declarative or procedural representation (Gardner, 1985, p. 161). The former codes knowledge as a set of stored facts or declarations, whereas the latter relies on sets of procedures or actions: "Proceduralists felt that human intelligence is best thought of as a set of activities that individuals know how to do; whatever knowledge is necessary can be embedded in the actual procedures for accomplishing things" (Gardner, 1985, p. 162).

This emphasis on activity-related knowledge raised questions about how such knowledge is organized and represented in mind, and led to interest in schemata, scripts, plans, and other information storage and processing concepts (see Sacerdoti, 1977; Shank & Abelson, 1977). This conceptual work joined the empirical program that compared expert and novice performances in such activities as chess playing and physics problem solving and eventually influenced the cognitive research on teaching. The term *expertise* took on precise meaning, rather than referring generally to the role, status, or general competence of a professional. Leinhardt (1990b, p. 147), for example, defines expertise as "a technical term that refers to working with speed, fluidity, flexibility, situationally encoded informational schemas, and mental models that permit larger chunks of information to be accessed and handled."

In the research that Leinhardt and her colleagues have conducted (see Leinhardt, 1988; Leinhardt & Greeno, 1986; Leinhardt & Smith, 1985; Leinhardt, Weidman, & Hammond, 1987), teachers' classroom knowledge is divided into two broad categories of subject matter and lesson structure. In the latter category, her research examines how expert teachers develop fluid activity structures represented mentally as schemata or "interrelated sets of organized actions" (Leinhardt & Greeno, 1986, p. 75). Using this framework, teachers' lessons may be divided into segments that are highly routinized. Lesson segments constitute the familiar stuff of classrooms: teacher presentation of information, guided practice, drill work, tutorials with individuals and small groups, or transitions between segments.

When expert and novice elementary teachers are compared in their management of these routine activities via videotaped classroom observation coupled with interviews, the research indicates that the experts are considerably more efficient in their routine practices and that they work in synch with their students. Experts choreograph their classrooms in the first days of the school year, efficiently teaching students how to participate in the various activities of the school day (Leinhardt et al., 1987). By contrast, novices constantly change the pattern of their activities and have few well-practiced routines to work with.

David Berliner and colleagues have conducted related studies within this general research program, focusing particularly on how expert teachers perceive and interpret the flux of classroom life (see Carter, Sabers, Cushing, Pinnegar, & Berliner, 1987; Sabers, Cushing, & Berliner, 1991). Their work also reveals that experts are more efficient in processing information, in selectively attending to events in classrooms, and in advancing interpretations about students and activities. Berliner concludes that expertise develops gradually, in stages (see Berliner, 1988), and with Leinhardt he argues that efficient routines are a necessary aspect of effective teaching. In their view, teaching is a highly complex activity occurring in an ill-structured, dynamic environment. To manage the complexity, they argue, teachers need efficient routines that help to "reduce the cognitive load and expand the teacher's facility to deal with unpredictable elements of the task" (Leinhardt & Greeno, 1986, p. 76). This perspective complements Schon's but focuses on the establishment of efficient routines as the necessary foundation for the strategic use of reflection in action.

How does the case idea fit with this interest in schemata, procedural knowledge, and the routinized aspects of teaching? Is there a role for cases in acquiring this type of expertise? The answers are by no means clear (for further discussion of this issue, see the exchange between Floden & Klinzing, 1990, and Lampert & Clark, 1990). Leinhardt, for example, speculates that a well-researched, carefully annotated library of videotaped expert lessons (i.e., cases) would be extremely helpful to novices in "building a rich taxonomy of lesson scripts that are known to be successful" (Leinhardt, 1988, p. 52). Cases would serve to display smoothly articulated classroom processes for modeling and emulation, rather than to provide material for inquiry. And Berliner (1988) wonders whether novices may have inadequate experience upon which to reflect. Better to defer the cultivation of reflective practice to the advanced beginner stage, he argues.

A Role for Cases in Learning Skills and Routines?

Case use in teacher education, based on the four kinds of encounter described above, emphasizes analysis, reflection, and inquiry more than acquisition of action routines, a development that reflects the contemporary interest in cognitive rather than behavioral accounts of complex human performances. But skills training, we suspect, still occupies a prominent place in many teacher education programs. The training model, one version of which has been set forth by Joyce and Showers (1988), relies on modeling, coaching, and guided practice as a time-honored method for acquiring skills and for improving skill-based performance, especially in such fields as athletics and the arts. Within this model, videotapes and other depictions and descriptions of performance serve two purposes—to portray exemplars and to supply feedback on the novice's own performance.

The training model in teacher education has a number of shortcomings that include the problem of transfer and sustained use of skills acquired initially in training; the lack of attention to conceptual underpinnings and rationales that allow practitioners to make flexible, adaptive uses of skills; and the integration of skills or instructional strategies into overall patterns of practice (see Kennedy, 1988). But these problems also point to certain conditions of effective training that include explanation and discussion of theory, rationale, and concepts and extensive follow-up in classrooms with peer coaches and mentors who themselves have mastered the particular skills and strategies (for review of the literature on training, see Cruickshank & Metcalf, 1990; Gleissman, 1984). Showers, Joyce, and Bennett (1987) note that

the purpose of providing training in any practice is not simply to generate the external visible teaching "moves" that bring that practice to bear in the instructional setting but to generate the conditions that enable the practice to be selected and used appropriately and integratively . . . a major, perhaps the major dimension of teaching skill is cognitive. (pp. 85–86)

In training-based approaches to teacher development, then, cases might serve as examples and models within the demonstration phase, and they might reveal the cognitive dimensions of skillful performance. Although many teacher educators supply skills training in such areas as classroom management and particular teaching strategies that involve performance mastery of rules and routines, the case idea is more at home in pursuit of aims framed in terms of decision making, problem solving, and inquiry. Cognitivists are interested in representing action in mind but have not yet demonstrated how cases can serve as tools for the efficient and effective learning of teaching routines.

THE FUTURE OF THE CASE IDEA

This review has drawn attention to the diversity of theory and practice surrounding the case idea in teacher education. In fields that have strong traditions and conventions of case use, the central questions are relatively settled. These include the form and content of cases, the methods through which cases are developed, their place in the codified knowledge of a profession, their role in the professional curriculum, and the character of teaching with cases. Notably absent, however, is a research literature that explores the nature of learning through cases.

Despite some early efforts to introduce the case method into teacher education, no parallel traditions emerged in this field, raising the question of whether the time is now ripe for such a development. A shift in the intellectual milieu of teacher education is under way, we have argued, that supports an interest in cases that would have been inconceivable even a decade or two ago given the dominant ideas about knowledge, science, educational practice, and their interrelationship. We have reviewed some of the leading ideas that support the turn to cases in teacher education, but it is fair to note the speculative and contingent character of the arguments and claims as they apply to the case idea. Teacher education is under mounting pressure today to demonstrate its value and its contribution to the larger reform agenda in education. These political pressures help create a hunger for new ideas and practices, including the case idea, and this fervor should induce a certain caution because a succession of enthusiasms has swept the field, leaving a legacy of disillusionment.

The future of the case idea, we suspect, rests more on development than research, or perhaps on research in the context of development. We mean that the central task ahead is to create and use rich and interesting case materials in a variety of settings for a variety of purposes, while simultaneously studying those uses. There is the familiar chicken and egg problem. To test an idea requires investment in development, training, and implementation across many sites and trials. To secure that investment requires advocacy for a bold idea before it has been tested. This is roughly the state of affairs with the case idea. Although its current status scarcely admits uniform conventions, codifications, or canons, local communities might pioneer shared uses and conceptions of cases, and then advertise their practices and results.

The current discussion of cases and case methods contains both a modest claim and a bolder implication, and around each research will be worthwhile. The modest claim is that cases and case methods provide appreciable advantages, over alternatives, for fostering some of the dispositions, knowledge, reasoning, and skills that are wanted in schoolteachers. The bolder implication is that the case idea provides a different way of framing and doing the business of teacher education that may significantly recast our ideas and practices. We conclude with some suggestions for research and development to test both claims.

A Modest Program of Research on the Case Idea

We identify three kinds of studies in pursuit of the modest claim. First are explorations that describe and assess the implementation of case methods. Second are comparisons of the case method with other forms of instruction. And third are examinations of the effects of variations in use of the case method.

Implementation Studies

Implementation studies rely on two sources of comparison. Argyris (1980) provides an example of one source that compares the espoused theory of the case method with the theory in use and with the actual teaching. His study of case teaching in a business seminar revealed significant discrepancies between the espoused theory and the theory in use. Implementation studies of this sort, based on interviews coupled with classroom observation, can identify some of the difficulties associated with case teaching, help in refining subsequent instruction, and provide a basis for assessing effects on learning. The factors that Argyris draws attention to, including the ways case teachers induce dependence on them and the mutual face-saving between students and teachers, are subtle but unmistakable violations of the teachers' expressed purposes and intentions. Close study of actual teaching with cases can contribute a literature useful in helping case teachers refine their practice.

Studies of classrooms considering the same text (or case) are a second source of comparison useful in revealing how the "instructional text" created through discourse influences learning from cases. A lively line of research has developed that employs multiple theoretical frameworks in the close analysis of text-based lessons (e.g., see Green & Harker, 1988). This form of research involves study of two or more teachers teaching the same text to similar groups of students, and then comparing lesson structures, discourse modes, participation patterns, and story recall outcomes. Sociolinguistic perspectives, semantic propositional analysis, and reader response theory are used to analyze the videotaped lesson, lesson transcripts, and story recall data. Such inquiry reveals the intricate interactions among the author's text, the teacher's instructional text, and the student's reconstructed text. Comparison across classrooms reveals marked differences in the lesson structure, norms of participation, information presented and turns allocated, question complexity, and stress on particular themes and ideas in the text (Green, Harker, & Golden, 1987). Learning from cases, this research suggests, will depend on the interaction among what the text presents, what the reader (or viewer) brings, and what the teacher does with the text in class.

Comparison With Alternatives

A second type of inquiry compares case-based teaching with other alternatives. A small literature has emerged that draws attention to the limitations and disadvantages of case-based teaching and to potential advantages of other forms of instruction (in addition to Argyris, 1980, see Leone, 1989). This work charges that a case curriculum is an inefficient means of conveying the codified knowledge useful to a complex practice; that single, vivid cases fail to help students see underlying issues and develop useful principles and generalizations; and that case discussions lack the impact of experiential learning, substituting instead the analysis of others' experiences. Our review has also raised questions about the relationship of case-based teaching to the acquisition and use of skills and routines in teaching.

This brief mention of the criticisms of the case method suggests research that compares case teaching with other instructional approaches to determine relative effectiveness. However, this type of research suffers from a number of well-known defects in experimental comparisons that include the substantial within-treatment variation relative to cross-treatment differences (characterized either as uneven implementation or as treatment by setting interactions) and the differences in aims and outcomes associated with different approaches to instruction. One solution to these difficulties is to study case teaching in particular settings using a range of outcomes, implementation measures, and data collection methods that allow a balanced appraisal, including attention to outcomes for which case methods may produce weak results. Similar studies of other instructional methods will reveal different patterns of outcomes and so allow some comparative judgments. The utility of such research lies in determining the relative advantages of case-based teaching judged against a range of desirable objectives.

Exploring Variations

A third type of inquiry explores patterns and effects of variations in curriculum and instruction with cases, within the context of design and development. In light of the diversity of practices that are emerging, case developers and users confront a range of choices about cases—their length, genre, medium, theoretical specification, and others—and about case encounters. What has influenced these choices to date have been the theoretical and value orientations of developers together with their purposes. But within particular development projects, investigators may begin experimenting with design variations, exploring effects on learners and learning. For example, in the context of elementary education, rather than teacher education, the Cognition and Technology Group at Vanderbilt prefers visual rather than textual formats for cases because they "allow students to develop pattern recognition skills (a major disadvantage of text is that it represents the output of the writer's pattern recognition processes)." They continue, "Video allows a more veridical representation of events than text; it is dynamic, visual, and spatial; and students can more easily form rich mental models of the problem situations" (p. 3). Claims such as these, for the importance of the case medium in this instance, are worth testing in the context of design and use.

The aim of such inquiry is to refine case methods based on modest evidence of effects. Theoretical hunches of various sorts, however, might guide developmental inquiry around such questions as the effects of literary properties of narrative on student learning; the relative merits of using many short vignettes versus extended "paradigm" cases; the effects of various kinds and degrees of contextual detail in written and video cases; the situating of cases in simulations, field experiences, liberal arts courses, and other contexts of use; the sequencing of theoretical material with case particulars; and the role of cases in introducing innovative practices into the repertoires of novices.

Toward the Bold Claim for the Case Idea

Research directed to the modest claim examines variations in the design and use of case methods in teacher education, as guided by theoretical hunches and as anchored in effects on learning. Various strategic comparisons will be useful for this kind of research: between espoused and actual case methods, across classrooms using the same cases, between case-based and other instructional approaches, and among variations in case designs and activities. But case advocates also have advanced a set of fundamental claims about knowledge in teaching, learning to teach, and the community of practice that deserve attention. These bold claims are difficult to study because the proposed practices are not available, but we conclude our review with some suggestions around four central issues: the character and significance of case knowledge; methods for identifying, testing, and using teachers' craft knowledge; the prospects for a curriculum of cases; and the creation of settings within which teachers, researchers, and teacher educators may jointly pursue the case idea.

Case Knowledge in Learning to Teach

The bold claim rests on a number of assumptions about knowledge in teaching and learning to teach that challenge reigning ideas. The central claim is that "cases in teaching are important . . . not simply because they convey the complexity of classroom life, but also because they are probably the form in which teachers' meanings are stored, conveyed, and brought to bear" (Doyle, 1990b, p. 356). Knowledge in teaching, this view asserts, is particularistic and situational, intimately connected to the contexts and events of teaching. Consequently, the most effective way to represent knowledge in teaching is through cases that capture both the routines and the problematic, unique situations that call for reflection, analysis, and continuing inquiry. This perspective draws insights from the theories of situated cognition and expertise but raises many questions. What kinds of cases situationally encode teaching knowledge? How does exposure to case-based knowledge help novices form principles, generalizations, and routines that are integrative and economical? What is the role of general cognitive skills versus domain-specific knowledge in learning to teach with cases, and what are the implications of this issue for the design of cases? What patterns of case-based instruction promote transfer of knowledge and skill to situations in teaching? How does case teaching modify novices' prior beliefs, values, and assumptions? How does this view of teaching knowledge accommodate the introduction of innovative practices through case teaching? Conceptual and empirical work are needed to address these and related questions.

Identifying Craft Knowledge

A second part of this program to reconceive teaching knowledge will explore methods for identifying, testing, and sharing teachers' craft or case knowledge. If the wisdom of practice in teaching is essentially ineffable, if tacit knowledge cannot be made explicit or tested for its validity and applicability, then there may be few implications for teacher education. The hope is to discover ways of tapping the wisdom of practice as a source of knowledge that could be represented in the curriculum of teacher education. Such methods might include encouraging teachers to construct case narratives and commentaries and to report on their responses to common problems of practice, videotaping and annotating classroom teaching that directly portrays craft wisdom, and constructing theoretically guided observer accounts of teaching that the teachers of record may subsequently modify, amplify, and interpret.

The Case-Based Curriculum

The most ambitious project in pursuit of the bold claim is to construct and test a curriculum of cases. This is a logical extension of the claim that knowledge in teaching is case based. Lee Shulman (in press) reflects this view in arguing that "the [teaching] field is itself a body of cases linked loosely by working principles, and case methods are the most valid way of representing that structure in teaching." Case methods, he asserts, constitute the appropriate strategy for transforming propositional knowledge into "narratives that motivate and educate." These assertions capture the bold implication of the case idea most directly and point to a curriculum of cases. The precedents are available in other fields, notably law, but that analogy is likely inapt for teacher education. There is another history to consult for leads: the experiments in other fields that break with curricular orthodoxy to fundamentally redesign the professional studies program. Medicine offers several examples, including the development of a problem-based curriculum in several universities around the world. The familiar and long-standing division into liberal and professional courses, the foundations-methods-clinical practice sequence, and the foundationalist construal of theory into practice are not so well-regarded or so obviously effective as to preclude program redesigns based on the central use of a variety of cases. The full test of the case idea in teacher education awaits such bold trials.

Collaborative Organization

Finally, the turn to cases in teacher education reflects an interest within the academy in drawing closer to practice and in exploring new means of connecting practice and research (for an expression of this trend, see the report from the National Academy of Education, 1991). Cases serve as a natural site for collaboration among researchers, teacher educators, and teachers, and case development may serve as a crucial activity in the formation of new professional communities. But this prospect's promise will depend on new institutional arrangements that support regular, long-term, collaborative work. Professional development schools (see Holmes Group, 1990) could serve as one site for such inquiry. This review has emphasized the importance of narrative, discourse, and forms of reasoning cultivated within communities of practice. We believe that the full exploration of the case idea will require as much attention to the formation of community as to the technical tasks of development and research. Social inventions will be necessary that bring teachers into productive relations with university colleagues, and so the efforts under way around the country to establish school-university partnerships may eventually support aggressive development of the case idea.

REFERENCES

Allen, A. (1990). *Cross-cultural counseling: The guidance project and the reluctant seniors* (Teaching cases in cross-cultural education No. 7). Fairbanks: Center for Cross-Cultural Studies, College of Rural Alaska, University of Alaska—Fairbanks.

Argyris, C. (1980). Some limitations of the case method: Experiences in a management development program. *Academy of Management Review, 5*, 291–298.

Ball, D. L. (1990a). The mathematical understandings that prospective teachers bring to teacher education. *Elementary School Journal, 90*, 449–466.

Ball, D. L. (1990b). *With an eye on the mathematical horizon: Dilemmas of teaching elementary school mathematics* (Craft Paper 90-3). East Lansing: Michigan State University, National Center for Research on Teacher Education.

Barnett, C. (1991, April). *Case methods: A promising vehicle for expanding the pedagogical knowledge base in mathematics.* Paper presented at the annual meeting of the American Educational Research Association, Chicago.

Berliner, D. C. (1985). Laboratory settings and the study of teacher education. *Journal of Teacher Education, 36*, 2–8.

Berliner, D. C. (1988). Implications of studies of expertise in pedagogy for teacher education and evaluation. In *New directions for teacher assessment: Proceedings of the 1988 ETS invitational conference* (pp. 39–67). Princeton, NJ: Educational Testing Service.

Bird, T. (1991, April). *Making conversations about teaching and learning in an introductory teacher education course.* Paper presented at the annual meeting of the American Educational Research Association, Chicago.

Bok, S. (1978). *Lying: Moral choice in public and private life.* New York: Vintage Books.

Bok, S. (1983). *Secrets: On the ethics of concealment and revelation.* New York: Random House.

Bransford, J., Franks, J., Vye, N., & Sherwood, R. (1989). New approaches to instruction: Because wisdom can't be told. In S. Vosniadou & A. Ortony (Eds.), *Similarities and analogical reasoning* (pp. 470–497). Cambridge, England: Cambridge University Press.

Brophy, J. (in press). *Advances in research on teaching, Vol. 2.* Greenwich, CT: JAI Press.

Broudy, H. S. (1990). Case studies—Why and how. *Teachers College Record, 91*, 449–459.

Brown, J. S., Collins, A., & Duguid, P. (1989). Situated cognition and the culture of learning. *Educational Researcher, 18*(1), 32–42.

Bruner, J. (1985). Narrative and paradigmatic modes of thought. In E. Eisner (Ed.), *Learning and teaching the ways of knowing: Eighty-fourth yearbook of the National Society for the Study of Education* (pp. 97–115). Chicago: University of Chicago Press.

Bruner, J. (1986). *Actual minds, possible worlds.* Cambridge, MA: Harvard University Press.

Bruner, J. (1990). *Acts of meaning.* Cambridge, MA: Harvard University Press.

Buchmann, M. (1989). *Breaking from experience in teacher education: When is it necessary, how is it possible?* East Lansing, MI: Institute for Research on Teaching.

Buchmann, M. (1991). Making new or making do: An inconclusive argument about teaching. *American Journal of Education, 99*, 279–297.

Buchmann, M., & Schwille, J. R. (1983). Education: The overcoming of experience. *American Journal of Education, 92*, 30–51.

Bussis, A. M., Chittenden, E. A., Amarel, M., & Klausner, E. (1985). *Inquiry into meaning.* Hillsdale, NJ: Erlbaum.

Carey, R. (1989). *Harassment in Lomavik: A case study* (Teaching cases in cross-cultural education No. 1). Fairbanks: Center for Cross-Cultural Studies, College of Rural Alaska, University of Alaska—Fairbanks.

Carter, K. (1988). Using cases to frame mentor-novice conversations about teaching. *Theory into Practice, 27,* 214–222.

Carter, K. (1990). Teachers' knowledge and learning to teach. In W. R. Houston (Ed.), *Handbook of research on teacher education* (pp. 291–310). New York: Macmillan.

Carter, K. (in press). Toward a cognitive conception of classroom management: A case of teacher comprehension. In J. H. Shulman (Ed.), *Case methods in teacher education.* New York: Teachers College Press.

Carter, K., Sabers, D., Cushing, K., Pinnegar, S., & Burliner, D. (1987). Processing and using information about students: A study of expert, novice, and postulant teachers. *Teaching and Teacher Education, 3,* 147–157.

Carter, K., & Unklesbay, R. (1989). Cases in teaching and law. *Journal of Curriculum Studies, 21,* 527–536.

Christensen, C. R., & Hansen, A. J. (1987). *Teaching and the case method.* Boston: Harvard Business School.

Clift, R., Houston, W. R., & Pugach, M. (1991). *Encouraging reflective practice in education: An examination of issues and programs.* New York: Teachers College Press.

Cognition and Technology Group at Vanderbilt. (1990). Anchored instruction and its relationship to situated cognition. *Educational Researcher, 19*(5), 2–10.

Coles, R. (1989). *The call of stories.* Boston: Houghton Mifflin.

Collins, A., Brown, J. S., & Newman, S. (1989). The new apprenticeship: Teaching students the craft of reading, writing, and mathematics. In L. B. Resnick (Ed.), *Knowing, learning, and instruction* (pp. 453–494). Hillsdale, NJ: Erlbaum.

Collins, A., & Stevens, A. L. (1982). Goals and strategies of inquiry teachers. In R. Glaser (Ed.), *Advances in instructional psychology, Vol. 2* (pp. 65–120). Hillsdale, NJ: Erlbaum.

Connelly, F. M., & Clandinin, J. (1990). Stories of experience and narrative inquiry. *Educational Researcher, 19*(4), 2–14.

Cruickshank, D. (1988). The uses of simulations in teacher preparation. *Simulation and Games, 19,* 133–156.

Cruickshank, D., & Metcalf, K. (1990). Training within teacher preparation. In W. R. Houston (Ed.), *Handbook of research on teacher education* (pp. 469–497). New York: Macmillan.

Dewey, J. (1929). *The quest for certainty: A study of the relation of knowledge and action.* New York: Minton, Balch.

Doyle, W. (1990a, Winter). Case methods in teacher education. *Teacher Education Quarterly,* pp. 7–15.

Doyle, W. (1990b). Classroom knowledge as a foundation for teaching. *Teachers College Record, 91,* 347–360.

Feiman-Nemser, S., & Buchmann, M. (1985). Pitfalls of experience in teacher preparation. *Teachers College Press, 87,* 53–65.

Feiman-Nemser, S., & Buchmann, M. (1989). Describing teacher education: A framework and illustrative findings from a longitudinal study of six students. *Elementary School Journal, 89,* 365–377.

Feinberg, W., & Soltis, J. F. (1985). *School and society.* New York: Teachers College Press.

Feldman, C. F., Bruner, J., Renderer, B., & Spitzer, S. (1990). Narrative comprehension. In B. Britton & A. Pellegrini (Eds.), *Narrative thought and narrative language* (pp. 1–78). Hillsdale, NJ: Erlbaum.

Fenstermacher, G. D., & Soltis, J. F. (1986). *Approaches to teaching.* New York: Teachers College Press.

Finley, A. F. (1988). *Malaise of the spirit: A case study.* Fairbanks: Center for Cross-Cultural Studies, College of Rural Alaska, University of Alaska—Fairbanks.

Finley, A. F. (1990). *Gender wars at John Adams High School: A case study* (Teaching cases in cross-cultural education No. 6). Fairbanks: Center for Cross-Cultural Studies, College of Rural Alaska, University of Alaska—Fairbanks.

Floden, R. E., Buchmann, M., & Schwille, J. R. (1987). Breaking with everyday experience. *Teachers College Record, 88,* 485–506.

Floden, R. E., & Klinzing, H. G. (1990). What can research on teacher thinking contribute to teacher preparation? A second opinion. *Educational Researcher, 19*(4), 15–20.

Florio-Ruane, S., & Clark, C. M. (1990, Winter). Using case studies to enrich field experiences. *Teacher Education Quarterly,* pp. 17–28.

Florio-Ruane, S., & Lensmire, T. (1990). Transforming future teachers' ideas about writing instruction. *Journal of Curriculum Studies, 22,* 277–289.

Gardner, H. (1985). *The mind's new science: A history of the cognitive revolution.* New York: Basic Books.

Geertz, C. (1981). *Local knowledge: Further essays in interpretive anthropology.* New York: Basic Books.

Gleissman, D. (1984). Changing teacher performance. In L. Katz & J. Raths (Eds.), *Advances in teacher education, Vol. 1* (pp. 95–112). Norwood, NJ: Ablex.

Gore, J. (in press). Practicing what we preach: Action research and the supervision of student teachers. In B. R. Tabachnick & K. Zeichner (Eds.), *Issues and practices in inquiry-oriented teacher education.* Philadelphia: Falmer Press.

Gore, J., & Zeichner, K. (1991). Action research and reflective teaching in preservice teacher education: A case study from the United States. *Teaching and Teacher Education, 7,* 119–136.

Green, J. L., & Harker, J. O. (Eds.). (1988). *Multiple perspective analyses of classroom discourse, Vol. XXVIII.* Norwood, NJ: Ablex.

Green, J. L., Harker, J. O., & Golden, J. M. (1987). Lesson construction: Differing views. In G. Noblit & W. Pink (Eds.), *Schooling in social context* (pp. 46–77). Norwood, NJ: Ablex.

Greenwood, G. E., & Parkay, F. W. (1989). *Case studies for teacher decision making.* New York: Random House.

Grimmett, P., & Erickson, G. (Eds.). (1988). *Reflection in teacher education.* New York: Teachers College Press.

Hollingsworth, S. (1989). Prior beliefs and cognitive change in learning to teach. *American Educational Research Journal, 26,* 160–189.

Holmes Group. (1990). *Tomorrow's schools.* East Lansing: Michigan State University College of Education.

Jonsen, A., & Toulmin, S. (1988). *The abuse of casuistry.* Berkeley: University of California Press.

Joyce, B., & Showers, B. (1988). *Student achievement through staff development.* New York: Longman.

Kennedy, M. (1988). Inexact sciences. Professional development and the education of expertise. In E. Z. Rothkopf (Ed.), *Review of research in education* (Vol. 14, pp. 133–167). Washington, DC: American Educational Research Association.

Kennedy, M. (1990). *A survey of recent literature on teachers' subject matter knowledge* (Issue Paper 90-3). East Lansing: Michigan State University, National Center for Research on Teacher Education.

Kennedy, M. (1991). *An agenda for research on teacher learning*. East Lansing: National Center for Research on Teacher Learning, Michigan State University.

Kleinfeld, J. (1990, Winter). The special virtues of the case method in preparing teachers for minority schools. *Teacher Education Quarterly*, pp. 43–52.

Kleinfeld, J. (1991, April). *Wrestling with the angel: What student teachers learn from writing cases*. Paper presented at the annual meeting of the American Educational Research Association, Chicago.

Kleinfeld, J. (in press). Learning to think like a teacher: The study of cases. In J. Shulman (Ed.), *Case methods in teacher education*. New York: Teachers College Press.

Kowalski, T. J., Weaver, R. A., & Hensen, K. T. (1990). *Case studies in teaching*. New York: Longman.

Laboskey, V. K. (in press). Case investigations: Preservice teacher research as an aid to reflection. In J. H. Shulman (Ed.), *Case methods in teacher education*. New York: Teachers College Press.

Lampert, M. (1985). Mathematics learning in context: The voyage of the Mimi. *Journal of Mathematical Behavior, 4*, 157–167.

Lampert, M. (1990). When the problem is not the question and the solution is not the answer: Mathematical knowing and teaching. *American Educational Research Journal, 27*, 29–63.

Lampert, M., & Ball, D. L. (1990). *Using hypermedia technology to support a new pedagogy of teacher education* (Issue Paper 90-5). East Lansing: National Center for Research on Teacher Education, Michigan State University.

Lampert, M., & Clark, C. (1990). Expert knowledge and expert thinking in teaching: A response to Floden and Klinzing. *Educational Researcher, 19*(4), 21–23.

Leck, G. M. (1990). Examining gender as a foundation within foundational studies. *Teachers College Record, 91*, 382–395.

Leinhardt, G. (1988). Expertise in instructional lessons: An example from fractions. In D. A. Grouws & R. J. Cooney (Eds.), *Perspectives on research on effective mathematics teaching* (pp. 47–66). Hillsdale, NJ: Erlbaum.

Leinhardt, G. (1990a). Capturing craft knowledge in teaching. *Educational Researcher, 19*(2), 18–25.

Leinhardt, G. (1990b). Situated knowledge and expertise in teaching. In J. Calderhead (Ed.), *Teachers' professional knowledge* (pp. 146–168). London: Falmer Press.

Leinhardt, G., & Greeno, J. G. (1986). The cognitive skill of teaching. *Journal of Educational Psychology, 78*, 75–95.

Leinhardt, G., & Smith, D. (1985). Expertise in mathematics instruction: Subject matter knowledge. *Journal of Educational Psychology, 77*, 247–271.

Leinhardt, G., Weidman, C., & Hammond, K. M. (1987). Introduction and integration of classroom routines by expert teachers. *Curriculum Inquiry, 17*, 135–176.

Leone, R. (1989). *Teaching management without cases*. Unpublished manuscript, Boston University School of Management.

Liston, D. P., & Zeichner, K. (1991). *Teacher education and the social conditions of schooling*. New York: Routledge.

Masoner, M. (1988). *An audit of the case study method*. New York: Praeger.

Mattingly, C. (1991). Normative reflections on practical actions: Two learning experiments in reflective storytelling. In D. Schon (Ed.), *The reflective turn: Case studies in and on educational practice* (pp. 235–257). New York: Teachers College Press.

McAninch, A. R. (1991). Casebooks for teacher education: The latest fad or lasting contribution? *Journal of Curriculum Studies, 23*, 345–355.

McCorduck, P. (1979). *Machines who think.* San Francisco: W. H. Freeman.

McDiarmid, G. W. (1990). What do prospective teachers learn in their liberal arts classes? *Theory into Practice, 29*, 21–29.

Merseth, K. (1991). *The case for cases.* Washington, DC: American Association for Higher Education.

Merseth, K. (in press). Cases for decision-making in teacher education. In J. H. Shulman (Ed.), *Case methods in teacher education.* New York: Teachers College Press.

National Academy of Education. (1991). *Research and the renewal of education.* Palo Alto, CA: Stanford University.

Neustadt, R. E., & May, E. R. (1986). *Thinking in time: The uses of history for decision makers.* New York: The Free Press.

Noffke, S., & Brennan, M. (in press). Action research and reflective student teaching at the University of Wisconsin—Madison: Issues and examples. In B. R. Tabachnick & K. Zeichner (Eds.), *Issues and practices in inquiry-oriented teacher education.* Philadelphia: Falmer Press.

Nozick, R. (1974). *Anarchy, state, and utopia.* New York: Basic Books.

Paine, L. (1989). *Orientations toward diversity: What do prospective teachers bring?* (Research Report 89-9). East Lansing: National Center for Research on Teacher Education, Michigan State University.

Paley, V. (1979). *White teacher.* Cambridge, MA: Harvard University Press.

Paley, V. (1984). *Boys and girls: Superheroes in the doll corner.* Chicago: University of Chicago Press.

Perkins, D. N., & Salomon, G. (1989). Are cognitive skills context-bound? *Educational Researcher, 18*(1), 16–25.

Peterson, P. L., Clark, C. M., & Dickson, W. P. (1990). Educational psychology as a foundation in teacher education: Reforming an old notion. *Teachers College Record, 91*, 322–346.

Phillips, D. C., & Soltis, J. F. (1985). *Perspectives on learning.* New York: Teachers College Press.

Posner, G. J., Strike, K. A., Hewson, P. W., & Gertzog, W. A. (1982). Accommodation of a scientific conception: Toward a theory of conceptual change. *Science Education, 66*, 211–227.

Rawls, J. (1971). *A theory of justice.* Cambridge, MA: Harvard University Press.

Richert, A. (in press). Writing cases: A vehicle for inquiry into the teaching process. In J. H. Shulman (Eds.), *Case methods in teacher education.* New York: Teachers College Press.

Rorty, R. (1979). *Philosophy and the mirror of nature.* Princeton, NJ: Princeton University Press.

Rorty, R. (1982). Method, social science, and social hope. In *Consequences of pragmatism* (pp. 191–216). Minneapolis: University of Minnesota Press.

Sabers, D., Cushing, K., & Berliner, D. (1991). Differences among teachers in a task characterized by simultaneity, multidimensionality, and immediacy. *American Educational Research Journal, 28*, 63–88.

Sacerdoti, E. D. (1977). *A structure for plans and behavior.* New York: Elsevier–North Holland.

Scheffler, I. (1974). *Four pragmatists*. London: Routledge & Kegan Paul.

Schon, D. (1983). *The reflective practitioner*. New York: Basic Books.

Schon, D. (1987). *Educating the reflective practitioner*. San Francisco: Jossey-Bass.

Schon, D. (Ed.). (1991). *The reflective turn. Case studies in and on educational practice*. New York: Teachers College Press.

Schwab, J. (1978). *Science, curriculum, and liberal education: Selected essays*. Chicago: University of Chicago Press.

Shank, R. C., & Abelson, R. (1977). *Scripts, plans, goals, and understanding*. Hillsdale, NJ: Erlbaum.

Showers, B., Joyce, B., & Bennett, B. (1987). Synthesis of research on staff development: A framework for future study and a state-of-the-art analysis. *Educational Leadership, 45*, 77–87.

Shulman, J. H. (Ed.). (in press). *Case methods in teacher education*. New York: Teachers College Press.

Shulman, J. H., & Colbert, J. A. (Eds.). (1987). *The mentor teacher casebook*. San Francisco: Far West Laboratory for Educational Research and Development.

Shulman, J. H., & Colbert, J. A. (Eds.). (1988). *The intern teacher casebook: Cases and commentaries*. San Francisco: Far West Laboratory for Educational Research and Development.

Shulman, L. S. (1986a). Paradigms and research programs in the study of teaching. In M. C. Wittrock (Ed.), *Handbook of research on teaching* (3rd ed., pp. 3–36). New York: Macmillan.

Shulman, L. S. (1986b). Those who understand: Knowledge growth in teaching. *Educational Researcher, 15*(2), 4–14.

Shulman, L. S. (1987). Knowledge and teaching: Foundations of the new reform. *Harvard Educational Review, 57*, 1–21.

Shulman, L. S. (1990). Reconnecting foundations to the substance of teacher education. *Teachers College Record, 91*, 300–310.

Shulman, L. S. (in press). Toward a pedagogy of cases. In J. H. Shulman (Ed.), *Case methods in teacher education*. New York: Teachers College Press.

Shuman, R. B. (1989). *Classroom encounters: Problems, case studies, solutions*. Washington, DC: National Education Association.

Silverman, R., Welty, W. M., & Lyon, S. (1991). *Case studies for teacher problem-solving*. New York: McGraw-Hill.

Soltis, J. F. (1990). A reconceptualization of educational foundations. *Teachers College Record, 91*, 311–321.

Spiro, R. J., Coulson, R. L., Feltovich, P. J., & Anderson, D. K. (1988). *Cognitive flexibility theory: Advanced knowledge acquisition in ill-structured domains* (Technical Report No. 441). Champaign: University of Illinois, Center for the Study of Reading.

Spiro, R. J., Feltovich, P. J., Coulson, R. L., & Anderson, D. K. (1989). Multiple analogies for complex concepts: Antidotes for analogy-induced misconception in advanced knowledge acquisition. In S. Vosniadou & A. Ortony (Eds.), *Similarity and analogical reasoning* (pp. 498–531). Cambridge, England: Cambridge University Press.

Spiro, R. J., & Jehng, J. (1990). Cognitive flexibility and hypertext: Theory and technology for the nonlinear and multidimensional traversal of complex subject matter. In D. Nix & R. Spiro (Eds.), *Cognition, education, and multimedia: Exploring ideas in high technology* (pp. 164–205). Hillsdale, NJ: Erlbaum.

Spiro, R. J., Vispoel, W. P., Schmitz, J. G., Samarapungavan, A., & Boerger, A. E. (1987). Knowledge acquisition for application: Cognitive flexibility and transfer in complex domains. In B. C. Britton (Ed.), *Executive control processes* (pp. 177–199). Hillsdale, NJ: Erlbaum.

Strike, K. A., & Soltis, J. F. (1985). *The ethics of teaching*. New York: Teachers College Press.

Sykes, G. (1989). Learning to teach with cases. *Colloquy, 2*(2), 7–13.

Tozer, S., Anderson, T. H., & Armbruster, B. B. (1990). Psychological and social foundations in teacher education: A thematic introduction. *Teachers College Record, 91*, 293–299.

Walker, D. F., & Soltis, J. F. (1986). *Curriculum and aims*. New York: Teachers College Press.

Walzer, M. (1979). *Just and unjust wars*. New York: Basic Books.

Wilson, S. M. (in press). A case concerning content: Using case studies to teach about subject matter. In J. H. Shulman (Ed.), *Case methods in teacher education*. New York: Teachers College Press.

Wittgenstein, L. (1953). *Philosophical investigations*. New York: Macmillan.

Zeichner, K. M., & Liston, D. (1987). Teaching student teachers to reflect. *Harvard Educational Review, 57*, 23–48.